Modern Control Theory

Modern Control Theory

KR Varmah MSc (Engg)

Professor
Department of Electrical and Electronics Engineering
Muthoot Institute of Technology and Science, Puthencruz
Kerala
Email: krvarmah@gmail.com

CBS

CBS Publishers & Distributors Pvt Ltd

New Delhi • Bengaluru • Chennai • Kochi • Kolkata • Mumbai
Bhopal • Bhubaneswar • Hyderabad • Jharkhand • Nagpur • Patna • Pune • Uttarakhand • Dhaka (Bangladesh)

Modern Control Theory

ISBN: 978-93-86217-76-9

First Edition: 2017
Reprint: 2020

Published by Satish Kumar Jain and produced by Varun Jain for

CBS Publishers & Distributors Pvt Ltd
4819/XI Prahlad Street, 24 Ansari Road, Daryaganj, New Delhi 110 002, India.
Ph: 23289259, 23266861, 23266867   Fax: 011-23243014   Website: www.cbspd.com
e-mail: delhi@cbspd.com; cbspubs@airtelmail.in.

Corporate Office: 204 FIE, Industrial Area, Patparganj, Delhi 110 092
Ph: 4934 4934   Fax: 4934 4935   e-mail: publishing@cbspd.com; publicity@cbspd.com

Branches

- **Bengaluru:** Seema House 2975, 17th Cross, K.R. Road,
  Banasankari 2nd Stage, Bengaluru 560 070, Karnataka
  Ph: +91-80-26771678/79   Fax: +91-80-26771680   e-mail: bangalore@cbspd.com
- **Chennai:** 7, Subbaraya Street, Shenoy Nagar, Chennai 600 030, Tamil Nadu
  Ph: +91-44-26680620, 26681266   Fax: +91-44-42032115   e-mail: chennai@cbspd.com
- **Kochi:** 42/1325, 1326, Power House Road, Opposite KSEB Power House,
  Ernakulam 682 018, Kochi, Kerala
  Ph: +91-484-4059061-65   Fax: +91-484-4059065   e-mail: kochi@cbspd.com
- **Kolkata:** 6/B, Ground Floor, Rameswar Shaw Road, Kolkata-700 014, West Bengal
  Ph: +91-33-22891126, 22891127, 22891128   e-mail: kolkata@cbspd.com
- **Mumbai:** 83-C, Dr E Moses Road, Worli, Mumbai-400018, Maharashtra
  Ph: +91-22-24902340/41   Fax: +91-22-24902342   e-mail: mumbai@cbspd.com

Representatives

Bhopal	0-8319310552	Bhubaneswar	0-9911037372	Hyderabad	0-9885175004
Jharkhand	0-9811541605	Nagpur	0-9421945513	Patna	0-9334159340
Pune	0-9623451994	Uttarakhand	0-9716462459	Dhaka (Bangladesh)	01912-003485

Printed at: JS Offset, Patparganj Industrial Area, Delhi, India

to

my parents and Gurus

Preface

All the theories and the methods employed in the analysis and design of linear control systems are not applicable to nonlinear systems. State space techniques are applicable, equally well, to both linear and nonlinear systems. This book introduces students to the modern control theory based on state variables and state space. It gives a basic approach to the design and analysis of continuous time control systems using state space representations.

The main objective of writing this second level book in control systems is the feeling that very few text books are available that cater to the needs of students: a book that presents the theories and explanations in a simple and lucid manner that help students to learn and imbibe the concepts without much difficulty. Academic interaction with my students and many years of my teaching the subject encouraged me in writing this book.

The topics are divided into ten chapters. Each chapter is divided into several sections, clearly numbered and labelled. The content of each chapter is well explained and many worked out examples are included to reinforce the theory. Each chapter ends with a summary of subject matter discussed. Many practice problems and review questions are included that enable the students to test their understanding of the subject.

Chapter 1 deals with fundamental concepts of state variables and state space. It also discusses the methods of formulation of state space models of physical systems. **Chapter 2** discusses methods of obtaining canonical forms of state models from the given transfer function of a system. **Chapter 3** explains the methods of solving state equations in both the s-domain and in time domain. **Chapter 4** gives a treatment on similarity transformations. It discusses the transformation of a given state model into canonical forms and also the transformation from a non-diagonal canonical form to diagonal canonical form. **Chapter 5** presents the concept of controllability and observability of a state model that represents a system. It also describes the method of transformation of a state model into controllable canonical form and to observable canonical form.

Chapter 6 discusses the technique of pole placement design of state controllers and state observers. **Chapter 7** presents an introduction to nonlinear systems and their properties. It also explains the method of deriving the describing function for a particular nonlinear element. The method of assessing the stability of limit cycles using describing function analysis is also discussed in this chapter. **Chapter 8** introduces the concept of phase plane and phase trajectory. It also discusses the construction of phase trajectories using both analytical and graphical methods and gives an insight into the nature of phase trajectories in the vicinity of singular points. **Chapter 9** introduces the concept of Lyapunov stability. It explains the stability analysis of both linear and nonlinear systems, using Lyapunov stability theorem. It also explains the method of generating a proper Lyapunov function for a given nonlinear system represented by state equations. **Chapter 10** gives an introduction to the concept of optimal control. It also discusses the method of optimization of quadratic performance measures using Lyapunov equation and Riccati equation.

Illustrations in abundance and many solved problems are included, for a better understanding of the theoretical concepts. It is expected that students using this book would have completed a basic course in linear control systems and mathematics through differential equations and Laplace transforms.

My parents and Gurus were a source of inspiration to my career as a teacher. I hereby acknowledge their profound love, affection and blessings and I dedicate this book at their feet. I also sincerely thank my wife Suja and daughter Ayana for their immense patience and support that helped me greatly, in completing this work.

Constructive criticism and suggestions for improvement are always welcome. I appeal to the users of this book to bring to my notice mistakes, if any, that might have crept in inadvertently.

KR Varmah

Contents

1 State Space and State Variables

1.1 INTRODUCTION

A physical system which is to be controlled is called a *plant*. The dynamics of a plant can be represented by ordinary differential equations with time t as independent variable. It can also be represented by many first-order differential equations, and the number of first-order differential equations depends on the order of the differential equation that describes the system. A differential equation is said to be linear if each term of the equation contains at most only the first power of the dependent variable or its derivatives. A system that can be described by a linear differential equation is said to be a linear system.

A linear system can be either a linear time-invariant system or a linear time-varying system. A linear time-invariant system has its parameters that do not change with time. A differential equation that represents such a system will have the coefficient of each term of it as constant. Most physical systems are time varying. For a linear time-varying system, many terms of its differential equation description will be a function of time.

A nonlinear differential equation contains higher powers of the dependent variable or its derivatives. Systems described by nonlinear differential equations are classified as nonlinear systems.

1.2 CLASSICAL VERSUS MODERN CONTROL THEORY

The analysis and design of linear control systems are carried out using root locus plots, Bode plots and Nyquist plots. These plots are drawn based on the transfer functions of linear systems. A transfer function establishes only the relationship between the input and the output. Transfer functions are used to analyse a system by determining the output response and error of the system for a set of given input test signals. This control system theory based on transfer function is called *classical* or *conventional control theory*. The classical control theory is based on a frequency domain approach.

This transfer function approach is not generally applicable to all types of control systems and hence these methods suffer from certain limitations. A few limitations are listed below:

(1) Conventional control theory is based on transfer functions derived from linear differential equations with constant coefficients. Hence, it is well suited for the analysis and design of linear time-invariant single-input single-output control systems and is not applicable to nonlinear, time-varying and multiple-input–multiple output systems.

(2) While deriving the transfer function, the initial conditions of the system are set to zero. Hence, in the methods based on transfer functions, initial conditions of a system cannot be included in the analysis and design.

(3) The transfer function approach does not give any information about the internal variables of the system.

(4) Frequency response methods such as Bode plots and Nyquist plots cannot be applied for the analysis and design of nonlinear and time varying systems.

(5) The system design is carried out on trial-and-error procedures and may not yield optimal control systems.

(6) The error analysis is based on certain test signals like unit impulse, unit step, unit ramp and unit parabolic inputs.

The state space technique is a time-domain approach applicable to the analysis and design of both linear and nonlinear control systems. It is based on certain chosen internal variables in a system. The control theory based on state space technique is also called *modern control theory*. Following are major advantages of the state space methods:

(1) It is applicable to both linear and nonlinear systems and time-invariant and time-varying systems.

(2) Multiple-input multiple-output systems can be represented and analysed in state space.

(3) Initial conditions can be incorporated in the analysis and design.

(4) The state space approach is generally based on certain internal variables in a system, and these variables, at any time, can be known.

(5) From the state model of a system, a discrete-time model can be obtained which can be analysed using a digital computer.

1.3 CONCEPT OF STATE AND STATE VARIABLES

Consider a linear time-invariant continuous time system with an input $u(t)$ and output response $y(t)$ as shown in Figure 1.1. The output response $y(t)$ is given by the convolution integral,

$$y(t) = \int_{-\infty}^{t} h(t - \theta) u(\theta) d\theta \qquad (1.1)$$

where $h(t)$ is the impulse response of the system, that is, the response of the system for an input of unit impulse signal.

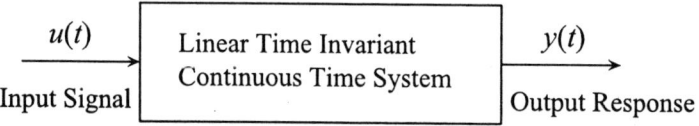

Figure 1.1 Input–Output relationship in a linear time-invariant (LTI) system

Consider that the input is applied at time $t = 0$. Then equation (1.1) can be written as

$$y(t) = \int_{-\infty}^{0} h(t-\theta)u(\theta)d\theta + \int_{0}^{t} h(t-\theta)u(\theta)d\theta$$

The first term on the right hand side is the response of the system at $t=0$, and this is the initial condition of the system. The second term represents the response of the system for an input $u(t)$. Hence, the output depends not only on the input but also on the initial condition. Then $y(t)$ can be written as

$$y(t) = y(0) + \int_{0}^{t} h(t-\theta)u(\theta)d\theta \tag{1.2}$$

The output $y(t)$ can be determined completely for a given input $u(t)$, if the initial condition $y(0)$ is known. In this equation, $y(0)$ is a state of the system at $t = 0$. The state of a system is the summary of the complete status of the system at a particular instant of time. At any particular time, the state of a system can be described by the values of a set of variables of the system. These variables that represent the state of a system are called *state variables*.

There may be many state variables that are required to represent the complete state of a system. The *state variables* of a dynamic system are the *minimum* set of variables that are necessary to specify the *state* of a system. Given the initial states at $t = t_0$ and the inputs for $t \geq t_0$, it is possible to determine the system response completely for any time $t \geq t_0$. If at least n variables are required to completely determine the output of a system, then these n variables are called the *state variables*. State variables need not be physical quantities. They need not even be measurable.

1.4 THE STATE VECTOR

If n state variables $x_1(t), x_2(t), x_3(t), -----, x_n(t)$ are necessary to completely determine the state of a system response, then these n state variables can be represented as a vector $X(t)$, and this vector is called a *state vector*.

$$X(t) = \begin{bmatrix} x_1(t) \\ x_2(t) \\ x_3(t) \\ -- \\ -- \\ x_n(t) \end{bmatrix}$$

1.5 THE STATE SPACE

The n-dimensional space whose co-ordinate axes are the $x_1 - axis$, $x_2 - axis$, $---$, $x_n - axis$ is called a *state space*. The variables $x_1(t), x_2(t), x_3(t), ------, x_n(t)$ are state variables. A state of a system can be represented by a point in the state space.

1.6 STATE MODEL FOR LINEAR SYSTEMS

A state model for a dynamic system consists of state equations and output equations. The state equations are first-order differential equations. In each state equation, the derivative of a state variable is expressed as a linear combination of the state variables chosen for the system and the system inputs. There will be as many first-order differential equations as there are state variables. The number of state equations is equal to the order of the system. The output equation is a set of equations for output variables which are linear combinations of the state variables and the inputs.

Consider a third-order linear time-invariant system, which is described by a third-order differential equation. The number of state variables required is three, which is the order of the system. Let the state variables be $x_1(t), x_2(t)$ and $x_3(t)$. Let there be two inputs $r_1(t)$ and $r_2(t)$ and two outputs $y_1(t)$ and $y_2(t)$. A set of three first-order differential equations can be written in the form

$$\frac{dx_1(t)}{dt} = a_{11} x_1(t) + a_{12} x_2(t) + a_{13} x_3(t) + b_{11} r_1(t) + b_{12} r_2(t) \tag{1.3}$$

$$\frac{dx_2(t)}{dt} = a_{21} x_1(t) + a_{22} x_2(t) + a_{23} x_3(t) + b_{21} r_1(t) + b_{22} r_2(t) \tag{1.4}$$

$$\frac{dx_3(t)}{dt} = a_{31} x_1(t) + a_{32} x_2(t) + a_{33} x_3(t) + b_{31} r_1(t) + b_{32} r_2(t) \tag{1.5}$$

These equations are linearly independent. The state equations (1.3), (1.4) and (1.5) can be written in the matrix form as

$$\begin{bmatrix} \dfrac{dx_1(t)}{dt} \\ \dfrac{dx_2(t)}{dt} \\ \dfrac{dx_3(t)}{dt} \end{bmatrix} = \begin{bmatrix} a_{11} & a_{12} & a_{13} \\ a_{21} & a_{22} & a_{23} \\ a_{31} & a_{32} & a_{33} \end{bmatrix} \begin{bmatrix} x_1(t) \\ x_2(t) \\ x_3(t) \end{bmatrix} + \begin{bmatrix} b_{11} & b_{12} \\ b_{21} & b_{22} \\ b_{31} & b_{32} \end{bmatrix} \begin{bmatrix} r_1(t) \\ r_2(t) \end{bmatrix} \tag{1.6}$$

The output equations are

$$y_1(t) = c_{11}\, x_1(t) + c_{12}\, x_2(t) + c_{13}\, x_3(t) + d_{11}\, r_1(t) + d_{12}\, r_2(t) \qquad (1.7)$$

$$y_2(t) = c_{21}\, x_1(t) + c_{22}\, x_2(t) + c_{23}\, x_3(t) + d_{21}\, r_1(t) + d_{22}\, r_2(t) \qquad (1.8)$$

The output equations (1.7) and (1.8) can be expressed in the matrix form as

$$\begin{bmatrix} y_1(t) \\ y_2(t) \end{bmatrix} = \begin{bmatrix} c_{11} & c_{12} & c_{13} \\ c_{21} & c_{22} & c_{23} \end{bmatrix} \begin{bmatrix} x_1(t) \\ x_2(t) \\ x_3(t) \end{bmatrix} + \begin{bmatrix} d_{11} & d_{12} \\ d_{21} & d_{22} \end{bmatrix} \begin{bmatrix} r_1(t) \\ r_2(t) \end{bmatrix}. \qquad (1.9)$$

The two equations (1.6) and (1.9) together are said to be the state model of the system.

For an n-th order system, the state equations are a set of n first-order differential equations. For an n-th order system with p inputs and q outputs, the state equation in the matrix form is

$$\begin{bmatrix} \dfrac{dx_1(t)}{dt} \\[4pt] \dfrac{dx_2(t)}{dt} \\[4pt] \dfrac{dx_3(t)}{dt} \\ \cdot \\ \cdot \\ \cdot \\ \dfrac{dx_n(t)}{dt} \end{bmatrix} = \begin{bmatrix} a_{11} & a_{12} & a_{13}\ldots\ldots a_{1n} \\ a_{21} & a_{22} & a_{23}\ldots\ldots a_{2n} \\ \cdot\cdot\cdot\cdot\cdot\cdot\cdot\cdot\cdot\cdot \\ \cdot\cdot\cdot\cdot\cdot\cdot\cdot\cdot\cdot\cdot \\ a_{n1} & a_{n2} & a_{n3}\ldots\ldots a_{nn} \end{bmatrix} \begin{bmatrix} x_1(t) \\ x_2(t) \\ x_3(t) \\ \cdot \\ \cdot \\ x_n(t) \end{bmatrix} + \begin{bmatrix} b_{11} & b_{12} & b_{13} & --- b_{1p} \\ b_{21} & b_{22} & b_{23} & --- b_{2p} \\ -- & -- & -- & -- \\ b_{n1} & b_{n2} & b_{n3} & --- b_{np} \end{bmatrix} \begin{bmatrix} r_1(t) \\ r_2(t) \\ -- \\ r_p(t) \end{bmatrix}. \qquad (1.10)$$

The output can be expressed in the matrix form as,

$$\begin{bmatrix} y_1(t) \\ y_2(t) \\ --- \\ y_q(t) \end{bmatrix} = \begin{bmatrix} c_{11} & c_{12} & --- c_{1n} \\ c_{21} & c_{22} & --- c_{2n} \\ & -- & \\ c_{q1} & c_{q2} & --- c_{qn} \end{bmatrix} \begin{bmatrix} x_1(t) \\ x_2(t) \\ x_3(t) \\ \cdots \\ \cdots \\ x_n(t) \end{bmatrix} + \begin{bmatrix} d_{11} & d_{12} & --- d_{1p} \\ d_{21} & d_{22} & --- d_{2p} \\ & -- & \\ d_{q1} & d_{q2} & --- d_{qp} \end{bmatrix} \begin{bmatrix} r_1(t) \\ r_2(t) \\ -- \\ r_p(t) \end{bmatrix} \qquad (1.11)$$

In general, these are written as,

$$\dot{X}(t) = A\ X(t) + B\ R(t) \; - \text{state equation}$$

$$Y(t) = C\ X(t) + D\ R(t) \; - \text{output equation}$$

where $\dot{X}(t)$ represents derivative of state variables, $X(t)$ is the state vector and $R(t)$ is the input vector representing multiple inputs; $Y(t)$ is the output vector.

$$R(t) = \begin{bmatrix} r_1(t) \\ r_2(t) \\ r_3(t) \\ .... \\ .... \\ r_p(t) \end{bmatrix} \quad and \quad Y(t) = \begin{bmatrix} y_1(t) \\ y_2(t) \\ y_3(t) \\ .... \\ .... \\ y_q(t) \end{bmatrix}$$

A is called the *system matrix*, B is the *input matrix*, C is the *output matrix* and D is called the *input–output coupling matrix* or *feed forward matrix*. The coupling matrix represents direct coupling between the input and the output. For an n-th order system with p number of inputs and q number of outputs, A is an $n \times n$ matrix, B is an $n \times p$ matrix, C is a $q \times n$ matrix and D is a $q \times p$ matrix.

A *state equation* and the *output equation* together make the *state model* for a system. A state model is not unique for a given system. State model for a system depends on the choice of state variables. As such, there can be many state models for the same system. The advantage of first-order equations is that, in addition to the input–output characteristics, the internal characteristics of the system are represented by these equations.

For time-varying systems, a state equation will be a function of time, the independent variable. They are represented in terms of n state variables and p inputs as

$$\dot{x}_1 = f_1(x_1, x_2, x_3, ---, x_n, r_1, r_2, r_3, ---, r_p, t)$$

$$\dot{x}_2 = f_2(x_1, x_2, x_3, ---, x_n, r_1, r_2, r_3, ---, r_p, t)$$

$$...$$

$$\dot{x}_n = f_n(x_1, x_2, x_3, ---, x_n, r_1, r_2, r_3, ---, r_p, t).$$

The general form of a state equation is

$$\begin{bmatrix} \dfrac{dx_1(t)}{dt} \\ \dfrac{dx_2(t)}{dt} \\ \dfrac{dx_3(t)}{dt} \\ . \\ . \\ \dfrac{dx_n(t)}{dt} \end{bmatrix} = \begin{bmatrix} a_{11}(t) & a_{12}(t) & a_{13}(t) ..... a_{1n}(t) \\ a_{21}(t) & a_{22}(t) & a_{23}(t) ..... a_{2n}(t) \\ & - - - - - \\ & - - - - - \\ a_{n1}(t) & a_{n2}(t) & a_{n3}(t) ..... a_{nn}(t) \end{bmatrix} \begin{bmatrix} x_1(t) \\ x_2(t) \\ x_3(t) \\ . \\ . \\ x_n(t) \end{bmatrix} + \begin{bmatrix} b_{11}(t) & b_{12}(t) & b_{13}(t) --- b_{1p}(t) \\ b_{21}(t) & b_{22}(t) & b_{23}(t) --- b_{2p}(t) \\ & - - - - - - \\ b_{n1}(t) & b_{n2}(t) & b_{n3}(t) --- b_{np}(t) \end{bmatrix} \begin{bmatrix} r_1(t) \\ r_2(t) \\ -- \\ r_p(t) \end{bmatrix},$$

$$\dot{X} = A(t)X(t) + B(t)R(t) .$$

The general form of an output equation is

$$
\begin{bmatrix} y_1(t) \\ y_2(t) \\ --- \\ y_q(t) \end{bmatrix} =
\begin{bmatrix} c_{11}(t) & c_{12}(t) & --- c_{1n}(t) \\ c_{21}(t) & c_{22}(t) & --- c_{2n}(t) \\ & --- \\ c_{q1}(t) & c_{q2}(t) & --- c_{qn}(t) \end{bmatrix}
\begin{bmatrix} x_1(t) \\ x_2(t) \\ x_3(t) \\ \cdots \\ \cdots \\ x_n(t) \end{bmatrix} +
\begin{bmatrix} d_{11}(t) & d_{12}(t) --- d_{1p}(t) \\ d_{21}(t) & d_{22}(t) --- d_{2p}(t) \\ & --- \\ d_{q1}(t) & d_{q2}(t) --- d_{qp}(t) \end{bmatrix}
\begin{bmatrix} r_1(t) \\ r_2(t) \\ -- \\ r_p(t) \end{bmatrix} ,
$$

$$Y(t) = C(t)X(t) + D(t)R(t) .$$

1.7 THE STATE DIAGRAM

A pictorial representation of the state equation and the output equation is called the *state diagram* and can be drawn from the state model:

$$\dot{X}(t) = A\ X(t) + B\ R(t)$$

$$Y(t) = C\ X(t) + D\ R(t)$$

The state diagram is shown in Figure 1.2.

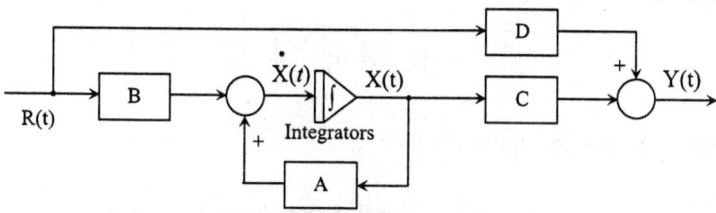

Figure 1.2 The State Diagram

1.8 STATE MODEL FROM DIFFERENTIAL EQUATIONS

(A) Differential Equations without any Derivative of Input

The method of obtaining state model from differential equations is explained by means of some examples.

Example 1.1 The differential equation that describes a linear time-invariant system is

$$\frac{d^2 y(t)}{dt^2} + 6\frac{dy(t)}{dt} + 8y(t) = r(t) .$$

Obtain a state model for the system and draw the state diagram.

Solution: Choose a set of two state variables

$$x_1 = y \qquad and \quad x_2 = \dot{x}_1 = \dot{y}$$

$$\text{So that} \quad \dot{x}_1 = x_2 . \tag{A}$$

From the given differential equation,

$$\frac{d^2 y(t)}{dt^2} = -8y(t) - 6\frac{dy(t)}{dt} + r(t)$$

$$\ddot{y} = \dot{x}_2 = -8x_1 - 6x_2 + r(t) . \tag{B}$$

Equations (A) and (B) are state equations, and in matrix form, they can be written as

$$\begin{bmatrix} \dot{x}_1 \\ \dot{x}_2 \end{bmatrix} = \begin{bmatrix} 0 & 1 \\ -8 & -6 \end{bmatrix} \begin{bmatrix} x_1 \\ x_2 \end{bmatrix} + \begin{bmatrix} 0 \\ 1 \end{bmatrix} r(t) .$$

The output equation is

$$y = \begin{bmatrix} 1 & 0 \end{bmatrix} \begin{bmatrix} x_1 \\ x_2 \end{bmatrix} .$$

The state diagram is shown in Figure E1.1.

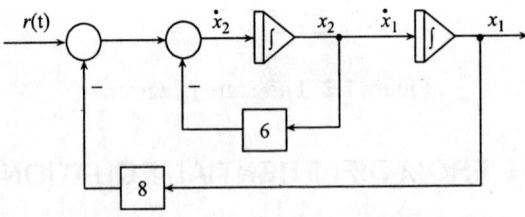

Figure E 1.1 State diagram for Example 1.1

Example 1.2 A third-order LTI system is represented by the differential equation

$$\frac{d^3y(t)}{dt^2}+6\frac{d^2y(t)}{dt^2}+11\frac{dy(t)}{dt}+6y(t)=3r(t).$$

Represent the system in a state space model and draw the state diagram.

Solution: Choose a set of three state variables

$$x_1=y \qquad x_2=\dot{x}_1=\dot{y} \ \text{ and } \ x_3=\dot{x}_2=\ddot{y},$$

so that $\dot{x}_1=x_2$　　　　　　　　　　　　　　　　(A)

and $\dot{x}_2=x_3$　　　　　　　　　　　　　　　　(B)

From the given differential equation,

$$\frac{d^3y(t)}{dt^2}=-6y(t)-11\frac{dy(t)}{dt}-6\frac{d^2y(t)}{dt^2}+3r(t)$$

$$\dddot{y}=\dot{x}_3=-6x_1-11x_2-6x_3+3r(t)\qquad\qquad(C)$$

Equations (A), (B) and (C) are state equations, and in matrix form, they can be written as

$$\begin{bmatrix}\dot{x}_1\\\dot{x}_2\\\dot{x}_3\end{bmatrix}=\begin{bmatrix}0&1&0\\0&0&1\\-6&-11&-6\end{bmatrix}\begin{bmatrix}x_1\\x_2\\x_3\end{bmatrix}+\begin{bmatrix}0\\0\\3\end{bmatrix}r(t).$$

The output equation is

$$y=\begin{bmatrix}1&0&0\end{bmatrix}\begin{bmatrix}x_1\\x_2\\x_3\end{bmatrix}.$$

The state diagram is shown in Figure E 1.2.

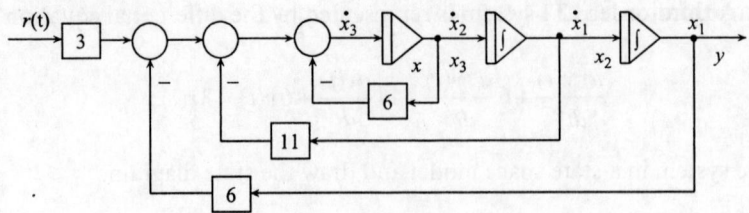

Figure E 1.2 State diagram for Example 1.2

(B) Differential Equations Involving Derivatives of Input

A derivative of input cannot be a part of the state equation. A method to avoid a term with a derivative of input in the state equation is to choose the state variables in a different form.

Consider a second-order system represented by the differential equation

$$\ddot{y} + a_1 \dot{y} + a_0 y = b_2 \ddot{r} + b_1 \dot{r} + b_0 r \,. \tag{1.12}$$

Choose a set of two state variables as

$$x_1 = y - k_0 r \tag{1.13}$$

$$x_2 = \dot{x}_1 - k_1 r \,. \tag{1.14}$$

From equation (1.14),

$$\dot{x}_1 = x_2 + k_1 r \,. \tag{A}$$

From equation (1.13),

$$y = x_1 + k_0 r \,. \tag{1.15}$$

Differentiating equation (1.15),

$$\dot{y} = \dot{x}_1 + k_0 \dot{r} \,. \tag{1.16}$$

Combining equations (A) and (1.16),

$$\dot{y} = (x_2 + k_1 r) + k_0 \dot{r} \,. \tag{1.17}$$

Differentiating equation (1.17),

$$\ddot{y} = \dot{x}_2 + k_1 \dot{r} + k_0 \ddot{r} \tag{1.18}$$

Substituting equations (1.15), (1.17) and (1.18) in the given differential equation,

$$(\ddot{x}_2 + k_1 \dot{r} + k_0 \ddot{r}) + a_1 (\dot{x}_2 + k_1 r + k_0 \dot{r}) + a_0 (x_1 + k_0 r) = b_2 \ddot{r} + b_1 \dot{r} + b_0 r$$

$$\dot{x}_2 + a_1 x_2 + a_0 x_1 = (b_2 - k_0) \ddot{r} + (b_1 - k_1 - a_1 k_0) \dot{r} + (b_0 - a_1 k_1 - a_0 k_0) r. \tag{1.19}$$

For the state equation to be independent of derivatives of input,

$$(b_2 - k_0) = 0 \ \ or \ \ k_0 = b_2 \tag{1.20}$$

$$(b_1 - k_1 - a_1 k_0) = 0 \ \ or \ \ k_1 = b_1 - a_1 k_0 \tag{1.21}$$

$$k_2 = (b_0 - a_1 k_1 - a_0 k_0) \ . \tag{1.22}$$

From equation (1.19),

$$\dot{x}_2 = -a_1 x_2 - a_0 x_1 + k_2 r \tag{B}$$

Equations (A) and (B) are state equations and they can be kept in the matrix form as

$$\begin{bmatrix} \dot{x}_1 \\ \dot{x}_2 \end{bmatrix} = \begin{bmatrix} 0 & 1 \\ -a_0 & -a_1 \end{bmatrix} \begin{bmatrix} x_1 \\ x_2 \end{bmatrix} + \begin{bmatrix} k_1 \\ k_2 \end{bmatrix} r \ .$$

The output equation is

$$y = x_1 + k_0 r = \begin{bmatrix} 1 & 0 \end{bmatrix} \begin{bmatrix} x_1 \\ x_2 \end{bmatrix} + k_0 r \ .$$

Example 1.3

Represent the system in the state space model, the differential equation description of which is $\ddot{y} + 5 \dot{y} + 4y = 3r(t) + 2\dot{r}(t)$. Choose the state variables as $x_1 = y - k_0 r$ and $x_2 = \dot{x}_1 - k_1 r$.

Solution:

$$x_1 = y - k_0 r \tag{1}$$

$$x_2 = \dot{x}_1 - k_1 r \tag{2}$$

From equation (2),

$$\dot{x}_1 = x_2 + k_1 r . \tag{A}$$

From equation (1),

$$y = x_1 + k_0 r . \tag{3}$$

Differentiating this equation,

$$\dot{y} = \dot{x}_1 + k_0 \dot{r} . \tag{4}$$

Combining equations (A) and (4),

$$\dot{y} = (x_2 + k_1 r) + k_0 \dot{r} . \tag{5}$$

Differentiating this equation,

$$\ddot{y} = \dot{x}_2 + k_1 \dot{r} + k_0 \ddot{r} . \tag{6}$$

Substituting equations (3), (5) and (6) in the given differential equation,

$$\dot{x}_2 + k_1 \dot{r} + k_0 \ddot{r} + 5(x_2 + k_1 r + k_0 \dot{r}) + 4(x_1 + k_0 r) = 3r + 2\dot{r}$$

$$\dot{x}_2 = -4x_1 - 5x_2 - k_0 \ddot{r} + (2 - k_1 - 5k_0)\dot{r} + (3 - 5k_1 - 4k_0)r . \tag{7}$$

For the state equation to be independent of derivatives of input,

$$k_0 = 0 .$$

$$(2 - k_1 - 5k_0) = 0 \quad or \quad k_1 = 2$$

$$k_2 = (3 - 5k_1 - 4k_0) = -7 .$$

Equation (7) becomes

$$\dot{x}_2 = -4x_1 - 5x_2 - 7r . \tag{B}$$

Equations (A) and (B) are state equations and they can be kept in the matrix form as

$$\begin{bmatrix} \dot{x}_1 \\ \dot{x}_2 \end{bmatrix} = \begin{bmatrix} 0 & 1 \\ -4 & -5 \end{bmatrix} \begin{bmatrix} x_1 \\ x_2 \end{bmatrix} + \begin{bmatrix} 2 \\ -7 \end{bmatrix} r .$$

The output equation is

$$y = x_1 + k_0\, r = x_1 = \begin{bmatrix} 1 & 0 \end{bmatrix} \begin{bmatrix} x_1 \\ x_2 \end{bmatrix}.$$

Example 1.4 Represent the system in the state space model, the differential equation description of which is

$$\dddot{y} + 6\,\ddot{y} + 5\,\dot{y} + 4y = 3r(t) + 2\dot{r}(t) + \ddot{r}.$$

Choose the state variables as

$$x_1 = y - k_0\, r,\quad x_2 = \dot{x}_1 - k_1\, r \quad \text{and} \quad x_3 = \dot{x}_2 - k_2\, r.$$

Solution:

$$x_1 = y - k_0\, r \tag{1}$$

$$x_2 = \dot{x}_1 - k_1\, r \tag{2}$$

$$\text{and}\ \ x_3 = \dot{x}_2 - k_2\, r \tag{3}$$

From equation (2),

$$\dot{x}_1 = x_2 + k_1\, r \tag{A}$$

This is a state equation.
From equation (3),

$$\dot{x}_2 = x_3 + k_2\, r \tag{B}$$

This is also a state equation.
From equation (1),

$$y = x_1 + k_0\, r. \tag{4}$$

Differentiating this equation,

$$\dot{y} = \dot{x}_1 + k_0\, \dot{r} = x_2 + k_1\, r + k_0\, \dot{r}. \tag{5}$$

Differentiating this equation,

$$\ddot{y} = \dot{x}_3 + k_2\, r + k_1\, \dot{r} + k_0\, \ddot{r}. \tag{6}$$

Differentiating this equation again,

$$\dddot{y} = \dot{x_3} + k_2 \, \dot{r} + k_1 \, \ddot{r} + k_0 \, \dddot{r} \ . \tag{7}$$

Substituting equations (4), (5), (6) and (7) in the given differential equation,

$$(\dot{x_3} + k_2 \, \dot{r} + k_1 \, \ddot{r} + k_0 \, \dddot{r}) + 6(x_3 + k_2 \, r + k_1 \, \dot{r} + k_0 \, \ddot{r}) + 5(x_2 + k_1 \, r + k_0 \, \dot{r}) + 4(x_1 + k_0 \, r)$$

$$= 3r(t) + 2\dot{r}(t) + \ddot{r} \ .$$

$$\dot{x_3} = -4x_1 - 5x_2 - 6x_3 + (3 - 4k_0 - 5k_1 - 6k_2)r + (2 - 5k_0 - 6k_1 - k_2)\dot{r}$$

$$+ (1 - 6k_0 - k_1)\ddot{r} - k_0 \, \dddot{r} \ . \tag{8}$$

For the state equation to be independent of the derivatives of input r, the coefficients of $\dot{r}$, $\ddot{r}$ and $\dddot{r}$ in equation (8) are to be zero.

Hence $k_0 = 0$ $\qquad (1 - 6k_0 - k_1) = 0 \ or \ k_1 = 1$.

$$2 - 5k_0 - 6k_1 - k_2 = 0 \qquad\qquad or \ k_2 = -4 \ .$$

$$k_3 = 3 - 4k_0 - 5k_1 - 6k_2 = 22 \ .$$

From equation (8),

$$\dot{x_3} = -4x_1 - 5x_2 - 6x_3 + 22r \ . \tag{C}$$

Keeping equations A, B and C in matrix form,

$$\begin{bmatrix} \dot{x_1} \\ \dot{x_2} \\ \dot{x_3} \end{bmatrix} = \begin{bmatrix} 0 & 1 & 0 \\ 0 & 0 & 1 \\ -4 & -5 & -6 \end{bmatrix} \begin{bmatrix} x_1 \\ x_2 \\ x_3 \end{bmatrix} + \begin{bmatrix} 1 \\ -4 \\ 22 \end{bmatrix} r(t) \ .$$

The output equation is

$$y = x_1 + k_0 \, r = x_1 = \begin{bmatrix} 1 & 0 & 0 \end{bmatrix} \begin{bmatrix} x_1 \\ x_2 \\ x_3 \end{bmatrix} .$$

1.9 STATE MODEL FROM A STATE DIAGRAM

For a system, state equations can be written down from a given state diagram. The method is to choose the output of every integrator as a state variable. The input to an integrator is the derivative of a state variable, and its equation can be written down from the state diagram.

Example 1. 5 A state diagram for a dynamic system is shown in Figure E 1.5. Write down the state equations and the output equation.

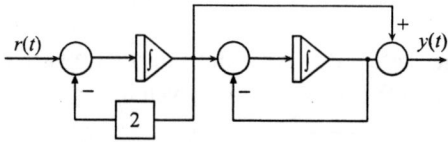

Figure E 1.5 for Example 1.5

Solution: There are two integrators and hence two state variables. The output of each integrator is taken as a state variable. The state variables and their derivatives are marked in the state diagram given in Figure E 1.5(a).

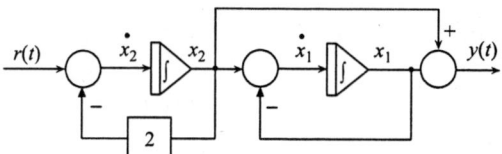

Figure E 1.5 (a) for Example E 1.5

From the state diagram,

$$\dot{x}_1 = -x_1 + x_2 \tag{A}$$

$$\text{and } \dot{x}_2 = -2x_2 + r(t). \tag{B}$$

Keeping equations (A) and (B) in matrix form,

$$\begin{bmatrix} \dot{x}_1 \\ \dot{x}_2 \end{bmatrix} = \begin{bmatrix} -1 & 1 \\ 0 & -2 \end{bmatrix} \begin{bmatrix} x_1 \\ x_2 \end{bmatrix} + \begin{bmatrix} 0 \\ 1 \end{bmatrix} r(t).$$

The output equation is

$$y(t) = x_1 + x_2 = \begin{bmatrix} 1 & 1 \end{bmatrix} \begin{bmatrix} x_1 \\ x_2 \end{bmatrix}.$$

Example 1. 6 A state diagram for a dynamic system is shown in Figure E 1.6. Write down the state equations and the output equation.

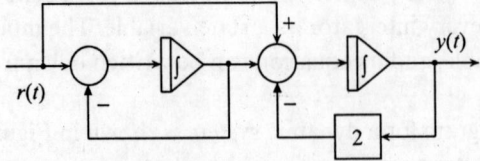

Figure E 1.6 for Example 1.6

Solution: The output of each integrator is taken to be a state variable. The state variables and their derivatives are marked in the state diagram given in Figure E 1.6(a).

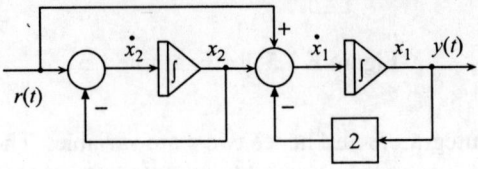

Figure E 1.6 (a) for Example 1.6

From the state diagram,

$$\dot{x}_1 = -2x_1 + x_2 + r(t) \tag{A}$$

$$\text{and } \dot{x}_2 = -x_2 + r(t). \tag{B}$$

Equations (A) and (B) are state equations and they can be kept in the matrix form,

$$\begin{bmatrix} \dot{x}_1 \\ \dot{x}_2 \end{bmatrix} = \begin{bmatrix} -2 & 1 \\ 0 & -1 \end{bmatrix} \begin{bmatrix} x_1 \\ x_2 \end{bmatrix} + \begin{bmatrix} 1 \\ 1 \end{bmatrix} r(t).$$

The output equation is

$$y(t) = x_1 = \begin{bmatrix} 1 & 0 \end{bmatrix} \begin{bmatrix} x_1 \\ x_2 \end{bmatrix}.$$

Example 1.7 A state diagram of a dynamic system is shown in Figure E 1.7. Write down the state equations and the output equation.

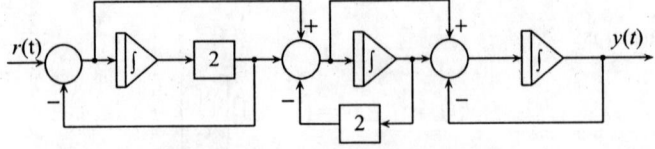

Figure E 1.7 for Example E 1.7

Solution: The output of each integrator is taken to be a state variable. The state variables and their derivatives are marked in the state diagram given in Figure E 1.7(a).

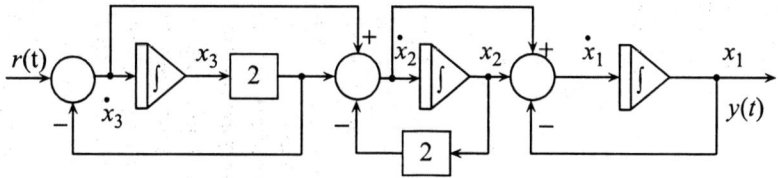

Figure E 1.7 (a) for Example E 1.7

From the state diagram,

$$\dot{x}_1 = -x_1 + x_2 + \dot{x}_2 \tag{1}$$

$$\dot{x}_2 = -2x_2 + 2x_3 + \dot{x}_3 \tag{2}$$

$$\dot{x}_3 = -2x_3 + r(t). \tag{3}$$

Combining equations (2) and (3),

$$\dot{x}_2 = -2x_2 + 2x_3 - 2x_3 + r(t) = -2x_2 + r(t). \tag{4}$$

Combining equations (1) and (4),

$$\dot{x}_1 = -x_1 + x_2 - 2x_2 + r(t) = -x_1 - x_2 + r(t). \tag{5}$$

Equations (3), (4) and (5) are state equations and can be represented in matrix form,

$$\begin{bmatrix} \dot{x}_1 \\ \dot{x}_2 \\ \dot{x}_3 \end{bmatrix} = \begin{bmatrix} -1 & -1 & 0 \\ 0 & -2 & 0 \\ 0 & 0 & -2 \end{bmatrix} \begin{bmatrix} x_1 \\ x_2 \\ x_3 \end{bmatrix} + \begin{bmatrix} 1 \\ 1 \\ 1 \end{bmatrix} r(t).$$

The output equation is

$$y(t) = x_1 = \begin{bmatrix} 1 & 0 & 0 \end{bmatrix} \begin{bmatrix} x_1 \\ x_2 \\ x_3 \end{bmatrix}.$$

1.10 STATE MODEL FOR NONLINEAR SYSTEMS

Differential equations for nonlinear systems generally contain higher powers of derivatives or products of derivatives. For such a system, the state equation and the output equation cannot be kept in matrix form. As an example, consider a nonlinear time-invariant system described by

$$\frac{d^3y}{dt^3} + a\frac{d^2y}{dt^2}\frac{dy}{dt} + b\frac{d^2y}{dt^2} + c\left(\frac{dy}{dt}\right)^2 + y = 0.$$

This can be written as

$$\overset{\cdots}{y} + a(\overset{\cdot\cdot}{y})(\overset{\cdot}{y}) + b\overset{\cdot\cdot}{y} + c(\overset{\cdot}{y})^2 + y = 0.$$

Let $x_1 = y$, $x_2 = \overset{\cdot}{x_1} = \overset{\cdot}{y}$ and $x_3 = \overset{\cdot}{x_2} = \overset{\cdot\cdot}{y}$ be three state variables. Then the two state equations are

$$\overset{\cdot}{x_1} = x_2$$

$$\text{and } \overset{\cdot}{x_2} = x_3.$$

The third state equation can be obtained from the given differential equation as

$$\frac{d^3y}{dt^3} = -y - c\left(\frac{dy}{dt}\right)^2 - b\frac{d^2y}{dt^2} - a\frac{d^2y}{dt^2}\frac{dy}{dt}.$$

Or

$$\overset{\cdot}{x_3} = -x_1 - c(x_2)^2 - bx_3 - a x_3 x_2.$$

It can be observed that these equations cannot be kept in matrix form. The general form of the state and output equations are

$$\overset{\cdot}{X} = f\left(X(t), R(t)\right)$$

and

$$Y(t) \quad g\left(X(t), R(t)\right).$$

For a nonlinear time-varying system, the state and the output equations are also functions of time. The general form is

$$\overset{\cdot}{X} = f\left(X(t), R(t), t\right)$$

and

$$Y(t) = g\left(X(t), R(t), t\right).$$

Example 1.8 A dynamic system is represented by the differential equation

$$\frac{d^2y}{dt^2} + a\frac{dy}{dt} + (1 + b^2 y^2)y = 0$$

Represent the system in a state space model.

Solution: Let $x_1 = y$ and $x_2 = \dot{y} = \dot{x}_1$ be the two state variables.

$$\dot{x}_1 = x_2 \tag{A}$$

From the given differential equation

$$\frac{d^2 y}{dt^2} = -a\frac{dy}{dt} - (1 + b^2 y^2)y$$

$$\ddot{y} = -a\dot{y} - (1 + b^2 y^2)y = -a\dot{y} - y - b^2 y^3$$

This can be written as

$$\dot{x}_2 = -ax_2 - x_1 - b^2 x_1^3 \tag{B}$$

Equations (A) and (B) are state equations but they cannot be kept in matrix form.

1.11 LINEARIZATION

A nonlinear system can be linearized about its operating state. In order to obtain a linear mathematical model for a nonlinear system, it must be assumed that the variables deviate only slightly from a given state. The method of linearization is based on Taylor's series. Consider a function $y = f(x)$. If the operating point is (x_0, y_0), using Taylor's series

$$f(x) = f(x_0) + f'(x_0)(x - x_0) + \frac{1}{2!}f''(x_0)(x - x_0)^2 + ----$$

When x is very close to x_0 or in other words if the variation $(x - x_0)$ is very small, higher order terms can be neglected so that as an approximation

$$f(x) = f(x_0) + f'(x_0)(x - x_0)$$

$$\text{or}\quad y = y_0 + f'(x_0)(x - x_0)$$

This shows that $(y - y_0)$ is proportional to $(x - x_0)$. This equation gives a linear mathematical model for a nonlinear system.

Linearization with Two Variables

Let $z = f(x, y)$ be a nonlinear equation where x and y are two variables. Let (x_0, y_0) be an operating point. Using Taylor's series expansion and neglecting higher degree terms

$$(z-z_0) = \frac{\partial f}{\partial x}\bigg|_{(x_0,y_0)}(x-x_0) + \frac{\partial f}{\partial y}\bigg|_{(x_0,y_0)}(y-y_0)$$

$$(z-z_0) = k_1(x-x_0) + k_2(y-y_0)$$

Let

$$\delta z = (z-z_0),\ \delta x = (x-x_0)\ \text{and}\ \delta y = (y-y_0)$$

Hence

$$\delta z = \begin{bmatrix} k_1 & k_2 \end{bmatrix}\begin{bmatrix} \delta x \\ \delta y \end{bmatrix}$$

$$= J(x_0,y_0)\begin{bmatrix} \delta x \\ \delta y \end{bmatrix} \tag{1.23}$$

where

$$J(x_0,y_0) = \begin{bmatrix} k_1 & k_2 \end{bmatrix} = \begin{bmatrix} \dfrac{\partial f}{\partial x}\bigg|_{(x_0,y_0)} & \dfrac{\partial f}{\partial y}\bigg|_{(x_0,y_0)} \end{bmatrix}$$

Equation (1.23) represents a linear model for the nonlinear system $z = f(x, y)$, about the point (x_0, y_0). The variables in this linearized model do not represent the actual values but the deviations from the values at the steady state operating point.

Linearization of State Equations

Consider the state equations for a time-invariant second-order system.

$$\dot{x}_1 = f_1(x_1, x_2)\ \text{and}\ \dot{x}_2 = f_2(x_1, x_2)$$

Let $\{x_1(0), x_2(0)\}$ be an operating point.

$$\begin{bmatrix} \dot{\delta x_1} \\ \dot{\delta x_2} \end{bmatrix} = J\{x_1(0), x_2(0)\}\begin{bmatrix} \delta x_1 \\ \delta x_2 \end{bmatrix} = \begin{bmatrix} k_1 & k_2 \\ k_3 & k_4 \end{bmatrix}\begin{bmatrix} \delta x_1 \\ \delta x_2 \end{bmatrix}, \tag{1.24}$$

where

$$\dot{\delta x_1} = \dot{x}_1 - \dot{x}_1(0) \qquad \dot{\delta x_2} = \dot{x}_2 - \dot{x}_2(0)$$

$$\delta x_1 = x_1 - x_1(0) \quad \delta x_2 = x_2 - x_2(0).$$

Equation (1.24) is a set of linearized state equations in terms of small variations in state variables x_1 and x_2.

$$J\{x_1(0), x_2(0)\} = \begin{bmatrix} k_1 & k_2 \\ k_3 & k_4 \end{bmatrix} = \begin{bmatrix} \dfrac{\partial f_1}{\partial x_1} & \dfrac{\partial f_1}{\partial x_2} \\ \dfrac{\partial f_2}{\partial x_1} & \dfrac{\partial f_2}{\partial x_2} \end{bmatrix}_{\{x_1(0),\ x_2(0)\}} \tag{1.25}$$

Using equation (1.25), the Eigen values of the linearized state model can be determined by solving the equation

$$\begin{vmatrix} \lambda - k_1 & -k_2 \\ -k_3 & \lambda - k_4 \end{vmatrix} = 0 \,.$$

Example 1.9 Linearize the equation $\dot{y} + 2y^3 - 2 = 0$ about the point y = 0.5.

Solution: Replace the nonlinear term by a variable z. Let $z = 2y^3$
The linearized model for the system is

$$(z - z_0) = \frac{dz}{dy}\bigg|_{y=0.5} (y - y_0) \tag{1}$$

$$z_0 = 2(y_0)^3 = 2(0.5)^3 = 0.25$$

$$\frac{dz}{dy}\bigg|_{y=0.5} = 6y^2\bigg|_{y=0.5} = 1.5 \,.$$

Substituting these values in equation (1),

$$(z - 0.25) = 1.5(y - 0.5) \text{ or } z = 1.5y\text{-}0.5 \,.$$

The given nonlinear equation can be linearized about the point y = 0.5 as

$$\dot{y} + 1.5y - 0.5 - 2 = 0 \ \text{ or } \ \dot{y} + 1.5y - 2.5 = 0 \,.$$

Example 1.10 Linearize the differential equation $\ddot{x} + 2x^2 \dot{x} + 3(\dot{x})^2 + x = 0$ about the operating point $(x, \dot{x}) = (0.5, 1)$.

Solution:

$$\ddot{x} + 2x^2 \dot{x} + 3(\dot{x})^2 + x = 0 \,. \tag{1}$$

Let $y = \dot{x}$. The given differential equation is

$$\dot{y} + 2(x)^2 y + 3(y)^2 + x = 0$$

The operating state is $(x_0, y_0) = (0.5, 1)$
Let the nonlinear term be written as

$$z = 2x^2 \dot{x} + 3(\dot{x})^2 \,.$$

$$= 2(x)^2 y + 3(y)^2 \quad z_0 = 2(0.5)^2 1 + 3(1)^2 = 3.5 \,.$$

$$\frac{\partial z}{\partial x}\bigg|_{(0.5,1)} = 4xy\big|_{(0.5,1)} = 2 \qquad \frac{\partial z}{\partial y}\bigg|_{(0.5,1)} = 2(x)^2 + 6y\big|_{(0.5,1)} = 6.5 \,.$$

$$z - z_0 = 2(x - x_0) + 6.5(y - y_0)$$

$$(z - 3.5) = 2(x - 0.5) + 6.5(y - 1).$$

Or $$z = 2x + 6.5y - 4.$$ (2)

Substituting equation (2) in equation (1),

$$\ddot{x} + [2x + 6y - 4] + x = 0$$

$$\ddot{x} + 6\dot{x} + 3x - 4 = 0.$$

Example 1.11 A nonlinear equation is represented by the differential equation
$$\ddot{x} + 2x^2\,\dot{x} + 3(\dot{x})^2 + x = 0.$$

Obtain a state model for the system about the operating point $(x, \dot{x}) = (0.5, 1)$.

Solution: $\ddot{x} + 2x^2\,\dot{x} + 3(\dot{x})^2 + x = 0$

Let $$x_1 = x \ \ and \ \ x_2 = \dot{x}.$$

The given differential equation now becomes

$$\dot{x}_2 + 2(x_1)^2 x_2 + 3(x_2)^2 + x_1 = 0.$$

The operating state is $$(x_{10}, x_{20}) = (0.5, 1).$$
The state equations are

$$\dot{x}_1 = x_2 = f_1(x_1, x_2)$$ (1)

$$\dot{x}_2 = -2(x_1)^2 x_2 - 3(x_2)^2 - x_1 = f_2(x_1, x_2)$$ (2)

$$\frac{\partial f_1}{\partial x_1} = 0 \qquad\qquad \frac{\partial f_1}{\partial x_2} = 1.$$

$$\left.\frac{\partial f_2}{\partial x_1}\right|_{(0.5,1)} = (-4x_1 x_2 - 1)\big|_{(0.5,1)} = -3 \qquad \left.\frac{\partial f_2}{\partial x_2}\right|_{(0.5,1)} = -2(x_1)^2 - 6x_2\big|_{(0.5,1)} = -6.5.$$

$$\dot{x}_1(0) = f_1(x_{10}, x_{20}) = x_{20} = 1$$

$$\dot{x}_2(0) = f_2(x_{10}, x_{20}) = -2(0.5)^2 1 - 3(1)^2 - 0.5 = -4.$$

The linearized state equations are

$$
\begin{bmatrix} \dot{x}_1 - \dot{x}_1(0) \\ \dot{x}_2 - \dot{x}_2(0) \end{bmatrix} = J(x_{10}, x_{20}) \begin{bmatrix} x_1 - x_{10} \\ x_2 - x_{20} \end{bmatrix}
$$

$$
\begin{bmatrix} \dot{x}_1 - 1 \\ \dot{x}_2 + 4 \end{bmatrix} = \begin{bmatrix} 0 & 1 \\ -3 & -6.5 \end{bmatrix} \begin{bmatrix} x_1 - 0.5 \\ x_2 - 1 \end{bmatrix}.
$$

The Eigen values of the linearized model can be determined as

$$
\begin{vmatrix} \lambda & -1 \\ 3 & \lambda + 6.5 \end{vmatrix} = 0 \qquad\qquad \lambda^2 + 6.5\lambda + 3 = 0 .
$$

The Eigen values are $\lambda = -0.5\ and\ \lambda = -6$.

1.12 STATE VARIABLES FOR PHYSICAL SYSTEMS

Electrical System

The Capacitor

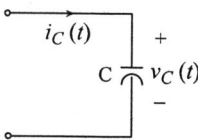

Figure 1.3 State variable for a capacitor

Consider a capacitor as shown in Figure 1.3. The current through the capacitor is

$$
i_C(t) = C \frac{dv_C(t)}{dt} .
$$

The voltage across the capacitor is

$$
v_C(t) = \frac{1}{C} \int_{-\infty}^{t} i_C(t)\, dt
$$

$$
= \frac{1}{C} \int_{-\infty}^{0} i_C(t)\, dt + \frac{1}{C} \int_{0}^{t} i_C(t)\, dt .
$$

The first term on the right hand side is the initial voltage across the capacitor at $t = 0$. It indicates the state of the capacitor at $t = 0$.

$$v_C(t) = v_C(0) + \frac{1}{C}\int_0^t i_C(t)\,dt.$$ (1.26)

Comparing equations (1.2) and (1.26), it can be concluded that the capacitor voltage can be taken as a state variable. Since the charge on a capacitor is given by $q(t) = C v_C(t)$, the charge on the capacitor $q(t)$ can also be taken as a state variable.

The Inductor

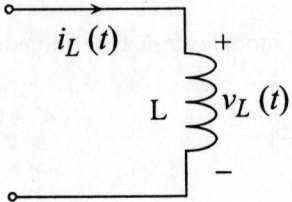

Figure 1.4 State variable for an inductor

Consider an inductor as shown in Figure 1.4. The voltage across the inductor is

$$v_L(t) = L\frac{di_L(t)}{dt}.$$

The current through the inductor is

$$i_L(t) = \frac{1}{L}\int_{-\infty}^t v_L(t)\,dt$$

$$= \frac{1}{L}\int_{-\infty}^0 v_L(t)\,dt + \frac{1}{L}\int_0^t v_L(t)\,dt.$$

The first term on the right hand side is the initial current through the inductor at $t = 0$. It indicates the state of the inductor at $t = 0$.

$$i_L(t) = i_L(0) + \frac{1}{L}\int_0^t v_L(t)\,dt.$$ (1.27)

Comparing equations (1.2) and (1.27), it can be concluded that the inductor current can be taken as state variable. Since the electromagnetic flux linkage is $\psi(t) = L i_L(t)$, the flux linkage can also be taken as a state variable.

Note that inductors and capacitors are energy-storing elements. State variables are associated with energy-storing elements. A resistor absorbs energy and dissipates it as heat. It cannot store energy. Hence, no state variable is associated with a resistor.

Mechanical Translational System

Just as a resistor, a capacitor and an inductor form the elements of a basic electrical circuit, mass, linear spring and a damper form the elements of a basic mechanical translational system.

The Mass

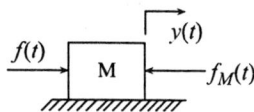

Figure 1.5 State variable for a mass M

Consider a mass M moving on a plane, as shown in Figure 1.5. The externally applied force that causes the motion is $f(t)$. The reactive force is $f_M(t)$, which is the force due to acceleration of the mass. If $y(t)$ is the displacement of mass and $u(t)$ is the velocity of mass, the dynamic equation of motion is

$$f(t) = f_M(t) = M\frac{d^2y}{dt^2} = M\frac{du(t)}{dt}.$$

The velocity of mass can be found out as

$$u(t) = \frac{1}{M}\int_{-\infty}^{t} f(t)\,dt$$

$$= \frac{1}{M}\int_{-\infty}^{0} f(t)\,dt + \frac{1}{M}\int_{0}^{t} f(t)\,dt.$$

The first term on the right hand side is the initial velocity of mass at $t = 0$. It indicates the state of the mass at $t = 0$.

$$u(t) = u(0) + \frac{1}{M}\int_{0}^{t} f(t)\,dt. \tag{1.28}$$

Comparing equations (1.2) and (1.28), it can be concluded that the velocity of mass can be taken as a state variable.

Since $u(t) = \frac{d}{dt}y(t)$, the displacement of mass can also be taken as a state variable.

The Spring

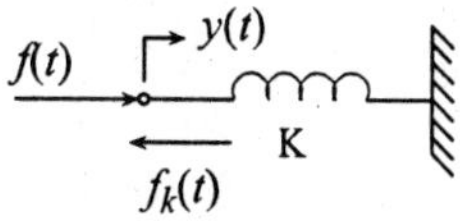

Figure 1.6 State variable for a linear spring

Consider a mechanical spring, which is considered to be linear, as shown in Figure 1.6. The externally applied force is $f(t)$. The force of reaction of the spring is proportional to the displacement and the constant of proportionality is the spring constant K. If $f_K(t)$ is the reactive force,

$$f(t) = f_K(t) = Ky(t) = K \int_{-\infty}^{t} u(t)$$

$$f_K(t) = K \int_{-\infty}^{0} u(t)\,dt + K \int_{0}^{t} u(t)\,dt .$$

The first term on the right hand side is the initial value of the spring force at $t = 0$. It indicates the state of the spring at $t = 0$.

$$f_K(t) = f_K(0) + K \int_{0}^{t} u(t)\,dt . \tag{1.29}$$

Comparing equations (1.2) and (1.29), it can be concluded that the spring force can be taken as state variable. Since the spring force is proportional to the displacement, displacement of spring can also be taken as a state variable.

A mass in motion stores kinetic energy, and a spring, stretched or compressed, stores potential energy. Thus, state variables are associated with a mass and a spring. A mechanical damper or a dash pot absorbs energy and dissipates it as heat. It cannot store energy. Hence, no state variable is associated with a dashpot.

Mechanical Rotational System

Rotating Shaft

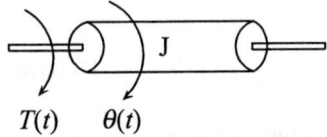

$$T(t) \qquad \theta(t)$$

Figure 1.7 State Variable for the inertia of shaft

Consider a rotating shaft with moment of inertia J, shown in Figure 1.7. If $T(t)$ is the input torque to the shaft and $\theta(t)$ is the angular displacement, then the torque due to acceleration of the shaft $T_J(t)$ is

$$T(t) = T_J(t) = J \frac{d^2\theta(t)}{dt^2} = J \frac{d\omega(t)}{dt},$$

where $\omega(t) = \dfrac{d\theta(t)}{dt}$ is the angular speed of rotation,

$$\omega(t) = \frac{1}{J} \int_{-\infty}^{t} T(t)\,dt$$

$$\omega(t) = \frac{1}{J} \int_{-\infty}^{0} T(t)\,dt + \frac{1}{J} \int_{0}^{t} T(t)\,dt \ .$$

The term on the right hand side is the initial value of the angular speed at $t = 0$. It indicates the state of the shaft at $t = 0$.

$$\omega(t) = \omega(0) + \frac{1}{J} \int_{0}^{t} T(t)\,dt \tag{1.30}$$

Comparing equation (1.2) and (1.30), it can be understood that the angular speed of the rotating shaft can be taken as a state variable. Since $\omega(t) = \dfrac{d\theta(t)}{dt}$, the angular displacement of shaft can also be taken as a state variable.

Torsion Spring

Consider a torsion spring shown in Figure 1.8. The externally applied torque is $T(t)$. The reaction torque $T_K(t)$ of the spring is equal to and opposite of the applied torque.

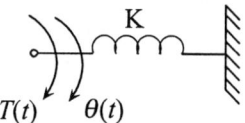

$$T(t) \qquad \theta(t)$$

Figure 1.8 State variable for a torsion spring

$$T(t) = T_K(t) = K\theta(t) = K \int_{-\infty}^{t} \omega(t)$$

$$T_K(t) = K \int_{-\infty}^{0} \omega(t) + K \int_{0}^{t} \omega(t)$$

The first term on the right hand side is the initial value of the spring reaction torque at $t = 0$. It indicates the state of the spring at $t = 0$.

$$T_K(t) = T_K(0) + K \int_{0}^{t} \omega(t) \tag{1.31}$$

Comparing equation (1.2) and (1.31), it can be concluded that for a torsion spring, the spring torque can be taken as a state variable. Since the spring torque is proportional to the angular displacement, the angular displacement can also be taken as a state variable.

A rotating mass stores kinetic energy, and torsion spring stores potential energy. A damper or a dash pot in a rotating system absorbs energy and dissipates it as heat. It cannot store energy. Hence, no state variable is associated with a damper.

The variables for electrical and mechanical systems are listed below in Table 1.1.

System	State Variables
Electrical	Inductor current, magnetic flux linkage Capacitor voltage, capacitor charge
Mechanical (translational)	Displacement of mass, velocity of mass Displacement of spring, spring force
Mechanical (rotational)	Angular displacement, angular velocity Reaction torque of torsion spring

Table 1.1 State Variables for Different Systems

1.13 CHOICE OF STATE VARIABLES

The number of state variables required for a system is equal to the order of the system, that is, the order of the differential equation that describes the system. The choice of state variables for a system is not unique. Many sets of state variables may be chosen, each containing the minimum number required, and each set results in a different state model.

(1) For an electrical network, inductor currents and capacitor voltages can be chosen as state variables.

(2) For mechanical translational system, displacement and velocity of mass, displacement of spring and spring force qualify to be state variables.

(3) For mechanical rotational system, angular displacement and angular velocity of rotating shaft and the angular displacement and the spring torque can be taken as state variables.

(4) As a general guidance, after writing simple derivative equation for each energy storing element, each differentiated variable can be chosen as a state variable.

(5) The state variables chosen must be linearly independent.

1.14 PHYSICAL VARIABLE MODEL AND PHASE VARIABLE MODEL

Inductor current, capacitor voltage, mechanical displacement, velocity and spring force are physical variables. If some of these are selected as state variables, then the resulting state model is called a *physical variable model*.

For a physical system, consider that a particular quantity is chosen as a state variable and labelled as x_1. If the successive derivatives of this variable are also taken as state variables, the state model thus obtained is said to be in *phase variable form*. For example, consider a third-order system. Three state variables are necessary to obtain a state model. For this system, consider that the variable x_1 is chosen as a state variable. Then $x_2 = \dfrac{dx_1}{dt}$ and $x_3 = \dfrac{dx_2}{dt}$ are also taken as state variables required to represent the system. If $r_1(t)$ and $r_2(t)$ are the two inputs to the system, then the three state equations can be written in the form,

$$\frac{dx_1}{dt} = x_2, \quad \frac{dx_2}{dt} = x_3$$

$$\frac{dx_3}{dt} = a_1\, x_1 + a_2\, x_2 + a_3\, x_3 + b_1\, r_1 + b_2\, r_2.$$

This set of equations are said to be in phase variable form. These equations are denoted in matrix form as

$$\begin{bmatrix} \dfrac{dx_1(t)}{dt} \\[2mm] \dfrac{dx_2(t)}{dt} \\[2mm] \dfrac{dx_3(t)}{dt} \end{bmatrix} = \begin{bmatrix} 0 & 1 & 0 \\ 0 & 0 & 1 \\ a_1 & a_2 & a_3 \end{bmatrix} \begin{bmatrix} x_1(t) \\ x_2(t) \\ x_3(t) \end{bmatrix} + \begin{bmatrix} 0 & 0 \\ 0 & 0 \\ b_1 & b_2 \end{bmatrix} \begin{bmatrix} r_1(t) \\ r_2(t) \end{bmatrix}.$$

Example 1.12 An electrical network is shown in Figure E 1.12. Select a set of proper state variables and write down a state equation, in physical variable form, to represent the system.

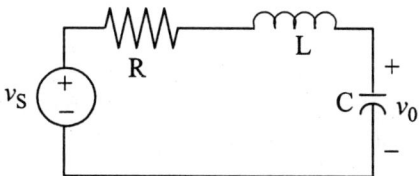

Figure E 1.12

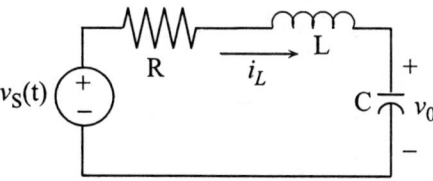

Figure E 1.12(a)

Solution There are two energy-storing elements, the inductor and the capacitor. The order of the system is two and two state variables are required to get a state model. The current through the inductor and the voltage across the capacitor are taken as state variables. Writing the KVL equation, with respect to Figure E 1.12(a),

$$v_S(t) = R\, i_L(t) + L\frac{di_L(t)}{dt} + v_C(t) \qquad (1)$$

Or
$$\frac{di_L(t)}{dt} = -\frac{R}{L}i_L(t) - \frac{1}{L}v_C(t) + \frac{1}{L}v_S(t). \qquad (A)$$

The capacitor current is the same as the inductor current and is given by,

$$C\frac{dv_C(t)}{dt} = i_L(t) \qquad (2)$$

Or
$$\frac{dv_C(t)}{dt} = \frac{1}{C}i_L(t). \qquad (B)$$

Equations (A) and (B) are state equations and can be represented in the matrix form as,

$$\begin{bmatrix} \dfrac{di_L(t)}{dt} \\[2mm] \dfrac{dv_C(t)}{dt} \end{bmatrix} = \begin{bmatrix} -\dfrac{R}{L} & -\dfrac{1}{L} \\[2mm] \dfrac{1}{C} & 0 \end{bmatrix} \begin{bmatrix} i_L(t) \\[1mm] v_C(t) \end{bmatrix} + \begin{bmatrix} \dfrac{1}{L} \\[1mm] 0 \end{bmatrix} v_S(t) \qquad \text{State equation}$$

The output equation is

$$v_0(t) = v_C(t) = \begin{bmatrix} 0 & 1 \end{bmatrix}\begin{bmatrix} i_L(t) \\ v_C(t) \end{bmatrix}.$$

Example 1.13 For the network of Example 1.12, obtain a state model in phase variable form. Select the capacitor voltage as a state variable.

Solution The network is shown in Figure E 1.13.

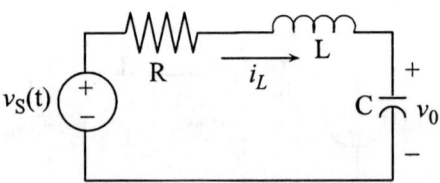

Figure E 1.13 Example 1.13

The capacitor voltage $x_1 = v_c(t)$ and $x_2 = \dot{x}_1 = \dfrac{dv_C(t)}{dt}$ are taken as the two state variables.

$$\dot{x}_1 = x_2 \qquad (A)$$

Writing the KVL equation,

$$v_S(t) = R\, i_L(t) + L\frac{di_L(t)}{dt} + v_C(t). \tag{1}$$

The inductor current is the same as the capacitor current.

$$i_L(t) = C\frac{dv_C(t)}{dt} = C\,x_2 \tag{2}$$

$$\frac{di_L(t)}{dt} = C\,\dot{x}_2 \tag{3}$$

From equations (1), (2) and (3), $v_S(t) = R\,C\,x_2 + LC\,\dot{x}_2 + x_1$

$$\dot{x}_2 = -\frac{1}{LC}x_1 - \frac{R}{L}x_2 + \frac{1}{LC}v_S(t). \tag{B}$$

Keeping equations (A) and (B) in the matrix form, the state model is

$$\begin{bmatrix} \dot{x}_1 \\ \dot{x}_2 \end{bmatrix} = \begin{bmatrix} 0 & 1 \\ -\dfrac{1}{LC} & -\dfrac{R}{L} \end{bmatrix} \begin{bmatrix} x_1 \\ x_2 \end{bmatrix} + \begin{bmatrix} 0 \\ \dfrac{1}{LC} \end{bmatrix} v_S(t).$$

Taking capacitor voltage as output,

$$v_0(t) = v_C(t) = \begin{bmatrix} 1 & 0 \end{bmatrix} \begin{bmatrix} x_1 \\ x_2 \end{bmatrix}.$$

Example 1.14 For the network of Example 1.12, obtain a state model in phase variable form, selecting the inductor current as a state variable.

Solution: The network is shown in Figure E 1.14.

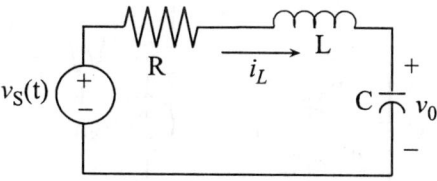

Figure E1.14 Example 1.14

Let $x_1 = \int i_L(t)\,dt$ and the inductor current $x_2 = i_L(t)$ be two state variables. Then

$$\dot{x}_1 = i_L(t) = x_2 \tag{A}$$

Writing the KVL equation,

$$v_S(t) = Ri_L + L\frac{di_L}{dt} + \frac{1}{C}\int i_L\, dt \tag{1}$$

$$\dot{x_2} = \frac{di_L}{dt}. \tag{2}$$

From equation (1),

$$v_S(t) = R\,x_2 + L\,\dot{x_2} + \frac{1}{C}x_1$$

Or

$$\dot{x_2} = -\frac{1}{LC}x_1 - \frac{R}{L}x_2 + \frac{1}{L}v_s. \tag{B}$$

Expressing equations (A) and (B) in the matrix form, the state equation is

$$\begin{bmatrix} \dot{x_1} \\ \dot{x_2} \end{bmatrix} = \begin{bmatrix} 0 & 1 \\ -\dfrac{1}{LC} & -\dfrac{R}{L} \end{bmatrix} \begin{bmatrix} x_1 \\ x_2 \end{bmatrix} + \begin{bmatrix} 0 \\ \dfrac{1}{L} \end{bmatrix} v_S.$$

If the capacitor voltage is the output, the output equation is

$$v_0(t) = v_C(t) = \frac{1}{C}\int i_L\, dt = \frac{1}{C}x_1,$$

or

$$v_0(t) = \begin{bmatrix} \dfrac{1}{C} & 0 \end{bmatrix} \begin{bmatrix} x_1 \\ x_2 \end{bmatrix}.$$

Example 1.15 An electrical network is shown in Figure E1.15. Select a set of proper state variables and write down state equations, in physical variable form, to represent the system.

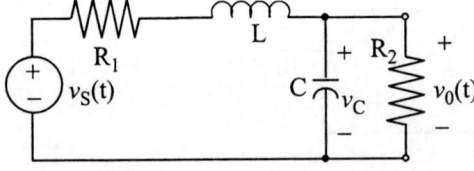

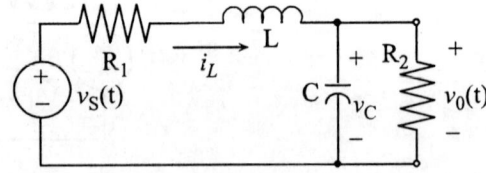

Figure E1.15 Example 1.15   **Figure E1.15(a)** Example 1.15

Solution: The current through the inductor $i_L(t)$ and the voltage cross capacitor $v_C(t)$ are taken as state variables.

Writing the KVL equation, with respect to Figure E 1.15(a)

$$v_S(t) = R_1 i_L(t) + L\frac{di_L(t)}{dt} + v_C(t).$$

Or

$$\frac{di_L(t)}{dt} = -\frac{R_1}{L}i_L(t) - \frac{1}{L}v_C(t) + \frac{1}{L}v_S(t).$$ (A)

The capacitor current is given by,

$$C\frac{dv_C(t)}{dt} = i_L(t) - \frac{v_C(t)}{R_2}.$$

Or

$$\frac{dv_C(t)}{dt} = \frac{1}{C}i_L(t) - \frac{v_C(t)}{R_2 C}.$$ (B)

Equations (A) and (B) are state equations and can be represented in the matrix form as,

$$\begin{bmatrix} \dfrac{di_L(t)}{dt} \\ \dfrac{dv_c(t)}{dt} \end{bmatrix} = \begin{bmatrix} -\dfrac{R_1}{L} & -\dfrac{1}{L} \\ \dfrac{1}{C} & -\dfrac{1}{R_2 C} \end{bmatrix} \begin{bmatrix} i_L(t) \\ v_C(t) \end{bmatrix} + \begin{bmatrix} \dfrac{1}{L} \\ 0 \end{bmatrix} v_S(t),$$

The output equation is

$$v_O(t) = v_C(t) = \begin{bmatrix} 0 & 1 \end{bmatrix}\begin{bmatrix} i_L(t) \\ v_C(t) \end{bmatrix}.$$

Example 1.16 For the network of Example 1.15, obtain a state model in phase variable form, selecting the capacitor voltage as a state variable.

Solution: The network is shown in Figure E1.16. The state variables selected are

$$x_1 = v_C(t) \ and \ x_2 = \dot{v_C}(t).$$

Hence $\dot{x_1} = x_2$.   (A)

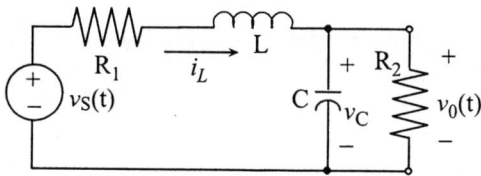

Figure E 1.16 for Example 1.16

Writing the KVL equation, with respect to Figure E 1.16,

$$v_S(t) = R_1 i_L(t) + L \frac{di_L(t)}{dt} + v_C(t) \tag{1}$$

$$i_L(t) = C \frac{dv_C(t)}{dt} + \frac{v_C(t)}{R_2} . \tag{2}$$

Differentiating equation (2),

$$\frac{di_L(t)}{dt} = C \frac{d^2 v_C(t)}{dt^2} + \frac{1}{R_2} \frac{dv_C(t)}{dt} \tag{3}$$

Substituting equations (2) and (3) in equation (1),

$$v_S(t) = R_1 \left(C \dot{v_C}(t) + \frac{1}{R_2} v_C(t) \right) + L \left(C \ddot{v_C}(t) + \frac{1}{R_2} \dot{v_C}(t) \right) + v_C(t) . \tag{4}$$

Rearranging equation (4),

$$\ddot{v_C}(t) = -\left(\frac{R_1 + R_2}{R_2 LC} \right) v_C(t) - \left(\frac{R_1}{L} + \frac{1}{R_2 C} \right) \dot{v_C}(t) + \frac{1}{LC} v_S(t) . \tag{B}$$

Keeping equations (A) and (B) in the matrix form, the state equation is

$$\begin{bmatrix} \dot{v_C} \\ \ddot{v_C} \end{bmatrix} = \begin{bmatrix} 0 & 1 \\ -\dfrac{(R_1 + R_2)}{R_2 LC} & -\dfrac{(R_1 R_2 C + L)}{R_2 LC} \end{bmatrix} \begin{bmatrix} v_C(t) \\ \dot{v_C}(t) \end{bmatrix} + \begin{bmatrix} 0 \\ \dfrac{1}{LC} \end{bmatrix} v_S(t) .$$

The output equation is

$$v_O(t) = v_C(t) = \begin{bmatrix} 1 & 0 \end{bmatrix} \begin{bmatrix} v_C(t) \\ \dot{v_C}(t) \end{bmatrix} .$$

Example 1.17 Represent the network shown in Figure E 1.17 in the physical variable form of the state model.

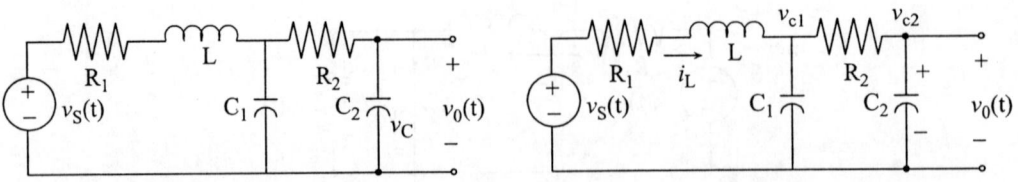

Figure E 1.17 for Example 1.17   Figure E 1.17 (a) for Example 1.17

Solution: Select the inductor current i_L and the capacitor voltages v_{C1} and v_{C2}, shown in Figure E 1.17(a), as state variables. Writing the KVL equation,

$$v_S(t) = R_1\, i_L + L\,\frac{di_L}{dt} + v_{C1}\,. \tag{1}$$

From equation (1), $\dfrac{di_L}{dt} = -\dfrac{R_1}{L} i_L - \dfrac{1}{L} v_{C1} + \dfrac{1}{L} v_S(t)\,.$ (A)

The inductor current $i_L = C_1\, \dot{v}_{C1} + \dfrac{v_{C1} - v_{C2}}{R_2}\,.$ (2)

From equation (2), $\dot{v}_{C1} = \dfrac{1}{C_1} i_L - \dfrac{1}{R_2 C_1} v_{C1} + \dfrac{1}{R_2 C_1} v_{C2}\,.$ (B)

The current through the capacitor C_2 is

$$\frac{v_{C1} - v_{C2}}{R_2} = C_2\, \dot{v}_{C2}\,. \tag{3}$$

From equation (3), $\dot{v}_{C2} = \dfrac{1}{R_2 C_2} v_{C1} - \dfrac{1}{R_2 C_2} v_{C2}\,.$ (C)

Equations (A), (B) and (C) are state equations and keeping them in the matrix form,

$$
\begin{bmatrix} \dot{i}_L \\ \dot{v}_{C1} \\ \dot{v}_{C2} \end{bmatrix}
=
\begin{bmatrix}
-\dfrac{R_1}{L} & -\dfrac{1}{L} & 0 \\[2mm]
\dfrac{1}{C_1} & -\dfrac{1}{R_2 C_1} & \dfrac{1}{R_2 C_1} \\[2mm]
0 & \dfrac{1}{R_2 C_2} & -\dfrac{1}{R_2 C_2}
\end{bmatrix}
\begin{bmatrix} i_L \\ v_{C1} \\ v_{C2} \end{bmatrix}
+
\begin{bmatrix} \dfrac{1}{L} \\ 0 \\ 0 \end{bmatrix}
v_S(t)\,.
$$

If the capacitor voltage v_{C2} is taken as the output

$$v_O(t) = v_{C2}(t) = \begin{bmatrix} 0 & 0 & 1 \end{bmatrix} \begin{bmatrix} i_L(t) \\ v_{C1}(t) \\ v_{C2}(t) \end{bmatrix}\,.$$

Example 1.18 An electrical network is shown in Figure E1.18. Represent it in a state model in physical variable form.

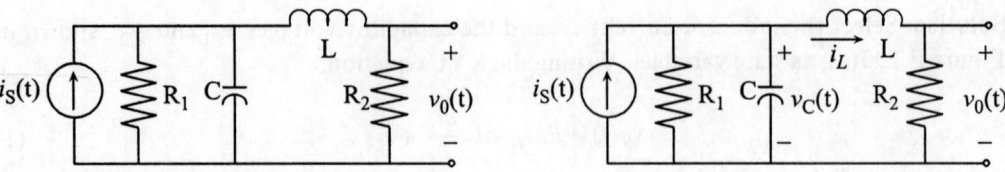

Figure E 1.18 for Example 1.18   **Figure E 1.18** (a) for Example 1.18

Solution: There are two energy-storing elements, and hence two state variables are to be selected. Choose the current through the inductor i_L and the voltage across the capacitor v_C, shown in Figure E 1.18 (a). Writing down the KCL equation,

$$i_S(t) = \frac{v_C}{R_1} + C\frac{dv_C}{dt} + i_L. \qquad (1)$$

From equation (1), $\dfrac{dv_C}{dt} = -\dfrac{1}{R_1C}v_C - \dfrac{1}{C}i_L + \dfrac{1}{C}i_S(t).$ \qquad (A)

The capacitor voltage is $v_C = L\dfrac{di_L}{dt} + R_2 i_L.$ \qquad (2)

From equation (2), $\dfrac{di_L}{dt} = -\dfrac{R_2}{L}i_L + \dfrac{1}{L}v_C.$ \qquad (B)

Equations (A) and (B) are state equations and can be written in the matrix form as

$$\begin{bmatrix} \dfrac{dv_C(t)}{dt} \\ \dfrac{di_L(t)}{dt} \end{bmatrix} = \begin{bmatrix} -\dfrac{1}{R_1C} & -\dfrac{1}{C} \\ \dfrac{1}{L} & -\dfrac{R_2}{L} \end{bmatrix} \begin{bmatrix} v_C(t) \\ i_L(t) \end{bmatrix} + \begin{bmatrix} \dfrac{1}{C} \\ 0 \end{bmatrix} i_S(t).$$

If the voltage across the resistor R_2 is taken as output, then the output equation is

$$v_0(t) = i_L R_2. \qquad (3)$$

From equation (3),

$$v_0(t) = \begin{bmatrix} 0 & R_2 \end{bmatrix} \begin{bmatrix} v_C(t) \\ i_L(t) \end{bmatrix}.$$

Example 1.19 For the network of Example 1.18, obtain a state model in its phase variable form. Select the current through the inductor as a state variable.

Solution: The network is re-drawn as shown in Figure E 1.19.

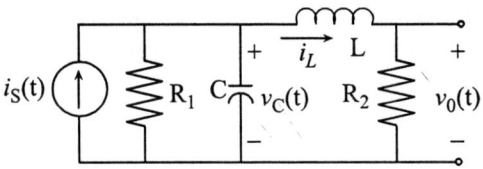

Figure E 1.19 Example 1.19

Select $x_1 = i_L$ and $x_2 = \dot{x}_1 = \dfrac{di_L}{dt}$ as two state variables.

Hence $\dot{x}_1 = x_2$. (A)

Writing KCL equation,

$$i_S = \frac{v_C}{R_1} + C\frac{dv_C}{dt} + i_L. \tag{1}$$

The capacitor voltage $v_C = L\dfrac{di}{dt} + R_2\, i_L = L\, x_2 + R_2\, x_1$. (2)

Differentiating equation (2),

$$\dot{v}_C = L\,\dot{x}_2 + R_2\,\dot{x}_1 = L\,\dot{x}_2 + R_2\, x_2. \tag{3}$$

Substituting equations (2) and (3) in equation (1),

$$i_S(t) = \frac{1}{R_1}(L\, x_2 + R_2\, x_1) + C(L\,\dot{x}_2 + R_2\, x_2) + i_L.$$

$$= \frac{1}{R_1}(L\, x_2 + R_2\, x_1) + C(L\,\dot{x}_2 + R_2\, x_2) + x_1. \tag{4}$$

Re-arranging equation (4),

$$\dot{x}_2 = -\frac{(R_1 + R_2)}{R_1 LC}x_1 - \frac{(L + R_1 R_2 C)}{R_1 LC}x_2 + \frac{1}{LC}i_S(t). \tag{B}$$

Equations (A) and (B) are state equations and can be expressed in the matrix form as

$$\begin{bmatrix} \dot{x}_1 \\ \dot{x}_2 \end{bmatrix} = \begin{bmatrix} 0 & 1 \\ -\dfrac{(R_1 + R_2)}{R_1 LC} & -\dfrac{(L + R_1 R_2 C)}{R_1 LC} \end{bmatrix} \begin{bmatrix} x_1 \\ x_2 \end{bmatrix} + \begin{bmatrix} 0 \\ \dfrac{1}{LC} \end{bmatrix} i_S(t).$$

The output equation is $v_0(t) = i_L R_2 = x_1 R_2 = \begin{bmatrix} R_2 & 0 \end{bmatrix} \begin{bmatrix} x_1 \\ x_2 \end{bmatrix}$.

Example 1.20 Write down the necessary equations for the electrical network shown in Figure E 1.20 and represent it in a state model. Use the physical variables as the state variables. Also draw the corresponding state diagram.

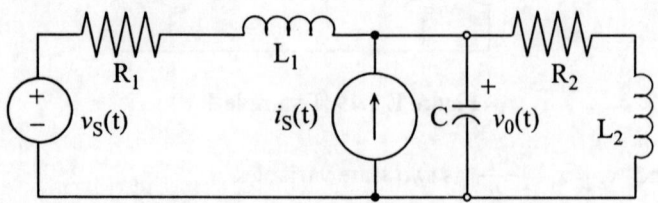

Figure E 1.20 for Example 1.20

Solution: The network is drawn in Figure E1.20 (a) showing the inductor currents and the capacitor voltage. Let the state variables be $x_1 = i_{L1}$, $x_2 = v_C$ and $x_3 = i_{L2}$.

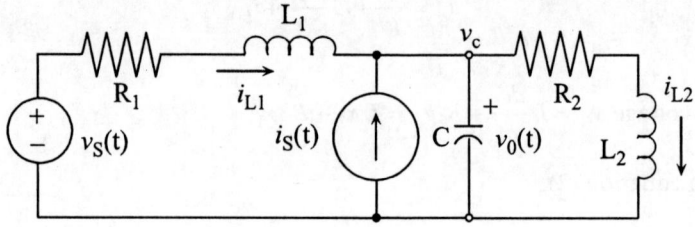

Figure E 1.20 (a) for Example 1.20

Writing the KVL equation,

$$v_S(t) = R_1 i_{L1} + L_1 \frac{di_{L1}}{dt} + v_C = R_1 x_1 + L_1 \dot{x}_1 + x_2 . \tag{1}$$

From equation (1), $\dot{x}_1 = -\frac{R_1}{L_1} x_1 - \frac{1}{L_1} x_2 + \frac{1}{L_1} v_S(t)$. \hfill (A)

Writing the KCL equation,

$$i_{L1} + i_S(t) = C \dot{v}_C + i_{L2} .$$

Or $$x_1 + i_S(t) = C \dot{x}_2 + x_3 . \tag{2}$$

From equation (2), $\dot{x}_2 = \frac{1}{C} x_1 - \frac{1}{C} x_3 + \frac{1}{C} i_S(t)$. \hfill (B)

The capacitor voltage $v_C = R_2 i_{L2} + L_2 \frac{di_{L2}}{dt}$.

Or $$x_2 = R_2 x_3 + L_2 \dot{x}_3 . \tag{3}$$

From equation (3), $\dot{x}_3 = \dfrac{1}{L_2}x_2 - \dfrac{R_2}{L_2}x_3$. (C)

Equations (A), (B) and (C) are state equations and keeping them in the matrix form,

$$\begin{bmatrix} \dot{x}_1 \\ \dot{x}_2 \\ x_3 \end{bmatrix} = \begin{bmatrix} -\dfrac{R_1}{L_1} & -\dfrac{1}{L_1} & 0 \\ \dfrac{1}{C} & 0 & -\dfrac{1}{C} \\ 0 & \dfrac{1}{L_2} & -\dfrac{R_2}{L_2} \end{bmatrix}\begin{bmatrix} x_1 \\ x_2 \\ x_3 \end{bmatrix} + \begin{bmatrix} \dfrac{1}{L_1} & 0 \\ 0 & \dfrac{1}{C} \\ 0 & 0 \end{bmatrix}\begin{bmatrix} v_S(t) \\ i_S(t) \end{bmatrix}.$$

If the capacitor voltage is taken as output, then the output equation is

$$v_0(t) = v_C(t) = x_2 = \begin{bmatrix} 0 & 1 & 0 \end{bmatrix}\begin{bmatrix} x_1 \\ x_2 \\ x_3 \end{bmatrix}.$$

The state diagram is shown in Figure 1.20 (b).

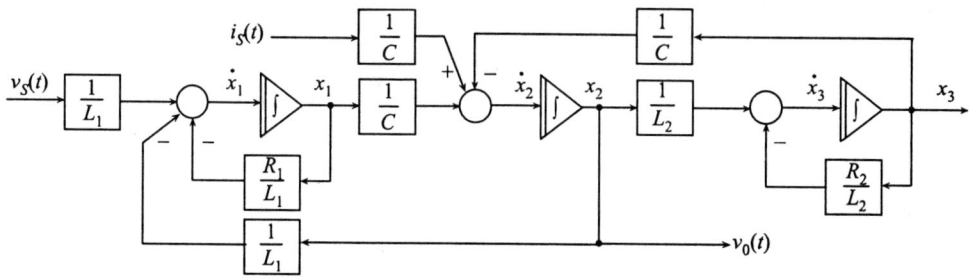

Figure E1.20 (b) State diagram for Example E 1.20

Example 1.21 Realize a state model for the mechanical system shown in Figure E1.21.

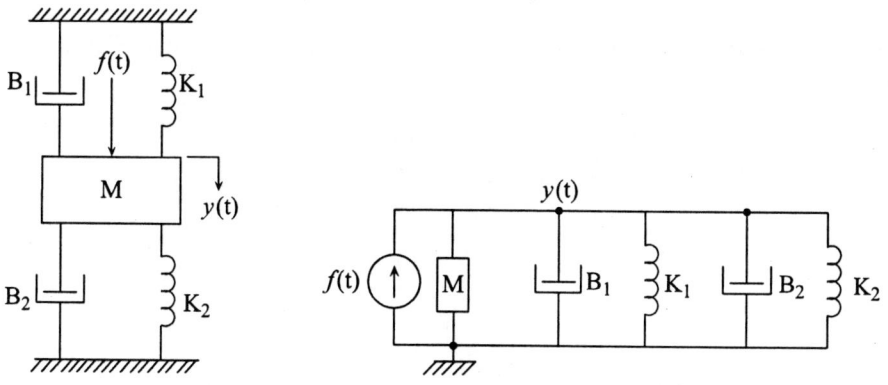

Figure E 1.21 for Example 1.21

Figure E 1.21 (a) for Example 1.21

Solution: For the convenience of analysis, an equivalent mechanical diagram is drawn as shown in Figure E 1.21(a). From this diagram, the equation of motion is

$$f(t) = M\ddot{y}(t) + (B_1 + B_2)\dot{y}(t) + (K_1 + K_2)y(t). \tag{1}$$

The variables y and $\dot{y}$ are, respectively, the displacement and the velocity of mass, and these are selected as the two state variables.

$$x_1 = y \ and \ x_2 = \dot{y}$$

so that $\dot{x}_1 = x_2$ (A)

Equation (1) can be written as

$$f(t) = M\dot{x}_2 + (B_1 + B_2)x_2 + (K_1 + K_2)x_1. \tag{2}$$

Rearranging equation (2),

$$\dot{x}_2 = -\frac{(K_1 + K_2)}{M}x_1 - \frac{(B_1 + B_2)}{M}x_2 + \frac{1}{M}f(t). \tag{B}$$

Equations (A) and (B) are state equations and keeping them in matrix form,

$$\begin{bmatrix} \dot{x}_1 \\ \dot{x}_2 \end{bmatrix} = \begin{bmatrix} 0 & 1 \\ -\dfrac{(K_1 + K_2)}{M} & -\dfrac{(B_1 + B_2)}{M} \end{bmatrix} \begin{bmatrix} x_1 \\ x_2 \end{bmatrix} + \begin{bmatrix} 0 \\ \dfrac{1}{M} \end{bmatrix} f(t).$$

If the velocity of mass is taken as the output, then the output equation is

$$v_0(t) = \dot{y} = x_2 = \begin{bmatrix} 0 & 1 \end{bmatrix} \begin{bmatrix} x_1 \\ x_2 \end{bmatrix}.$$

Example 1.22 Obtain a state model for an armature-controlled DC motor and draw the state diagram.

Solution: The schematic diagram of a DC motor driving a shaft is shown in Figure E 1.22.

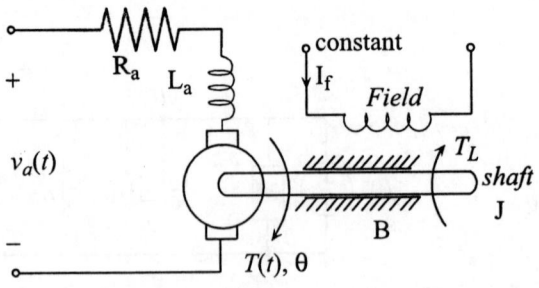

Figure E 1.22 for Example E 1.22

Solution For an electromechanical system, a set of three equations are to be written: one for the electrical circuit, another for the mechanical system and the third equation relating these two.

R_a – Resistance of armature circuit, Ω

L_a – Inductance of armature circuit, H

$i_a(t)$ – Instantaneous armature current, A

$v_a(t)$ – Applied voltage across the armature, V

e_b – Back emf of the motor, V

I_f – Constant field current, A

J – Moment of inertia of rotating parts, Kg-m²

B – Coefficient of viscous friction, N-sec/radian

θ – Angular displacement of the shaft, radians

T_L – External load torque on the motor, N.

The equation for the electrical circuit (armature circuit) is

$$v_a(t) = R_a i_a(t) + L_a \frac{d i_a(t)}{dt} + e_b(t) . \tag{1}$$

The back emf is proportional to the angular speed and is given by

$$e_b(t) = k_b \frac{d\theta}{dt} , \tag{2}$$

where k_b is a constant of proportionality.

Combining equations (1) and (2),

$$v_a(t) = R_a i_a(t) + L_a \frac{d i_a(t)}{dt} + k_b \frac{d\theta}{dt} . \tag{3}$$

The angular velocity is

$$\omega = \frac{d\theta}{dt} \text{ rad/sec.} \tag{4}$$

Hence, equation (3) can be written as

$$v_a(t) = R_a i_a(t) + L_a \frac{d i_a(t)}{dt} + k_b \, \omega . \tag{5}$$

The torque developed by the motor has to overcome the torque due to angular acceleration, the torque lost to overcome the viscous friction and the externally applied load torque. Hence, the equation for the mechanical system is

$$T(t) = J \frac{d^2\theta}{dt^2} + B \frac{d\theta}{dt} + T_L = J \frac{d\omega}{dt} + B\omega + T_L . \tag{6}$$

The torque developed by the motor is proportional to the armature current when the field flux is constant. Hence,

$$T(t) = k_t \, i_a(t) \,, \tag{7}$$

where k_t is a constant.
Combining equations (6) and (7),

$$k_t \, i_a(t) = J\frac{d\omega}{dt} + B\omega + T_L \,. \tag{8}$$

Select the armature current $i_a(t)$ and the angular speed ω as state variables. From equation (5),

$$\frac{d i_a(t)}{dt} = -\frac{R_a}{L_a}i_a(t) - \frac{k_b}{L_a}\omega + \frac{1}{L_a}v_a(t) \,. \tag{A}$$

From equation (8),

$$\frac{d\omega}{dt} = \frac{k_t}{J}i_a(t) - \frac{B}{J}\omega - \frac{T_L}{J} \,. \tag{B}$$

Equations (A) and (B) are state equations and can be expressed in the matrix form as

$$\begin{bmatrix} \dfrac{d i_a(t)}{dt} \\ \dfrac{d\omega}{dt} \end{bmatrix} = \begin{bmatrix} -\dfrac{R_a}{L_a} & -\dfrac{k_b}{L_a} \\ \dfrac{k_t}{J} & -\dfrac{B}{J} \end{bmatrix} \begin{bmatrix} i_a(t) \\ \omega \end{bmatrix} + \begin{bmatrix} \dfrac{1}{L_a} & 0 \\ 0 & \dfrac{-1}{J} \end{bmatrix} \begin{bmatrix} v_a(t) \\ T_L \end{bmatrix} .$$

If the angular speed is taken as output

$$v_0(t) = \omega = \begin{bmatrix} 0 & 1 \end{bmatrix} \begin{bmatrix} i_a(t) \\ \omega \end{bmatrix} .$$

The state diagram is shown in Figure E 1.22(a).

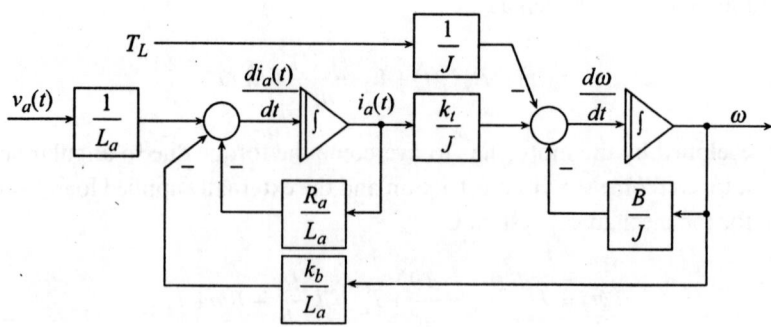

Figure E 1.22 (a) State model for an armature-controlled DC motor

Example 1.23 Represent a field-controlled DC motor in a state space model.

Solution: The schematic diagram of a field-controlled DC motor is shown in Figure E 1.23.

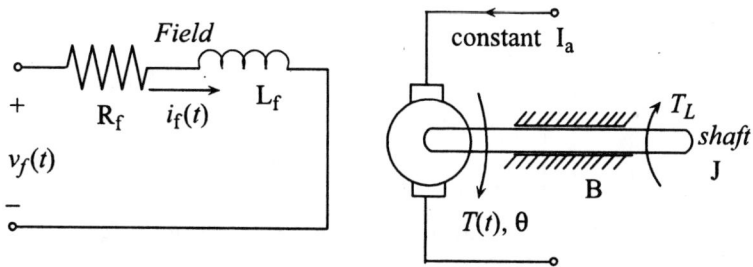

Figure E 1.23 for Example E 1.23

Equation for the field circuit is

$$R_f\, i_f(t) + L_f\, \frac{di_f(t)}{dt} = v_f(t). \tag{1}$$

The electromagnetic torque developed by the motor is proportional to the field current when the armature current is constant. Hence,

$$T(t) = k_f\, i_f(t). \tag{2}$$

The torque developed by the motor has to overcome the torque due to acceleration, the torque required to overcome the frictional torque and to counterbalance the externally applied load torque on the motor shaft. Thus,

$$T(t) = J\frac{d^2\theta}{dt^2} + B\frac{d\theta}{dt} + T_L = J\frac{d\omega}{dt} + B\omega + T_L, \tag{3}$$

where $\omega = \dfrac{d\theta}{dt}$ is the angular speed of rotation.

Combining equations (2) and (3),

$$J\frac{d\omega}{dt} + B\omega + T_L = k_f\, i_f(t). \tag{4}$$

Select the field current $i_f(t)$ and the angular speed of the shaft as two state variables. From equation (1),

$$\frac{di_f(t)}{dt} = -\frac{R_f}{L_f} i_f(t) + \frac{1}{L_f} v_f(t). \tag{A}$$

From equation (4),

$$\frac{d\omega}{dt} = \frac{k_f}{J} i_f(t) - \frac{B}{J}\omega - \frac{T_L}{J}. \tag{B}$$

Equations (A) and (B) are state equations and they can be expressed in the matrix form as

$$\begin{bmatrix} \dfrac{di_f(t)}{dt} \\[2ex] \dfrac{d\omega}{dt} \end{bmatrix} = \begin{bmatrix} -\dfrac{R_f}{L_f} & 0 \\[2ex] \dfrac{k_f}{J} & -\dfrac{B}{J} \end{bmatrix} \begin{bmatrix} i_f(t) \\[1ex] \omega \end{bmatrix} + \begin{bmatrix} \dfrac{1}{L_f} & 0 \\[2ex] 0 & -\dfrac{1}{J} \end{bmatrix} \begin{bmatrix} v_f(t) \\[1ex] T_L \end{bmatrix}.$$

If the angular speed of the motor is taken as output, then the output equation is

$$y(t) = \omega = \begin{bmatrix} 0 & 1 \end{bmatrix} \begin{bmatrix} i_f(t) \\ \omega \end{bmatrix}.$$

The state diagram is shown in Figure E1.23(a).

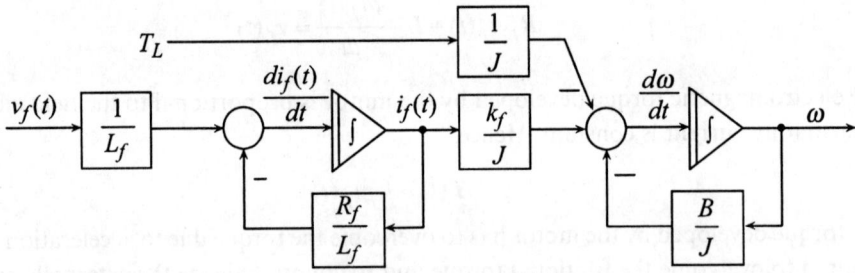

Figure E1.23 (a) State model for a field-controlled DC motor

1.15 NON-UNIQUENESS OF A SET OF STATE VARIABLES

For a given system, there can be a number of sets of state variables. In many of the earlier examples, it was shown that a system can be represented in a state model either in terms of physical variables or in terms of phase variables. There can be different sets of physical variables and so also different sets of phase variables. These variables need not have physical existence. For example, in an electrical system, if i_L is an inductor current and v_C is a capacitor voltage, then $x = i_L + v_C$ can be a state variable. Clearly, this variable has no physical entity. In general, if $x_1, x_2, x_3, ---, x_n$ are state variables, then $z_1, z_2, z_3, ---, z_n$ given by

$$z_1 = f_1(x_1, x_2, x_3, ---, x_n)$$

$$z_2 = f_2(x_1, x_2, x_3, ---, x_n)$$

$$...$$

$$z_n = f_n(x_1, x_2, x_3, ---, x_n)$$

can also be selected as state variables to represent the same system; but they must be linearly independent. The selection of a set of state variables affects the analytical and computational effort and complexity. A set of state variables even if they are not measurable or not physical variables may greatly reduce the computational effort.

SUMMARY

- The conventional approach to the analysis and design are by using root locus, Bode plots and Nyquist plots. They are applicable only to linear time-invariant systems and are not applicable to nonlinear and time-varying systems.
- The conventional method is based on open-loop and closed-loop transfer functions and does not throw any light on the internal variables of the system.
- Transfer function approach is applicable only to single-input single-output systems.
- State space method of analysis and design of a control system is a time-domain method.
- State space method is also applicable to nonlinear systems, time-varying systems and multiple-input/multipleoutput systems.
- A set of state variables of a dynamic system is a set of a minimum number of variables with which the system can be modelled and analysed completely.
- The number of state variables required to represent a system is equal to the order of the system.
- In a state equation, the first derivative of a state variable is expressed in terms of the state variables selected and the inputs to the system.
- The state equations are linearly independent.
- Output equation expresses the output variable in terms of the state variable selected and the inputs to the system.
- Variables associated with energy-storing elements qualify to be state variables.
- For an electrical network, inductor current and capacitor voltage are taken as state variables.
- For a mechanical translational system, displacement and velocity of mass, displacement of spring and spring force can be taken as state variables.
- For a mechanical rotational system, angular displacement and angular velocity of a shaft, displacement of torsion spring and spring torque can be taken as state variables.
- A physical variable model is a state model in which the state variables are physical variables.
- In a phase variable model, the state variables are successive derivatives of the first state variable selected.
- State variables need not be physical variables.
- State diagram is a pictorial representation of the state equation and the output equation.
- A nonlinear differential equation representing a system can be linearized using Taylor's series expansion.

PRACTICE PROBLEMS

PP 1.1 Obtain a physical variable form of state model for the electrical network shown in Figure PP1.1.

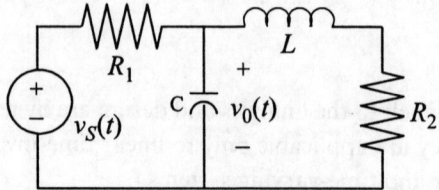

Figure PP 1.1

PP 1.2 For the network of PP1.1, obtain a state model in the phase variable form and draw the corresponding state diagram.

PP 1.3 An electrical network is shown in Figure PP1.3. Obtain a state model selecting physical variables as the state variables and draw the corresponding state diagram.

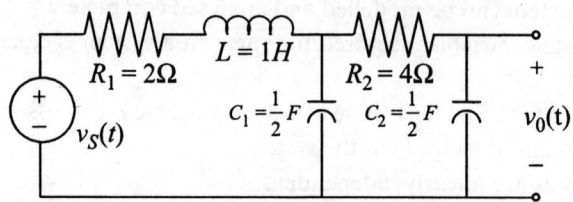

Figure PP 1.3

PP 1.4 For the network of PP1.3, obtain a state model in the phase variable form and draw the state diagram.

PP 1.5 An electrical network is shown in Figure P1.5. Obtain a state model in the physical variable form.

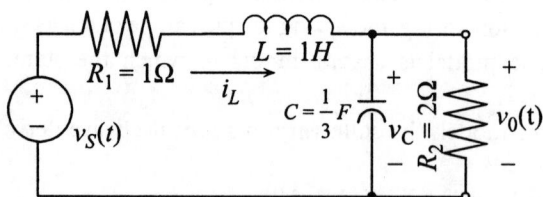

Figure PP1.5

PP 1.6 For the problem of PP1.5, obtain a state model in the phase variable form.

PP 1.7 An electrical network is shown in Figure P1.7. Obtain a state model in the physical variable form.

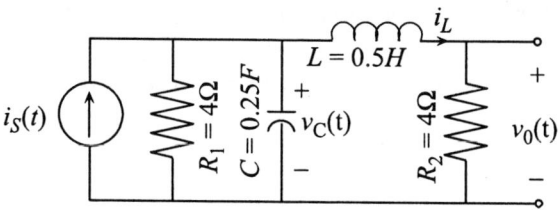

Figure PP1.7

PP 1.8 For the problem of PP1.7, obtain a state model in the phase variable form.

PP 1.9 For a field-controlled DC motor, obtain a state space representation in the phase variable form and draw the state diagram.

PP 1.10 For an armature-controlled DC motor, obtain a state space representation in the phase variable form and draw the state diagram.

PP 1.11 Realize a state model for the mechanical system shown in Figure PP1.11 using physical variables and draw the state diagram.

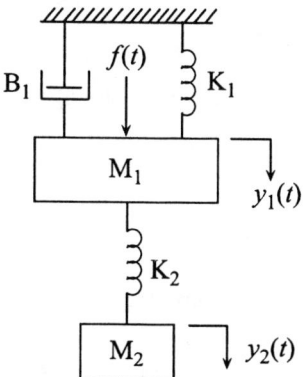

Figure PP 1.11

PP 1.12 Draw a state diagram for the mechanical system shown in Figure PP 1.11, selecting a phase variable form of state model.

PP 1.13 Realize a state model in the physical variable form for the mechanical rotational system shown in Figure PP 1.13 and draw the state diagram.

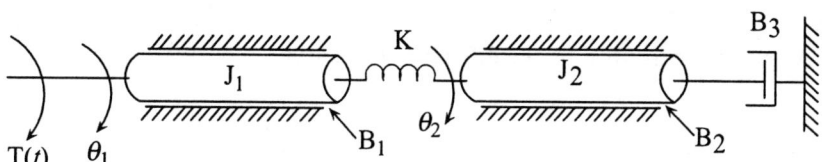

Figure PP 1.13

PP 1.14 Obtain a state model for the mechanical rotational system shown in Figure PP1.13, using phase variables

PP 1.15 For the state diagram shown in Figure PP 1.15, write down the state equations and the output equation.

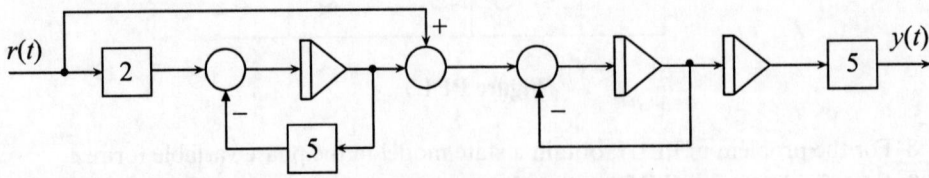

Figure PP 1.15

PP 1.16 A system is described by the differential equation

$$\ddot{y} + 3\dot{y} + 6y = 2r .$$

Obtain a state model for the system.

PP 1.17 A system is described by the differential equation

$$\ddot{y} + 3\dot{y} + 6y = 2r + \dot{r} .$$

Obtain a state model for the system.

PP 1.18 A system is described by the differential equation

$$\dddot{y} + 4\ddot{y} + 5\dot{y} + 2y = 3r .$$

Obtain a state model for the system.

PP 1.19 A system is described by the differential equation

$$\dddot{y} + 4\ddot{y} + 5\dot{y} + 2y = 3r + 2\dot{r} .$$

Obtain a state model for the system.

PP 1.20 A system is described by the differential equation

$$\dddot{y} + 5\ddot{y} + 7\dot{y} + 3y = 3r + 2\dot{r} + \ddot{r} .$$

Obtain a state model for the system.

PP 1.21 The differential equation description for a dynamic system is

$$\frac{d^2y}{dt^2} + 2y^2\frac{dy}{dt} + \left(\frac{dy}{dt}\right)^2 + y = 0.$$

Linearize the system about the equilibrium state $(y_0, \dot{y}_0) = (1,1)$ and obtain a state model for the system.

PP 1.22 Obtain a state model for the nonlinear system described by the differential equation

$$\ddot{y} + 2y^2\,\dot{y} + 3(\dot{y})^2 + y = 0$$

and linearize it about the point (0, 0).

PP 1.23 Obtain a state model for the nonlinear system described by the differential equation

$$\ddot{y} + 2y(\dot{y})^2 + 3y^3 + y = 0$$

and linearize it about the point (0, 0).

REVIEW QUESTIONS

1.1   List the demerits of classical control theory.
1.2   Enumerate the merits of modern control theory.
1.3   Compare and contrast the modern control theory and the classical control theory.
1.4   Explain the concept of state, state variable and state space.
1.5   Show that state variables are associated with energy-storing elements.
1.6   Bring out the differences between the physical variable and phase variable forms of state models.
1.7   State variables need not be physical variables. Discuss.
1.8   Explain the steps involved in obtaining a state model for a given dynamic system.
1.9   State variables must be linearly independent. Explain.
1.10 What are the factors that influence the choice of state variables in a dynamic system?

2   State Models from Transfer Functions

2.1 TRANSFER MATRIX

Consider an n-th order system with p inputs and q outputs. The state model is

$$\dot{X} = A\,X + B\,R \quad \text{: state equation}$$
$$Y = C\,X + D\,R \quad \text{: output equation}$$

where X = state vector with n state variables

    R = input vector with p inputs
    Y = output vector with q outputs
    A = system matrix, $n \times n$
    B = input matrix, $n \times p$
    C = output matrix, $q \times n$
    D = coupling matrix, $q \times p$.

Taking Laplace transform of the state equation,

$$s\,X(s) - X(0) = A\,X(s) + B\,R(s)$$

$$[s\,I - A]X(s) = X(0) + B\,R(s) \tag{2.1}$$

where I is the identity matrix.

Assume that the initial state X(0) is zero. Then from equation (2.1),

$$[s\,I - A]X(s) = B\,R(s)$$

$$X(s) = [s\,I - A]^{-1}\,B\,R(s) \tag{2.2}$$

Taking Laplace transform of the output equation

$$Y(s) = C\,X(s) + D\,R(s)$$

Using equations (2.2) and (2.1)

$$Y(s) = C[sI-A]^{-1}BR(s) + DR(s)$$

$$= \left[C[sI-A]^{-1}B + D\right]R(s)$$

Or

$$Y(s) = T_M(s)R(s)$$

The transfer matrix is

$$T_M(s) = C[sI-A]^{-1}B + D \qquad (2.3)$$

This is the transfer matrix.

If D=0, the *Transfer matrix* is that matrix which pre-multiplying the input vector yields the output vector, when the initial states are all zeros.

Example 2.1 A system is represented by the state equation as

$$\begin{bmatrix} \dot{x}_1 \\ \dot{x}_2 \end{bmatrix} = \begin{bmatrix} -1 & 1 \\ 0 & -2 \end{bmatrix}\begin{bmatrix} x_1 \\ x_2 \end{bmatrix} + \begin{bmatrix} 1 & -1 \\ 0 & 1 \end{bmatrix}\begin{bmatrix} r_1 \\ r_2 \end{bmatrix}$$

$$\begin{bmatrix} y_1 \\ y_2 \end{bmatrix} = \begin{bmatrix} 1 & 0 \\ 1 & -1 \end{bmatrix}\begin{bmatrix} x_1 \\ x_2 \end{bmatrix}.$$

Derive the transfer matrix.

Solution: The transfer matrix, from equation (2.3) is

$$T_M(s) = C[sI-A]^{-1}B + D$$

$$A = \begin{bmatrix} -1 & 1 \\ 0 & -2 \end{bmatrix} \qquad B = \begin{bmatrix} 1 & -1 \\ 0 & 1 \end{bmatrix}$$

$$C = \begin{bmatrix} 1 & 0 \\ 1 & -1 \end{bmatrix} \qquad D = 0$$

$$[sI-A]^{-1} = \begin{bmatrix} s+1 & -1 \\ 0 & s+2 \end{bmatrix}^{-1} = \frac{1}{(s+1)(s+2)}\begin{bmatrix} s+2 & 1 \\ 0 & s+1 \end{bmatrix}$$

$$T_M(s) = C[sI-A]^{-1}B + D = \frac{1}{(s+1)(s+2)}\begin{bmatrix} 1 & 0 \\ 1 & -1 \end{bmatrix}\begin{bmatrix} s+2 & 1 \\ 0 & s+1 \end{bmatrix}\begin{bmatrix} 1 & -1 \\ 0 & 1 \end{bmatrix} + [0]$$

$$= \frac{1}{(s+1)(s+2)}\begin{bmatrix} s+2 & 1 \\ s+2 & -s \end{bmatrix}\begin{bmatrix} 1 & -1 \\ 0 & 1 \end{bmatrix}$$

$$= \frac{1}{(s+1)(s+2)} \begin{bmatrix} s+2 & -(s+1) \\ s+2 & -2(s+1) \end{bmatrix}$$

$$= \begin{bmatrix} \dfrac{1}{s+1} & -\dfrac{1}{s+2} \\ \dfrac{1}{s+1} & -\dfrac{2}{s+2} \end{bmatrix}.$$

This is the transfer matrix and

$$Y(s) = \begin{bmatrix} \dfrac{1}{s+1} & -\dfrac{1}{s+2} \\ \dfrac{1}{s+1} & -\dfrac{2}{s+2} \end{bmatrix} R(s).$$

2.2 TRANSFER FUNCTION FROM THE STATE EQUATION

Transfer function is a special case of transfer matrix when the system is a single-input/single-output system. It is the ratio of the Laplace transform of the output to the Laplace transform of the input, with all initial states being zero.

Consider a linear time-invariant single-input/single-output system shown in Figure 2.1 described by a state model,

$$\dot{X} = A X + B r \quad : \text{state equation}$$

$$y = C X + D r \quad : \text{output equation}$$

where $r(t)$ is the single input, $y(t)$ is the single output and X(t) is the state vector.

$$X(t) = \begin{bmatrix} x_1(t) \\ x_2(t) \\ x_3(t) \\ --- \\ x_n(t) \end{bmatrix}$$

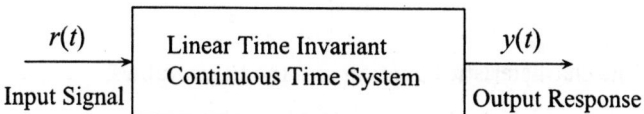

Figure 2.1 Linear time-invariant system

Taking Laplace transform of the state equation, with zero initial states

$$s X(s) = A X(s) + B R(s)$$

$$[s\,\mathrm{I}- \mathrm{A}]X(s) = \mathrm{B}\,R(s)$$

$$X(s) = [s\,\mathrm{I}- \mathrm{A}]^{-1}\,\mathrm{B}\,R(s) \tag{2.4}$$

where I is the identity matrix.

Taking Laplace transform of the output equation

$$Y(s) = C\,X(s) + D\,R(s) \tag{2.5}$$

Combining equations (2.4) and (2.5)

$$Y(s) = C[s\,\mathrm{I}- \mathrm{A}]^{-1}\,\mathrm{B}\,R(s) + D\,R(s)$$

The transfer function is

$$\frac{Y(s)}{R(s)} = C[s\,\mathrm{I}- \mathrm{A}]^{-1}\,\mathrm{B} + D \tag{2.6}$$

Since this is a single-input-single-output system, this is a ratio of a numerator polynomial and a denominator polynomial in terms of s.

$$\frac{Y(s)}{R(s)} = C\,\frac{Adj\,[s\,\mathrm{I}- \mathrm{A}]}{Det\,[s\,\mathrm{I}- \mathrm{A}]}\,\mathrm{B} + D \cdot \tag{2.7}$$

With D = 0, the transfer function is

$$\frac{Y(s)}{R(s)} = C\,\frac{Adj\,[s\,\mathrm{I}- \mathrm{A}]}{Det\,[s\,\mathrm{I}- \mathrm{A}]}\,\mathrm{B} \cdot \tag{2.8}$$

Eigen Values

The equation $Det\,[s\,\mathrm{I}- \mathrm{A}] = 0$ is called the *characteristic equation* and the roots of this characteristic equation are the poles of the transfer function. In matrix terminology, these roots are known as *Eigen values*. These are the critical values of the complex frequency s. The Eigen values decide the stability of a system. There is a system response corresponding to every Eigen value. Eigen values decide the nature of the natural response of a system.

Example 2.2 For a linear time-invariant system, the system matrix is

$$A = \begin{bmatrix} 0 & 1 \\ -2 & -3 \end{bmatrix}$$

Determine the characteristic equation and the Eigen values.

Solution: The characteristic equation is

$$\left| s\,\mathrm{I}- \mathrm{A} \right| = 0$$

$$\left| s\begin{bmatrix} 1 & 0 \\ 0 & 1 \end{bmatrix} - \begin{bmatrix} 0 & 1 \\ -2 & -3 \end{bmatrix} \right| = \left| \begin{matrix} s & -1 \\ 2 & s+3 \end{matrix} \right| = 0$$

This simplifies to $s^2 + 3s + 2 = 0$.

This is the characteristic equation. Solving this equation, we get $s = -1$ *and* $s = -2$ and the Eigen values are $s = -1$ *and* $s = -2$.

2.3 STATE TRANSITION MATRIX

The matrix $[s\,I - A]^{-1}$ is called the *resolvent matrix* or the *state transition matrix* in the s-domain and is denoted as $\Phi(s)$. The state transition matrix in the time domain is obtained by taking the inverse Laplace transform of $\Phi(s)$ and can be expressed as

$$\phi(t) = L^{-1}\left\{[s\,I - A]^{-1}\right\}.$$

It is known that

$$[s\,I - A]^{-1} = \frac{I}{s} + \frac{A}{s^2} + \frac{A^2}{s^3} + \frac{A^3}{s^4} + \dots.$$

Taking inverse Laplace transforms

$$L^{-1}\{[s\,I - A]^{-1}\} = I + A\,t + \frac{A^2 t^2}{2!} + \frac{A^3 t^3}{3!} + \dots = e^{A\,t}.$$

This is called the *matrix exponential* or the *state transition matrix*. The importance of state transition matrix will be discussed in Chapter 3.

Example 2.3 Show that $\dfrac{d}{dt}e^{A\,t} = A\,e^{A\,t}$ and evaluate $\dot{\phi}(0)$.

Solution: The state transition matrix is defined as

$$e^{A\,t} = I + A\,t + \frac{A^2 t^2}{2!} + \frac{A^3 t^3}{3!} + \frac{A^4 t^4}{4!} + \dots + \frac{A^k t^k}{k!} + \dots$$

Differentiating each term on the right hand side,

$$\frac{d}{dt}e^{A\,t} = \frac{d}{dt}\left[I + A\,t + \frac{A^2 t^2}{2!} + \frac{A^3 t^3}{3!} + \frac{A^4 t^4}{4!} + \dots + \frac{A^k t^k}{k!} + \dots \right]$$

$$= A + A^2 t + \frac{A^3 t^2}{2!} + \frac{A^4 t^3}{3!} + \dots \tag{1}$$

$$= A\left[I + A\,t + \frac{A^2 t^2}{2!} + \frac{A^3 t^3}{3!} + \frac{A^4 t^4}{4!} + \dots + \frac{A^k t^k}{k!} + \dots \right]$$

$$= A\,e^{A\,t}.$$

Equation (1) can also be written as

$$= \left[I + A t + \frac{A^2 t^2}{2!} + \frac{A^3 t^3}{3!} + \frac{A^4 t^4}{4!} + \dots + \frac{A^k t^k}{k!} + \dots \right] A = e^{A t} A$$

Hence

$$\frac{d}{dt} e^{A t} = A e^{A t} = e^{A t} A$$

$$\dot{\phi}(t) = \frac{d}{dt} e^{A t} = A e^{A t}$$

$$\dot{\phi}(0) = A I = A .$$

Example 2.4 Evaluate $\int_0^t e^{A \theta} d\theta$.

Solution: $\int_0^t e^{A \theta} d\theta = \int_0^t \left[I + A \theta + \frac{A^2 \theta^2}{2!} + \frac{A^3 \theta^3}{3!} + \frac{A^4 \theta^4}{4!} + \dots + \frac{A^k \theta^k}{k!} + \dots \right] d\theta$

Integrating term by term on the right hand side,

$$\int_0^t e^{A \theta} d\theta = I \int_0^t d\theta + A \int_0^t \theta d\theta + \frac{1}{2!} A^2 \int_0^t \theta^2 d\theta + \frac{1}{3!} A^3 \int_0^t \theta^3 d\theta + \dots$$

$$= I t + \frac{A t^2}{2!} + \frac{A^2 t^3}{3!} + \frac{A^3 t^4}{4!} + \dots$$

$$A \int_0^t e^{A \theta} d\theta = A t + \frac{A^2 t^2}{2!} + \frac{A^3 t^3}{3!} + \frac{A^4 t^4}{4!} + \dots$$

$$I + A \int e^A d = I + A t + \frac{A^2 t^2}{2!} + \frac{A^3 t^3}{3!} + \frac{A^4 t^4}{4!} + \dots = e^A .$$

If A^{-1} exists,

$$\int_0^t e^{A \theta} d\theta = A^{-1} [e^{A t} - I].$$

Example 2.5 For the system matrix of Example 2.2, determine the state transition matrix and show that $\dot{\phi}(0) = A$.

Solution:

$$A = \begin{bmatrix} 0 & 1 \\ -2 & -3 \end{bmatrix}.$$

$$\Phi(s) = [s\,I - A]^{-1} = \begin{bmatrix} s & -1 \\ 2 & s+3 \end{bmatrix}^{-1} = \frac{1}{s(s+3)+2}\begin{bmatrix} s+3 & 1 \\ -2 & s \end{bmatrix} = \frac{1}{s^2+3s+2}\begin{bmatrix} s+3 & 1 \\ -2 & s \end{bmatrix}$$

$$= \begin{bmatrix} \dfrac{s+3}{(s+1)(s+2)} & \dfrac{1}{(s+1)(s+2)} \\[3mm] \dfrac{-2}{(s+1)(s+2)} & \dfrac{s}{(s+1)(s+2)} \end{bmatrix}$$

$$= \begin{bmatrix} \dfrac{2}{s+1} - \dfrac{1}{s+2} & \dfrac{1}{s+1} - \dfrac{1}{s+2} \\[3mm] \dfrac{-2}{s+1} + \dfrac{2}{s+2} & \dfrac{-1}{s+1} + \dfrac{2}{s+2} \end{bmatrix}.$$

Taking inverse Laplace transforms, the state transition matrix is

$$\phi(t) = e^{At} = \begin{bmatrix} 2e^{-t} - e^{-2t} & e^{-t} - e^{-2t} \\ -2e^{-t} + 2e^{-2t} & -e^{-t} + 2e^{-2t} \end{bmatrix}$$

$$\dot{\phi}(t) = \begin{bmatrix} -2e^{-t} + 2e^{-2t} & -e^{-t} + 2e^{-2t} \\ 2e^{-t} - 4e^{-2t} & e^{-t} - 4e^{-2t} \end{bmatrix}$$

$$\dot{\phi}(0) = \begin{bmatrix} 0 & 1 \\ -2 & -3 \end{bmatrix} = A.$$

Example 2.6 The state equation for a system is

$$\dot{X} = \begin{bmatrix} -1 & 0 \\ 0 & -2 \end{bmatrix} X + \begin{bmatrix} 1 \\ 1 \end{bmatrix} r(t).$$

Find the resolvent matrix and the state transition matrix.

Solution: The resolvent matrix is

$$\Phi(s) = [s\,I - A]^{-1} = \begin{bmatrix} s+1 & 0 \\ 0 & s+2 \end{bmatrix}^{-1} = \frac{1}{(s+1)(s+2)}\begin{bmatrix} s+2 & 0 \\ 0 & s+1 \end{bmatrix} = \begin{bmatrix} \dfrac{1}{s+1} & 0 \\[3mm] 0 & \dfrac{1}{s+2} \end{bmatrix}.$$

Taking inverse Laplace transforms, the state transition matrix is

$$\phi(t) = e^{At} = \begin{bmatrix} e^{-t} & 0 \\ 0 & e^{-2t} \end{bmatrix}.$$

Example 2.7 The state equation for a system is

$$\dot{X} = \begin{bmatrix} -2 & 1 \\ 0 & -2 \end{bmatrix} X + \begin{bmatrix} 1 \\ 1 \end{bmatrix} r(t).$$

Find the state transition matrix.

Solution: The resolvent matrix is

$$\Phi(s) = [s\,I - A]^{-1} = \begin{bmatrix} s+2 & -1 \\ 0 & s+2 \end{bmatrix}^{-1} = \frac{1}{(s+2)^2}\begin{bmatrix} s+2 & 1 \\ 0 & s+2 \end{bmatrix} = \begin{bmatrix} \dfrac{1}{s+2} & \dfrac{1}{(s+2)^2} \\ 0 & \dfrac{1}{s+2} \end{bmatrix}.$$

Taking inverse Laplace transforms, the state transition matrix is

$$\phi(t) = e^{At} = \begin{bmatrix} e^{-2t} & t\,e^{-2t} \\ 0 & e^{-2t} \end{bmatrix}.$$

Example 2.8 For the given system matrix, determine the state transition matrix.

$$A = \begin{bmatrix} -2 & 1 & 0 \\ 0 & -2 & 0 \\ 0 & 0 & -3 \end{bmatrix}.$$

Solution: The resolvent matrix is

$$\Phi(s) = [s\,I - A]^{-1} = \begin{bmatrix} s+2 & -1 & 0 \\ 0 & s+2 & 0 \\ 0 & 0 & s+3 \end{bmatrix}^{-1}$$

$$= \frac{1}{(s+2)^2(s+3)}\begin{bmatrix} +(s+2)(s+3) & -\{-(s+3)\} & +0 \\ -0 & +(s+2)(s+3) & -0 \\ +0 & -0 & +(s+2)^2 \end{bmatrix}$$

$$= \begin{bmatrix} \dfrac{1}{(s+2)} & \dfrac{1}{(s+2)^2} & 0 \\ 0 & \dfrac{1}{(s+2)} & 0 \\ 0 & 0 & \dfrac{1}{(s+3)} \end{bmatrix}.$$

Taking inverse Laplace transforms,

$$\phi(t) = e^{At} = \begin{bmatrix} e^{-2t} & te^{-2t} & 0 \\ 0 & e^{-2t} & 0 \\ 0 & 0 & e^{-3t} \end{bmatrix}.$$

2.4 PROPERTIES OF STATE TRANSITION MATRIX

The state transition matrix is

$$\phi(t) = e^{At} = L^{-1}\left\{ [sI - A]^{-1} \right\}$$

1. It can be expressed as an infinite series as

$$e^{At} = I + At + \frac{A^2 t^2}{2!} + \frac{A^3 t^3}{3!} + \dots + \frac{A^k t^k}{k!} + \dots = \sum_{k=0}^{\infty} \frac{A^k t^k}{k!}$$

This converges absolutely for all values of t.

2. $\phi(0) = I$ is the identity matrix

3. $\dot{\phi}(0) = A$ is the system matrix.

4. The state transition matrix is non-singular, and its inverse exists for all t:

$$\left[\phi(t) \right]^{-1} = \left[e^{At} \right]^{-1} = \left[e^{-At} \right] = \phi(-t).$$

5. $\phi(t)\phi(-t) = e^{At} e^{-At} = I$

The state transition matrix for a system can be determined using properties (2) and (3). This is illustrated in Example 2.12.

Example 2.9 For the state variable description of a system given by

$$\dot{X} = \begin{bmatrix} -1 & 0 \\ 1 & -2 \end{bmatrix} X + \begin{bmatrix} 1 \\ 0 \end{bmatrix} r(t)$$

$$y(t) = \begin{bmatrix} 1 & 1 \end{bmatrix} X.$$

Draw a state diagram and determine the Eigen values, state transition matrix and the transfer function.

Solution: The state equations are

$$\dot{x}_1 = -x_1 + r(t) \tag{A}$$

$$\dot{x}_2 = x_1 - 2x_2 \tag{B}$$

The output equation is

$$y = x_1 + x_2 .$$

Using these equations, the state diagram can be drawn as shown in Figure E2.9.

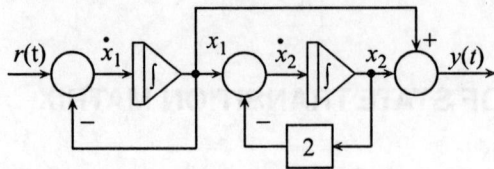

Figure E 2.9 The state diagram for Example 2.9

The Eigen values are obtained by solving the equation

$$|sI - A| = 0 \qquad or \qquad \begin{vmatrix} s+1 & 0 \\ -1 & s+2 \end{vmatrix} = 0 .$$

The characteristic equation is

$$(s+1)(s+2) = 0 .$$

The Eigen values are $s = -1$ and $s = -2$.
   The state transition matrix in the s-domain is

$$\Phi(s) = [sI - A]^{-1} = \begin{bmatrix} s+1 & 0 \\ -1 & s+2 \end{bmatrix}^{-1} = \frac{1}{(s+1)(s+2)} \begin{bmatrix} s+2 & 0 \\ 1 & s+1 \end{bmatrix}$$

$$= \begin{bmatrix} \dfrac{1}{s+1} & 0 \\ \dfrac{1}{(s+1)(s+2)} & \dfrac{1}{s+2} \end{bmatrix} = \begin{bmatrix} \dfrac{1}{s+1} & 0 \\ \dfrac{1}{(s+1)} - \dfrac{1}{(s+2)} & \dfrac{1}{s+2} \end{bmatrix} .$$

The state transition in the time domain is

$$\phi(t) = e^{At} = L^{-1} \begin{bmatrix} \dfrac{1}{s+1} & 0 \\ \dfrac{1}{(s+1)} - \dfrac{1}{(s+2)} & \dfrac{1}{s+2} \end{bmatrix} = \begin{bmatrix} e^{-t} & 0 \\ e^{-t} - e^{-2t} & e^{-2t} \end{bmatrix} .$$

The transfer function is

$$\frac{Y(s)}{R(s)} = C[sI - A]^{-1} B = \frac{1}{(s+1)(s+2)} \begin{bmatrix} 1 & 1 \end{bmatrix} \begin{bmatrix} s+2 & 0 \\ 1 & s+1 \end{bmatrix} \begin{bmatrix} 1 \\ 0 \end{bmatrix} = \frac{s+3}{(s+1)(s+2)}.$$

Example 2.10 A state model for an LTI system is given by

$$\dot{X} = \begin{bmatrix} 0 & 1 \\ -5 & -2 \end{bmatrix} X + \begin{bmatrix} 1 \\ 6 \end{bmatrix} r(t)$$

$$y(t) = \begin{bmatrix} 1 & 1 \end{bmatrix} X.$$

Determine the (a) characteristic equation, (b) Eigen values, (c) state transition matrix and (d) transfer function.

Solution: The characteristic equation is

$$\begin{vmatrix} s & -1 \\ 5 & (s+2) \end{vmatrix} = 0 \ or \ s^2 + 2s + 5 = 0.$$

The Eigen values are $s = -1 \pm j2$.
The state transition matrix in the s-domain or the resolvent matrix is

$$\Phi(s) = [sI - A]^{-1} = \begin{bmatrix} s & -1 \\ 5 & s+2 \end{bmatrix}^{-1} = \frac{1}{(s^2 + 2s + 5)} \begin{bmatrix} s+2 & 1 \\ -5 & s \end{bmatrix}$$

$$= \begin{bmatrix} \dfrac{s+2}{s^2 + 2s + 5} & \dfrac{1}{s^2 + 2s + 5} \\ -\dfrac{5}{s^2 + 2s + 5} & \dfrac{s}{s^2 + 2s + 5} \end{bmatrix}$$

$$= \begin{bmatrix} \dfrac{s+1}{(s+1)^2 + 2^2} + \dfrac{1}{2}\dfrac{2}{(s+1)^2 + 2^2} & \dfrac{1}{2}\dfrac{2}{(s+1)^2 + 2^2} \\ -\dfrac{5}{2}\dfrac{2}{(s+1)^2 + 2^2} & \dfrac{s+1}{(s+1)^2 + 2^2} - \dfrac{1}{2}\dfrac{2}{(s+1)^2 + 2^2} \end{bmatrix}$$

Taking inverse Laplace transforms, the state transition matrix in the time domain is

$$\phi(t) = e^{At} = \begin{bmatrix} e^{-t}\cos 2t + \dfrac{1}{2}e^{-t}\sin 2t & \dfrac{1}{2}e^{-t}\sin 2t \\ -\dfrac{5}{2}e^{-t}\sin 2t & e^{-t}\cos 2t - \dfrac{1}{2}e^{-t}\sin 2t \end{bmatrix}.$$

The transfer function is

$$\frac{Y(s)}{R(s)} = C[sI - A]^{-1} B = C\,\Phi(s)B$$

$$= \frac{1}{(s^2 + 2s + 5)} \begin{bmatrix} 1 & 1 \end{bmatrix} \begin{bmatrix} s+2 & 1 \\ -5 & s \end{bmatrix} \begin{bmatrix} 1 \\ 6 \end{bmatrix}$$

$$= \frac{1}{(s^2 + 2s + 5)} \begin{bmatrix} 1 & 1 \end{bmatrix} \begin{bmatrix} s+8 \\ 6s-5 \end{bmatrix}$$

$$= \frac{7s+3}{(s^2 + 2s + 5)}.$$

Example 2.11 Draw the state diagram and find the state transition matrix and the transfer function of a system represented by the state model

$$\dot{X} = \begin{bmatrix} 0 & 2 & 1 \\ 0 & -2 & 2 \\ -1 & -1 & -4 \end{bmatrix} X + \begin{bmatrix} 0 \\ 0 \\ 2 \end{bmatrix} r$$

$$y(t) = \begin{bmatrix} 1 & 0 & 0 \end{bmatrix} X .$$

Solution: The state equations are

$$\dot{x}_1 = 2x_2 + x_3 \qquad \dot{x}_2 = -2x_2 + 2x_3 \qquad \dot{x}_3 = -x_1 - x_2 - 4x_3 + 2r .$$

The output equation is $y = x_1$.

Using these equations, the state diagram can be drawn, keeping the output of an integrator as a state variable. The state diagram is shown in Figure E2.11.

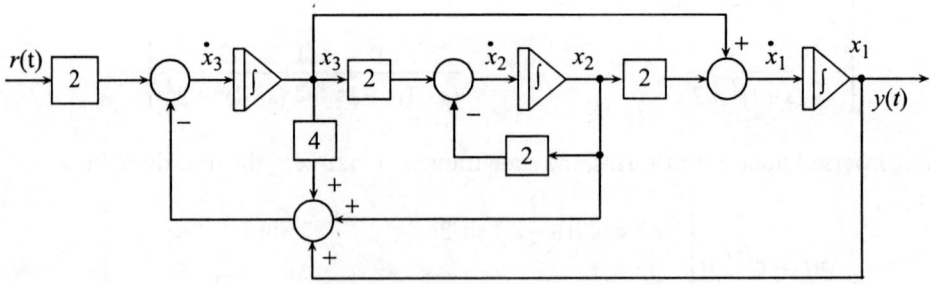

Figure E 2.11 The state diagram for Example 2.11

The state transition matrix is

$$\Phi(s) = [sI - A]^{-1} = \begin{bmatrix} s & -2 & -1 \\ 0 & s+2 & -2 \\ 1 & 1 & s+4 \end{bmatrix}^{-1} .$$

$$Det[sI - A] = s[(s+2)(s+4)+2] + 4 + s + 2 = s^3 + 6s^2 + 11s + 6$$

$$\Phi(s) = \frac{1}{s^3 + 6s^2 + 11s + 6} \begin{bmatrix} +(s^2+6s+8+2) & -(-2s-8+1) & +(4+s+2) \\ -(2) & +(s^2+4s+1) & -(-2s) \\ +\{-(s+2)\} & -(s+2) & +(s^2+2s) \end{bmatrix}$$

$$= \frac{1}{s^3 + 6s^2 + 11s + 6} \begin{bmatrix} (s^2+6s+10) & 2s+7 & (s+6) \\ -2 & (s^2+4s+1) & 2s \\ -(s+2) & -(s+2) & (s^2+2s) \end{bmatrix} .$$

The transfer function is

$$\frac{Y(s)}{R(s)} = C\,\Phi(s)\,B$$

$$= \frac{1}{s^3 + 6s^2 + 11s + 6} [1 \quad 0 \quad 0] \begin{bmatrix} (s^2+6s+10) & 2s+7 & (s+6) \\ -2 & (s^2+4s+1) & 2s \\ -(s+2) & -(s+2) & (s^2+2s) \end{bmatrix} \begin{bmatrix} 0 \\ 0 \\ 2 \end{bmatrix}$$

$$= \frac{2(s+6)}{s^3 + 6s^2 + 11s + 6} .$$

Example 2.12 A system matrix for a linear time-invariant system is given as

$$A = \begin{bmatrix} -1 & 1 \\ 0 & -3 \end{bmatrix} .$$

Determine the state transition matrix using its properties.

Solution: The Eigen values are obtained by solving the equation

$$|sI - A| = 0 \qquad or \qquad \begin{vmatrix} s+1 & -1 \\ 0 & s+3 \end{vmatrix} = 0 .$$

Or

$$(s+1)(s+3) = 0 .$$

The Eigen values are $s = -1$ and $s = -3$. Each element of the state transition matrix will be a polynomial in terms of e^{-t} and e^{-3t}. Hence

$$\phi(t) = e^{At} = \begin{bmatrix} a_1 e^{-t} + a_2 e^{-3t} & b_1 e^{-t} + b_2 e^{-3t} \\ c_1 e^{-t} + c_2 e^{-3t} & d_1 e^{-t} + d_2 e^{-3t} \end{bmatrix}.$$

Using the property $\phi(0) = I$,

$$\begin{bmatrix} a_1 + a_2 & b_1 + b_2 \\ c_1 + c_2 & d_1 + d_2 \end{bmatrix} = \begin{bmatrix} 1 & 0 \\ 0 & 1 \end{bmatrix}. \tag{1}$$

Using the property $\dot{\phi}(0) = A$,

$$\begin{bmatrix} -a_1 - 3a_2 & -b_1 - 3b_2 \\ -c_1 - 3c_2 & -d_1 - 3d_2 \end{bmatrix} = \begin{bmatrix} -1 & 1 \\ 0 & -3 \end{bmatrix}. \tag{2}$$

From equations (1) and (2),

$a_1 + a_2 = 1$ $a_1 + 3a_2 = 1$ Solving these equations $a_1 = 1$ $a_2 = 0$

$b_1 + b_2 = 0$ $b_1 + 3b_2 = -1$ Solving these equations $b_1 = \dfrac{1}{2}$ $b_2 = -\dfrac{1}{2}$

$c_1 + c_2 = 0$ $c_1 + 3c_2 = 0$ Solving these equations $c_1 = 0$ $c_2 = 0$

$d_1 + d_2 = 1$ $d_1 + 3d_2 = 3$ Solving these equations $d_1 = 0$ $d_2 = 1$

Hence, the state transition matrix is

$$\phi(t) = e^{At} = \begin{bmatrix} e^{-t} & \dfrac{1}{2}e^{-t} - \dfrac{1}{2}e^{-3t} \\ 0 & e^{-3t} \end{bmatrix}.$$

2.5 REALIZATION OF STATE MODELS FROM THE TRANSFER FUNCTION

As mentioned earlier, a state model obtained for a system is not unique and depends on the choice of state variables. The state variables can be physical variables or phase variables or any other set of variables which even may not have physical entity. Hence, for a given system, many state models can be derived representing the same system. Given the transfer function of a system, many state models can be realized using different methods of processing the transfer function.

For an n-th order system, the system matrix is of size $n \times n$. If there are p inputs and q outputs, the input matrix B is of size $n \times p$ and the output matrix C is of size $q \times n$. The input–output coupling matrix is of size $q \times p$. Hence, the number of elements in the

A, B, C, and D matrices is equal to $n^2 + np + qn + qp$. This number is very large and can be significantly reduced by deriving state models from transfer functions using certain standard and accepted methods. Such state models with matrices having minimum number of non-zero elements are called *canonical forms*. State models in canonical forms are very convenient in the analysis and design of physical systems. Canonical state models have certain advantages over other state models. In certain canonical forms, the state transition matrix can be written down by inspection.

The realization of the following four canonical forms will be discussed here:

1. Controllable canonical form
2. Observable canonical form
3. Diagonal canonical form
4. Jordan canonical form.

2.6 CONTROLLABLE CANONICAL FORM

The transfer function of a physical system is the ratio of polynomials in terms of the variable s. Usually the degree of the numerator polynomial is less than the degree of the denominator polynomial and such a transfer function is said to be *strictly proper*.

Consider the transfer function of a third-order system given as

$$\frac{Y(s)}{R(s)} = \frac{b_2 s^2 + b_1 s + b_0}{s^3 + a_2 s^2 + a_1 s + a_0}.$$

This can be written as

$$\frac{Y(s)}{R(s)} = \frac{b_2 s^2 + b_1 s + b_0}{s^3 + a_2 s^2 + a_1 s + a_0} \frac{W(s)}{W(s)}$$

and can be represented by a block diagram shown in Figure 2. 2.

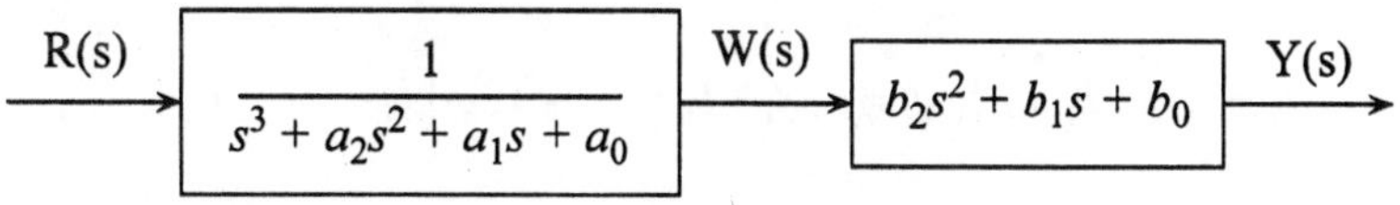

Figure 2.2 Realization into controllable canonical form

$W(s)$ is an intermediate variable. From the block diagram,

$$\frac{W(s)}{R(s)} = \frac{1}{s^3 + a_2 s^2 + a_1 s + a_0}$$

$$(s^3 + a_2 s^2 + a_1 s + a_0) W(s) = R(s)$$

Taking inverse Laplace transforms,

$$\dddot{w} + a_2\,\ddot{w} + a_1\,\dot{w} + a_0 w = r(t) . \tag{2.9}$$

Choose $x_1 = w$, $x_2 = \dot{x}_1 = \dot{w}$ and $x_3 = \dot{x}_2 = \ddot{w}$ as the three state variables.

$$\dot{x}_1 = x_2 \tag{2.10}$$

$$\dot{x}_2 = x_3 \tag{2.11}$$

From equation (2.9),

$$\dddot{w} = -a_0 w - a_1\,\dot{w} - a_2\,\ddot{w} + r(t)$$

Or

$$\dot{x}_3 = -a_0 x_1 - a_1 x_2 - a_2 x_3 + r(t) \tag{2.12}$$

Equations (2.10), (2.11) and (2.12) are state equations and can be expressed in the matrix form as

$$\begin{bmatrix} \dot{x}_1 \\ \dot{x}_2 \\ \dot{x}_3 \end{bmatrix} = \begin{bmatrix} 0 & 1 & 0 \\ 0 & 0 & 1 \\ -a_0 & -a_1 & -a_2 \end{bmatrix} \begin{bmatrix} x_1 \\ x_2 \\ x_3 \end{bmatrix} + \begin{bmatrix} 0 \\ 0 \\ 1 \end{bmatrix} r .$$

The output equation, from the block diagram, is

$$Y(s) = (b_2 s^2 + b_1 s + b_0) W(s)$$

Taking inverse Laplace transforms,

$$y(t) = b_2\,\ddot{w} + b_1\,\dot{w} + b_0\,w = b_0 x_1 + b_1 x_2 + b_2 x_3 = \begin{bmatrix} b_0 & b_1 & b_2 \end{bmatrix} \begin{bmatrix} x_1 \\ x_2 \\ x_3 \end{bmatrix} . \tag{2.13}$$

Note: It can be observed that in the system matrix A, the elements of all the rows except the last are either 0 or 1. The last row contains the coefficients of the denominator polynomial of the transfer function, with negative sign. The elements of the input matrix B are all 0 except the last row, which is 1. The elements of the output matrix C are the coefficients of the numerator polynomial of the transfer function.

The state diagram can be drawn as shown in figure 2.3.

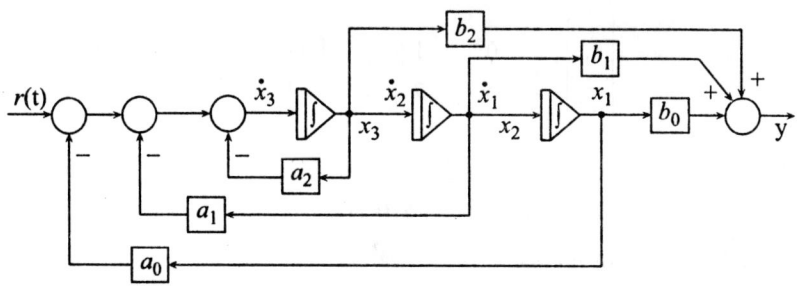

Figure 2.3 State diagram – controllable canonical form

Example 2.13 The transfer function of a dynamic system is

$$\frac{Y(s)}{R(s)} = \frac{2s+5}{s^2+7s+10}.$$

Obtain the state model in the controllable canonical form.

Solution: The transfer function can be written as

$$\frac{Y(s)}{R(s)} = \frac{2s+5}{s^2+7s+10}\frac{W(s)}{W(s)}$$

$$(s^2+7s+10)W(s) = R(s).$$

Taking inverse Laplace transforms,

$$\ddot{w}+7\dot{w}+10w = r(t). \tag{1}$$

Choose $x_1 = w$ and $x_2 = \dot{x_1} = \dot{w}$ as the two state variables.

$$\dot{x_1} = x_2 \tag{A}$$

From equation (1),

$$\ddot{w} = -7\dot{w}-10w+r(t)$$

Or

$$\dot{x_2} = -10x_1 - 7x_2 + r(t). \tag{B}$$

Keeping equations (A) and (B) in the matrix form,

$$\begin{bmatrix} \dot{x}_1 \\ \dot{x}_2 \end{bmatrix} = \begin{bmatrix} 0 & 1 \\ -10 & -7 \end{bmatrix} \begin{bmatrix} x_1 \\ x_2 \end{bmatrix} + \begin{bmatrix} 0 \\ 1 \end{bmatrix} r.$$

The output equation is

$$Y(s) = (2s+5)W(s).$$

Taking inverse Laplace transforms,

$$y(t) = 2\dot{w} + 5w = 5x_1 + 2x_2 = \begin{bmatrix} 5 & 2 \end{bmatrix} \begin{bmatrix} x_1 \\ x_2 \end{bmatrix}.$$

Example 2.14 The transfer function of a dynamic system is

$$\frac{Y(s)}{R(s)} = \frac{10(s^2 + 4s + 8)}{s^3 + 7s^2 + 10s}.$$

Derive the state model in controllable canonical form.

Solution: The transfer function can be written as

$$\frac{Y(s)}{R(s)} = \frac{10(s^2 + 4s + 8)}{s^3 + 7s^2 + 10s} \frac{W(s)}{W(s)}$$

$$(s^3 + 7s^2 + 10s)W(s) = R(s).$$

Taking inverse Laplace transforms,

$$\dddot{w} + 7\ddot{w} + 10\dot{w} = r(t) \qquad (1)$$

Choose $x_1 = w$, $x_2 = \dot{x}_1 = \dot{w}$ and $x_3 = \dot{x}_2 = \ddot{w}$ as the three state variables.

$$\dot{x}_1 = x_2 \qquad (A)$$

$$\dot{x}_2 = x_3 \qquad (B)$$

From equation (1),

$$\dddot{w} = -7\ddot{w} - 10\dot{w} + r(t)$$

Or

$$\dot{x}_3 = -10x_2 - 7x_3 + r(t). \qquad (C)$$

Keeping equations (A), (B) and (C) in the matrix form,

$$\begin{bmatrix} \dot{x_1} \\ \dot{x_2} \\ \dot{x_3} \end{bmatrix} = \begin{bmatrix} 0 & 1 & 0 \\ 0 & 0 & 1 \\ 0 & -10 & -7 \end{bmatrix} \begin{bmatrix} x_1 \\ x_2 \\ x_3 \end{bmatrix} + \begin{bmatrix} 0 \\ 0 \\ 1 \end{bmatrix} r .$$

The output equation is

$$Y(s) = (10s^2 + 40s + 80)W(s) .$$

Taking inverse Laplace transforms,

$$y(t) = 10\ddot{w} + 40\dot{w} + 80\,w = 10\,x_3 + 40\,x_2 + 80\,x_1 = \begin{bmatrix} 80 & 40 & 10 \end{bmatrix} \begin{bmatrix} x_1 \\ x_2 \\ x_3 \end{bmatrix} .$$

Example 2.15 Derive the state model in its controllable canonical form for a system with transfer function

$$\frac{Y(s)}{R(s)} = \frac{5s^2 + 14s + 8}{s^3 + 9s^2 + 23s + 15} .$$

Solution: The transfer function can be written as

$$\frac{Y(s)}{R(s)} = \frac{5s^2 + 14s + 8}{s^3 + 9s^2 + 23s + 15} \frac{W(s)}{W(s)}$$

$$(s^3 + 9s^2 + 23s + 15)W(s) = R(s) . \tag{1}$$

Taking inverse Laplace transforms of equation (1),

$$\dddot{w} + 9\ddot{w} + 23\dot{w} + 15w = r(t) . \tag{2}$$

Choose $x_1 = w$, $x_2 = \dot{x_1} = \dot{w}$ and $x_3 = \dot{x_2} = \ddot{w}$ as the three state variables.

$$\dot{x_1} = x_2 \tag{A}$$

$$\dot{x_2} = x_3 \tag{B}$$

From equation (2),

$$\dddot{w} = -9\ddot{w} - 23\dot{w} - 15w + r(t)$$

Or

$$\dot{x_3} = -15x_1 - 23x_2 - 9x_3 + r(t) . \tag{C}$$

Keeping equations (A), (B) and (C) in the matrix form,

$$
\begin{bmatrix} \dot{x}_1 \\ \dot{x}_2 \\ \dot{x}_3 \end{bmatrix} = \begin{bmatrix} 0 & 1 & 0 \\ 0 & 0 & 1 \\ -15 & -23 & -9 \end{bmatrix} \begin{bmatrix} x_1 \\ x_2 \\ x_3 \end{bmatrix} + \begin{bmatrix} 0 \\ 0 \\ 1 \end{bmatrix} r .
$$

The output equation is

$$Y(s) = (5s^2 + 14s + 8)W(s).$$

Taking inverse Laplace transforms,

$$
y(t) = 5\ddot{w} + 14\dot{w} + 8w = 8x_1 + 14x_2 + 5x_3 = \begin{bmatrix} 8 & 14 & 5 \end{bmatrix} \begin{bmatrix} x_1 \\ x_2 \\ x_3 \end{bmatrix}.
$$

2.7 OBSERVABLE CANONICAL FORM

Consider a third-order system with transfer function

$$\frac{Y(s)}{R(s)} = \frac{b_2 s^2 + b_1 s + b_0}{s^3 + a_2 s^2 + a_1 s + a_0} .$$

After cross multiplication, this can be written as

$$(s^3 + a_2 s^2 + a_1 s + a_0) Y(s) = (b_2 s^2 + b_1 s + b_0)R(s).$$

Taking inverse Laplace transforms and using the symbol $D = \dfrac{d}{dt}$,

$$(D^3 + a_2 D^2 + a_1 D + a_0)y = (b_2 D^2 + b_1 D + b_0)r$$

$$D^3 y = D^2(b_2 r - a_2 y) + D(b_1 r - a_1 y) + (b_0 r - a_0 y)$$

Or

$$y = D^{-1}(b_2 r - a_2 y) + D^{-2}(b_1 r - a_1 y) + D^{-3}(b_0 r - a_0 y). \tag{2.14}$$

Let the three state variables be $x_1 = y$, x_2 and x_3.
Hence, from equation (2.14),

$$x_1 = D^{-1}(b_2 r - a_2 x_1) + D^{-2}(b_1 r - a_1 x_1) + D^{-3}(b_0 r - a_0 x_1).$$

Differentiating this equation,

$$\dot{x}_1 = (b_2 r - a_2 x_1) + D^{-1}(b_1 r - a_1 x_1) + D^{-2}(b_0 r - a_0 x_1). \tag{2.15}$$

Let

$$x_2 = D^{-1}(b_1 r - a_1 x_1) + D^{-2}(b_0 r - a_0 x_1). \tag{2.16}$$

Hence, from equation (2.15) and (2.16),

$$\dot{x}_1 = b_2 r - a_2 x_1 + x_2. \tag{2.17}$$

Differentiating equation (2.16),

$$\dot{x}_2 = (b_1 r - a_1 x_1) + D^{-1}(b_0 r - a_0 x_1). \tag{2.18}$$

Let

$$x_3 = D^{-1}(b_0 r - a_0 x_1). \tag{2.19}$$

Hence, from equation (2.18) and (2.19),

$$\dot{x}_2 = b_1 r - a_1 x_1 + x_3 \tag{2.20}$$

Differentiating equation (2.19),

$$\dot{x}_3 = b_0 r - a_0 x_1. \tag{2.21}$$

Keeping equations (2.17), (2.20) and (2.21) in the matrix form,

$$\begin{bmatrix} \dot{x}_1 \\ \dot{x}_2 \\ \dot{x}_3 \end{bmatrix} = \begin{bmatrix} -a_2 & 1 & 0 \\ -a_1 & 0 & 1 \\ -a_0 & 0 & 0 \end{bmatrix} \begin{bmatrix} x_1 \\ x_2 \\ x_3 \end{bmatrix} + \begin{bmatrix} b_2 \\ b_1 \\ b_0 \end{bmatrix} r.$$

The output equation is

$$y = x_1 = \begin{bmatrix} 1 & 0 & 0 \end{bmatrix} \begin{bmatrix} x_1 \\ x_2 \\ x_3 \end{bmatrix}.$$

Note: It can be observed that in the system matrix, the elements of the first column are the coefficients of the denominator polynomial of the transfer function with negative sign and the elements of the other columns are either 0 or 1. The elements of the column matrix B are the coefficients of the numerator polynomial of the transfer function. The first element of the output row matrix C is '1' and all other elements are 0.

The state diagram is shown in Figure 2.4.

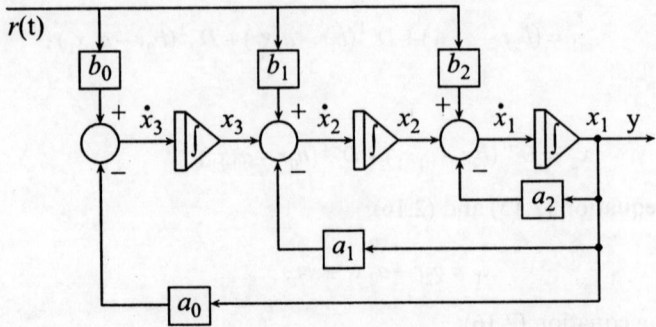

Figure 2.4 State diagram – observable canonical form

Example 2.16 Derive the state model in its observable canonical form for a system with transfer function

$$\frac{Y(s)}{R(s)} = \frac{2s+5}{s^2+7s+10}.$$

Solution: Rearranging the transfer function,

$$(s^2+7s+10)Y(s) = (2s+5)R(s).$$

Taking inverse Laplace transforms,

$$\ddot{y}+7\dot{y}+10y = 2\dot{r}+5r.$$

This can be written in terms of the differential operator as

$$(D^2+7D+10)y = (2D+5)r$$

$$D^2y = D(2r-7y)+(5r-10y)$$

$$y = D^{-1}(2r-7y)+D^{-2}(5r-10y). \tag{1}$$

Let $x_1 = y$ *and* x_2 be the two state variables.
    Equation (1) can be written as

$$x_1 = D^{-1}(2r-7x_1)+D^{-2}(5r-10x_1). \tag{2}$$

Differentiating equation (2),

$$\dot{x}_1 = (2r-7x_1)+D^{-1}(5r-10x_1). \tag{3}$$

Let

$$x_2 = D^{-1}(5r-10x_1). \tag{4}$$

Combining equations (3) and (4),

$$\dot{x}_1 = (2r - 7x_1) + x_2 .$$ (A)

Differentiating equation (4),

$$\dot{x}_2 = 5r - 10x_1 .$$ (B)

Keeping equations (A) and (B) in the matrix form, the state equation is

$$\begin{bmatrix} \dot{x}_1 \\ \dot{x}_2 \end{bmatrix} = \begin{bmatrix} -7 & 1 \\ -10 & 0 \end{bmatrix} \begin{bmatrix} x_1 \\ x_2 \end{bmatrix} + \begin{bmatrix} 2 \\ 5 \end{bmatrix} r .$$

The output equation is

$$y = x_1 = \begin{bmatrix} 1 & 0 \end{bmatrix} \begin{bmatrix} x_1 \\ x_2 \end{bmatrix}.$$

Example 2.17 Derive the state model in its observable canonical form for a system with transfer function

$$\frac{Y(s)}{R(s)} = \frac{5s^2 + 14s + 8}{s^3 + 9s^2 + 23s + 15}.$$

Solution: The transfer function can be written as

$$(s^3 + 9s^2 + 23s + 15)Y(s) = (5s^2 + 14s + 8)R(s) .$$

Taking inverse Laplace transforms,

$$\dddot{y} + 9\ddot{y} + 23\dot{y} + 15y = 5\ddot{r} + 14\dot{r} + 8r .$$

In terms of the differential operator D, this can be written as

$$D^3 y + 9D^2 y + 23Dy + 15y = 5D^2 r + 14Dr + 8r .$$

From this equation,

$$y = D^{-1}(5r - 9y) + D^{-2}(14r - 23y) + D^{-3}(8r - 15y) ,$$ (1)

Let $x_1 = y$, x_2 *and* x_3 be the three state variables.
From equation (1),

$$x_1 = D^{-1}(5r - 9x_1) + D^{-2}(14r - 23x_1) + D^{-3}(8r - 15x_1) .$$ (2)

Differentiating equation (2),

$$\dot{x}_1 = (5r - 9x_1) + D^{-1}(14r - 23x_1) + D^{-2}(8r - 15x_1) .$$ (3)

Let

$$x_2 = D^{-1}(14r - 23x_1) + D^{-2}(8r - 15x_1).$$ (4)

Combining equations (3) and (4),

$$\dot{x}_1 = 5r - 9x_1 + x_2.$$ (A)

Differentiating equation (4),

$$\dot{x}_2 = (14r - 23x_1) + D^{-1}(8r - 15x_1).$$ (5)

Let

$$x_3 = D^{-1}(8r - 15x_1).$$ (6)

Combining equations (5) and (6),

$$\dot{x}_2 = 14r - 23x_1 + x_3.$$ (B)

Differentiating equation (6),

$$\dot{x}_3 = 8r - 15x_1.$$ (C)

Equations (A), (B) and (C) are state equations and keeping them in the matrix form

$$\begin{bmatrix} \dot{x}_1 \\ \dot{x}_2 \\ \dot{x}_3 \end{bmatrix} = \begin{bmatrix} -9 & 1 & 0 \\ -23 & 0 & 1 \\ -15 & 0 & 0 \end{bmatrix} \begin{bmatrix} x_1 \\ x_2 \\ x_3 \end{bmatrix} + \begin{bmatrix} 5 \\ 14 \\ 8 \end{bmatrix} r.$$

The output equation is

$$y = x_1 = \begin{bmatrix} 1 & 0 & 0 \end{bmatrix} \begin{bmatrix} x_1 \\ x_2 \\ x_3 \end{bmatrix}.$$

2.8 DIAGONAL CANONICAL FORM

This canonical form of the state model is obtained using the partial fraction expansion of the given transfer function, provided the Eigen values are distinct.

Consider a third-order system with transfer function

$$\frac{Y(s)}{R(s)} = \frac{b_2 s^2 + b_1 s + b_0}{s^3 + a_2 s^2 + a_1 s + a_0}.$$

Let the three Eigen values be distinct and let them be $s = -\lambda_1$ $s = -\lambda_2$ and $s = -\lambda_3$.
Hence the transfer function can be written as

$$\frac{Y(s)}{R(s)} = \frac{b_2 s^2 + b_1 s + b_0}{(s + \lambda_1)(s + \lambda_2)(s + \lambda_3)} .$$

Let the partial fraction expansion be

$$\frac{Y(s)}{R(s)} = \frac{k_1}{(s + \lambda_1)} + \frac{k_2}{(s + \lambda_2)} + \frac{k_3}{(s + \lambda_3)} .$$

Rearranging this transfer function,

$$Y(s) = \frac{k_1}{(s + \lambda_1)} R(s) + \frac{k_2}{(s + \lambda_2)} R(s) + \frac{k_3}{(s + \lambda_3)} R(s) \qquad (2.22)$$

Or

$$Y(s) = k_1 X_1(s) + k_2 X_2(s) + k_3 X_3(s) . \qquad (2.23)$$

These equations are represented by means of a block diagram as shown in Figure 2.5. Since
the blocks are in parallel, this method of realization is also called *parallel realization*.

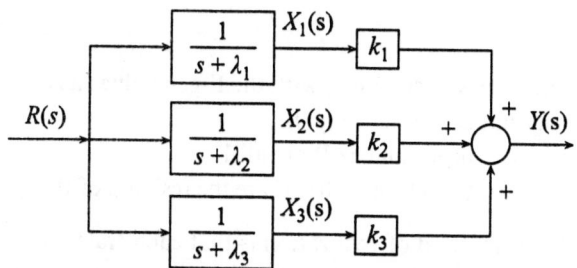

Figure 2.5 Block diagram – Diagonal form

From equations (2.22) and (2.23),

$$X_1(s) = \frac{R(s)}{s + \lambda_1} \qquad X_2(s) = \frac{R(s)}{s + \lambda_2} \qquad X_3(s) = \frac{R(s)}{s + \lambda_3} .$$

$$(s + \lambda_1) X_1(s) = R(s) .$$

Taking inverse Laplace transforms,

$$\dot{x}_1 + \lambda_1 x_1 = r(t) \qquad\qquad \dot{x}_1 = -\lambda_1 x_1 + r(t) \qquad (2.24)$$

$$(s + \lambda_2) X_2(s) = R(s) .$$

Taking inverse Laplace transforms,

$$\dot{x}_2 + \lambda_2 x_2 = r(t) \qquad \dot{x}_2 = -\lambda_2 x_2 + r(t) \qquad (2.25)$$

and $(s + \lambda_3)X_3(s) = R(s)$.

Taking inverse Laplace transforms,

$$\dot{x}_3 + \lambda_3 x_3 = r(t) \qquad \dot{x}_3 = -\lambda_3 x_3 + r(t) \qquad (2.26)$$

Equations (2.24), (2.25) and (2.26) are the state equations and keeping them in the matrix form

$$
\begin{bmatrix} \dot{x}_1 \\ \dot{x}_2 \\ \dot{x}_3 \end{bmatrix} =
\begin{bmatrix} -\lambda_1 & 0 & 0 \\ 0 & -\lambda_2 & 0 \\ 0 & 0 & -\lambda_3 \end{bmatrix}
\begin{bmatrix} x_1 \\ x_2 \\ x_3 \end{bmatrix} +
\begin{bmatrix} 1 \\ 1 \\ 1 \end{bmatrix} r.
$$

Taking inverse Laplace transform of equation (2.23), the output equation is

$$y(t) = k_1 x_1 + k_2 x_2 + k_3 x_3 = \begin{bmatrix} k_1 & k_2 & k_3 \end{bmatrix} \begin{bmatrix} x_1 \\ x_2 \\ x_3 \end{bmatrix}.$$

Note:

1. The system matrix is diagonal, with the Eigen values as the diagonal elements and all the off-diagonal elements being 0.
2. The elements of input matrix B are all 1.
3. The elements of the output matrix C are the residues of the transfer function at its poles.

The advantage of this method of realization is that each state equation is a first-order differential equation in only one variable. Thus these equations can be solved independently. Hence, these equations are said to be decoupled and the state model is said to be in decoupled form. The state diagram is shown in Figure 2.6.

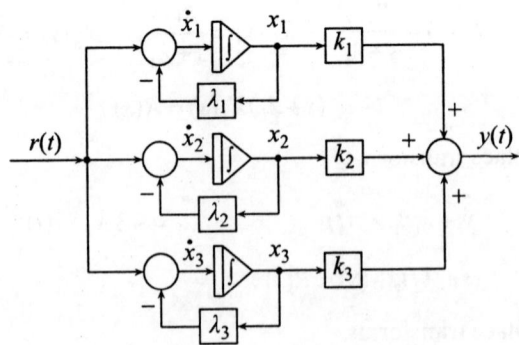

Figure 2.6 State diagram – Diagonal canonical form

Example 2.18 Derive the state model in its diagonal canonical form for a system with transfer function

$$\frac{Y(s)}{R(s)} = \frac{2s+5}{s^2+7s+10}.$$

Solution: The denominator polynomial can be factorized as

$$s^2+7s+10 = (s+2)(s+5).$$

Hence, the Eigen values are $s = -2$ *and* $s = -5$.

Expanding the transfer function in partial fractions,

$$\frac{Y(s)}{R(s)} = \frac{2s+5}{s^2+7s+10} = \frac{(1/3)}{s+2} + \frac{(5/3)}{s+5}$$

$$Y(s) = \frac{(1/3)}{s+2}R(s) + \frac{(5/3)}{s+5}R(s) = \frac{1}{3}X_1(s) + \frac{5}{3}X_2(s),\qquad (1)$$

where

$$X_1(s) = \frac{R(s)}{s+2} \quad and \quad X_2(s) = \frac{R(s)}{s+5}$$

$$(s+2)X_1(s) = R(s).$$

Taking inverse Laplace transforms,

$$\dot{x}_1 + 2x_1 = r(t) \qquad \text{or } \dot{x}_1 = -2x_1 + r(t) \qquad (A)$$

$$(s+5)X_2(s) = R(s).$$

Taking inverse Laplace transforms,

$$\dot{x}_2 + 5x_2 = r(t) \quad \text{or} \quad \dot{x}_2 = -5x_2 + r(t).\qquad (B)$$

Equations (A) and (B) are the state equations and keeping them in the matrix form,

$$\begin{bmatrix} \dot{x}_1 \\ \dot{x}_2 \end{bmatrix} = \begin{bmatrix} -2 & 0 \\ 0 & -5 \end{bmatrix}\begin{bmatrix} x_1 \\ x_2 \end{bmatrix} + \begin{bmatrix} 1 \\ 1 \end{bmatrix}r(t).$$

From equation (1),

$$Y(s) = \frac{1}{3}X_1(s) + \frac{5}{3}X_2(s).$$

Taking inverse Laplace transforms,

$$y(t) = \frac{1}{3}x_1 + \frac{5}{3}x_2 = \begin{bmatrix} \frac{1}{3} & \frac{5}{3} \end{bmatrix} \begin{bmatrix} x_1 \\ x_2 \end{bmatrix}.$$

Example 2.19 Derive the state model in its diagonal canonical form and draw the state diagram for a system with transfer function

$$\frac{Y(s)}{R(s)} = \frac{5s^2 + 14s + 8}{s^3 + 9s^2 + 23s + 15}.$$

Solution:

The denominator polynomial can be factorized as

$$s^3 + 9s^2 + 23s + 15 = (s+1)(s+3)(s+5).$$

Hence, the Eigen values are $s = -1$, $s = -3$ and $s = -5$.

Expanding the transfer function in partial fractions

$$\frac{Y(s)}{R(s)} = \frac{5s^2 + 14s + 8}{s^3 + 9s^2 + 23s + 15} = \frac{-(1/8)}{s+1} - \frac{(11/4)}{s+3} + \frac{(63/8)}{s+5}$$

$$Y(s) = -\frac{1}{8}\frac{R(s)}{s+1} - \frac{11}{4}\frac{R(s)}{s+3} + \frac{63}{8}\frac{R(s)}{s+5}. \tag{1}$$

Let

$$Y(s) = -\frac{1}{8}X_1(s) - \frac{11}{4}X_2(s) + \frac{63}{8}X_3(s). \tag{2}$$

From equations (1) and (2),

$$X_1(s) = \frac{R(s)}{s+1} \qquad\qquad X_2(s) = \frac{R(s)}{s+3} \qquad\qquad X_3(s) = \frac{R(s)}{s+5}$$

$$(s+1)X_1(s) = R(s).$$

Taking inverse Laplace transforms,

$$\dot{x}_1 + x_1 = r(t) \qquad or \quad \dot{x}_1 = -x_1 + r(t) \tag{A}$$

$$(s+3)X_2(s) = R(s).$$

Taking inverse Laplace transforms,

$$\dot{x}_2 + 3x_2 = r(t) \qquad or \quad \dot{x}_2 = -3x_2 + r(t) \tag{B}$$

$$(s+5)X_3(s) = R(s).$$

Taking inverse Laplace transforms,

$$\dot{x}_3 + 5x_3 = r(t) \quad or \quad \dot{x}_3 = -5x_3 + r(t).$$ (C)

Keeping equations (A), (B) and (C) in the matrix form,

$$\begin{bmatrix} \dot{x}_1 \\ \dot{x}_2 \\ \dot{x}_3 \end{bmatrix} = \begin{bmatrix} -1 & 0 & 0 \\ 0 & -3 & 0 \\ 0 & 0 & -5 \end{bmatrix} \begin{bmatrix} x_1 \\ x_2 \\ x_3 \end{bmatrix} + \begin{bmatrix} 1 \\ 1 \\ 1 \end{bmatrix} r.$$

From equation (2), the output equation is

$$Y(s) = -\frac{1}{8}X_1(s) - \frac{11}{4}X_2(s) + \frac{63}{8}X_3(s).$$

Taking inverse Laplace transforms,

$$y(t) = -\frac{1}{8}x_1 - \frac{11}{4}x_2 + \frac{63}{8}x_3 = \begin{bmatrix} -\frac{1}{8} & -\frac{11}{4} & \frac{63}{8} \end{bmatrix} \begin{bmatrix} x_1 \\ x_2 \\ x_3 \end{bmatrix}.$$

The state diagram is shown in Figure E 2.19.

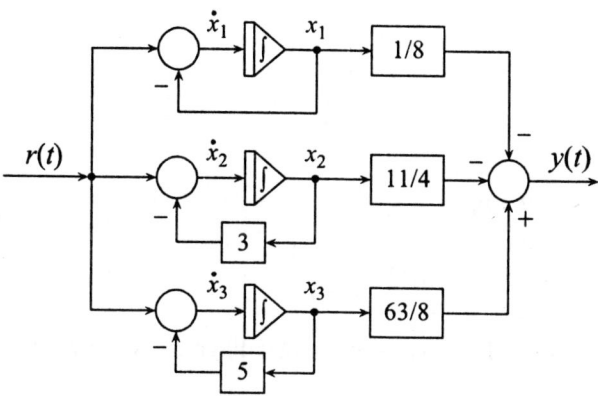

Figure E 2.19 State diagram for Example 2.19

2.9 JORDAN CANONICAL FORM

The Jordan canonical form of a state model is also obtained by the partial fraction expansion of the given transfer function. But in this case some of the Eigen values are repeated. The diagonal canonical form is a special case of Jordan canonical form.

Consider a third-order system with transfer function

$$\frac{Y(s)}{R(s)} = \frac{b_2 s^2 + b_1 s + b_0}{s^3 + a_2 s^2 + a_1 s + a_0}.$$

Let the three Eigen values be $s = -\lambda_1$, $s = -\lambda_1$ and $s = -\lambda_2$. Note that $s = -\lambda_1$ is a repeated Eigen value. Hence the transfer function can be written as

$$\frac{Y(s)}{R(s)} = \frac{b_2 s^2 + b_1 s + b_0}{(s + \lambda_1)^2 (s + \lambda_2)}.$$

Let the partial fraction expansion be

$$\frac{Y(s)}{R(s)} = \frac{k_{11}}{(s + \lambda_1)^2} + \frac{k_{12}}{(s + \lambda_1)} + \frac{k_2}{(s + \lambda_2)}$$

$$Y(s) = k_{11} \frac{R(s)}{(s + \lambda_1)^2} + k_{12} \frac{R(s)}{(s + \lambda_1)} + k_2 \frac{R(s)}{(s + \lambda_2)} \tag{2.27}$$

Or

$$Y(s) = k_{11} X_1(s) + k_{12} X_2(s) + k_2 X_3(s). \tag{2.28}$$

From equations (2.27) and (2.28),

$$X_1(s) = \frac{R(s)}{(s + \lambda_1)^2} \tag{2.29}$$

$$X_2(s) = \frac{R(s)}{(s + \lambda_1)} \tag{2.30}$$

$$X_3(s) = \frac{R(s)}{(s + \lambda_2)}. \tag{2.31}$$

From equations (2.29) and (2.30),

$$X_1(s) = \frac{X_2(s)}{(s + \lambda_1)}. \tag{2.32}$$

These equations are represented by means of a block diagram as shown in Figure 2.7.

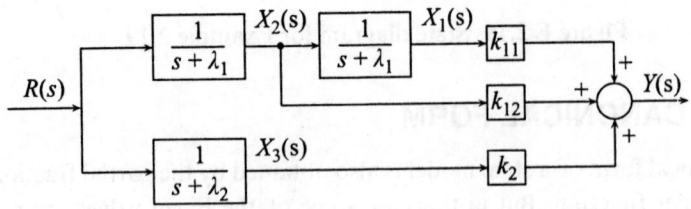

Figure 2.7 Block diagram – Jordan canonical form

From equation (2.32),

$$X_1(s) = \frac{X_2(s)}{s + \lambda_1} \qquad \text{or } (s + \lambda_1)X_1(s) = X_2(s).$$

Taking inverse Laplace transforms,

$$\dot{x}_1 + \lambda_1 x_1 = x_2 \qquad \text{or } \dot{x}_1 = -\lambda_1 x_1 + x_2. \qquad (2.33)$$

From equation (2.30),

$$X_2(s) = \frac{R(s)}{s + \lambda_1} \qquad \text{or } (s + \lambda_1)X_2(s) = R(s).$$

Taking inverse Laplace transforms,

$$\dot{x}_2 + \lambda_1 x_2 = r(t) \qquad \text{or } \dot{x}_2 = -\lambda_1 x_2 + r(t). \qquad (2.34)$$

From equation (2.31)

$$X_3(s) = \frac{R(s)}{s + \lambda_2} \qquad \text{or } (s + \lambda_2)X_3(s) = R(s)$$

Taking inverse Laplace transforms,

$$\dot{x}_3 + \lambda_2 x_3 = r(t) \qquad \text{or } \dot{x}_3 = -\lambda_2 x_3 + r(t). \qquad (2.35)$$

Keeping equations (2.33), (2.34) and (2.35) in the matrix form, the state equation is

$$\begin{bmatrix} \dot{x}_1 \\ \dot{x}_2 \\ \dot{x}_3 \end{bmatrix} = \begin{bmatrix} -\lambda_1 & 1 & 0 \\ 0 & -\lambda_1 & 0 \\ 0 & 0 & -\lambda_2 \end{bmatrix} \begin{bmatrix} x_1 \\ x_2 \\ x_3 \end{bmatrix} + \begin{bmatrix} 0 \\ 1 \\ 1 \end{bmatrix} r.$$

Taking inverse Laplace transforms of equation (2.28), the output equation is

$$y(t) = k_{11}x_1 + k_{12}x_2 + k_2 x_3 = \begin{bmatrix} k_{11} & k_{12} & k_2 \end{bmatrix} \begin{bmatrix} x_1 \\ x_2 \\ x_3 \end{bmatrix}.$$

The state diagram is shown in Figure 2.8.

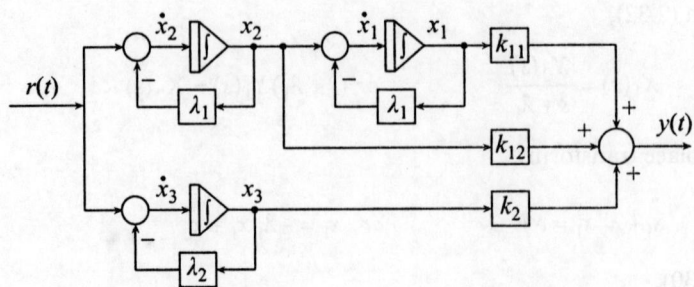

Figure 2.8 State diagram – Jordan canonical form

Note:

1. The sub-matrix $\begin{bmatrix} -\lambda_1 & 1 \\ 0 & -\lambda_1 \end{bmatrix}$ in the system matrix is called a *Jordan block*.

2. The diagonal elements of the system matrix are the Eigen values.
3. All the elements below the diagonal are 0.
4. The element in the previous row and just above the repeated Eigen value is 1and all other elements above the diagonal are 0.
5. The element of the output matrix B corresponding to the last row of the Jordan block is 1 and all the previous elements corresponding to each row of the Jordan block are 0.
6. The element of the output matrix B corresponding to the non-repeated Eigen value in the system matrix is 1.

Example 2.20 The transfer function of a system is

$$\frac{Y(s)}{R(s)} = \frac{2s+5}{(s+1)^2}.$$

Obtain a state model in the Jordan canonical form and draw the state diagram.

Solution: The Eigen values are repeated and are equal to $s = -1$. Expanding in partial fraction,

$$\frac{Y(s)}{R(s)} = \frac{2s+5}{(s+1)^2} = \frac{3}{(s+1)^2} + \frac{2}{(s+1)}.$$

Or

$$Y(s) = \frac{3}{(s+1)^2} R(s) + \frac{2}{(s+1)} R(s). \tag{1}$$

Let

$$Y(s) = 3X_1(s) + 2X_2(s). \tag{2}$$

From equations (1) and (2),

$$X_1(s) = \frac{R(s)}{(s+1)^2}$$ (3)

$$X_2(s) = \frac{R(s)}{s+1}.$$ (4)

From equations (3) and (4),

$$X_1(s) = \frac{X_2(s)}{s+1}.$$ (5)

From equation (5),

$$(s+1)X_1(s) = X_2(s).$$

Taking inverse Laplace transforms,

$$\dot{x}_1 + x_1 = x_2 \qquad or \quad \dot{x}_1 = -x_1 + x_2.$$ (A)

From equation (4),

$$(s+1)X_2(s) = R(s).$$

Taking inverse Laplace transforms,

$$\dot{x}_2 + x_2 = r(t) \qquad or \quad \dot{x}_2 = -x_2 + r(t).$$ (B)

Keeping equations (A) and (B) in the matrix form,

$$\begin{bmatrix} \dot{x}_1 \\ \dot{x}_2 \end{bmatrix} = \begin{bmatrix} -1 & 1 \\ 0 & -1 \end{bmatrix} \begin{bmatrix} x_1 \\ x_2 \end{bmatrix} + \begin{bmatrix} 0 \\ 1 \end{bmatrix} r(t).$$

Taking inverse Laplace transform of equation (2), the output equation is

$$y(t) = 3x_1 + 2x_2 = \begin{bmatrix} 3 & 2 \end{bmatrix} \begin{bmatrix} x_1 \\ x_2 \end{bmatrix}.$$

The state diagram is shown in Figure E2.20.

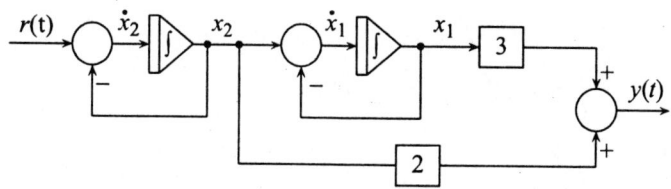

Figure E2.20 State diagram for Example 2.20

Example 2.21 For a system of order six, the six Eigen values are $s = \lambda_1$ repeated thrice, $s = \lambda_2$ repeated twice and $s = \lambda_3$. Write down the state equation in its Jordan canonical form and comment on the structure of the state equation.

Solution: The six Eigen values are $s = \lambda_1$ repeated thrice, $s = \lambda_2$ repeated twice and $s = \lambda_3$. The state equation in the Jordan canonical form can be written down as

$$
\begin{bmatrix} \dot{x}_1 \\ \dot{x}_2 \\ \dot{x}_3 \\ \dot{x}_4 \\ \dot{x}_5 \\ \dot{x}_6 \end{bmatrix} = \begin{bmatrix} -\lambda_1 & 1 & 0 & 0 & 0 & 0 \\ 0 & -\lambda_1 & 1 & 0 & 0 & 0 \\ 0 & 0 & -\lambda_1 & 0 & 0 & 0 \\ 0 & 0 & 0 & -\lambda_2 & 1 & 0 \\ 0 & 0 & 0 & 0 & -\lambda_2 & 0 \\ 0 & 0 & 0 & 0 & 0 & -\lambda_3 \end{bmatrix} \begin{bmatrix} x_1 \\ x_2 \\ x_3 \\ x_4 \\ x_5 \\ x_6 \end{bmatrix} + \begin{bmatrix} 0 \\ 0 \\ 1 \\ 0 \\ 1 \\ 1 \end{bmatrix} r
$$

There are two Jordan blocks as expected. In the output matrix, the element that corresponds to the last row of each Jordan block and the element that corresponds to the non-repeated Eigen values are 1 and all other elements are 0. This is indicated by the arrows.

Example 2.22 Obtain a state model in Jordan canonical for a system with transfer function

$$
\frac{Y(s)}{R(s)} = \frac{2(s^2 + 2s + 3)}{(s+1)^2(s+3)}.
$$

Also draw the state diagram.

Solution: Expanding the given transfer function in partial fractions

$$
\frac{Y(s)}{R(s)} = \frac{2(s^2 + 2s + 3)}{(s+1)^2(s+3)} = \frac{2}{(s+1)^2} - \frac{1}{(s+1)} + \frac{3}{(s+3)}
$$

$$
Y(s) = \frac{2R(s)}{(s+1)^2} - \frac{R(s)}{(s+1)} + \frac{3R(s)}{(s+3)}. \tag{1}
$$

Let

$$
Y(s) = 2X_1(s) - X_2(s) + 3X_3(s). \tag{2}
$$

From equations (1) and (2),

$$
X_1(s) = \frac{R(s)}{(s+1)^2} \tag{3}
$$

$$X_2(s) = \frac{R(s)}{(s+1)} \tag{4}$$

$$X_3(s) = \frac{R(s)}{(s+3)}. \tag{5}$$

From equations (3) and (4),

$$X_1(s) = \frac{X_2(s)}{(s+1)} \qquad or \ \ (s+1)X_1(s) = X_2(s).$$

Taking inverse Laplace transforms,

$$\dot{x}_1 + x_1 = x_2 \qquad or \ \ \dot{x}_1 = -x_1 + x_2. \tag{A}$$

From equation (4),

$$(s+1)X_2(s) = R(s).$$

Taking inverse Laplace transforms,

$$\dot{x}_2 + x_2 = r(t) \qquad or \ \ \dot{x}_2 = -x_2 + r(t). \tag{B}$$

From equation (5),

$$X_3(s) = \frac{R(s)}{(s+3)} \qquad or \ \ (s+3)X_3(s) = R(s).$$

Taking inverse Laplace transforms,

$$\dot{x}_3 + 3x_3 = r(t) \qquad or \ \ \dot{x}_3 = -3x_3 + r(t). \tag{C}$$

Keeping equation (A), (B) and (C) in the matrix form

$$\begin{bmatrix} \dot{x}_1 \\ \dot{x}_2 \\ \dot{x}_3 \end{bmatrix} = \begin{bmatrix} -1 & 1 & 0 \\ 0 & -1 & 0 \\ 0 & 0 & -3 \end{bmatrix} \begin{bmatrix} x_1 \\ x_2 \\ x_3 \end{bmatrix} + \begin{bmatrix} 0 \\ 1 \\ 1 \end{bmatrix} r(t).$$

Taking inverse Laplace transforms of equation (2), the output equation is

$$y(t) = 2x_1 - x_2 + 3x_3 = \begin{bmatrix} 2 & -1 & 3 \end{bmatrix} \begin{bmatrix} x_1 \\ x_2 \\ x_3 \end{bmatrix}.$$

The state diagram is shown in Figure E2.22.

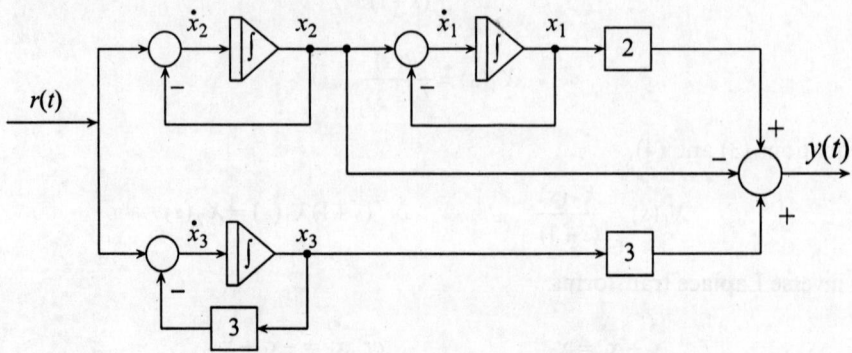

Figure E2.22 State diagram for Example 2.22

Example 2.23 Obtain a state model in Jordan canonical for a system with transfer function

$$\frac{Y(s)}{R(s)} = \frac{16}{(s+1)^2(s+3)(s+5)}.$$

Also draw the state diagram.

Solution: Expanding the given transfer function in partial fractions

$$\frac{Y(s)}{R(s)} = \frac{16}{(s+1)^2(s+3)(s+5)} = \frac{2}{(s+1)^2} - \frac{(3/2)}{(s+1)} + \frac{2}{(s+3)} - \frac{(1/2)}{(s+5)}$$

$$Y(s) = 2\frac{R(s)}{(s+1)^2} - \frac{3}{2}\frac{R(s)}{(s+1)} + 2\frac{R(s)}{(s+3)} - \frac{1}{2}\frac{R(s)}{(s+5)}. \tag{1}$$

Let

$$Y(s) = 2X_1(s) - \frac{3}{2}X_2(s) + 2X_3(s) - \frac{1}{2}X_4(s). \tag{2}$$

From equations (1) and (2),

$$X_1(s) = \frac{R(s)}{(s+1)^2} \tag{3}$$

$$X_2(s) = \frac{R(s)}{(s+1)} \tag{4}$$

$$X_3(s) = \frac{R(s)}{(s+3)} \tag{5}$$

$$X_4(s) = \frac{R(s)}{(s+5)}. \tag{6}$$

From equations (3) and (4),

$$X_1(s) = \frac{X_2(s)}{(s+1)} \tag{7}$$

From equation (7),

$$(s+1)X_1(s) = X_2(s).$$

Taking inverse Laplace transforms,

$$\dot{x}_1 + x_1 = x_2 \qquad\qquad \text{or } \dot{x}_1 = -x_1 + x_2. \tag{A}$$

From equation (4),

$$(s+1)X_2(s) = R(s).$$

Taking inverse Laplace transforms,

$$\dot{x}_2 + x_2 = r(t) \qquad\qquad \text{or } \dot{x}_2 = -x_2 + r(t). \tag{B}$$

From equation (5),

$$(s+3)X_3(s) = R(s).$$

Taking inverse Laplace transforms,

$$\dot{x}_3 + 3x_3 = r(t) \qquad\qquad \text{or } \dot{x}_3 = -3x_3 + r(t). \tag{C}$$

From equation (6),

$$(s+5)X_4(s) = R(s).$$

Taking inverse Laplace transforms,

$$\dot{x}_4 + 5x_4 = r(t) \qquad\qquad \text{or } \dot{x}_4 = -5x_4 + r(t). \tag{D}$$

Keeping equations (A), (B), (C) and (D) in the matrix form, the state equation is

$$\begin{bmatrix} \dot{x}_1 \\ \dot{x}_2 \\ \dot{x}_3 \\ \dot{x}_4 \end{bmatrix} = \begin{bmatrix} -1 & 1 & 0 & 0 \\ 0 & -1 & 0 & 0 \\ 0 & 0 & -3 & 0 \\ 0 & 0 & 0 & -5 \end{bmatrix} \begin{bmatrix} x_1 \\ x_2 \\ x_3 \\ x_4 \end{bmatrix} + \begin{bmatrix} 0 \\ 1 \\ 1 \\ 1 \end{bmatrix} r(t).$$

Taking inverse Laplace transforms of equation (2), the output equation is

$$y(t) = 2x_1 - \frac{3}{2}x_2 + 2x_3 - \frac{1}{2}x_4 = \begin{bmatrix} 2 & -\frac{3}{2} & 2 & -\frac{1}{2} \end{bmatrix} \begin{bmatrix} x_1 \\ x_2 \\ x_3 \\ x_4 \end{bmatrix}.$$

The state diagram is shown in Figure E2.23.

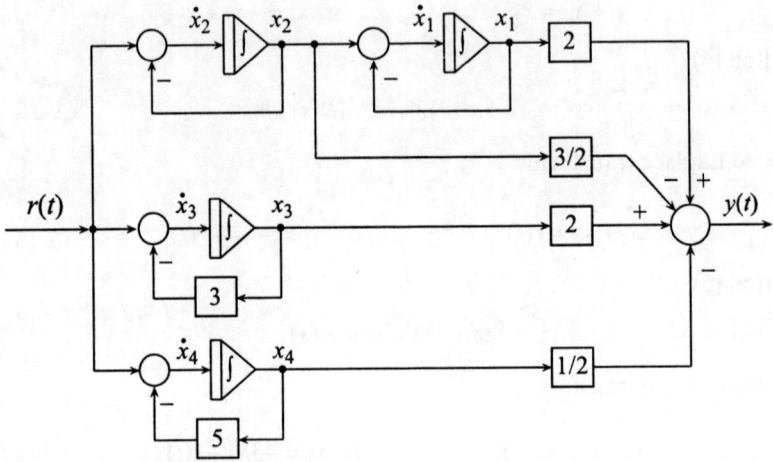

Figure E2.23 State diagram for Example 2.23

2.10 CASCADE REALIZATION

This method of realization is achieved by writing the given transfer function as a product of simple ratios of polynomials. The degree of the numerator polynomial or the denominator polynomial is at most two. Generally, the denominator polynomial is of first or second degree and the numerator is either a first-degree polynomial or a constant term. These factors can be represented by a series of blocks in a block diagram. The output of each block is taken as a state variable. A differential equation can be obtained using each block and, hence, a state equation. The method is illustrated by means of some examples.

Example 2.24 The transfer function of a second-order system given by

$$\frac{Y(s)}{R(s)} = \frac{k(s+b)}{(s+a_1)(s+a_2)}.$$

Obtain a state model by cascade realization.

Solution:

$$\frac{Y(s)}{R(s)} = \frac{k(s+b)}{(s+a_1)(s+a_2)}.$$

Splitting the right hand side into two ratios,

$$\frac{Y(s)}{R(s)} = \frac{k}{(s+a_1)} \frac{(s+b)}{(s+a_2)}.$$

These factors can be represented by block diagram as shown in Figure E2.24.

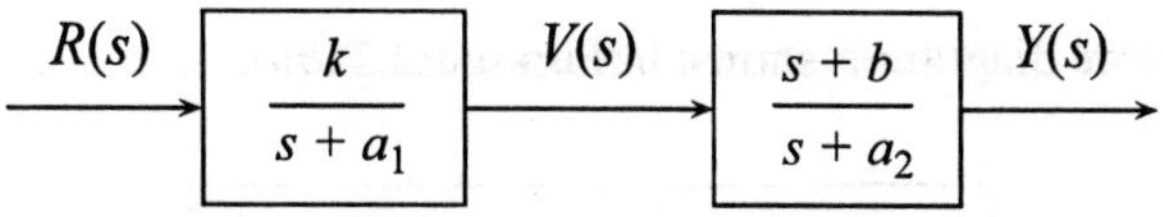

Figure E2.24 Cascade realization for Example 2.24

From the block diagram

$$\frac{V(s)}{R(s)} = \frac{k}{s+a_1} \qquad \qquad or \;\; (s+a_1)V(s) = k\,R(s).$$

Taking inverse Laplace transforms,

$$\overset{\bullet}{v} + a_1\, v = k\, r(t) \qquad\qquad or \;\; \overset{\bullet}{v} = -a_1\, v + k\, r(t). \tag{1}$$

Also

$$\frac{Y(s)}{V(s)} = \frac{s+b}{s+a_2} \qquad\qquad or \;\; (s+a_2)Y(s) = (s+b)V(s).$$

Taking inverse Laplace transforms,

$$\overset{\bullet}{y} + a_2\, y = \overset{\bullet}{v} + b\,v \qquad\qquad or \;\; \overset{\bullet}{y} = -a_2\, y + b\,v + \overset{\bullet}{v}. \tag{2}$$

Select the output of the blocks y and v as state variables. Hence $x_1 = y$ and $x_2 = v$ are chosen as state variables.

From equation (2),

$$\overset{\bullet}{x_1} = -a_2\, x_1 + b\,x_2 + \overset{\bullet}{x_2}. \tag{3}$$

From equation (1),

$$\overset{\bullet}{x_2} = -a_1\, x_2 + k\, r(t). \tag{A}$$

Substituting equation (A) in equation (3),

$$\overset{\bullet}{x_1} = -a_2\, x_1 + b\,x_2 - a_1\, x_2 + k\, r(t) = -a_2\, x_1 + (b-a_1)x_2 + k\,r(t). \tag{B}$$

Keeping equations (A) and (B) in the matrix form,

$$\begin{bmatrix} \dot{x}_1 \\ \dot{x}_2 \end{bmatrix} = \begin{bmatrix} -a_2 & (b-a_1) \\ 0 & -a_1 \end{bmatrix} \begin{bmatrix} x_1 \\ x_2 \end{bmatrix} + \begin{bmatrix} k \\ k \end{bmatrix} r(t).$$

The output equation is

$$y(t) = x_1 = \begin{bmatrix} 1 & 0 \end{bmatrix} \begin{bmatrix} x_1 \\ x_2 \end{bmatrix}.$$

The corresponding state diagram is shown in Figure E2.24 (a).

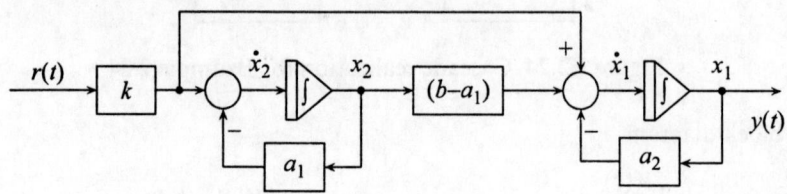

Figure E2.24 (a) State diagram of the cascade realization for Example 2.24

Example 2.25 For the example of 2.24, obtain an alternate state model by cascade realization.

Solution:

$$\frac{Y(s)}{R(s)} = \frac{k(s+b)}{(s+a_1)(s+a_2)}$$

$$= \frac{k}{(s+a_1)} \frac{(s+b)}{(s+a_2)}.$$

These factors can be represented by a block diagram as shown in Figure E2.25.

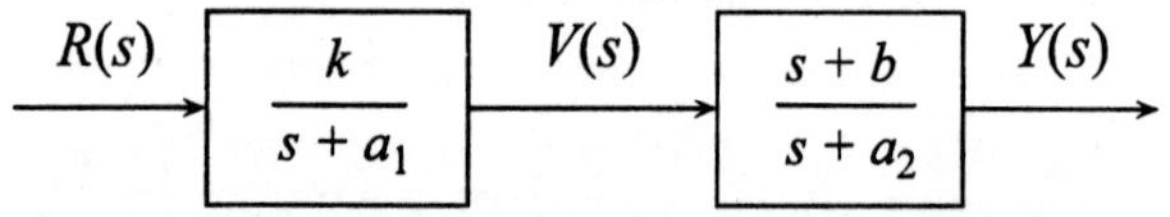

$$R(s) \rightarrow \boxed{\frac{k}{s+a_1}} \xrightarrow{V(s)} \boxed{\frac{s+b}{s+a_2}} \rightarrow Y(s)$$

Figure E2.25 Cascade realization for Example 2.25

$$\frac{V(s)}{R(s)} = \frac{k}{s+a_1} \qquad\qquad or \ (s+a_1)V(s) = k\,R(s).$$

Taking inverse Laplace transforms,

$$\dot{v} + a_1 v = k\,r(t) \qquad\qquad or \ \dot{v} = -a_1 v + k\,r(t). \qquad\qquad (1)$$

Choose $x_1 = v$ as a state variable. Hence, equation (1) can be written as

$$\dot{x}_1 = -a_1 x_1 + kr. \tag{A}$$

$$\frac{Y(s)}{V(s)} = \frac{s+b}{s+a_2} = 1 + \frac{(b-a_2)}{s+a_2}$$

$$Y(s) = V(s) + \frac{(b-a_2)}{s+a_2} V(s).$$

Let

$$Y(s) = V(s) + W(s). \tag{2}$$

where

$$W(s) = \frac{(b-a_2)}{s+a_2} V(s) \qquad\qquad (s+a_2)W(s) = (b-a_2)V(s).$$

Taking inverse Laplace transforms,

$$\dot{w} + a_2 w = (b-a_2)v \qquad\qquad or \ \dot{w} = -a_2 w + (b-a_2)v. \tag{3}$$

Select $x_2 = w$ as the second state variable. Hence, equation (3) can be written as

$$\dot{x}_2 = (b-a_2)x_1 - a_2 x_2. \tag{B}$$

Keeping equations (A) and (B) in the matrix form

$$\begin{bmatrix} \dot{x}_1 \\ \dot{x}_2 \end{bmatrix} = \begin{bmatrix} -a_1 & 0 \\ (b-a_2) & -a_2 \end{bmatrix} \begin{bmatrix} x_1 \\ x_2 \end{bmatrix} + \begin{bmatrix} k \\ 0 \end{bmatrix} r(t).$$

Taking inverse Laplace transforms of equation (2) the output equation is

$$y = v + w = x_1 + x_2 = \begin{bmatrix} 1 & 1 \end{bmatrix} \begin{bmatrix} x_1 \\ x_2 \end{bmatrix}.$$

The corresponding state diagram is shown in Figure E2.25(a).

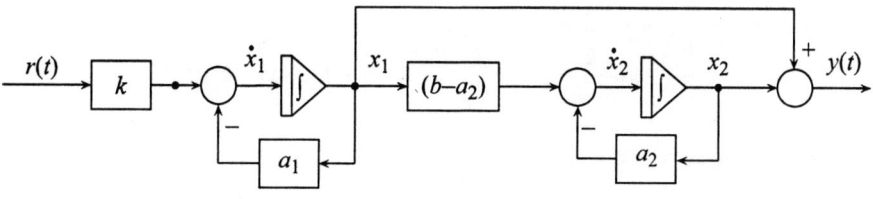

Figure E2.25 (a) Alternate form of state diagram – cascade realization

Example 2.26 For the example 2.24, obtain another state model by cascade realization.

Solution:

$$\frac{Y(s)}{R(s)} = \frac{k(s+b)}{(s+a_1)(s+a_2)} \ .$$

Rearranging the factor terms of the transfer function,

$$\frac{Y(s)}{R(s)} = \frac{(s+b)}{(s+a_2)} \frac{k}{(s+a_1)} \ .$$

These factors can be represented by a block diagram as shown in Figure E2.26.

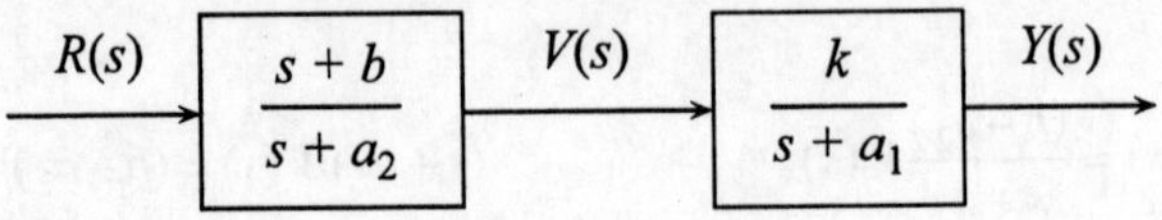

Figure E2.26 Rearrangement of blocks – cascade realization

From the block diagram

$$\frac{V(s)}{R(s)} = \frac{s+b}{s+a_2} = 1 + \frac{(b-a_2)}{(s+a_2)}$$

$$V(s) = R(s) + \frac{(b-a_2)}{(s+a_2)} R(s) = R(s) + W(s) \ . \tag{1}$$

$$W(s) = \frac{(b-a_2)}{(s+a_2)} R(s) \ or \ (s+a_2)W(s) = (b-a_2)R(s) \ .$$

Taking inverse Laplace transforms,

$$\dot{w} + a_2 \, w = (b-a_2)r(t) \qquad\qquad or \ \dot{w} = -a_2 \, w + (b-a_2)r(t) \ . \tag{2}$$

Choose $x_1 = w$ as a state variable. Hence, equation (2) can be written as

$$\dot{x}_1 = -a_2 \, x_1 + (b-a_2)r(t) \ . \tag{A}$$

Also

$$\frac{Y(s)}{V(s)} = \frac{k}{s+a_1} \qquad\qquad or \ (s+a_1)Y(s) = k \, V(s) \ .$$

Taking inverse Laplace transforms,

$$\dot{y} + a_1 \, y = k \, v \qquad\qquad or \ \dot{y} = -a_1 \, y + k \, v \ . \tag{3}$$

Taking inverse Laplace transforms of equation (1),

$$v = r + w .$$ (4)

Combining equations (3) and (4),

$$\dot{y} = -a_1 y + k(r + w) .$$ (5)

Select $x_2 = y$ as another state variable. Thus, from equation (5),

$$\dot{x}_2 = -a_1 x_2 + k(r + x_1) = kx_1 - a_1 x_2 + kr .$$ (B)

Keeping equations (A) and (B) in the matrix form,

$$\begin{bmatrix} \dot{x}_1 \\ \dot{x}_2 \end{bmatrix} = \begin{bmatrix} -a_2 & 0 \\ k & -a_1 \end{bmatrix} \begin{bmatrix} x_1 \\ x_2 \end{bmatrix} + \begin{bmatrix} (b - a_2) \\ k \end{bmatrix} r(t) .$$

The output equation is

$$y = x_2 = \begin{bmatrix} 0 & 1 \end{bmatrix} \begin{bmatrix} x_1 \\ x_2 \end{bmatrix} .$$

The state diagram is shown in Figure E2.26 (a).

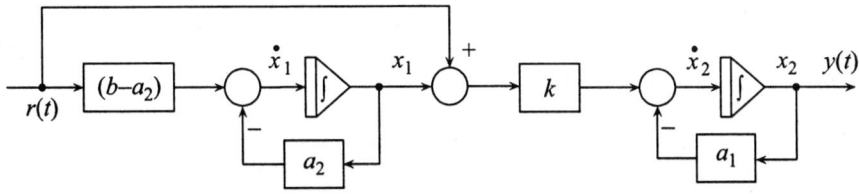

Figure E2.26 (a) State diagram – cascade realization

The three state models obtained for the same system of Example 2.24 are different. It can
be seen that the cascade realization does not give rise to a unique state model. Hence, the
state model obtained by the cascade realization does not belong to the class of canonical
forms.

Example 2.27 The transfer function of a system is

$$\frac{Y(s)}{R(s)} = \frac{5(s + 3)}{s(s + 1)(s + 5)} .$$

Obtain a state model by cascade realization.

Solution: The given transfer function can be written in a factored form as

$$\frac{Y(s)}{R(s)} = \frac{5}{(s+5)} \frac{(s+3)}{(s+1)} \frac{1}{s}.$$

This is represented by a block diagram as shown in Figure E2.27.

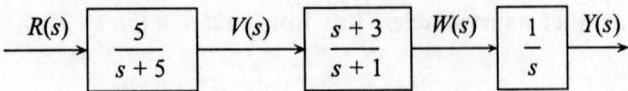

Figure E2.27 Block diagram for Example 2.27

Consider $\dfrac{V(s)}{R(s)} = \dfrac{5}{s+5}$.

From this factor,

$$(s+5)V(s) = 5R(s).$$

Taking inverse Laplace transforms,

$$\dot{v} + 5v = 5r(t) \qquad\qquad or \ \dot{v} = -5v + 5r(t). \tag{1}$$

Now

$$\frac{W(s)}{V(s)} = \frac{(s+3)}{(s+1)} \qquad\qquad or \ (s+1)W(s) = (s+3)V(s).$$

Taking inverse Laplace transforms,

$$\dot{w} + w = \dot{v} + 3v \qquad\qquad or \ \dot{w} = -w + \dot{v} + 3v. \tag{2}$$

Also

$$\frac{Y(s)}{W(s)} = \frac{1}{s} \qquad\qquad or \ sY(s) = W(s).$$

Taking inverse Laplace transforms,

$$\dot{y} = w. \tag{3}$$

Select $x_1 = y$, $x_2 = w$ and $x_3 = v$ as state variables. Hence, from equation (3),

$$\dot{x}_1 = x_2. \tag{A}$$

From equation (2),

$$\dot{x}_2 = -x_2 + \dot{x}_3 + 3x_3. \tag{4}$$

From equation (1),

$$\dot{x}_3 = -5x_3 + 5r(t).$$ (B)

Substituting equation (B) in equation (4),

$$\dot{x}_2 = -x_2 + \{-5x_3 + 5r(t)\} + 3x_3 = -x_2 - 2x_3 + 5r(t).$$ (C)

Equations (A), (B) and (C) are the state equations. Keeping them in the matrix form,

$$\begin{bmatrix} \dot{x}_1 \\ \dot{x}_2 \\ \dot{x}_3 \end{bmatrix} = \begin{bmatrix} 0 & 1 & 0 \\ 0 & -1 & -2 \\ 0 & 0 & 5 \end{bmatrix} \begin{bmatrix} x_1 \\ x_2 \\ x_3 \end{bmatrix} + \begin{bmatrix} 0 \\ 5 \\ 5 \end{bmatrix} r(t).$$

The output equation is

$$y = x_1 = \begin{bmatrix} 1 & 0 & 0 \end{bmatrix} \begin{bmatrix} x_1 \\ x_2 \\ x_3 \end{bmatrix}.$$

The state diagram is shown in Figure E 2.27(a).

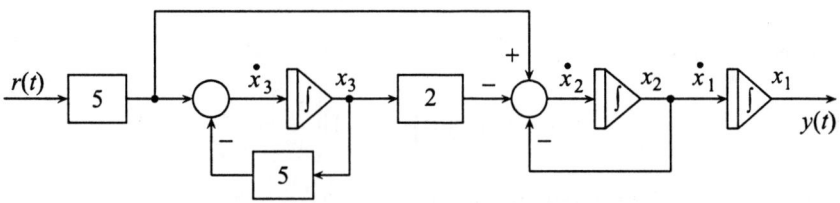

Figure E 2.27(a) The state diagram for Example 2.27

Example 2.28 A state diagram of a linear time-invariant system is shown in Figure E 2.28. Realize a state model for the system and hence find the transfer function.

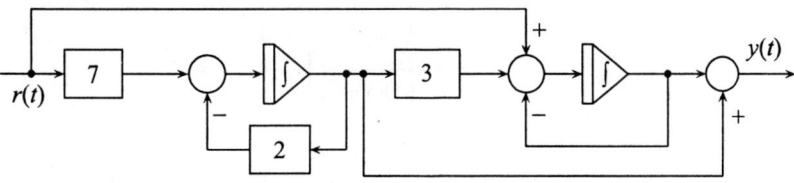

Figure E 2.28 The state diagram for Example 2.28

Solution: The output of each integrator is taken as a state variable. The given state diagram is redrawn as shown in Figure E 2.28 (a), selecting the output of each integrator as state variables.

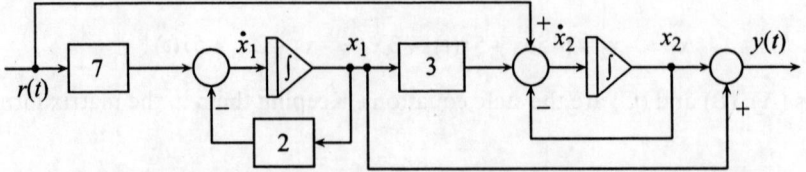

Figure E 2.28(a) The redrawn state diagram for Example 2.28

From the state diagram, the state equations can be written as

$$\dot{x}_1 = -2x_1 + 7r(t) \tag{1}$$

$$\dot{x}_1 = -x_2 + 3x_1 + r(t). \tag{2}$$

These are the state equations and keeping them in the matrix form,

$$\begin{bmatrix} \dot{x}_1 \\ \dot{x}_2 \end{bmatrix} = \begin{bmatrix} -2 & 0 \\ 3 & -1 \end{bmatrix}\begin{bmatrix} x_1 \\ x_2 \end{bmatrix} + \begin{bmatrix} 7 \\ 1 \end{bmatrix}r(t).$$

The output equation is

$$y = x_1 + x_2 = \begin{bmatrix} 1 & 1 \end{bmatrix}\begin{bmatrix} x_1 \\ x_2 \end{bmatrix}.$$

The transfer function of the system is given by

$$C[sI - A]^{-1}B = \begin{bmatrix} 1 & 1 \end{bmatrix}\begin{bmatrix} s+2 & 0 \\ -3 & s+1 \end{bmatrix}^{-1}\begin{bmatrix} 7 \\ 1 \end{bmatrix}$$

$$= \frac{1}{(s+1)(s+2)}\begin{bmatrix} 1 & 1 \end{bmatrix}\begin{bmatrix} s+1 & 0 \\ 3 & s+2 \end{bmatrix}\begin{bmatrix} 7 \\ 1 \end{bmatrix}$$

$$= \frac{1}{(s+1)(s+2)}\begin{bmatrix} s+4 & s+2 \end{bmatrix}\begin{bmatrix} 7 \\ 1 \end{bmatrix}$$

$$= \frac{8s+30}{(s+1)(s+2)}.$$

SUMMARY

- The transfer matrix is that matrix which pre-multiplying the input vector yields the output vector.
- The transfer function is a special case of transfer matrix. It is defined with respect to a single-input/single-output (SISO) system. It is defined as the ratio of the Laplace transform of the output to the Laplace transform of the input, with all zero initial states.
- Eigen values are the system poles. They are the roots of the characteristic equation $|sI - A| = 0$ where A is the system matrix.
- State transition matrix in the s-domain is called the resolvent matrix: $[sI - A]^{-1}$
- The inverse Laplace transform of this is the state transition matrix in the time domain.
- A state model with the matrices A, B, C and D which have the minimum number of non-zero elements is said to be in canonical form.
- The different canonical forms are controllable canonical form, observable canonical form, diagonal canonical form and Jordan canonical form.
- Diagonal form of a state model is associated with a system with distinct Eigen values.
- In the diagonal canonical form of the state model, the system matrix is diagonal with the diagonal elements as the Eigen values and all the off-diagonal elements as zeros.
- Diagonal canonical form is also called parallel form or decoupled form.
- In the diagonal canonical form, each state equation is a first-order differential equation in only one state variable and the input. The equations are said to be decoupled. The state equations can be solved independently.
- Systems with repeated Eigen values can be expressed in Jordan canonical forms.
- The diagonal and Jordan canonical forms are obtained by the method of partial fraction expansion of the transfer function.
- The state model obtained using cascade realization is not unique. More than one state model can be realized by this method from the same transfer function.
- For obtaining a state model from a given state diagram, the variable at the output of each integrator is taken as a state variable and the equation for the variable at the input of each integrator is written down.

PRACTICE PROBLEMS

PP 2.1 Determine the transfer matrix of a multiple-input/multiple-output system represented by the state model

$$\begin{bmatrix} \dot{x}_1 \\ \dot{x}_2 \end{bmatrix} = \begin{bmatrix} 0 & 1 \\ -2 & -3 \end{bmatrix} \begin{bmatrix} x_1 \\ x_2 \end{bmatrix} + \begin{bmatrix} 1 & 2 \\ 0 & 1 \end{bmatrix} \begin{bmatrix} r_1(t) \\ r_2(t) \end{bmatrix}$$

$$\begin{bmatrix} y_1(t) \\ y_2(t) \end{bmatrix} = \begin{bmatrix} 2 & 1 \\ 1 & 0 \end{bmatrix} \begin{bmatrix} x_1 \\ x_2 \end{bmatrix}.$$

PP 2.2 Draw a state diagram for the system in problem PP2.1.

PP 2.3 A state variable description of a system is given by

$$\begin{bmatrix} \dot{x}_1 \\ \dot{x}_2 \end{bmatrix} = \begin{bmatrix} -1 & 0 \\ 1 & -2 \end{bmatrix} \begin{bmatrix} x_1 \\ x_2 \end{bmatrix} + \begin{bmatrix} 1 \\ 0 \end{bmatrix} r(t) \qquad y(t) = \begin{bmatrix} 1 & 1 \end{bmatrix} \begin{bmatrix} x_1 \\ x_2 \end{bmatrix}.$$

Find the Eigen values, state transition matrix and the transfer function. Also draw the state diagram.

PP 2.4 The system matrix of a system is

$$A = \begin{bmatrix} -1 & 0 \\ 0 & -2 \end{bmatrix}.$$

Find the state transition matrix.

PP 2.5 The system matrix of a system is

$$A = \begin{bmatrix} -1 & 1 \\ 0 & -1 \end{bmatrix}$$

Find the state transition matrix.

PP 2.6 If the system matrix of a system is

$$A = \begin{bmatrix} 0 & 2 & 1 \\ 0 & -2 & 2 \\ -1 & -1 & -4 \end{bmatrix}.$$

Find the Eigen values and the state transition matrix in the *s*-domain.

PP 2.7 For the system of PP2.6, find the state transition matrix in time domain.

PP 2.8 A state variable description of a system is given by

$$\dot{X} = \begin{bmatrix} 0 & 2 & 1 \\ 0 & -2 & 2 \\ -1 & -1 & -4 \end{bmatrix} X + \begin{bmatrix} 0 \\ 0 \\ 2 \end{bmatrix} r \qquad y = \begin{bmatrix} 1 & 0 & 0 \end{bmatrix} X.$$

Find the transfer function and draw the state diagram.

PP 2.9 The transfer function of a system is given by

$$\frac{Y(s)}{R(s)} = \frac{2s+3}{s^2 + 4s + 3}.$$

Obtain a state model in controllable canonical form and draw the state diagram.

PP 2.10 For the system of PP2.9 obtain a state model in observable canonical form and draw the corresponding state diagram.

PP 2.11 For the system of PP2.9 obtain a state model in diagonal canonical form and draw the corresponding state diagram.

PP 2.12 Realize a state model in Jordan canonical form for a system described by the transfer function

$$\frac{Y(s)}{R(s)} = \frac{2s+5}{s^2 + 4s + 4}.$$

Also draw the corresponding state diagram.

PP 2.13 The transfer function of a system is given by

$$\frac{Y(s)}{R(s)} = \frac{3s^2 + s + 2}{s^3 + 7s^2 + 14s + 8}.$$

Obtain a state model in controllable canonical form and draw the state diagram.

PP 2.14 For the system of PP2.13, obtain a state model in observable canonical form and draw the corresponding state diagram.

PP 2.15 The transfer function of a system is

$$\frac{Y(s)}{R(s)} = \frac{3s^2 + 2s + 6}{s^3 + 7s^2 + 14s + 8}.$$

Obtain a state model in diagonal form and draw the corresponding state diagram.

PP 2.16 A linear system has the transfer function

$$\frac{Y(s)}{R(s)} = \frac{4s^2 + 3s + 5}{(s+2)(s+1)^2}.$$

Obtain a state model in Jordan canonical form and draw the state diagram.

PP 2.17 A state diagram for a linear system is shown in Figure PP2.17. Obtain a state model and find the transfer function of the system.

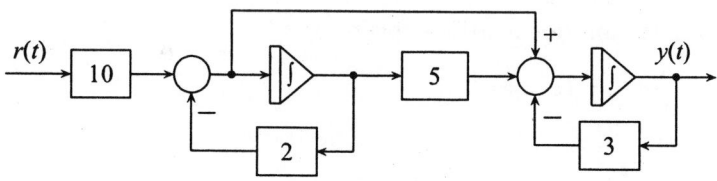

Figure PP2.17 Practice problem P2.17

PP 2.18 A linear system is represented by a block diagram as shown in Figure PP2.18. Obtain a state model by realising each block.

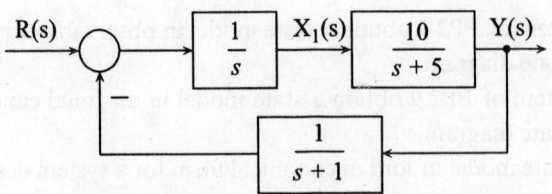

Figure PP2.18 Practice problem PP2.18

PP 2.19 Obtain a state model for a system, the transfer function of which is given as

$$\frac{Y(s)}{R(s)} = \frac{2s^3 + 5s^2 + 4s + 7}{s^3 + 6s^2 + 11s + 6}$$ [*Hint : Choose one state variable* $x_1 = y - 2r$].

PP 2.20 A linear time-invariant system is represented by the state equation

$$\dot{X} = \begin{bmatrix} 0 & 1 \\ -8 & -6 \end{bmatrix} X + \begin{bmatrix} 2 \\ 1 \end{bmatrix} r.$$

Determine the state transition matrix using its properties.

PP 2.21 A state model of a system is

$$\dot{X} = \begin{bmatrix} -a & 1 \\ -b & 0 \end{bmatrix} X + \begin{bmatrix} k_1 \\ k_2 \end{bmatrix} r \qquad y(t) = \begin{bmatrix} 1 & 0 \end{bmatrix} X.$$

Find the value of a, b, k_1 and k_2 if the transfer function is

$$\frac{Y(s)}{R(s)} = \frac{5(s+3)}{s^2 + 6s + 16}.$$

REVIEW QUESTIONS

2.1   Define transfer matrix and derive an expression for it.

2.2   Differentiate between transfer matrix and transfer function.

2.3   Define Eigen value. What is its significance?

2.4   What is a state transition matrix? Explain.

2.5   Show that $\dfrac{d}{dt} e^{At} = A e^{At} = e^{At} A$

2.6   Evaluate $\displaystyle\int_0^t e^{A\theta} d\theta$

2.7   What is meant by canonical form of state model?

2.8   List a few canonical form of state model.

2.9   What are the features of a controllable canonical form of state model?

2.10 What are the features of an observable canonical form of state model?

2.11 What are the features of a diagonal canonical form of state model?

2.12 What are the advantages of a diagonal canonical form of state model?

2.13 What are the features of a Jordan canonical form of state model?

2.14 Explain the method of cascade realization for obtaining a state model.

2.15 Given a state diagram of a system, explain the method of writing down its state model.

2.16 Given a block diagram of a system, explain the method of obtaining its state model.

2.17 List the properties of state transition matrix.

2.18 Discuss the method of obtaining state transition matrix using its properties.

Solution of State Equations

3

3.1 SOLUTION OF STATE EQUATION IN S-DOMAIN

Consider that a system is represented by a state equation

$$\dot{X}(t) = A\,X(t) + B\,R(t).$$

Taking Laplace transform of the state equation and assuming $X(0)$ as initial state vector,

$$s\,X(s) - X(0) = A\,X(s) + B\,R(s)$$

$$[s\,I - A]\,X(s) = X(0) + B\,R(s)$$

$$X(s) = [s\,I - A]^{-1}\,X(0) + [s\,I - A]^{-1}\,B\,R(s). \qquad (3.1)$$

Taking inverse Laplace transforms, the state vector $X(t)$ can be obtained. The output that can be obtained from the equation is

$$Y(t) = C\,X(t) + D\,R(t).$$

Taking Laplace transforms,

$$Y(s) = C\,X(s) + D\,R(s). \qquad (3.2)$$

Substituting equation (3.1) in equation (3.2),

$$Y(s) = C[s\,I - A]^{-1}\,X(0) + \{C[s\,I - A]^{-1}\,B + D\}\,R(s). \qquad (3.3)$$

Taking inverse Laplace transforms, the output $y(t)$ can be obtained. The first term on the right hand side of equation (3.3) is generated by the initial state $X(0)$ only and is independent of the input $r(t)$. Hence, it is called the Zero Input Response (ZIR). The second term on the right hand side of equation (3.3) is due to the input $r(t)$ only and is independent of the initial state $X(0)$. Hence, it is called the Zero State Response (ZSR). If $D = 0$, then

$$Y(s) = C[s\,I - A]^{-1}\,X(0) + C[s\,I - A]^{-1}\,B\,R(s).$$

Zero Input Response: $Y_{ZIR}(s) = C[sI-A]^{-1} X(0)$

Zero State Response: $Y_{ZSR}(s) = C[sI-A]^{-1} B\ R(s)$.

Unit Impulse Response

Consider a SISO system. For unit impulse input, $r(t) = \delta(t)$ so that $R(s) = 1$.

Hence, the zero state response of the system for unit impulse input is

$$Y_{ZSR}(s) = C[sI-A]^{-1} B.$$

Response of the system is

$$Y(s) = Y_{ZIR}(s) + Y_{ZSR}(s) = C[sI-A]^{-1} X(0) + C[sI-A]^{-1} B = C[sI-A]^{-1} [X(0)+B].$$

Taking inverse Laplace transforms, the response $y(t)$ can be obtained.

Unit Step Response

Consider a SISO system. For unit step input, $r(t) = 1$ so that $R(s) = \dfrac{1}{s}$.

Hence, the zero state response of the system for unit step input is

$$Y_{ZSR}(s) = C[sI-A]^{-1} B\left(\frac{1}{s}\right).$$

Response of the system is

$$Y(s) = C[sI-A]^{-1} X(0) + C[sI-A]^{-1} B\left(\frac{1}{s}\right)$$

$$= C[sI-A]^{-1} X(0) + C\left\{\frac{1}{s}[sI-A]^{-1}\right\}B \tag{3.4}$$

$$Y_{ZIR}(s) = C[sI-A]^{-1} X(0) \tag{3.5}$$

$$Y_{ZSR}(s) = C\left\{\frac{1}{s}[sI-A]^{-1}\right\}B . \tag{3.6}$$

Taking inverse Laplace transform of equation (3.5) and equation (3.6), the ZIR and the ZSR in time domain can be obtained for a unit step input and hence the total response.

Example 3.1 The system matrix of a linear time invariant system is given as $A = \begin{bmatrix} 0 & 1 \\ -3 & -4 \end{bmatrix}$ and the initial state is $X(0) = \begin{bmatrix} 2 \\ -1 \end{bmatrix}$. Determine the state response of the system for zero input.

Solution: The resolvent matrix is

$$\Phi(s) = [sI-A]^{-1} = \begin{bmatrix} s & -1 \\ 3 & s+4 \end{bmatrix}^{-1} = \frac{1}{s^2+4s+3}\begin{bmatrix} s+4 & 1 \\ -3 & s \end{bmatrix}.$$

The state response for zero input is

$$X_{ZIR}(s) = [sI - A]^{-1} X(0)$$

$$= \frac{1}{s^2 + 4s + 3} \begin{bmatrix} s+4 & 1 \\ -3 & s \end{bmatrix} \begin{bmatrix} 2 \\ -1 \end{bmatrix} = \begin{bmatrix} \dfrac{2s+7}{(s+1)(s+3)} \\ -\dfrac{(s+6)}{(s+1)(s+3)} \end{bmatrix}.$$

The time domain solution can be obtained by taking inverse Laplace transform of this, using partial fraction expansion.

$$X_{ZIR}(s) = \begin{bmatrix} \dfrac{(5/2)}{s+1} - \dfrac{(1/2)}{s+3} \\ \dfrac{-(5/2)}{s+1} + \dfrac{(3/2)}{s+3} \end{bmatrix}.$$

Taking inverse Laplace transform of each element,

$$\begin{bmatrix} x_1(t) \\ x_2(t) \end{bmatrix} = \begin{bmatrix} \dfrac{5}{2}e^{-t} - \dfrac{1}{2}e^{-3t} \\ -\dfrac{5}{2}e^{-t} + \dfrac{3}{2}e^{-3t} \end{bmatrix}.$$

Example 3.2 A linear time invariant system is described by the state equation

$$\dot{X} = \begin{bmatrix} -1 & 2 \\ -1 & -3 \end{bmatrix} X + \begin{bmatrix} 1 \\ 1 \end{bmatrix} r(t) \qquad y(t) = \begin{bmatrix} 1 & 1 \end{bmatrix} X$$

If the initial state vector is $X(0) = \begin{bmatrix} 2 \\ 0 \end{bmatrix}$, find the Zero Input Response and the Zero State Response due to a unit impulse input. Also find the total response.

Solution:

$$\dot{X} = \begin{bmatrix} -1 & 2 \\ -1 & -3 \end{bmatrix} X + \begin{bmatrix} 1 \\ 1 \end{bmatrix} r(t) \qquad y(t) = \begin{bmatrix} 1 & 1 \end{bmatrix} X \qquad X(0) = \begin{bmatrix} 2 \\ 0 \end{bmatrix}.$$

The resolvent matrix is

$$\Phi(s) = [sI - A]^{-1} = \begin{bmatrix} s+1 & -2 \\ 1 & s+3 \end{bmatrix}^{-1} = \frac{1}{s^2 + 4s + 5} \begin{bmatrix} s+3 & 2 \\ -1 & s+1 \end{bmatrix}.$$

$$C[sI - A]^{-1} = \frac{1}{(s+2)^2 + 1} \begin{bmatrix} 1 & 1 \end{bmatrix} \begin{bmatrix} s+3 & 2 \\ -1 & s+1 \end{bmatrix}$$

$$= \frac{1}{(s+2)^2 + 1}\begin{bmatrix} s+2 & s+3 \end{bmatrix}$$

The zero input response is

$$Y_{ZIR}(s) = C[sI-A]^{-1} X(0)$$

$$= \frac{1}{(s+2)^2 + 1}\begin{bmatrix} s+2 & s+3 \end{bmatrix}\begin{bmatrix} 2 \\ 0 \end{bmatrix} = \frac{2s+4}{(s+2)^2 + 1} = \frac{2(s+2)}{(s+2)^2 + 1}.$$

Taking inverse Laplace transforms, the time domain solution is

$$y_{ZIR}(t) = 2e^{-2t} \cos t .$$

For a unit impulse input $r(t) = \delta(t)$, $R(s) = 1$. The zero state response is

$$Y_{ZSR}(s) = C[sI-A]^{-1} B .$$

$$= \frac{1}{(s+2)^2 + 1}\begin{bmatrix} s+2 & s+3 \end{bmatrix}\begin{bmatrix} 1 \\ 1 \end{bmatrix} = \frac{2s+5}{(s+2)^2 + 1}$$

$$= \frac{2(s+2)+1}{(s+2)^2 + 1} = \frac{2(s+2)}{(s+2)^2 + 1} + \frac{1}{(s+2)^2 + 1}.$$

Taking inverse Laplace transforms,

$$y_{ZSR}(t) = 2e^{-2t} \cos t + e^{-2t} \sin t .$$

Total response of the system is

$$y(t) = 2e^{-2t} \cos t + 2e^{-2t} \cos t + e^{-2t} \sin t = 4e^{-2t} \cos t + e^{-2t} \sin t .$$

Example 3.3 For the system of Example 3.2, determine the Zero Input Response and the Zero State Response due to a unit step input. Also find the total response.

Solution:

$$\dot{X} = \begin{bmatrix} -1 & 2 \\ -1 & -3 \end{bmatrix} X + \begin{bmatrix} 1 \\ 1 \end{bmatrix} r(t) \qquad y(t) = \begin{bmatrix} 1 & 1 \end{bmatrix} X \qquad X(0) = \begin{bmatrix} 2 \\ 0 \end{bmatrix}.$$

The resolvent matrix is

$$\Phi(s) = [sI-A]^{-1} = \begin{bmatrix} s+1 & -2 \\ 1 & s+3 \end{bmatrix}^{-1} = \frac{1}{s^2 + 4s + 5}\begin{bmatrix} s+3 & 2 \\ -1 & s+1 \end{bmatrix}.$$

$$C[sI-A]^{-1} = \frac{1}{(s+2)^2+1}[1 \quad 1]\begin{bmatrix} s+3 & 2 \\ -1 & s+1 \end{bmatrix} = \frac{1}{(s+2)^2+1}[s+2 \quad s+3]$$

The zero input response is

$$Y_{ZIR}(s) = C[sI-A]^{-1}X(0)$$

$$= \frac{1}{(s+2)^2+1}[s+2 \quad s+3]\begin{bmatrix} 2 \\ 0 \end{bmatrix} = \frac{2s+4}{(s+2)^2+1} = \frac{2(s+2)}{(s+2)^2+1}.$$

Taking inverse Laplace transforms, the time domain solution is

$$y_{ZIR}(t) = 2e^{-2t}\cos t.$$

The input is $r(t) = 1$, a unit step input so that $R(s) = 1/s$. The zero state response is

$$Y_{ZSR}(s) = C[sI-A]^{-1}B\,R(s)$$

$$= \frac{1}{(s+2)^2+1}[s+2 \quad s+3]\begin{bmatrix} 1 \\ 1 \end{bmatrix}\frac{1}{s}$$

$$= \frac{2s+5}{s\{(s+2)^2+1\}} = \frac{1}{s} - \frac{s+2}{(s+2)^2+1}.$$

Taking inverse Laplace transforms, the zero state response is

$$y_{ZSR}(t) = 1 - e^{-2t}\cos t$$

The total response is

$$y(t) = y_{ZIR}(t) + y_{ZSR}(t) = 2e^{-2t}\cos t + 1 - e^{-2t}\cos t$$

$$= 1 + e^{-2t}\cos t$$

Example 3.4 A linear time-invariant system is described by the state equation

$$\dot{X} = \begin{bmatrix} -3 & 1 \\ -2 & 0 \end{bmatrix}X + \begin{bmatrix} 0 \\ 1 \end{bmatrix}r(t) \qquad y(t) = [1 \quad 0]X$$

If the initial state vector is $X(0) = \begin{bmatrix} 1 \\ -1 \end{bmatrix}$, find the Zero Input Response and the Zero State Response due to a unit impulse input. Also find the total response.

Solution:

$$\dot{X} = \begin{bmatrix} -3 & 1 \\ -2 & 0 \end{bmatrix}X + \begin{bmatrix} 0 \\ 1 \end{bmatrix}r(t) \qquad y(t) = [1 \quad 0]X \quad X(0) = \begin{bmatrix} 1 \\ -1 \end{bmatrix}.$$

The resolvent matrix is

$$\Phi(s) = [sI - A]^{-1} = \begin{bmatrix} s+3 & -1 \\ 2 & s \end{bmatrix}^{-1} = \frac{1}{s^2+3s+2}\begin{bmatrix} s & 1 \\ -2 & s+3 \end{bmatrix}$$

$$C[sI - A]^{-1} = \frac{1}{s^2+3s+2}\begin{bmatrix} 1 & 0 \end{bmatrix}\begin{bmatrix} s & 1 \\ -2 & s+3 \end{bmatrix} = \frac{1}{(s+1)(s+2)}\begin{bmatrix} s & 1 \end{bmatrix}.$$

The ZIR is

$$Y_{ZIR} = C[sI - A]^{-1}X(0)$$

$$= \frac{1}{(s+1)(s+2)}\begin{bmatrix} s & 1 \end{bmatrix}\begin{bmatrix} 1 \\ -1 \end{bmatrix} = \frac{s-1}{(s+1)(s+2)}$$

$$= \frac{-2}{s+1} + \frac{3}{s+2}.$$

Taking inverse Laplace transforms, the ZIR is

$$y_{ZIR}(t) = -2e^{-t} + 3e^{-2t}.$$

For a unit impulse input $r(t) = \delta(t)$, so that $R(s) = 1$. The zero state response is

$$Y_{ZSR}(s) = C[sI - A]^{-1}B$$

$$= \frac{1}{(s+1)(s+2)}\begin{bmatrix} s & 1 \end{bmatrix}\begin{bmatrix} 0 \\ 1 \end{bmatrix} = \frac{1}{(s+1)(s+2)}$$

$$= \frac{1}{(s+1)} - \frac{1}{(s+2)}.$$

Taking inverse Laplace transforms,

$$y_{ZSR}(t) = e^{-t} - e^{-2t}.$$

Total response of the system is

$$y(t) = (-2e^{-t} + 3e^{-2t}) + (e^{-t} - e^{-2t}) = -e^{-t} + 2e^{-2t}.$$

Example 3.5 For the system of Example 3.4, determine the Zero Input Response, the Zero State Response due to a unit step input. Also find the total response.

Solution:

$$\dot{X} = \begin{bmatrix} -3 & 1 \\ -2 & 0 \end{bmatrix}X + \begin{bmatrix} 0 \\ 1 \end{bmatrix}r(t) \qquad y(t) = \begin{bmatrix} 1 & 0 \end{bmatrix}X \qquad X(0) = \begin{bmatrix} 1 \\ -1 \end{bmatrix}.$$

The resolvent matrix is

$$\Phi(s) = \left[sI - A\right]^{-1} = \begin{bmatrix} s+3 & -1 \\ 2 & s \end{bmatrix}^{-1} = \frac{1}{s^2 + 3s + 2} \begin{bmatrix} s & 1 \\ -2 & s+3 \end{bmatrix}.$$

$$C\left[sI - A\right]^{-1} = \frac{1}{s^2 + 3s + 2} \begin{bmatrix} 1 & 0 \end{bmatrix} \begin{bmatrix} s & 1 \\ -2 & s+3 \end{bmatrix} = \frac{1}{(s+1)(s+2)} \begin{bmatrix} s & 1 \end{bmatrix}.$$

The ZIR is

$$Y_{ZIR} = C[sI - A]^{-1} X(0)$$

$$= \frac{1}{(s+1)(s+2)} \begin{bmatrix} s & 1 \end{bmatrix} \begin{bmatrix} 1 \\ -1 \end{bmatrix} = \frac{s-1}{(s+1)(s+2)}$$

$$= \frac{s-1}{s^2 + 3s + 2} = \frac{-2}{s+1} + \frac{3}{s+2}.$$

Taking inverse Laplace transforms, the ZIR is

$$y_{ZIR}(t) = -2e^{-t} + 3e^{-2t}.$$

The input is $r(t) = 1$, a unit step input so that $R(s) = 1/s$. The zero state response is

$$Y_{ZSR} = C[sI - A]^{-1} B R(s).$$

$$= \frac{1}{s(s+1)(s+2)} \begin{bmatrix} s & 1 \end{bmatrix} \begin{bmatrix} 0 \\ 1 \end{bmatrix}$$

$$= \frac{1}{s(s+1)(s+2)} = \frac{1}{2}\frac{1}{s} - \frac{1}{(s+1)} + \frac{1}{2}\frac{1}{s+2}.$$

Taking inverse Laplace transforms, the ZSR is

$$y_{ZSR}(t) = \frac{1}{2} - e^{-t} + \frac{1}{2}e^{-2t}.$$

Total response is

$$y(t) = y_{ZIR}(t) + y_{ZSR}(t) = (-2e^{-t} + 3e^{-2t}) + (\frac{1}{2} - e^{-t} + \frac{1}{2}e^{-2t})$$

$$= \frac{1}{2} - 3e^{-t} + \frac{7}{2}e^{-2t}.$$

Example 3.6 A state model for a linear system is given by

$$\dot{X} = \begin{bmatrix} 0 & 1 & 0 \\ 0 & 0 & 1 \\ -18 & -21 & -8 \end{bmatrix} X + \begin{bmatrix} 0 \\ 0 \\ 1 \end{bmatrix} r(t) \qquad y(t) = \begin{bmatrix} 4 & 1 & 0 \end{bmatrix} X.$$

Determine the response due to a unit step input, given that $X(0) = \begin{bmatrix} 1 & 0 & 0 \end{bmatrix}^T$.

Solution:

$$\dot{X} = \begin{bmatrix} 0 & 1 & 0 \\ 0 & 0 & 1 \\ -18 & -21 & -8 \end{bmatrix} X + \begin{bmatrix} 0 \\ 0 \\ 1 \end{bmatrix} r(t)$$

$$y(t) = \begin{bmatrix} 4 & 1 & 0 \end{bmatrix} X \qquad\qquad X(0) = \begin{bmatrix} 1 & 0 & 0 \end{bmatrix}^T .$$

The resolvent matrix is

$$[sI - A]^{-1} = \begin{bmatrix} s & -1 & 0 \\ 0 & s & -1 \\ 18 & 21 & s+8 \end{bmatrix}^{-1}$$

$$= \frac{1}{s^3 + 8s^2 + 21s + 18} \begin{bmatrix} +(s^2 + 8s + 21) & -\{-(s+8)\} & +1 \\ -18 & +s(s+8) & -\{-s\} \\ +\{-18s\} & -\{21s+18\} & +s^2 \end{bmatrix}$$

$$= \frac{1}{(s+2)(s+3)^2} \begin{bmatrix} (s^2 + 8s + 21) & (s+8) & 1 \\ -18 & s(s+8) & s \\ -18s & -(21s+18) & s^2 \end{bmatrix}.$$

$$C[sI - A]^{-1} = \frac{1}{(s+2)(s+3)^2} \begin{bmatrix} 4 & 1 & 0 \end{bmatrix} \begin{bmatrix} (s^2 + 8s + 21) & (s+8) & 1 \\ -18 & s(s+8) & s \\ -18s & -(21s+18) & s^2 \end{bmatrix}$$

$$= \frac{1}{(s+2)(s+3)^2} \begin{bmatrix} (4s^2 + 32s + 66) & (s^2 + 12s + 32) & (s+4) \end{bmatrix}.$$

The ZIR is

$$Y_{ZIR} = C[sI - A]^{-1} X(0)$$

$$= \frac{1}{s(s+2)(s+3)^2} \begin{bmatrix} (4s^2 + 32s + 66) & (s^2 + 12s + 32) & (s+4) \end{bmatrix} \begin{bmatrix} 1 \\ 0 \\ 0 \end{bmatrix}$$

$$= \frac{4(s^2 + 8s + 16.5)}{(s+2)(s+3)^2}$$

$$= \frac{18}{s+2} - \frac{14}{s+3} - \frac{6}{(s+3)^2} .$$

Taking inverse Laplace transforms, the ZIR is

$$y_{ZIR}(t) = 18e^{-2t} - 14e^{-3t} - 6te^{-3t}.$$

The input is $r(t) = 1$, a unit step input so that $R(s) = 1/s$. The zero state response is

$$Y_{ZSR} = C[sI - A]^{-1} B R(s)$$

$$= \frac{1}{s(s+2)(s+3)^2} \left[(4s^2 + 32s + 66) \quad (s^2 + 12s + 32) \quad (s+4) \right] \begin{bmatrix} 0 \\ 0 \\ 1 \end{bmatrix}$$

$$= \frac{s+4}{s(s+2)(s+3)^2}$$

$$= \frac{2}{9}\frac{1}{s} - \frac{1}{(s+2)} + \frac{7}{9}\frac{1}{(s+3)} + \frac{1}{3}\frac{1}{(s+3)^2}.$$

Taking inverse Laplace transforms, the ZSR is

$$y_{ZSR}(t) = \frac{2}{9} - e^{-2t} + \frac{7}{9}e^{-3t} + \frac{1}{3}te^{-3t}.$$

The complete response is

$$y(t) = y_{ZIR}(t) + y_{ZSR}(t) = \left(18e^{-2t} - 14e^{-3t} - 6te^{-3t} \right) + \left(\frac{2}{9} - e^{-2t} + \frac{7}{9}e^{-3t} + \frac{1}{3}te^{-3t} \right)$$

$$= \frac{2}{9} + 17e^{-2t} - \frac{119}{9}e^{-3t} - \frac{17}{3}te^{-3t}.$$

3.2 SOLUTION OF STATE EQUATION IN TIME-DOMAIN

The state equation is

$$\dot{X}(t) = A X(t) + B R(t)$$

or

$$\dot{X}(t) - A X(t) = B R(t).$$

Multiplying both sides by e^{-At},

$$e^{-At}\dot{X}(t) - e^{-At} A X(t) = e^{-At} B R(t).$$

This can be written as

$$\frac{d}{dt}\left[e^{-At} X(t) \right] = e^{-At} B R(t).$$

Integrating both sides with respect to t, within limits 0 to t,

$$e^{-At} X(t)\Big|_0^t = \int_0^t e^{-A\theta} B\ R(\theta)\,d\theta$$

$$e^{-At} X(t) - X(0) = \int_0^t e^{-A\theta} B\ R(\theta)\,d\theta$$

$$e^{-At} X(t) = X(0) + \int_0^t e^{-A\theta} B\ R(\theta)\,d\theta.$$

From this equation, the solution for the state vector is

$$X(t) = e^{At} X(0) + \int_0^t e^{A(t-\theta)} B\ R(\theta)\,d\theta. \tag{3.7}$$

The first term on the right hand side is the ZIR and the second term is the ZSR. The output is

$$Y(t) = C X(t) + D\ R(t).$$

With D = 0, the output equation is

$$Y(t) = C X(t) = C e^{At} X(0) + C \int_0^t e^{A(t-\theta)} B\ R(\theta)\,d\theta.$$

For a SISO system, the output equation is

$$y(t) = C e^{At} X(0) + C \int_0^t e^{A(t-\theta)} B\ r(\theta)\,d\theta \tag{3.8}$$

$$y(t) = y_{ZIR}(t) + y_{ZSR}(t)$$

$$y_{ZIR}(t) = C e^{At} X(0) \tag{3.9}$$

and

$$y_{ZSR}(t) = C \int_0^t e^{A(t-\theta)} B\ r(\theta)\,d\theta = \int_0^t C e^{A(t-\theta)} B\ r(\theta)\,d\theta. \tag{3.10}$$

Consider equation (3.7) with zero input. Then this equation can be written as

$$X(t) = e^{At} X(0).$$

This shows that the initial state X(0) is driven to a state X(t) at any time t and this transition in state is carried out by the matrix exponential e^{At}. Because of this property, e^{At} is known as state transition matrix.

3.2.1 Unit Impulse Response

For unit impulse input, $r(t) = \delta(t)$. Hence, the response of the system for unit impulse input is

$$y(t) = C e^{At} X(0) + \int_0^t C e^{A(t-\theta)} B \ \delta(\theta) d\theta .$$

$$= C \ e^{At} X(0) + C \ e^{At} B = C \ e^{At} [X(0) + B] \tag{3.11}$$

3.2.2 Unit Step Response

For unit step input, $r(t) = 1; t \geq 0$. Hence, the response of the system for unit step input is

$$y(t) = C e^{At} X(0) + \int_0^t C e^{A(t-\theta)} B \ d\theta$$

$$= C e^{At} X(0) + C[-A^{-1}] e^{At} [e^{-At} - I] B$$

$$= C e^{At} X(0) + C A^{-1} e^{At} [I - e^{-At}] B$$

$$= C e^{At} X(0) + C A^{-1} [e^{At} - I] B \tag{3.12}$$

$$= Y_{ZIR}(s) + Y_{ZSR}(s)$$

Example 3.7 For the system of Example 3.4, find the state transition matrix and hence the unit impulse response.

Solution:

$$\dot{X} = \begin{bmatrix} -3 & 1 \\ -2 & 0 \end{bmatrix} X + \begin{bmatrix} 0 \\ 1 \end{bmatrix} r(t) \qquad y(t) = \begin{bmatrix} 1 & 0 \end{bmatrix} X \quad X(0) = \begin{bmatrix} 1 \\ -1 \end{bmatrix} .$$

The resolvent matrix is

$$\Phi(s) = [sI - A]^{-1} = \begin{bmatrix} s+3 & -1 \\ 2 & s \end{bmatrix}^{-1} = \frac{1}{s^2 + 3s + 2} \begin{bmatrix} s & 1 \\ -2 & s+3 \end{bmatrix}$$

$$= \begin{bmatrix} \dfrac{s}{(s+1)(s+2)} & \dfrac{1}{(s+1)(s+2)} \\ \dfrac{-2}{(s+1)(s+2)} & \dfrac{s+3}{(s+1)(s+2)} \end{bmatrix} = \begin{bmatrix} \dfrac{-1}{(s+1)} + \dfrac{2}{(s+2)} & \dfrac{1}{(s+1)} - \dfrac{1}{(s+2)} \\ \dfrac{-2}{(s+1)} + \dfrac{2}{(s+2)} & \dfrac{2}{(s+1)} - \dfrac{1}{(s+2)} \end{bmatrix} .$$

The state transition matrix is

$$e^{At} = \begin{bmatrix} -e^{-t} + 2e^{-2t} & e^{-t} - e^{-2t} \\ -2e^{-t} + 2e^{-2t} & 2e^{-t} - e^{-2t} \end{bmatrix} .$$

Response due to unit impulse input is

$$y(t) = C \ e^{At} X(0) + C \ e^{At} B$$

$$y(t) = C\, e^{At} \left[X(0) + B \right]$$

$$= \begin{bmatrix} 1 & 0 \end{bmatrix} \begin{bmatrix} -e^{-t} + 2e^{-2t} & e^{-t} - e^{-2t} \\ -2e^{-t} + 2e^{-2t} & 2e^{-t} - e^{-2t} \end{bmatrix} \left\{ \begin{bmatrix} 1 \\ -1 \end{bmatrix} + \begin{bmatrix} 0 \\ 1 \end{bmatrix} \right\}$$

$$= \begin{bmatrix} 1 & 0 \end{bmatrix} \begin{bmatrix} -e^{-t} + 2e^{-2t} & e^{-t} - e^{-2t} \\ -2e^{-t} + 2e^{-2t} & 2e^{-t} - e^{-2t} \end{bmatrix} \begin{bmatrix} 1 \\ 0 \end{bmatrix}$$

$$= -e^{-t} + 2e^{-2t}.$$

Example 3.8 A linear time-invariant system is represented by a state diagram shown in Figure E3.8. Determine the state transition matrix, the ZIR, the ZSR due to a unit step input and the complete response. Assume that $X(0)$ $\begin{bmatrix} 1 & 1 \end{bmatrix}$. Obtain the solution in the time domain.

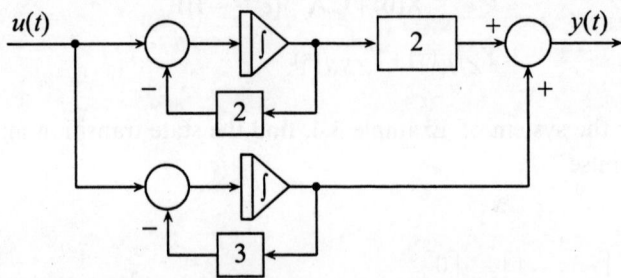

Figure E3.8 State diagram for Example 3.8

Solution: Select the output of each integrator as a state variable as shown in Figure E3.8 (a). From this diagram, the state equation and the output equation can be written as

$$\begin{bmatrix} \dot{x}_1 \\ \dot{x}_2 \end{bmatrix} = \begin{bmatrix} -2 & 0 \\ 0 & -3 \end{bmatrix} \begin{bmatrix} x_1 \\ x_2 \end{bmatrix} + \begin{bmatrix} 1 \\ 1 \end{bmatrix} u(t) \qquad y(t) = \begin{bmatrix} 2 & 1 \end{bmatrix} \begin{bmatrix} x_1 \\ x_2 \end{bmatrix}.$$

$$A = \begin{bmatrix} -2 & 0 \\ 0 & -3 \end{bmatrix} \qquad B = \begin{bmatrix} 1 \\ 1 \end{bmatrix} \quad C = \begin{bmatrix} 2 & 1 \end{bmatrix} \qquad X(0) = \begin{bmatrix} 1 & 1 \end{bmatrix}^T.$$

The system matrix A is diagonal and the Eigen values are $\lambda_1 = -2$ and $\lambda_2 = -3$.

$$A = \begin{bmatrix} -2 & 0 \\ 0 & -3 \end{bmatrix} \quad [sI - A]^{-1} = \begin{bmatrix} s+2 & 0 \\ 0 & s+3 \end{bmatrix}^{-1} = \frac{1}{(s+2)(s+3)} \begin{bmatrix} s+3 & 0 \\ 0 & s+2 \end{bmatrix} = \begin{bmatrix} \dfrac{1}{s+2} & 0 \\ 0 & \dfrac{1}{s+3} \end{bmatrix}.$$

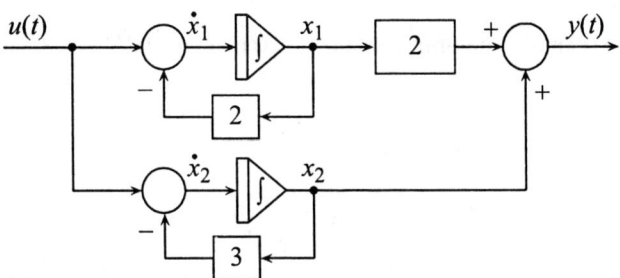

Figure E3.8(a) State diagram for Example 3.8

Taking inverse Laplace transforms, the state transition matrix is

$$e^{At} = L^{-1}\{[sI - A]^{-1}\} = \begin{bmatrix} e^{-2t} & 0 \\ 0 & e^{-3t} \end{bmatrix}.$$

The ZIR is

$$y_{ZIR}(t) = C e^{At} X(0)$$

$$= \begin{bmatrix} 2 & 1 \end{bmatrix} \begin{bmatrix} e^{-2t} & 0 \\ 0 & e^{-3t} \end{bmatrix} \begin{bmatrix} 1 \\ 1 \end{bmatrix} = 2e^{-2t} + e^{-3t}.$$

The input is $r(t) = 1$, unit step. Hence the ZSR is

$$y_{ZSR}(t) = C \int_0^t e^{A(t-\theta)} B \, u(\theta) d\theta$$

$$= \int_0^t \begin{bmatrix} 2 & 1 \end{bmatrix} \begin{bmatrix} e^{-2(t-\theta)} & 0 \\ 0 & e^{-3(t-\theta)} \end{bmatrix} \begin{bmatrix} 1 \\ 1 \end{bmatrix} d\theta$$

$$= \int_0^t \left(2e^{-2(t-\theta)} + e^{-3(t-\theta)} \right) d\theta$$

$$= 2e^{-2t} \int_0^t e^{2\theta} d\theta + e^{-3t} \int_0^t e^{3\theta} d\theta$$

$$= e^{-2t}(e^{2t} - 1) + \frac{1}{3} e^{-3t}(e^{3t} - 1)$$

$$= \frac{4}{3} - e^{-2t} - \frac{1}{3} e^{-3t} \; ; t \geq 0 \, .$$

Alternate method:

Zero state response can also be obtained as $y_{ZSR}(t) = CA^{-1}[e^{At} - I]B$

$$A = \begin{bmatrix} -2 & 0 \\ 0 & -3 \end{bmatrix} \qquad A^{-1} = \frac{1}{6}\begin{bmatrix} -3 & 0 \\ 0 & -2 \end{bmatrix}.$$

$$y_{ZSR}(t) = \begin{bmatrix} 2 & 1 \end{bmatrix}\frac{1}{6}\begin{bmatrix} -3 & 0 \\ 0 & -2 \end{bmatrix}\left\{\begin{bmatrix} e^{-2t} & 0 \\ 0 & e^{-3t} \end{bmatrix} - \begin{bmatrix} 1 & 0 \\ 0 & 1 \end{bmatrix}\right\}\begin{bmatrix} 1 \\ 1 \end{bmatrix}$$

$$= \frac{1}{6}\begin{bmatrix} -6 & -2 \end{bmatrix}\begin{bmatrix} e^{-2t} - 1 & 0 \\ 0 & e^{-3t} - 1 \end{bmatrix}\begin{bmatrix} 1 \\ 1 \end{bmatrix}$$

$$= \frac{1}{6}\begin{bmatrix} -6e^{-2t} + 6 & -2e^{-3t} + 2 \end{bmatrix}\begin{bmatrix} 1 \\ 1 \end{bmatrix}$$

$$= \frac{1}{6}\begin{bmatrix} 8 - 6e^{-2t} - 2e^{-3t} \end{bmatrix}$$

$$= \frac{4}{3} - e^{-2t} - \frac{1}{3}e^{-3t}.$$

The result is the same as that obtained earlier.

Example 3.9 A linear system is represented by the state model

$$\begin{bmatrix} \dot{x}_1 \\ \dot{x}_2 \end{bmatrix} = \begin{bmatrix} -1 & 1 \\ -1 & -3 \end{bmatrix}\begin{bmatrix} x_1 \\ x_2 \end{bmatrix} + \begin{bmatrix} 1 \\ 2 \end{bmatrix}r(t) \qquad y(t) = \begin{bmatrix} 2 & 1 \end{bmatrix}\begin{bmatrix} x_1 \\ x_2 \end{bmatrix} \qquad X(0) = \begin{bmatrix} 1 \\ -1 \end{bmatrix}.$$

Find the ZIR and the complete solution for the state equation for a unit step input.

Solution: The resolvent matrix is

$$\Phi(s) = [sI - A]^{-1} = \begin{bmatrix} s+1 & -1 \\ 1 & s+3 \end{bmatrix}^{-1} = \frac{1}{s^2 + 4s + 4}\begin{bmatrix} s+3 & 1 \\ -1 & s+1 \end{bmatrix}$$

$$= \begin{bmatrix} \dfrac{s+3}{(s+2)^2} & \dfrac{1}{(s+2)^2} \\ \dfrac{-1}{(s+2)^2} & \dfrac{s+1}{(s+2)^2} \end{bmatrix} = \begin{bmatrix} \dfrac{1}{(s+2)} + \dfrac{1}{(s+2)^2} & \dfrac{1}{(s+2)^2} \\ \dfrac{-1}{(s+2)^2} & \dfrac{1}{(s+2)} - \dfrac{1}{(s+2)^2} \end{bmatrix}.$$

Taking inverse Laplace transforms,

$$e^{At} = \begin{bmatrix} e^{-2t} + te^{-2t} & te^{-2t} \\ -te^{-2t} & e^{-2t} - te^{-2t} \end{bmatrix}.$$

$$y_{ZIR}(t) = C e^{At} X(0) = \begin{bmatrix} 2 & 1 \end{bmatrix} \begin{bmatrix} e^{-2t} + te^{-2t} & te^{-2t} \\ -te^{-2t} & e^{-2t} - te^{-2t} \end{bmatrix} \begin{bmatrix} 1 \\ -1 \end{bmatrix}$$

$$= \begin{bmatrix} 2 & 1 \end{bmatrix} \begin{bmatrix} e^{-2t} \\ -e^{-2t} \end{bmatrix} = e^{-2t}.$$

$$y_{ZSR}(t) = C A \begin{bmatrix} e^{At} - I \end{bmatrix} B$$

$$= \begin{bmatrix} 2 & 1 \end{bmatrix} \begin{bmatrix} -1 & 1 \\ -1 & -3 \end{bmatrix}^{-1} \begin{bmatrix} e^{-2t} + te^{-2t} - 1 & te^{-2t} \\ -te^{-2t} & e^{-2t} - te^{-2t} - 1 \end{bmatrix} \begin{bmatrix} 1 \\ 2 \end{bmatrix}$$

$$= \begin{bmatrix} 2 & 1 \end{bmatrix} \frac{1}{4} \begin{bmatrix} -3 & -1 \\ 1 & -1 \end{bmatrix} \begin{bmatrix} e^{-2t} + te^{-2t} - 1 & te^{-2t} \\ -te^{-2t} & e^{-2t} - te^{-2t} - 1 \end{bmatrix} \begin{bmatrix} 1 \\ 2 \end{bmatrix}$$

$$= \frac{1}{4} \begin{bmatrix} -5 & -3 \end{bmatrix} \begin{bmatrix} e^{-2t} + 3te^{-2t} - 1 \\ 2e^{-2t} - 3te^{-2t} - 2 \end{bmatrix} = -\frac{11}{4} e^{-2t} - \frac{3}{2} te^{-2t} + \frac{11}{4}.$$

Complete response is

$$y(t) = y_{ZIR}(t) + y_{ZSR}(t) = e^{-2t} + \frac{11}{4} - \frac{11}{4} e^{-2t} - \frac{3}{2} te^{-2t} = \frac{11}{4} - \frac{7}{4} e^{-2t} - \frac{3}{2} te^{-2t}$$

Example 3.10 A state model of a linear system is given as

$$\begin{bmatrix} \dot{x}_1 \\ \dot{x}_2 \end{bmatrix} = \begin{bmatrix} -1 & 2 \\ -1 & -3 \end{bmatrix} \begin{bmatrix} x_1 \\ x_2 \end{bmatrix} + \begin{bmatrix} 1 \\ 2 \end{bmatrix} r(t) \qquad y(t) = \begin{bmatrix} 1 & 1 \end{bmatrix} \begin{bmatrix} x_1 \\ x_2 \end{bmatrix} \qquad X(0) = \begin{bmatrix} 1 \\ 0 \end{bmatrix}.$$

Obtain, by the time domain method, the ZIR and the complete response for a unit impulse input.

Solution:

$$A = \begin{bmatrix} -1 & 2 \\ -1 & -3 \end{bmatrix} \qquad B = \begin{bmatrix} 1 \\ 2 \end{bmatrix} \qquad C = \begin{bmatrix} 1 & 1 \end{bmatrix} \qquad X(0) = \begin{bmatrix} 1 \\ 0 \end{bmatrix}.$$

The resolvent matrix is

$$\Phi(s) = \left[sI - A\right]^{-1} = \begin{bmatrix} s+1 & -2 \\ 1 & s+3 \end{bmatrix}^{-1} = \frac{1}{s^2 + 4s + 5} \begin{bmatrix} s+3 & 2 \\ -1 & s+1 \end{bmatrix}$$

$$= \frac{1}{(s+2)^2 + 1} \begin{bmatrix} s+3 & 2 \\ -1 & s+1 \end{bmatrix} = \begin{bmatrix} \dfrac{s+3}{(s+2)^2 + 1} & \dfrac{2}{(s+2)^2 + 1} \\ \dfrac{-1}{(s+2)^2 + 1} & \dfrac{s+1}{(s+2)^2 + 1} \end{bmatrix}$$

$$= \begin{bmatrix} \dfrac{s+2}{(s+2)^2 + 1} + \dfrac{1}{(s+2)^2 + 1} & \dfrac{2}{(s+2)^2 + 1} \\ \dfrac{-1}{(s+2)^2 + 1} & \dfrac{s+2}{(s+2)^2 + 1} - \dfrac{1}{(s+2)^2 + 1} \end{bmatrix}.$$

Taking inverse Laplace transforms,

$$e^{At} = \begin{bmatrix} e^{-2t}\cos t + e^{-2t}\sin t & 2e^{-2t}\sin t \\ -e^{-2t}\sin t & e^{-2t}\cos t - e^{-2t}\sin t \end{bmatrix}.$$

The zero input response is

$$y_{ZIR}(t) = C\,e^{At}\,X(0)$$

$$= \begin{bmatrix} 1 & 1 \end{bmatrix} \begin{bmatrix} e^{-2t}\cos t + e^{-2t}\sin t & 2e^{-2t}\sin t \\ -e^{-2t}\sin t & e^{-2t}\cos t - e^{-2t}\sin t \end{bmatrix} \begin{bmatrix} 1 \\ 0 \end{bmatrix}$$

$$= \begin{bmatrix} 1 & 1 \end{bmatrix} \begin{bmatrix} e^{-2t}\cos t + e^{-2t}\sin t \\ -e^{-2t}\sin t \end{bmatrix} = e^{-2t}\cos t.$$

The zero state response is

$$y_{ZSR}(t) = C\,e^{At}\,B = \begin{bmatrix} 1 & 1 \end{bmatrix} \begin{bmatrix} e^{-2t}\cos t + e^{-2t}\sin t & 2e^{-2t}\sin t \\ -e^{-2t}\sin t & e^{-2t}\cos t - e^{-2t}\sin t \end{bmatrix} \begin{bmatrix} 1 \\ 2 \end{bmatrix}$$

$$= \begin{bmatrix} 1 & 1 \end{bmatrix} \begin{bmatrix} e^{-2t}\cos t + e^{-2t}\sin t + 4e^{-2t}\sin t \\ -e^{-2t}\sin t + 2e^{-2t}\cos t - 2e^{-2t}\sin t \end{bmatrix}$$

$$= 3e^{-2t}\cos t + 2e^{-2t}\sin t.$$

Complete response is

$$y(t) = y_{ZIR}(t) + y_{ZSR}(t) = e^{-2t}\cos t + (3e^{-2t}\cos t + 2e^{-2t}\sin t) = 4e^{-2t}\cos t + 2e^{-2t}\sin t$$

Example 3.11 For the system of Example 3.10, obtain the complete solution for the state equation for a unit step input.

Solution:

$$\begin{bmatrix} \dot{x}_1 \\ \dot{x}_2 \end{bmatrix} = \begin{bmatrix} -1 & 2 \\ -1 & -3 \end{bmatrix}\begin{bmatrix} x_1 \\ x_2 \end{bmatrix} + \begin{bmatrix} 1 \\ 2 \end{bmatrix} r(t) \qquad y(t) = \begin{bmatrix} 1 & 1 \end{bmatrix}\begin{bmatrix} x_1 \\ x_2 \end{bmatrix} \qquad X(0) = \begin{bmatrix} 1 \\ 0 \end{bmatrix}.$$

The zero input response was obtained in Example 3.10 as

$$y_{ZIR}(t) = e^{-2t}\cos t.$$

The zero state response is

$$y_{ZSR}(t) = C A^{-1}\left[e^{At} - I\right] B$$

$$= \begin{bmatrix} 1 & 1 \end{bmatrix}\frac{1}{5}\begin{bmatrix} -3 & -2 \\ 1 & -1 \end{bmatrix}\begin{bmatrix} e^{-2t}\cos t + e^{-2t}\sin t - 1 & 2e^{-2t}\sin t \\ -e^{-2t}\sin t & e^{-2t}\cos t - e^{-2t}\sin t - 1 \end{bmatrix}\begin{bmatrix} 1 \\ 2 \end{bmatrix}$$

$$= \frac{1}{5}\begin{bmatrix} -2 & -3 \end{bmatrix}\begin{bmatrix} e^{-2t}\cos t + e^{-2t}\sin t - 1 + 4e^{-2t}\sin t \\ -e^{-2t}\sin t + 2e^{-2t}\cos t - 2e^{-2t}\sin t - 2 \end{bmatrix}$$

$$= \frac{1}{5}\begin{bmatrix} -2 & -3 \end{bmatrix}\begin{bmatrix} -1 + e^{-2t}\cos t + 5e^{-2t}\sin t \\ -2 + 2e^{-2t}\cos t - 3e^{-2t}\sin t \end{bmatrix}$$

$$= \frac{1}{5}\left[-2e^{-2t}\cos t - 2e^{-2t}\sin t + 2 - 8e^{-2t}\sin t + 3e^{-2t}\sin t - 6e^{-2t}\cos t + 6e^{-2t}\sin t + 6\right]$$

$$= \frac{8}{5} - \frac{8}{5}e^{-2t}\cos t - \frac{1}{5}e^{-2t}\sin t.$$

3.3 METHODS OF COMPUTING STATE TRANSITION MATRIX

For the solution of a state equation, the computation of state transition matrix e^{At} is necessary, especially in the time domain solution. Many methods are available for determining the state transition matrix:

1. Laplace transform method
2. Power series method
3. Using Sylvester's Interpolation Formula
4. Using Cayley–Hamilton Theorem
5. Diagonalization method

These methods are discussed here one by one.

3.3.1 Laplace Transform Method

From equation (3.5),

$$Y_{ZIR}(s) = C[sI - A]^{-1}X(0).$$

Taking inverse Laplace transforms,

$$y_{ZIR}(t) = L^{-1}\left\{C[sI - A]^{-1}X(0)\right\}$$

$$= C\left\{L^{-1}[sI - A]^{-1}\right\}X(0). \qquad (3.13)$$

From equation (3.9),

$$y_{ZIR}(t) = C e^{At}X(0). \qquad (3.14)$$

Comparing equations (3.13) and (3.14), the state transition matrix is

$$e^{At} = L^{-1}[sI - A]^{-1}. \qquad (3.15)$$

Example 3.12 From the result $\dfrac{d}{dt}e^{At} = A e^{At}$, devise a method of finding the state transition matrix.

Solution:

Let $F(t) = e^{At}$ $F(0) = e^{A(0)} = I$.

It is known that

$$\frac{d}{dt}(e^{At}) = A e^{At}$$

$$\frac{d}{dt}(F(t)) = A F(t).$$

Taking Laplace transforms,

$$s F(s) - F(0) = A F(s).$$

Or $s F(s) - I = A F(s)$ or $[sI - A]F(s) = I$.

From this, $F(s) = [sI - A]^{-1}$.

Taking inverse Laplace transforms,

$$F(t) = L^{-1}\left\{[sI - A]^{-1}\right\}$$

Hence, $e^{At} = L^{-1}\left\{[sI-A]^{-1}\right\}$.

3.3.2 Power Series Method

The state transition matrix e^{At} can be written in a power series as

$$e^{At} = I + At + \frac{(At)^2}{2!} + \frac{(At)^3}{3!} + \frac{(At)^4}{4!} + \ldots + \frac{(At)^k}{k!} + \ldots$$

$$= I + At + \frac{A^2 t^2}{2!} + \frac{A^3 t^3}{3!} + \frac{A^4 t^4}{4!} + \ldots + \frac{A^k t^k}{k!} + \ldots .$$

This method is practically very time consuming and is seldom used.

Example 3.13 Find the state transition matrix for $A = \begin{bmatrix} -1 & 0 \\ 0 & -2 \end{bmatrix}$

Solution: The state transition matrix is given by

$$e^{At} = I + At + \frac{A^2 t^2}{2!} + \frac{A^3 t^3}{3!} + \frac{A^4 t^4}{4!} + \ldots + \frac{A^k t^k}{k!} + \ldots$$

$$A = \begin{bmatrix} -1 & 0 \\ 0 & -2 \end{bmatrix} \qquad A^2 = \begin{bmatrix} -1 & 0 \\ 0 & -2 \end{bmatrix}\begin{bmatrix} -1 & 0 \\ 0 & -2 \end{bmatrix} = \begin{bmatrix} 1 & 0 \\ 0 & 4 \end{bmatrix}$$

$$A^3 = A\, A^2 = \begin{bmatrix} -1 & 0 \\ 0 & -2 \end{bmatrix}\begin{bmatrix} 1 & 0 \\ 0 & 4 \end{bmatrix} = \begin{bmatrix} -1 & 0 \\ 0 & -8 \end{bmatrix}$$

$$e^{At} = \begin{bmatrix} 1 & 0 \\ 0 & 1 \end{bmatrix} + \begin{bmatrix} -1 & 0 \\ 0 & -2 \end{bmatrix} t + \frac{1}{2!}\begin{bmatrix} 1 & 0 \\ 0 & 4 \end{bmatrix} t^2 + \frac{1}{3!}\begin{bmatrix} -1 & 0 \\ 0 & -8 \end{bmatrix} t^3 + \ldots$$

$$e^{At} = \begin{bmatrix} 1 - t + \frac{1}{2!}t^2 - \frac{1}{3!}t^3 + \ldots & 0 \\ 0 & 1 + (-2t) + \frac{1}{2!}(-2t)^2 + \frac{1}{3!}(-2t)^3 + \ldots \end{bmatrix}$$

Hence, $e^{At} = \begin{bmatrix} e^{-t} & 0 \\ 0 & e^{-2t} \end{bmatrix}$.

Example 3.14 Find the state transition matrix for $A = \begin{bmatrix} -2 & 1 \\ 0 & -2 \end{bmatrix}$.

Solution: The state transition matrix is given by

$$e^{At} = I + At + \frac{A^2 t^2}{2!} + \frac{A^3 t^3}{3!} + \frac{A^4 t^4}{4!} + \ldots + \frac{A^k t^k}{k!} + \ldots$$

$$A = \begin{bmatrix} -2 & 1 \\ 0 & -2 \end{bmatrix} \qquad A^2 = \begin{bmatrix} -2 & 1 \\ 0 & -2 \end{bmatrix}\begin{bmatrix} -2 & 1 \\ 0 & -2 \end{bmatrix} = \begin{bmatrix} 4 & -4 \\ 0 & 4 \end{bmatrix}$$

$$A^3 = A\,A^2 = \begin{bmatrix} -2 & 1 \\ 0 & -2 \end{bmatrix}\begin{bmatrix} 4 & -4 \\ 0 & 4 \end{bmatrix} = \begin{bmatrix} -8 & 12 \\ 0 & -8 \end{bmatrix}$$

$$e^{At} = \begin{bmatrix} 1 & 0 \\ 0 & 1 \end{bmatrix} + \begin{bmatrix} -2 & 1 \\ 0 & -2 \end{bmatrix}t + \frac{1}{2!}\begin{bmatrix} 4 & -4 \\ 0 & 4 \end{bmatrix}t^2 + \frac{1}{3!}\begin{bmatrix} -8 & 12 \\ 0 & -8 \end{bmatrix}t^3 + \cdots$$

$$= \begin{bmatrix} 1+(-2t)+\frac{1}{2!}(-2t)^2+\frac{1}{3!}(-2t)^3+\cdots & t+\frac{1}{2!}(-4t^2)+\frac{1}{3!}(12t^3)+\cdots \\[2mm] 0 & 1+(-2t)+\frac{1}{2!}(-2t)^2+\frac{1}{3!}(-2t)^3+\cdots \end{bmatrix}$$

$$= \begin{bmatrix} 1+(-2t)+\frac{1}{2!}(-2t)^2+\frac{1}{3!}(-2t)^3+\cdots & t\left(1+(-2t)+\frac{1}{2!}(-2t)^2+\cdots\right) \\[2mm] 0 & 1+(-2t)+\frac{1}{2!}(-2t)^2+\frac{1}{3!}(-2t)^3+\cdots \end{bmatrix}$$

Hence, $e^{At} = \begin{bmatrix} e^{-2t} & te^{-2t} \\ 0 & e^{-2t} \end{bmatrix}$.

Example 3.15 Find the state transition matrix for $A = \begin{bmatrix} -3 & 1 \\ -2 & 0 \end{bmatrix}$.

Solution: The state transition matrix is given by

$$e^{At} = I + At + \frac{A^2 t^2}{2!} + \frac{A^3 t^3}{3!} + \frac{A^4 t^4}{4!} + \cdots + \frac{A^k t^k}{k!} + \cdots.$$

$$A = \begin{bmatrix} -3 & 1 \\ -2 & 0 \end{bmatrix} \qquad A^2 = \begin{bmatrix} -3 & 1 \\ -2 & 0 \end{bmatrix}\begin{bmatrix} -3 & 1 \\ -2 & 0 \end{bmatrix} = \begin{bmatrix} 7 & -3 \\ 6 & -2 \end{bmatrix}$$

$$A^3 = A\,A^2 = \begin{bmatrix} -3 & 1 \\ -2 & 0 \end{bmatrix}\begin{bmatrix} 7 & -3 \\ 6 & -2 \end{bmatrix} = \begin{bmatrix} -15 & 7 \\ -14 & 6 \end{bmatrix}.$$

$$e^{At} = \begin{bmatrix} 1 & 0 \\ 0 & 1 \end{bmatrix} + \begin{bmatrix} -3 & 1 \\ -2 & 0 \end{bmatrix}t + \frac{1}{2!}\begin{bmatrix} 7 & -3 \\ 6 & -2 \end{bmatrix}t^2 + \frac{1}{3!}\begin{bmatrix} -15 & 7 \\ -14 & 6 \end{bmatrix}t^3 + \cdots$$

$$= \begin{bmatrix} 1-3t+\frac{7}{2}t^2-\frac{15}{6}t^3+\cdots & t-\frac{3}{2}t^2+\frac{7}{6}t^3+\cdots \\[2mm] -2t+3t^2-\frac{7}{3}t^3+\cdots & 1-2t-t^2+t^3+\cdots \end{bmatrix}.$$

It can be observed that it is very difficult to obtain a closed form of expression for each series in this matrix, and hence this method is not in practice.

3.3.3 Using Sylvester's Interpolation Formula

Case 1: System Matrix with Distinct Eigen Values

For an n-th order system, let the n Eigen values be $\lambda_1, \lambda_2, \lambda_3, ..., \lambda_n$. Then Sylvester's interpolation formula for computing the state transition matrix is given by

$$
\begin{vmatrix}
1 & 1 & 1 & 1 & \cdots & 1 & I \\
\lambda_1 & \lambda_2 & \lambda_3 & \lambda_4 & \cdots & \lambda_n & A \\
\lambda_1^2 & \lambda_2^2 & \lambda_3^2 & \lambda_4^2 & \cdots & \lambda_n^2 & A^2 \\
\lambda_1^3 & \lambda_2^3 & \lambda_3^3 & \lambda_4^3 & \cdots & \lambda_n^3 & A^3 \\
\cdots & & & & & & \\
\cdots & & & & & & \\
\lambda_1^{n-1} & \lambda_2^{n-1} & \lambda_3^{n-1} & \lambda_4^{n-1} & \cdots & \lambda_n^{n-1} & A^{n-1} \\
e^{\lambda_1 t} & e^{\lambda_2 t} & e^{\lambda_3 t} & e^{\lambda_4 t} & \cdots & e^{\lambda_n t} & e^{At}
\end{vmatrix} = 0 \tag{3.16}
$$

Solving this equation, the state transition matrix e^{At} can be determined.

Case 2: With Repeated Eigen values

For an n-th order system, assume that there are three repeated Eigen values and let these be $\lambda = \lambda_1$. For such a system matrix, Sylvester's interpolation formula can be written as

$$
\begin{vmatrix}
1 & 0 & 0 & 1 & \cdots & 1 & I \\
\lambda_1 & \dfrac{d}{d\lambda}(\lambda)\Big|_{\lambda=\lambda_1} & \dfrac{1}{2!}\dfrac{d^2}{d\lambda^2}(\lambda)\Big|_{\lambda=\lambda_1} & \lambda_3 & \cdots & \lambda_n & A \\
\lambda_1^2 & \dfrac{d}{d\lambda}(\lambda^2)\Big|_{\lambda=\lambda_1} & \dfrac{1}{2!}\dfrac{d^2}{d\lambda^2}(\lambda^2)\Big|_{\lambda=\lambda_1} & \lambda_3^2 & \cdots & \lambda_n^2 & A^2 \\
\lambda_1^3 & \dfrac{d}{d\lambda}(\lambda^3)\Big|_{\lambda=\lambda_1} & \dfrac{1}{2!}\dfrac{d^2}{d\lambda^2}(\lambda^3)\Big|_{\lambda=\lambda_1} & \lambda_3^3 & \cdots & \lambda_n^3 & A^3 \\
\cdots & & & & & & \\
\cdots & & & & & & \\
\lambda_1^{n-1} & \dfrac{d}{d\lambda}(\lambda^{n-1})\Big|_{\lambda=\lambda_1} & \dfrac{1}{2!}\dfrac{d^2}{d\lambda^2}(\lambda^{n-1})\Big|_{\lambda=\lambda_1} & \lambda_3^{n-1} & \cdots & \lambda_n^{n-1} & A^{n-1} \\
e^{\lambda_1 t} & \dfrac{d}{d\lambda}(e^{\lambda t})\Big|_{\lambda=\lambda_1} & \dfrac{1}{2!}\dfrac{d^2}{d\lambda^2}(e^{\lambda t})\Big|_{\lambda=\lambda_1} & e^{\lambda_3 t} & \cdots & e^{\lambda_n t} & e^{At}
\end{vmatrix} = 0 \quad (3.17)
$$

This reduces to

$$
\begin{vmatrix}
1 & 0 & 0 & 1 & \cdots & 1 & I \\
\lambda_1 & 1 & 0 & \lambda_3 & \cdots & \lambda_n & A \\
\lambda_1^2 & 2\lambda_1 & 1 & \lambda_3^2 & \cdots & \lambda_n^2 & A^2 \\
\lambda_1^3 & 3\lambda_1^2 & 3\lambda_1 & \lambda_3^3 & \cdots & \lambda_n^3 & A^3 \\
\cdots & & & & & & \\
\cdots & & & & & & \\
\lambda_1^{n-1} & (n-1)\lambda_1^{n-2} & \dfrac{(n-1)(n-2)\lambda_1^{n-3}}{2} & \lambda_3^{n-1} & \cdots & \lambda_n^{n-1} & A^{n-1} \\
e^{\lambda_1 t} & te^{\lambda_1 t} & \dfrac{t^2}{2}e^{\lambda_1 t} & e^{\lambda_3 t} & \cdots & e^{\lambda_n t} & e^{At}
\end{vmatrix} = 0 \qquad (3.18)
$$

Example 3.16 Find, using Sylvester's interpolation formula, the state transition matrix for

$$
A = \begin{bmatrix} -2 & 2 \\ 1 & -3 \end{bmatrix}.
$$

Solution: The Eigen values are obtained by solving the characteristic equation

$$
|\lambda I - A| = 0 \qquad or \qquad \begin{vmatrix} \lambda+2 & -2 \\ -1 & \lambda+3 \end{vmatrix} = 0.
$$

$$
\lambda^2 + 5\lambda + 4 = 0 \qquad\qquad or \;\; \lambda = -1 \; and \; \lambda = -4.
$$

The Eigen values are distinct. Hence, the Sylvester's interpolation formula is

$$
\begin{vmatrix} 1 & 1 & I \\ \lambda_1 & \lambda_2 & A \\ e^{\lambda_1 t} & e^{\lambda_2 t} & e^{At} \end{vmatrix} = \begin{vmatrix} 1 & 1 & I \\ -1 & -4 & A \\ e^{-t} & e^{-4t} & e^{At} \end{vmatrix} = 0.
$$

Expanding this determinant,

$$
-4e^{At} - e^{-4t}A + (e^{At} - e^{-4t}I) + e^{-t}(A + 4I) = 0
$$

$$
3e^{At} = -e^{-4t}(A+I) + e^{-t}(A+4I)
$$

$$
= -e^{-4t}\left\{ \begin{bmatrix} -2 & 2 \\ 1 & -3 \end{bmatrix} + \begin{bmatrix} 1 & 0 \\ 0 & 1 \end{bmatrix} \right\} + e^{-t}\left\{ \begin{bmatrix} -2 & 2 \\ 1 & -3 \end{bmatrix} + \begin{bmatrix} 4 & 0 \\ 0 & 4 \end{bmatrix} \right\}
$$

$$
= e^{-4t}\begin{bmatrix} 1 & -2 \\ -1 & 2 \end{bmatrix} + e^{-t}\begin{bmatrix} 2 & 2 \\ 1 & 1 \end{bmatrix}
$$

$$= \begin{bmatrix} e^{-4t} + 2e^{-t} & -2e^{-4t} + 2e^{-t} \\ -e^{-4t} + e^{-t} & 2e^{-4t} + e^{-t} \end{bmatrix}.$$

The state transition matrix is

$$e^{At} = \begin{bmatrix} \dfrac{2}{3}e^{-t} + \dfrac{1}{3}e^{-4t} & \dfrac{2}{3}e^{-t} - \dfrac{2}{3}e^{-4t} \\ \dfrac{1}{3}e^{-t} - \dfrac{1}{3}e^{-4t} & \dfrac{1}{3}e^{-t} + \dfrac{2}{3}e^{-4t} \end{bmatrix}.$$

Example 3.17 Find, using Sylvester's interpolation formula, the state transition matrix for

$$A = \begin{bmatrix} 0 & 2 \\ -2 & -4 \end{bmatrix}.$$

Solution: The Eigen values are obtained by solving the characteristic equation

$$|\lambda I - A| = 0 \qquad or \quad \begin{vmatrix} \lambda & -2 \\ 2 & \lambda + 4 \end{vmatrix} = 0.$$

$$\lambda^2 + 4\lambda + 4 = 0 \qquad or \ \lambda = -2 \ and \ \lambda = -2 \ ; repeated \ roots.$$

The Eigen values are repeated. Using the Sylvester's interpolation formula,

$$\begin{vmatrix} 1 & 0 & I \\ \lambda_1 & \dfrac{d}{d\lambda}(\lambda)\Big|_{\lambda=\lambda_1} & A \\ e^{\lambda_1 t} & \dfrac{d}{d\lambda}(e^{\lambda t})\Big|_{\lambda=\lambda_1} & e^{At} \end{vmatrix} = \begin{vmatrix} 1 & 0 & I \\ -2 & 1 & A \\ e^{-2t} & te^{-2t} & e^{At} \end{vmatrix} = 0.$$

Expanding this determinant,

$$e^{At} - te^{-2t} A + (-2te^{-2t} - e^{-2t})I = 0.$$

$$e^{At} = te^{-2t}(A + 2I) + e^{-2t}I$$

$$= te^{-2t}\left\{ \begin{bmatrix} 0 & 2 \\ -2 & -4 \end{bmatrix} + \begin{bmatrix} 2 & 0 \\ 0 & 2 \end{bmatrix} \right\} + e^{-2t}\begin{bmatrix} 1 & 0 \\ 0 & 1 \end{bmatrix}$$

$$= te^{-2t}\begin{bmatrix} 2 & 2 \\ -2 & -2 \end{bmatrix} + e^{-2t}\begin{bmatrix} 1 & 0 \\ 0 & 1 \end{bmatrix}$$

$$= \begin{bmatrix} e^{-2t} + 2t\,e^{-2t} & 2t\,e^{-2t} \\ -2t\,e^{-2t} & e^{-2t} - 2t\,e^{-2t} \end{bmatrix}.$$

3.3.4 Cayley–Hamilton Theorem

Cayley–Hamilton theorem states that every square matrix satisfies its own characteristic equation.

Consider a third order system with system matrix A. Its characteristic equation can be written as

$$P(\lambda) = \lambda^3 + a_2\lambda^2 + a_1\lambda + a_0 = 0.$$

The Cayley–Hamilton theorem states that

$$F(A) = A^3 + a_2\,A^2 + a_1\,A + a_0\,I = 0.$$

To illustrate this, consider a third-order system matrix

$$A = \begin{bmatrix} 0 & 1 & 0 \\ 0 & 0 & 1 \\ -6 & -11 & -6 \end{bmatrix}.$$

Its characteristic equation is

$$|\lambda I - A| = 0 \qquad or \qquad \begin{vmatrix} \lambda & -1 & 0 \\ 0 & \lambda & -1 \\ 6 & 11 & \lambda+6 \end{vmatrix} = 0.$$

Or $P(\lambda) = \lambda^3 + 6\lambda^2 + 11\lambda + 6 = 0$

$$A^2 = \begin{bmatrix} 0 & 1 & 0 \\ 0 & 0 & 1 \\ -6 & -11 & -6 \end{bmatrix}\begin{bmatrix} 0 & 1 & 0 \\ 0 & 0 & 1 \\ -6 & -11 & -6 \end{bmatrix} = \begin{bmatrix} 0 & 0 & 1 \\ -6 & -11 & -6 \\ 36 & 60 & 25 \end{bmatrix}$$

$$A^3 = \begin{bmatrix} 0 & 1 & 0 \\ 0 & 0 & 1 \\ -6 & -11 & -6 \end{bmatrix}\begin{bmatrix} 0 & 0 & 1 \\ -6 & -11 & -6 \\ 36 & 60 & 25 \end{bmatrix} = \begin{bmatrix} -6 & -11 & -6 \\ 36 & 60 & 25 \\ -150 & -239 & -90 \end{bmatrix}$$

$$P(A) = A^3 + 6A^2 + 11A + 6I = \begin{bmatrix} -6 & -11 & -6 \\ 36 & 60 & 25 \\ -150 & -239 & -90 \end{bmatrix} + \begin{bmatrix} 0 & 0 & 6 \\ -36 & -66 & -36 \\ 216 & 360 & 150 \end{bmatrix}$$

$$+\begin{bmatrix} 0 & 11 & 0 \\ 0 & 0 & 11 \\ -66 & -121 & -66 \end{bmatrix}+\begin{bmatrix} 6 & 0 & 0 \\ 0 & 6 & 0 \\ 0 & 0 & 6 \end{bmatrix}=\begin{bmatrix} 0 & 0 & 0 \\ 0 & 0 & 0 \\ 0 & 0 & 0 \end{bmatrix}.$$

This illustrates the Cayley–Hamilton theorem.

3.4 Applications of Cayley–Hamilton Theorem

3.4.1 Matrix Inversion

The Cayley–Hamilton theorem can be used to find the inverse of a square matrix.

For a system matrix of size $n \times n$, let the characteristic equation be

$$\lambda^n + a_{n-1}\lambda^{n-1} + a_{n-2}\lambda^{n-2} + ... + a_1\lambda + a_0 = 0. \tag{3.19}$$

Since the system matrix A satisfies its own characteristic equation,

$$A^n + a_{n-1}A^{n-1} + a_{n-2}A^{n-2} + ... + a_1 A + a_0 I = 0. \tag{3.20}$$

Assume that A^{-1} exists. Pre-multiplying equation (3.20) with A^{-1},

$$A^{n-1} + a_{n-1}A^{n-2} + a_{n-2}A^{n-3} + ... + a_1 I + a_0 A^{-1} = 0.$$

Hence, $A^{-1} = -\dfrac{1}{a_0}\left(A^{n-1} + a_{n-1}A^{n-2} + a_{n-2}A^{n-3} + ... + a_1 I\right)$ \hfill (3.21)

Example 3.18 Find the inverse of $A = \begin{bmatrix} 0 & 3 \\ -2 & -5 \end{bmatrix}$ using Cayley–Hamilton theorem.

Solution: The Eigen values are obtained by solving the characteristic equation

$$|\lambda I - A| = 0 \qquad or \begin{vmatrix} \lambda & -3 \\ 2 & \lambda+5 \end{vmatrix} = 0.$$

$$or \;\; \lambda^2 + 5\lambda + 6 = 0 \qquad\qquad or \;\; \lambda = -2 \; and \; \lambda = -3 .$$

The matrix A satisfies the characteristic equation and hence,

$$A^2 + 5A + 6I = 0 \;\; or \;\; A + 5I + 6A^{-1} = 0 .$$

$$A^{-1} = -\frac{1}{6}(A+5I) = -\frac{1}{6}\left\{\begin{bmatrix} 0 & 3 \\ -2 & -5 \end{bmatrix} + 5\begin{bmatrix} 1 & 0 \\ 0 & 1 \end{bmatrix}\right\} = \frac{1}{6}\begin{bmatrix} -5 & -3 \\ 2 & 0 \end{bmatrix}.$$

Example 3.19 Find the inverse of $A = \begin{bmatrix} -2 & 1 & 0 \\ -1 & -2 & 1 \\ 0 & 0 & -1 \end{bmatrix}$ using Cayley–Hamilton theorem.

Solution: The Eigen values are obtained by solving the characteristic equation

$$|\lambda I - A| = 0 \quad or \quad \begin{vmatrix} \lambda+2 & -1 & 0 \\ 1 & \lambda+2 & -1 \\ 0 & 0 & \lambda+1 \end{vmatrix} = 0,$$

$$(\lambda+2)(\lambda^2+3\lambda+2)+(\lambda+1)=0 \quad or \; \lambda^3+5\lambda^2+9\lambda+5=0.$$

The matrix equation is $A^3+5A^2+9A+5I=0 \quad or \quad A^2+5A+9I+5A^{-1}=0$.

$$A^{-1} = -\frac{1}{5}\left(A^2+5A+9I\right)$$

$$A^2 = \begin{bmatrix} -2 & 1 & 0 \\ -1 & -2 & 1 \\ 0 & 0 & -1 \end{bmatrix} \begin{bmatrix} -2 & 1 & 0 \\ -1 & -2 & 1 \\ 0 & 0 & -1 \end{bmatrix} = \begin{bmatrix} 3 & -4 & 1 \\ 4 & 3 & -3 \\ 0 & 0 & 1 \end{bmatrix}$$

$$A^{-1} = -\frac{1}{5}\left\{\begin{bmatrix} 3 & -4 & 1 \\ 4 & 3 & -3 \\ 0 & 0 & 1 \end{bmatrix} + 5\begin{bmatrix} -2 & 1 & 0 \\ -1 & -2 & 1 \\ 0 & 0 & -1 \end{bmatrix} + 9\begin{bmatrix} 1 & 0 & 0 \\ 0 & 1 & 0 \\ 0 & 0 & 1 \end{bmatrix}\right\}.$$

Hence, $A^{-1} = -\frac{1}{5}\begin{bmatrix} 2 & 1 & 1 \\ -1 & 2 & 2 \\ 0 & 0 & 5 \end{bmatrix}$.

3.4.2 Reduction of the Degree of a Matrix Polynomial

Using Cayley–Hamilton theorem, a matrix polynomial of degree n can be reduced to one with degree $(n–1)$ or less. Consider a system matrix of size $n \times n$. If λ represents any Eigen value, then the characteristic polynomial associated with the matrix A is

$$P(\lambda) = \lambda^n + c_{n-1}\lambda^{n-1} + c_{n-2}\lambda^{n-2} + ... + c_1\lambda + c_0.$$

Consider another polynomial $F(\lambda)$ of degree m; $m > n$.

$$F(\lambda) = \lambda^m + a_{m-1}\lambda^{m-1} + a_{m-2}\lambda^{m-2} + ... + a_1\lambda + a_0$$

$$\frac{F(\lambda)}{P(\lambda)} = Q(\lambda) + \frac{R(\lambda)}{P(\lambda)},$$

where $Q(\lambda)$ is a polynomial of degree $(m - n)$ and $R(\lambda)$ is the remainder polynomial of degree $(n - 1)$.

Now $F(\lambda) = P(\lambda)Q(\lambda) + R(\lambda)$. $\hspace{2cm}$ (3.22)

Also consider a matrix polynomial in terms of A of degree m. This can be written as

$$F(A) = A^m + a_{m-1} A^{m-1} + a_{m-2} A^{m-2} + ... + a_1 A + a_0 I.$$

In a similar way to equation (3.22), this matrix polynomial can be written as

$$F(A) = P(A)Q(A) + R(A). \hspace{2cm} (3.23)$$

At the Eigen values, $P(\lambda) = 0$, which is the characteristic equation.

By Cayley–Hamilton theorem, every square matrix satisfies its own characteristic equation. Hence, $P(A) = 0$

From equation (3.23),

$$F(A) = R(A)$$

The polynomial $R(A)$ is, at most, of degree $(m - 1)$ and hence F(A) can be written as

$$F(A) = R(A) = \alpha_0 I + \alpha_1 A + \alpha_2 A^2 + ... + \alpha_{n-1} A^{m-1}. \hspace{1cm} (3.24)$$

Example 3.20 If $A = \begin{bmatrix} 0 & 1 \\ -2 & 3 \end{bmatrix}$, compute $F(A) = A^4 + 4A^3 + 3A^2 + 2A + I$, using Cayley–Hamilton theorem.

Solution: The characteristic polynomial is

$$|\lambda I - A| = 0 \hspace{0.5cm} or \hspace{0.5cm} \begin{vmatrix} \lambda & -1 \\ 2 & \lambda - 3 \end{vmatrix} = 0 \hspace{0.5cm} or \hspace{0.3cm} \lambda^2 - 3\lambda + 2 = 0 .$$

The polynomial $F(A) = A^4 + 4A^3 + 3A^2 + 2A + I$ is of degree 4.
From the Cayley–Hamilton theorem,

$$A^2 - 3A + 2I = 0 \hspace{0.3cm} or \hspace{0.3cm} A^2 = 3A - 2I$$

$$A^3 = A A^2 = A(3A - 2I) = 3A^2 - 2A = 3(3A - 2I) - 2A = 7A - 6I$$

$$A^4 = A^2 A^2 = (3A - 2I)(3A - 2I) = 9A^2 - 12A + 4I$$

$$= 9(3A - 2I) - 12A + 4I = 15A - 14I .$$

Hence, $F(A) = (15A-14I)+4(7A-6I)+3(3A-2I)+2A+I = 54A-43I$.

Now F(A) is of degree 1.

$$F(A) = 54\begin{bmatrix} 0 & 1 \\ -2 & 3 \end{bmatrix} - 43\begin{bmatrix} 1 & 0 \\ 0 & 1 \end{bmatrix} = \begin{bmatrix} -43 & 54 \\ -108 & 119 \end{bmatrix}$$

3.4.3 Evaluation of the Analytic Functions of a Square Matrix

An analytic function of a square matrix is defined by a similar series as its scalar counterpart. For example,

$$\sin x = x - \frac{x^3}{3!} + \frac{x^5}{5!} - \frac{x^7}{7!} + \dots .$$

If A is a square matrix, then an analytic function similar to this series is

$$\sin A = A - \frac{A^3}{3!} + \frac{A^5}{5!} - \frac{A^7}{7!} + \dots .$$

Similarly $\cos x = 1 - \frac{x^2}{2!} + \frac{x^4}{4!} - \frac{x^6}{6!} + \dots$

$$\text{and } \cos A = I - \frac{A^2}{2!} + \frac{A^4}{4!} - \frac{A^6}{6!} + \dots$$

$$e^x = 1 + x + \frac{x^2}{2!} + \frac{x^3}{3!} + \frac{x^4}{4!} + \dots$$

$$\text{and } e^{At} = I + At + \frac{A^2 t^2}{2!} + \frac{A^3 t^3}{3!} + \frac{A^4 t^4}{4!} + \dots .$$

Computation of an analytic function of a square matrix can be carried out using equation (3.24) and its scalar counterpart. The method is discussed here. Consider that A is an $(n \times n)$ matrix and the distinct Eigen values are $\lambda_1, \lambda_2, \lambda_3, \dots, \lambda_n$.

From equation (3.24),

$$F(A) = R(A) = \alpha_0 I + \alpha_1 A + \alpha_2 A^2 + \dots + \alpha_{n-1} A^{n-1}.$$

The corresponding scalar equation is

$$f(\lambda) = R(\lambda) = \alpha_0 + \alpha_1 \lambda + \alpha_2 \lambda^2 + \dots + \alpha_{n-1} \lambda^{n-1}. \tag{3.25}$$

Equation (3.25) represents n simultaneous equations corresponding to n different Eigen values, and solving these, all the α coefficients $\alpha_0, \alpha_1, \alpha_2, ---, \alpha_{n-1}$ can be determined.

Substituting these values in equation (3.24), the analytical function of the matrix A can be determined. Selecting $F(A) = e^{At}$, the state transition matrix can be evaluated using this method.

Example 3.21 Evaluate the analytical function cos A of the matrix $A = \begin{bmatrix} 0 & 1 \\ -2 & -3 \end{bmatrix}$.

Solution: For the square matrix A, the Eigen values are obtained by solving

$$|\lambda I - A| = 0 \qquad or \quad \begin{vmatrix} \lambda & -1 \\ 2 & \lambda+3 \end{vmatrix} = 0 \qquad or \quad \lambda^2 + 3\lambda + 2 = 0 .$$

The Eigen values are $\lambda_1 = -1$ and $\lambda_2 = -2$.

Let $\cos A = \alpha_0 I + \alpha_1 A$. The corresponding scalar equations are $\cos \lambda_1 = \alpha_0 + \alpha_1 \lambda_1$ and $\cos \lambda_2 = \alpha_0 + \alpha_1 \lambda_2$.

Solving these equations,

$$\alpha_0 = \frac{\lambda_1 \cos(\lambda_2) - \lambda_2 \cos(\lambda_1)}{\lambda_1 - \lambda_2} = \frac{-\cos(-2) + 2\cos(-1)}{-1+2} = 1.4967$$

$$\alpha_1 = \frac{\cos(\lambda_1) - \cos(\lambda_2)}{\lambda_1 - \lambda_2} = \frac{\cos(-1) - \cos(-2)}{-1+2} = 0.9565 .$$

Hence, $\cos A = 1.4967 I + 0.9565 A$

$$= 1.4967 \begin{bmatrix} 1 & 0 \\ 0 & 1 \end{bmatrix} + 0.9565 \begin{bmatrix} 0 & 1 \\ -2 & -3 \end{bmatrix} = \begin{bmatrix} 1.4967 & 0.9565 \\ -1.913 & -1.3728 \end{bmatrix} .$$

Example 3.22 If A is a (2 × 2) matrix with distinct Eigen values, determine the state transition matrix using the Cayley–Hamilton theorem.

Solution: Let the Eigen values be λ_1 and λ_2 and the scalar polynomial be $f(\lambda) = e^{\lambda t}$.

$$f(\lambda_1) = e^{\lambda_1 t} = \alpha_0 + \alpha_1 \lambda_1$$

$$f(\lambda_2) = e^{\lambda_2 t} = \alpha_0 + \alpha_1 \lambda_2 .$$

Solving these equations,

$$\alpha_0 = \frac{\lambda_1 e^{\lambda_2 t} - \lambda_2 e^{\lambda_1 t}}{\lambda_1 - \lambda_2} \qquad and \quad \alpha_1 = \frac{e^{\lambda_1 t} - e^{\lambda_2 t}}{\lambda_1 - \lambda_2} .$$

Now the analytical function of the matrix to be evaluated is

$$f(A) = e^{At} = \alpha_0 I + \alpha_1 A$$

$$= \frac{\lambda_1 e^{\lambda_2 t} - \lambda_2 e^{\lambda_1 t}}{\lambda_1 - \lambda_2} \begin{bmatrix} 1 & 0 \\ 0 & 1 \end{bmatrix} + \frac{e^{\lambda_1 t} - e^{\lambda_2 t}}{\lambda_1 - \lambda_2} A.$$

Example 3.23 Evaluate the state transition matrix for $A = \begin{bmatrix} -2 & 2 \\ 1 & -3 \end{bmatrix}$ using Cayley–Hamilton theorem.

Solution: The Eigen values are obtained by solving

$$|\lambda I - A| = 0 \quad or \quad \begin{vmatrix} \lambda + 2 & -2 \\ -1 & \lambda + 3 \end{vmatrix} = 0 \quad or \quad \lambda^2 + 5\lambda + 4 = 0 .$$

The Eigen values are $\lambda_1 = -1$ and $\lambda_2 = -4$.

$$f(\lambda_1) = e^{\lambda_1 t} = \alpha_0 + \alpha_1 \lambda_1 \qquad e^{-t} = \alpha_0 - \alpha_1$$

$$f(\lambda_2) = e^{\lambda_2 t} = \alpha_0 + \alpha_1 \lambda_2 \qquad e^{-4t} = \alpha_0 - 4\alpha_1$$

Solving these equations,

$$\alpha_0 = \frac{4}{3}e^{-t} - \frac{1}{3}e^{-4t} \qquad and \quad \alpha_1 = \frac{1}{3}(e^{-t} - e^{-4t}).$$

Hence, the state transition matrix,

$$f(A) = e^{At} = \alpha_0 I + \alpha_1 A$$

$$= \alpha_0 \begin{bmatrix} 1 & 0 \\ 0 & 1 \end{bmatrix} + \alpha_1 \begin{bmatrix} -2 & 2 \\ 1 & -3 \end{bmatrix} = \begin{bmatrix} \alpha_0 - 2\alpha_1 & 2\alpha_1 \\ \alpha_1 & \alpha_0 - 3\alpha_1 \end{bmatrix}.$$

$$e^{At} = \begin{bmatrix} \frac{2}{3}e^{-t} + \frac{1}{3}e^{-4t} & \frac{2}{3}e^{-t} - \frac{2}{3}e^{-4t} \\ \frac{1}{3}e^{-t} - \frac{1}{3}e^{-4t} & \frac{1}{3}e^{-t} + \frac{2}{3}e^{-4t} \end{bmatrix}.$$

Example 3.24 If A is a (2 × 2) matrix with repeated Eigen values, determine the state transition matrix using the Cayley–Hamilton theorem.

Solution: Let the Eigen values be λ_1 and λ_1 and the scalar polynomial be $f(\lambda) = e^{\lambda t}$.

$$f(\lambda) = e^{\lambda t} = \alpha_0 + \alpha_1 \lambda \tag{1}$$

$$f(\lambda_1) = e^{\lambda_1 t} = \alpha_0 + \alpha_1 \lambda_1,$$ (2)

Differentiating equation (1) with respect to λ,

$$\frac{d}{d\lambda} f(\lambda) = \frac{d}{d\lambda} e^{\lambda t} = \frac{d}{d\lambda}(\alpha_0 + \alpha_1 \lambda)$$

$$te^{\lambda t} = \alpha_1 \qquad or \ \alpha_1 = te^{\lambda_1 t}.$$ (3)

Substituting equation (3) in equation (2),

$$\alpha_0 = e^{\lambda_1 t} - \lambda_1 te^{\lambda_1 t}$$ (4)

$$e^{At} = \alpha_0 I + \alpha_1 A$$

$$= \left[e^{\lambda_1 t} - \lambda_1 te^{\lambda_1 t} \right] I + te^{\lambda t} A$$

$$= \left[e^{\lambda_1 t} - \lambda_1 te^{\lambda_1 t} \right] \begin{bmatrix} 1 & 0 \\ 0 & 1 \end{bmatrix} + te^{\lambda t} A.$$

Example 3.25 Evaluate the state transition matrix for $A = \begin{bmatrix} -2 & 1 \\ 0 & -2 \end{bmatrix}$ using Cayley–Hamilton theorem.

Solution: The Eigen values are obtained by solving

$$|\lambda I - A| = 0 \quad or \quad \begin{vmatrix} \lambda + 2 & -1 \\ 0 & \lambda + 2 \end{vmatrix} = 0 \quad or \ (\lambda + 2)(\lambda + 2) = 0 .$$

The Eigen values are $\lambda_1 = -2$ and $\lambda_2 = -2$; *repeated Eigen values.*

$$f(\lambda) = e^{\lambda t} = \alpha_0 + \alpha_1 \lambda$$ (1)

$$f(\lambda_1) = e^{\lambda_1 t} = \alpha_0 + \alpha_1 \lambda_1 \qquad e^{-2t} = \alpha_0 - 2\alpha_1 .$$ (2)

To get another equation, differentiate equation (1) with respect to λ and substitute $\lambda = -2$.

$$\frac{d}{d\lambda} f(\lambda) \Big|_{\lambda = -2} = \frac{d}{d\lambda} e^{\lambda t} \Big|_{\lambda = -2} = \frac{d}{d\lambda}(\alpha_0 + \alpha_1 \lambda) \Big|_{\lambda = -2}$$

$$te^{-2t} = \alpha_1$$ (3)

From equations (2) and (3),

$$\alpha_0 = e^{-2t} + 2\alpha_1 = e^{-2t} + 2te^{-2t} .$$ (4)

The state transition matrix is given by

$$f(A) = e^{At} = \alpha_0 I + \alpha_1 A$$

$$= \alpha_0 \begin{bmatrix} 1 & 0 \\ 0 & 1 \end{bmatrix} + \alpha_1 \begin{bmatrix} -2 & 1 \\ 0 & -2 \end{bmatrix} = \begin{bmatrix} \alpha_0 - 2\alpha_1 & \alpha_1 \\ 0 & \alpha_0 - 2\alpha_1 \end{bmatrix}$$

$$= \begin{bmatrix} e^{-2t} & te^{-2t} \\ 0 & e^{-2t} \end{bmatrix}.$$

Example 3.26 Evaluate the state transition matrix for $A = \begin{bmatrix} -2 & 1 & 0 \\ 0 & -2 & 1 \\ 0 & 0 & -2 \end{bmatrix}$ by Cayley–Hamilton theorem.

Solution: The matrix A is in Jordan form. The Eigen values are $\lambda = -2, -2, -2$ repeated thrice.

$$f(\lambda) = e^{\lambda t} = \alpha_0 + \alpha_1 \lambda + \alpha_2 \lambda^2 \tag{1}$$

$$e^{-2t} = \alpha_0 - 2\alpha_1 + 4\alpha_2. \tag{2}$$

Differentiating equation (1) with respect to λ,

$$\frac{d}{d\lambda} e^{\lambda t}\bigg|_{\lambda=-2} = (\alpha_1 + 2\alpha_2 \lambda)\big|_{\lambda=-2}$$

$$te^{\lambda t}\bigg|_{\lambda=-2} = (\alpha_1 + 2\alpha_2 \lambda)\big|_{\lambda=-2}$$

$$te^{-2t} = \alpha_1 - 4\alpha_2. \tag{3}$$

Differentiating equation (1) with respect to λ twice,

$$\frac{d^2}{d\lambda^2} e^{\lambda t}\bigg|_{\lambda=-2} = \frac{d}{d\lambda}(\alpha_1 + 2\alpha_2 \lambda)\bigg|_{\lambda=-2}$$

$$t^2 e^{-2t} = 2\alpha_2. \tag{4}$$

$$A^2 = \begin{bmatrix} -2 & 1 & 0 \\ 0 & -2 & 1 \\ 0 & 0 & -2 \end{bmatrix} \begin{bmatrix} -2 & 1 & 0 \\ 0 & -2 & 1 \\ 0 & 0 & -2 \end{bmatrix} = \begin{bmatrix} 4 & -4 & 1 \\ 0 & 4 & -4 \\ 0 & 0 & 4 \end{bmatrix}$$

$$F(A) = e^{At} = \alpha_0 I + \alpha_1 A + \alpha_2 A^2$$

$$
= \alpha_0 \begin{bmatrix} 1 & 0 & 0 \\ 0 & 1 & 0 \\ 0 & 0 & 1 \end{bmatrix} + \alpha_1 \begin{bmatrix} -2 & 1 & 0 \\ 0 & -2 & 1 \\ 0 & 0 & -2 \end{bmatrix} + \alpha_2 \begin{bmatrix} 4 & -4 & 1 \\ 0 & 4 & -4 \\ 0 & 0 & 4 \end{bmatrix}
$$

$$
= \begin{bmatrix} \alpha_0 - 2\alpha_1 + 4\alpha_2 & \alpha_1 - 4\alpha_2 & \alpha_2 \\ 0 & \alpha_0 - 2\alpha_1 + 4\alpha_2 & \alpha_1 - 4\alpha_2 \\ 0 & 0 & \alpha_0 - 2\alpha_1 + 4\alpha_2 \end{bmatrix}
$$

$$
= \begin{bmatrix} e^{-2t} & te^{-2t} & \frac{1}{2}t^2 e^{-2t} \\ 0 & e^{-2t} & te^{-2t} \\ 0 & 0 & e^{-2t} \end{bmatrix}.
$$

Method of Diagonalization

The method of diagonalization for finding the state transition matrix is discussed in Chapter 4.

3.5 SPECIAL CASES OF DETERMINING THE STATE TRANSITION MATRIX

If the system matrix is either in the diagonal form or in the Jordan canonical form, then the state transition matrix can be written down by inspection.

3.5.1 System matrix in Diagonal form

Consider that for a third-order system, the state model is

$$
\begin{bmatrix} \dot{x}_1 \\ \dot{x}_2 \\ \dot{x}_3 \end{bmatrix} = \begin{bmatrix} -\lambda_1 & 0 & 0 \\ 0 & -\lambda_2 & 0 \\ 0 & 0 & -\lambda_3 \end{bmatrix} \begin{bmatrix} x_1 \\ x_2 \\ x_3 \end{bmatrix}.
$$

where $-\lambda_1$, $-\lambda_2$ and $-\lambda_3$ are the distinct Eigen values.

The corresponding differential equations are

$$
\dot{x}_1 = -\lambda_1 x_1 \tag{3.26}
$$

$$
\dot{x}_2 = -\lambda_2 x_2 \tag{3.27}
$$

$$\dot{x}_3 = -\lambda_3 \, x_3 \tag{3.28}$$

Solutions to these can be obtained as

$$x_1 = e^{-\lambda_1 t} x_1(0) \tag{3.29}$$

$$x_2 = e^{-\lambda_2 t} x_2(0) \tag{3.30}$$

$$x_3 = e^{-\lambda_3 t} x_3(0) . \tag{3.31}$$

Keeping in matrix form,

$$\begin{bmatrix} x_1 \\ x_2 \\ x_3 \end{bmatrix} = \begin{bmatrix} e^{-\lambda_1 t} & 0 & 0 \\ 0 & e^{-\lambda_2 t} & 0 \\ 0 & 0 & e^{-\lambda_3 t} \end{bmatrix} \begin{bmatrix} x_1(0) \\ x_2(0) \\ x_3(0) \end{bmatrix} . \tag{3.32}$$

This can be expressed as

$$X(t) = e^{At} X(0) \tag{3.33}$$

Comparing equations (3.32) and (3.33),

$$e^{At} = \begin{bmatrix} e^{-\lambda_1 t} & 0 & 0 \\ 0 & e^{-\lambda_2 t} & 0 \\ 0 & 0 & e^{-\lambda_3 t} \end{bmatrix} .$$

3.5.2 System matrix in Jordan form

Consider a state equation in Jordan canonical form

$$\dot{X} = J X . \tag{3.34}$$

For a third-order system, the state model in Jordan form can be written as

$$\begin{bmatrix} \dot{x}_1 \\ \dot{x}_2 \\ \dot{x}_3 \end{bmatrix} = \begin{bmatrix} -\lambda_1 & 1 & 0 \\ 0 & -\lambda_1 & 1 \\ 0 & 0 & -\lambda_1 \end{bmatrix} \begin{bmatrix} x_1 \\ x_2 \\ x_3 \end{bmatrix} .$$

with all the Eigen values at $\lambda = \lambda_1$.

The corresponding differential equations are

$$\dot{x}_1 = -\lambda_1 x_1 + x_2 \qquad (3.35)$$

$$\dot{x}_2 = -\lambda_1 x_2 + x_3 \qquad (3.36)$$

$$\dot{x}_3 = -\lambda_1 x_3 \qquad (3.37)$$

The solution to equation (3.37) is

$$x_3 = e^{-\lambda_1 t} x_3(0). \qquad (3.38)$$

Substituting equation (3.38) in equation (3.36),

$$\dot{x}_2 = -\lambda_1 x_2 + e^{-\lambda_1 t} x_3(0). \qquad (3.39)$$

Solving this first-order linear differential equation,

$$x_2 = e^{-\lambda_1 t} x_2(0) + t e^{-\lambda_1 t} x_3(0). \qquad (3.40)$$

Substituting equation (3.40) in equation (3.35),

$$\dot{x}_1 = -\lambda_1 x_1 + e^{-\lambda_1 t} x_2(0) + t e^{-\lambda_1 t} x_3(0). \qquad (3.41)$$

Solving this first-order linear differential equation,

$$x_1 = e^{-\lambda_1 t} x_1(0) + t e^{-\lambda_1 t} x_2(0) + \frac{1}{2} t^2 e^{-\lambda_1 t} x_3(0). \qquad (3.42)$$

Keeping equations (3.38), (3.40) and (3.42) in matrix form,

$$\begin{bmatrix} x_1 \\ x_2 \\ x_3 \end{bmatrix} = \begin{bmatrix} e^{-\lambda_1 t} & t e^{-\lambda_1 t} & \frac{1}{2} t^2 e^{-\lambda_1 t} \\ 0 & e^{-\lambda_1 t} & t e^{-\lambda_1 t} \\ 0 & 0 & e^{-\lambda_1 t} \end{bmatrix} \begin{bmatrix} x_1(0) \\ x_2(0) \\ x_3(0) \end{bmatrix} \qquad (3.43)$$

This can be written as

$$X(t) = e^{Jt} X(0). \qquad (3.44)$$

Comparing equations (3.43) and (3.44), the state transition matrix is given by

$$e^{Jt} = \begin{bmatrix} e^{-\lambda_1 t} & t e^{-\lambda_1 t} & \frac{1}{2} t^2 e^{-\lambda_1 t} \\ 0 & e^{-\lambda_1 t} & t e^{-\lambda_1 t} \\ 0 & 0 & e^{-\lambda_1 t} \end{bmatrix}.$$

This can be written as

$$
e^{Jt} = \begin{bmatrix} e^{\lambda_1 t} & \dfrac{d}{d\lambda}e^{\lambda t}\Big|_{\lambda=\lambda_1} & \dfrac{1}{2!}\dfrac{d^2}{d\lambda^2}e^{\lambda t}\Big|_{\lambda=\lambda_1} \\ 0 & e^{\lambda_1 t} & \dfrac{d}{d\lambda}e^{\lambda t}\Big|_{\lambda=\lambda_1} \\ 0 & 0 & e^{\lambda_1 t} \end{bmatrix}.
$$

In general for an n-th order system, with n repeated Eigen values,

$$
e^{At} = \begin{bmatrix} e^{-\lambda_1 t} & te^{-\lambda_1 t} & \dfrac{1}{2!}t^2 e^{-\lambda_1 t} & \dfrac{1}{3!}t^3 e^{-\lambda_1 t} & \cdots & \dfrac{1}{n!}t^n e^{-\lambda_1 t} \\ 0 & e^{-\lambda_1 t} & te^{-\lambda_1 t} & \dfrac{1}{2!}t^2 e^{-\lambda_1 t} & \cdots & \dfrac{1}{(n-1)!}t^{n-1} e^{-\lambda_1 t} \\ 0 & 0 & e^{-\lambda_1 t} & te^{-\lambda_1 t} & \cdots & \dfrac{1}{(n-2)!}t^{n-2} e^{-\lambda_1 t} \\ \cdots & & & & & \\ \cdots & & & & & \\ 0 & 0 & 0 & 0 & \cdots & e^{-\lambda_1 t} \end{bmatrix}.
$$

Example 3.27 Repeat Example 3.25, considering that A is in Jordan canonical form.

Solution: $A = \begin{bmatrix} -2 & 1 \\ 0 & -2 \end{bmatrix}$

The Eigen values are $\lambda = -2$, repeated twice.

The state transition matrix is given by

$$
e^{At} = \begin{bmatrix} e^{\lambda_1 t} & \dfrac{d}{d\lambda}e^{\lambda t}\Big|_{\lambda=-2} \\ 0 & e^{\lambda_1 t} \end{bmatrix} = \begin{bmatrix} e^{-2t} & te^{-2t} \\ 0 & e^{-2t} \end{bmatrix}.
$$

Example 3.28 Determine the state transition matrix for the system matrix

$$
A = \begin{bmatrix} -2 & 1 & 0 \\ 0 & -2 & 1 \\ 0 & 0 & -2 \end{bmatrix}.
$$

Solution: The given matrix is Jordan canonical form with repeated Eigen values $\lambda = \lambda_1 = -2$, repeated thrice. In this case, the state transition matrix is given by

$$
e^{At} =
\begin{bmatrix}
e^{\lambda_1 t} & \left.\dfrac{d}{d\lambda}e^{\lambda t}\right|_{\lambda=\lambda_1} & \left.\dfrac{1}{2!}\dfrac{d^2}{d\lambda^2}e^{\lambda t}\right|_{\lambda=\lambda_1} \\
0 & e^{\lambda_1 t} & \left.\dfrac{d}{d\lambda}e^{\lambda t}\right|_{\lambda=\lambda_1} \\
0 & 0 & e^{\lambda_1 t}
\end{bmatrix}
=
\begin{bmatrix}
e^{-2t} & te^{-2t} & \dfrac{1}{2}t^2 e^{-2t} \\
0 & e^{-2t} & te^{-2t} \\
0 & 0 & e^{-2t}
\end{bmatrix}.
$$

SUMMARY

- A state equation can be solved using Laplace transform method.
- A state equation can be solved by actual integration.
- For a SISO system, the zero input response is $y_{ZIR}(t) = C\,e^{At}\,X(0)$.
- For a SISO system, the zero state response due to a unit impulse input is $y_{ZSR}(t) = C\,e^{At}\,B$.
- For a SISO system, the zero state response due to a unit step input is $y_{ZSR}(t) = C\,A^{-1}[e^{At} - I]B$.
- The resolvent matrix of a square matrix A is $[sI - A]^{-1}$.
- The state transition matrix for any square matrix can be determined by using Laplace transform method, power series method, Sylvester's interpolation formula, Cayley–Hamilton theorem or by diagonalization method.
- Power series method of finding the state transition matrix of a square matrix is very time consuming and quite often may not yield a closed form of solution.
- Cayley–Hamilton theorem states that every square matrix satisfies its own characteristic equation.
- Cayley–Hamilton theorem can be used to find the inverse of a square matrix, if it exists.
- Using Cayley–Hamilton theorem, a matrix polynomial of degree n can be reduced to one of degree $(n - 1)$ or less.
- An analytic function of a square matrix is defined by a similar transcendental series as its scalar counterpart.
- Cayley–Hamilton theorem can be used to evaluate an analytical function of a square matrix.
- Cayley–Hamilton theorem can be used to evaluate the state transition matrix of a square matrix.

PRACTICE PROBLEMS

PP 3.1 A state model for a linear time invariant system is

$$\dot{X} = \begin{bmatrix} -1 & 2 \\ 0 & -3 \end{bmatrix} X + \begin{bmatrix} 1 \\ -1 \end{bmatrix} r \qquad y = \begin{bmatrix} 1 & 2 \end{bmatrix} X \qquad X(0) = \begin{bmatrix} 1 \\ 0 \end{bmatrix}$$

Find the ZIR using Laplace transform method.

PP 3.2 For the PP3.1 determine the ZSR for a unit impulse input, using Laplace transform method.

PP 3.3 For the problem 3.1, determine the ZSR for a unit step input, using Laplace transform method.

PP 3.4 A state model for a linear time-invariant system is

$$\dot{X} = \begin{bmatrix} -2 & 1 \\ -2 & -5 \end{bmatrix} X + \begin{bmatrix} 1 \\ 2 \end{bmatrix} r \qquad y = \begin{bmatrix} 2 & -1 \end{bmatrix} X \qquad X(0) = \begin{bmatrix} 1 \\ -1 \end{bmatrix}$$

Find the ZIR, ZSR due to a unit step input and the total response using Laplace transform method.

PP 3.5 A system is represented by a state model

$$\dot{X} = \begin{bmatrix} -1 & 1 \\ -2 & -3 \end{bmatrix} X + \begin{bmatrix} 2 \\ 1 \end{bmatrix} r \qquad y = \begin{bmatrix} 1 & 2 \end{bmatrix} X \qquad X(0) = \begin{bmatrix} 1 \\ 1 \end{bmatrix}.$$

Determine the unit impulse response by Laplace transform method.

PP 3.6 For the system of PP3.5, determine the unit step response by Laplace transform method.

PP 3.7 A linear system is represented by a state model

$$\dot{X} = \begin{bmatrix} -1 & 1 \\ -1 & -3 \end{bmatrix} X + \begin{bmatrix} 1 \\ 1 \end{bmatrix} r \qquad y = \begin{bmatrix} 2 & 1 \end{bmatrix} X \qquad X(0) = \begin{bmatrix} 1 \\ 0 \end{bmatrix}.$$

Determine the response of the system to a unit step input by Laplace transform method.

PP 3.8 A state space model for a system is

$$\dot{X} = \begin{bmatrix} -6 & 1 & 0 \\ -11 & 0 & 1 \\ -6 & 0 & 0 \end{bmatrix} X + \begin{bmatrix} 3 \\ 2 \\ 1 \end{bmatrix} r \qquad y = \begin{bmatrix} 1 & 0 & 0 \end{bmatrix} X \qquad X(0) = \begin{bmatrix} 1 \\ 1 \\ -1 \end{bmatrix}.$$

Find the unit step response by Laplace transform method.

PP 3.9 The state model of a system in diagonal canonical form is

$$\dot{X} = \begin{bmatrix} -1 & 0 \\ 0 & -2 \end{bmatrix} X + \begin{bmatrix} 1 \\ 1 \end{bmatrix} r \qquad y = \begin{bmatrix} 2 & -1 \end{bmatrix} X \qquad X(0) = \begin{bmatrix} 1 \\ 0 \end{bmatrix}.$$

Find the state transition matrix and hence find the unit step response by time domain method.

PP 3.10 A linear system is represented by a state model

$$\dot{X} = \begin{bmatrix} -3 & 1 \\ -2 & 0 \end{bmatrix} X + \begin{bmatrix} 2 \\ 1 \end{bmatrix} r \qquad y = [1 \quad 1] X \qquad X(0) = \begin{bmatrix} 1 \\ 0 \end{bmatrix}.$$

Find the state transition matrix and hence find the unit impulse response by time domain method.

PP 3.11 For the system of PP3.10, determine the unit step response by using time domain method.

PP 3.12 A linear system is represented by a state model

$$\dot{X} = \begin{bmatrix} -1 & 1 \\ -2 & -1 \end{bmatrix} X + \begin{bmatrix} 1 \\ 1 \end{bmatrix} r \qquad y = [1 \quad 0] X \qquad X(0) = \begin{bmatrix} 0 \\ 1 \end{bmatrix}.$$

Find the state transition matrix and hence find the unit step response by time domain method.

PP 3.13 For a square matrix $A = \begin{bmatrix} -1 & 1 \\ 0 & -2 \end{bmatrix}$, find the state transition matrix using the properties

$$\frac{d}{dt} e^{At} = A e^{At} \qquad and \quad e^{A(0)} = I.$$

PP 3.14 For a square matrix $A = \begin{bmatrix} 0 & 5 \\ -1 & -2 \end{bmatrix}$, find the state transition matrix by using (a) Laplace transform method (b) Sylvester's interpolation formula and (c) Cayley–Hamilton theorem.

PP 3.15 For a square matrix $A = \begin{bmatrix} -4 & 1 \\ -4 & 0 \end{bmatrix}$, find the state transition matrix by using (a) Laplace transform method (b) Sylvester's interpolation formula and (c) Cayley–Hamilton theorem.

PP 3.16 For a square matrix $A = \begin{bmatrix} 0 & -8 \\ 1 & -6 \end{bmatrix}$, find the state transition matrix by using (a) Laplace transform method (b) Sylvester's interpolation formula and (c) Cayley–Hamilton theorem.

PP 3.17 For a square matrix $A = \begin{bmatrix} 0 & -8 \\ 1 & -6 \end{bmatrix}$, find its inverse using Cayley–Hamilton theorem.

PP 3.18 Find the inverse of the given matrix using Cayley–Hamilton theorem.

$$A = \begin{bmatrix} -5 & -6 & 0 \\ 2 & 2 & 0 \\ 0 & 0 & -3 \end{bmatrix}$$

PP 3.19 If $A = \begin{bmatrix} 0 & -8 \\ 1 & -6 \end{bmatrix}$, compute $f(A) = A^4 + 3A^3 + 2A^2 + A + I$ using Cayley–Hamilton theorem.

PP 3.20 Compute the analytical functions (a) sin A and (b) cos A for $A = \begin{bmatrix} -1 & 1 \\ 0 & -2 \end{bmatrix}$

PP 3.21 Find the inverse of the given matrix using Cayley–Hamilton theorem.

$$A = \begin{bmatrix} -1 & 1 & 0 \\ 0 & -1 & 1 \\ 0 & 0 & -1 \end{bmatrix}$$

PP 3.22 If $A = \begin{bmatrix} 0 & -3 & 0 \\ 3 & 0 & 0 \\ 0 & 0 & -1 \end{bmatrix}$, find the state transition matrix using Cayley–Hamilton

theorem. Verify the result by Laplace transform method.

REVIEW QUESTIONS

3.1   Why is e^{At} called a matrix exponential?

3.2   Why is e^{At} called the state transition matrix?

3.3   A single-input/single-output (SISO) system is represented by the state model

$$\dot{X} = AX + Br \qquad\qquad y = CX$$

Derive the expression for the zero input response (ZIR).

3.4   A SISO system is represented by the state model

$$\dot{X} = AX + Br \qquad\qquad y = CX$$

Derive the expression for the zero state response (ZSR).

3.5   Show that for a SISO system, defined by

$$\dot{X} = AX + Br \qquad\qquad y = CX ,$$

the zero input response is given by

$$y_{ZIR}(t) = C\, e^{At}\, X(0) .$$

3.6   Show that for a SISO system, defined by

$$\dot{X} = AX + Br \qquad\qquad y = CX ,$$

the zero state response due to a unit impulse input is given by

$$y_{ZSR}(t) = C e^{At}\, B .$$

3.7 Show that for a SISO system, defined by

$$\dot{X} = AX + Br \qquad y = CX,$$

the zero state response due to a unit step input is given by

$$y_{ZSR}(t) = CA^{-1}[e^{At} - I]B.$$

3.8 What is meant by a resolvent matrix of a square matrix?

3.9 Explain the method of obtaining the state transition matrix of a square matrix by Laplace transform method. Illustrate the method by means of an example.

3.10 Explain the method of obtaining the state transition matrix of a square matrix by power series method. Illustrate the method by means of an example.

3.11 Explain the method of obtaining the state transition matrix of a square matrix by using Sylvester's interpolation formula. Illustrate the method by means of an example.

3.12 State and explain the Cayley–Hamilton theorem.

3.13 Explain the method of finding the inverse of a nonsingular matrix by Cayley–Hamilton theorem.

3.14 Using Cayley–Hamilton theorem, explain how a matrix polynomial of degree n can be reduced to one of degree $(n-1)$ or less.

3.15 What is meant by an analytic function of a square matrix?

3.16 Explain the method to evaluate an analytic function of a square matrix.

3.17 Explain the method of obtaining the state transition matrix of a square matrix, with distinct Eigen values, by using Cayley–Hamilton theorem. Illustrate the method by means of an example.

3.18 Explain the method of obtaining the state transition matrix of a square matrix, with repeated Eigen values, by using Cayley–Hamilton theorem. Illustrate the method by means of an example.

4 Similarity Transformations

4.1 SIMILARITY TRANSFORMATION

As mentioned in Chapter 2, canonical models have certain advantages over other state models. Often, it is convenient to represent the system in canonical forms for the simplicity of design and analysis. A non-canonical form of state model can be converted into different canonical forms by linear transformation. These systems are called *similar systems* and the method of obtaining these is called *similarity transformation*. The method is based on selecting a transformation matrix that transforms the given system matrix into a canonical form.

4.2 METHOD OF SIMILARITY TRANSFORMATION

Two $(n \times n)$ matrices A_X and A_Z are said to be similar if a non-singular matrix P exists such that

$$A_Z = [P^{-1} A_X P]$$

The matrix A_Z is said to be obtained from A_X by a similarity transformation and P is called the transformation matrix. It was discussed in Chapter 2 that different state models can be obtained from the same transfer function representing a system. Hence, a state model for a system is not unique. Consider the matrix equation

$$X = PZ.$$

The matrix P transforms a vector Z into a vector X. This transformation is valid only when the inverse of P exists. Let the state model for a SISO system be

$$\dot{X} = AX + Br \qquad (4.1)$$

and

$$y = CX \qquad (4.2)$$

Consider that this state model is to be converted into another state model and let the transformation be $X(t) = P Z(t)$, where Z(t) is the new state vector and P is the transformation matrix.

$$\dot{X} = P \dot{Z} \qquad (4.3)$$

From equation (4.1) and (4.3)

$$P \dot{Z} = APZ + Br$$

$$\dot{Z} = [P^{-1}AP]Z + [P^{-1}B]r = A_z Z + B_z r \qquad (4.4)$$

$A_z = [P^{-1}AP]$ is the new system matrix and $B_z = [P^{-1}B]$ is the new input matrix.

$$y = [CP]Z \qquad (4.5)$$

$C_z = [CP]$ is the new output matrix.

Equations (4.4) and (4.5) show the non-uniqueness of a state model. The set of equations (4.1) and (4.2) and the set (4.4) and (4.5) represent the same system. A system can have different state space representations, but these models have the same transfer function and Eigen values.

4.3 EIGEN VECTORS

Consider a square matrix A and let λ represents its Eigen values. A column vector P that satisfies the relation $AP = \lambda P$ is called the Eigen vector. For every Eigen value, there is an associated Eigen vector. Eigen vectors can be determined by any one of the following methods.

Method 1: By solving the equation $AP = \lambda P$

For an Eigen value λ_i,

$$AP = \lambda_i P$$

This gives

$$\left[\lambda_i I - A \right] P_i = \left[0 \right]$$

Consider a (3 × 3) system matrix with distinct Eigen values λ_1, λ_2 and λ_3. Let the Eigen vector associated with these Eigen values be

$$P_1 = \begin{bmatrix} p_{11} \\ p_{21} \\ p_{31} \end{bmatrix}, \; P_2 = \begin{bmatrix} p_{12} \\ p_{22} \\ p_{32} \end{bmatrix} \; and \; P_3 = \begin{bmatrix} p_{13} \\ p_{23} \\ p_{33} \end{bmatrix}.$$

Solving the equation

$$[\lambda I - A]_{\lambda = \lambda_1} \begin{bmatrix} p_{11} \\ p_{21} \\ p_{31} \end{bmatrix} = \begin{bmatrix} 0 \\ 0 \\ 0 \end{bmatrix},$$

$P_1 = \begin{bmatrix} p_{11} \\ p_{21} \\ p_{31} \end{bmatrix}$ can be obtained.

Similarly $P_2 = \begin{bmatrix} p_{12} \\ p_{22} \\ p_{32} \end{bmatrix}$ can be obtained by solving the equation

$$[\lambda I - A]_{\lambda = \lambda_2} \begin{bmatrix} p_{12} \\ p_{22} \\ p_{32} \end{bmatrix} = \begin{bmatrix} 0 \\ 0 \\ 0 \end{bmatrix}.$$

In a similar way, $P_3 = \begin{bmatrix} p_{13} \\ p_{23} \\ p_{33} \end{bmatrix}$ can be obtained by solving the equation

$$[\lambda I - A]_{\lambda = \lambda_2} \begin{bmatrix} p_{13} \\ p_{23} \\ p_{33} \end{bmatrix} = \begin{bmatrix} 0 \\ 0 \\ 0 \end{bmatrix}.$$

$$P = [P_1 \quad P_2 \quad P_3] = \begin{bmatrix} p_{11} & p_{12} & p_{13} \\ p_{21} & p_{22} & p_{23} \\ p_{31} & p_{32} & p_{33} \end{bmatrix}.$$

Method 2: Select any column of {adjoint $[\lambda_i I - A]$}.

This is the same as the cofactors of any one row of $[\lambda_i I - A]$, where λ_i represents the Eigen values. Then for each Eigen value, the corresponding Eigen vector can be found out. In the Eigen vector obtained, the common factor, if any, in the column elements, can be removed. These methods are illustrated in the following examples.

Example 4.1

For a square matrix $A = \begin{bmatrix} -4 & 2 \\ -1 & -1 \end{bmatrix}$, determine the Eigen values and the associated Eigen vectors.

Solution: The Eigen values are obtained by solving the equation

$$|\lambda I - A| = 0 \quad or \quad \begin{vmatrix} \lambda + 4 & -2 \\ 1 & \lambda + 1 \end{vmatrix} = 0 \quad or \quad \lambda^2 + 5\lambda + 6 = 0 .$$

The Eigen values are $\lambda_1 = -2$ and $\lambda_2 = -3$

The Eigen vectors are determined from

$$[\lambda I - A]P = [0] \quad or \quad \begin{bmatrix} \lambda + 4 & -2 \\ 1 & \lambda + 1 \end{bmatrix} P = [0] .$$

Let the Eigen vector associated with the Eigen value $\lambda_1 = -2$ be $P_1 = \begin{bmatrix} p_{11} \\ p_{21} \end{bmatrix}$. Then

$$\begin{bmatrix} \lambda + 4 & -2 \\ 1 & \lambda + 1 \end{bmatrix}_{\lambda = -2} \begin{bmatrix} p_{11} \\ p_{21} \end{bmatrix} = [0] .$$

These equations are

$$2 p_{11} - 2 p_{21} = 0 \quad and \quad p_{11} - p_{21} = 0 .$$

These equations cannot be uniquely solved to get the values of p_{11} and p_{21}. The method is to assign any value for p_{11} or p_{21} and get the other value.

Choose $p_{11} = 1$ so that $p_{21} = 1$. Hence $P_1 = \begin{bmatrix} 1 \\ 1 \end{bmatrix}$.

Let the Eigen vector associated with the Eigen value $\lambda_2 = -3$ be $P_2 = \begin{bmatrix} p_{12} \\ p_{22} \end{bmatrix}$.

$$\begin{bmatrix} \lambda + 4 & -2 \\ 1 & \lambda + 1 \end{bmatrix}_{\lambda = -3} \begin{bmatrix} p_{12} \\ p_{22} \end{bmatrix} = [0]$$

These equations are

$$p_{12} - 2 p_{22} = 0 \quad and \quad p_{12} - 2 p_{22} = 0 .$$

These equations cannot be uniquely solved to get the values of p_{12} and p_{22}. Assigning any value for p_{12} or p_{22}, the other value can be obtained.

Choose $p_{12} = 2$ so that $p_{22} = 1$. Hence $P_2 = \begin{bmatrix} 2 \\ 1 \end{bmatrix}$

Example 4.2 For a square matrix $A = \begin{bmatrix} 0 & 1 \\ -2 & -3 \end{bmatrix}$, determine the Eigen values and the associated Eigen vectors.

Solution: The Eigen values are obtained by solving the equation

$$|\lambda I - A| = 0 \qquad or \begin{vmatrix} \lambda & -1 \\ 2 & \lambda+3 \end{vmatrix} = 0 \qquad or \quad \lambda^2 + 3\lambda + 2 = 0 .$$

The Eigen values are $\lambda_1 = -1 \ and \ \lambda_2 = -2$.
The Eigen vectors are determined from

$$[\lambda I - A]P = [0] \quad or \quad \begin{bmatrix} \lambda & -1 \\ 2 & \lambda+3 \end{bmatrix} P = [0]$$

Let the Eigen vector associated with the Eigen value $\lambda_1 = -1$ be $P_1 = \begin{bmatrix} p_{11} \\ p_{21} \end{bmatrix}$.

$$\begin{bmatrix} \lambda & -1 \\ 2 & \lambda+3 \end{bmatrix}_{\lambda=-1} \begin{bmatrix} p_{11} \\ p_{21} \end{bmatrix} = [0]$$

These equations are

$$-p_{11} - p_{21} = 0 \text{ and } 2p_{11} + 2p_{21} = 0$$

Choose $p_{11} = 1$ so that $p_{21} = -1$. Hence $P_1 = \begin{bmatrix} 1 \\ -1 \end{bmatrix}$.

Let the Eigen vector associated with the Eigen value $\lambda_2 = -2$ be

$$P_2 = \begin{bmatrix} p_{12} \\ p_{22} \end{bmatrix}.$$

$$\begin{bmatrix} \lambda & -1 \\ 2 & \lambda+3 \end{bmatrix}_{\lambda=-2} \begin{bmatrix} p_{12} \\ p_{22} \end{bmatrix} = [0].$$

These equations are

$$-2p_{12} - p_{22} = 0 \text{ and } 2p_{12} + p_{22} = 0$$

Choose $p_{12} = 1$ so that $p_{22} = -2$. Hence $P_2 = \begin{bmatrix} 1 \\ -2 \end{bmatrix}$.

Example 4.3 For a square matrix $A = \begin{bmatrix} 0 & 1 & 0 \\ 3 & 0 & 2 \\ -12 & -7 & -6 \end{bmatrix}$, determine the Eigen values and the associated Eigen vectors.

Solution: The Eigen values are obtained by solving the equation

$$|\lambda I - A| = 0 \qquad or \begin{vmatrix} \lambda & -1 & 0 \\ -3 & \lambda & -2 \\ 12 & 7 & \lambda+6 \end{vmatrix} = 0$$

$$\lambda(\lambda^2 + 6\lambda + 14) + (-3\lambda - 18 + 24) = 0$$

$$\lambda^3 + 6\lambda^2 + 11\lambda + 6 = 0$$

$$(\lambda+1)(\lambda+2)(\lambda+3) = 0 .$$

The Eigen values are $\lambda = -1$, $\lambda = -2$ and $\lambda = -3$.

The Eigen vectors can be determined using the cofactors of any one row of $\left[\lambda_i I - A\right]$.

$$[\lambda I - A] = \begin{bmatrix} \lambda & -1 & 0 \\ -3 & \lambda & -2 \\ 12 & 7 & \lambda+6 \end{bmatrix} .$$

The cofactors of the first row of $[\lambda I - A]$ are

$$P_i = \begin{bmatrix} +(\lambda^2 + 6\lambda + 14) \\ -(-3\lambda+6) \\ +(-21-12\lambda) \end{bmatrix} = \begin{bmatrix} \lambda^2 + 6\lambda + 14 \\ 3\lambda - 6 \\ -12\lambda - 21 \end{bmatrix}$$

For $\lambda = -1$, $P_1 = \begin{bmatrix} 9 \\ -9 \\ -9 \end{bmatrix}$, this can be reduced to $P_1 = \begin{bmatrix} 1 \\ -1 \\ -1 \end{bmatrix}$.

For $\lambda = -2$, $P_2 = \begin{bmatrix} 6 \\ -12 \\ 3 \end{bmatrix}$, this can be reduced to $P_2 = \begin{bmatrix} 2 \\ -4 \\ 1 \end{bmatrix}$.

For $\lambda = -3$, $P_3 = \begin{bmatrix} 5 \\ -15 \\ 15 \end{bmatrix}$, this can be reduced to $P_3 = \begin{bmatrix} 1 \\ -3 \\ 3 \end{bmatrix}$.

4.4 GENERALIZED EIGEN VECTORS

When the Eigen values of a square matrix are distinct, the matrix is said to have a full set of Eigen vectors. When the Eigen values are repeated, the matrix will not have a full set of Eigen vectors. Consider a (3×3) square matrix with the Eigen values $\lambda = \lambda_1, \lambda_2, \lambda_2$, with one repeated Eigen value. Corresponding to $\lambda = \lambda_1$ and $\lambda = \lambda_2$, two Eigen vectors can be determined using a method explained earlier. There is an additional Eigen vector corresponding to the repeated Eigen value and is called a *generalized Eigen vector*.

Let P_i = Cofactors of any one row of $\left[\lambda I - A\right]$

$$P_1 = [P_i]_{\lambda=\lambda_1}$$

$$P_2 = [P_i]_{\lambda=\lambda_2} .$$

The generalized Eigen vector corresponding to the repeated Eigen value $\lambda = \lambda_2$ can be obtained as

$$P_3 = \frac{d}{d\lambda}[P_i]_{\lambda=\lambda_2}.$$

If there are m repeated Eigen values, the Eigen vector and the remaining generalized Eigen vectors can be determined as given below:

$$P_i = \text{Cofactors of any one row of } [\lambda I - A]$$

$$P_1 = [P_i]_{\lambda=\lambda_1}$$

$$P_2 = \frac{d}{d\lambda}[P_i]_{\lambda=\lambda_1}$$

$$P_3 = \frac{1}{2!}\frac{d^2}{d\lambda^2}[P_i]_{\lambda=\lambda_1}$$

$$\cdots$$
$$\cdots$$

$$P_m = \frac{1}{(m-1)!}\frac{d^{m-1}}{d\lambda^{m-1}}[P_i]_{\lambda=\lambda_1}.$$

$P_2, P_3, P_4, \ldots P_m$ are generalized Eigen vectors.
Generalized Eigen vectors can also be found using the following relations:

$$P_1 = [P_i]_{\lambda=\lambda_1}$$

$$[\lambda I - A]_{\lambda=\lambda_1} P_2 = -P_1$$

$$[\lambda I - A]_{\lambda=\lambda_1} P_3 = -P_2$$

$$[\lambda I - A]_{\lambda=\lambda_1} P_4 = -P_3$$

$$\cdots$$
$$\cdots$$

$$[\lambda I - A]_{\lambda=\lambda_1} P_m = -P_{m-1}.$$

Example 4.4 For a square matrix $A = \begin{bmatrix} -1 & 2 \\ 0 & -1 \end{bmatrix}$, determine the Eigen values and the associated Eigen vectors.

Solution: The Eigen values are obtained by solving the equation

$$|\lambda I - A| = 0 \qquad or \begin{vmatrix} \lambda+1 & -2 \\ 0 & \lambda+1 \end{vmatrix} = 0 \qquad or \quad (\lambda+1)(\lambda+1) = 0 .$$

The Eigen values are $\lambda_1 = -1$ and $\lambda_2 = -1$, with repeated Eigen values.

Let the Eigen vector associated with one Eigen value $\lambda_1 = -1$ be $P_1 = \begin{bmatrix} p_{11} \\ p_{21} \end{bmatrix}$.
Cofactors of the second row of $[\lambda I - A]$ are

$$P_i = \begin{bmatrix} -(-2) \\ +(\lambda+1) \end{bmatrix} = \begin{bmatrix} 2 \\ \lambda+1 \end{bmatrix}$$

$$P_1 = \begin{bmatrix} 2 \\ \lambda+1 \end{bmatrix}_{\lambda=-1} = \begin{bmatrix} 2 \\ 0 \end{bmatrix}.$$

The generalized Eigen vector corresponding to the repeated Eigen value $\lambda_2 = -1$ is

$$P_2 = \begin{bmatrix} p_{12} \\ p_{22} \end{bmatrix} = \frac{d}{d\lambda}[P_i] = \frac{d}{d\lambda}\begin{bmatrix} 2 \\ \lambda+1 \end{bmatrix}_{\lambda=-1} = \begin{bmatrix} 0 \\ 1 \end{bmatrix}.$$

Example 4.5 For a square matrix $A = \begin{bmatrix} -3 & 2 \\ -2 & 1 \end{bmatrix}$, determine the Eigen values and the associated Eigen vectors.

Solution: The Eigen values are obtained by solving the equation

$$|\lambda I - A| = 0 \qquad or \begin{vmatrix} \lambda+3 & -2 \\ 2 & \lambda-1 \end{vmatrix} = 0 \qquad or \quad \lambda^2 + 2\lambda + 1 = 0 .$$

The Eigen values are $\lambda_1 = -1$ and $\lambda_2 = -1$, with repeated Eigen values.

One Eigen vector corresponding to $\lambda_1 = -1$ is obtained as the cofactors of any one row of

$$|\lambda I - A| = \begin{bmatrix} \lambda+3 & -2 \\ 2 & \lambda-1 \end{bmatrix}.$$

Cofactors of the first row

$$P_i = \begin{bmatrix} +(\lambda-1) \\ -(2) \end{bmatrix} = \begin{bmatrix} \lambda-1 \\ -2 \end{bmatrix}$$

$$P_1 = \begin{bmatrix} \lambda-1 \\ -2 \end{bmatrix}_{\lambda=-1} = \begin{bmatrix} -2 \\ -2 \end{bmatrix}.$$

The generalized Eigen vector corresponding to $\lambda_2 = -1$ is

$$P_2 = \frac{d}{d\lambda}\begin{bmatrix} \lambda-1 \\ -2 \end{bmatrix}_{\lambda=-1} = \begin{bmatrix} 1 \\ 0 \end{bmatrix}.$$

Example 4.6 For a square matrix $A = \begin{bmatrix} -5 & 1 \\ -9 & 1 \end{bmatrix}$, determine the Eigen values and the associated Eigen vectors.

Solution: The Eigen values are obtained by solving the equation

$$|\lambda I - A| = 0 \qquad or \begin{vmatrix} \lambda + 5 & -1 \\ 9 & \lambda - 1 \end{vmatrix} = 0 \qquad or \ \lambda^2 + 4\lambda + 4 = 0 \ \ or \ (\lambda + 2)^2 = 0 .$$

The Eigen values are $\lambda_1 = -2 \, and \ \lambda_2 = -2$, with repeated Eigen values.

$$[\lambda I - A] = \begin{bmatrix} \lambda + 5 & -1 \\ 9 & \lambda - 1 \end{bmatrix}$$

Cofactors of the first row of $[\lambda I - A]$ are

$$P_i = \begin{bmatrix} +(\lambda - 1) \\ -(9) \end{bmatrix} = \begin{bmatrix} \lambda - 1 \\ -9 \end{bmatrix}.$$

The Eigen vector associated with Eigen value $\lambda_1 = -2$ is

$$P_1 = \begin{bmatrix} p_{11} \\ p_{21} \end{bmatrix} = \begin{bmatrix} \lambda - 1 \\ -9 \end{bmatrix}_{\lambda = -2} = \begin{bmatrix} -3 \\ -9 \end{bmatrix}.$$

The generalized Eigen vector corresponding to $\lambda_2 = -2$ can be obtained by solving the equation

$$\begin{bmatrix} \lambda + 5 & -1 \\ 9 & \lambda - 1 \end{bmatrix}_{\lambda = -2} [P_2] = [-P_1].$$

$$\begin{bmatrix} \lambda + 5 & -1 \\ 9 & \lambda - 1 \end{bmatrix}_{\lambda = -2} \begin{bmatrix} p_{12} \\ p_{22} \end{bmatrix} = \begin{bmatrix} -p_{11} \\ -p_{21} \end{bmatrix}$$

$$\begin{bmatrix} 3 & -1 \\ 9 & -3 \end{bmatrix} \begin{bmatrix} p_{12} \\ p_{22} \end{bmatrix} = \begin{bmatrix} 3 \\ 9 \end{bmatrix}$$

$$3p_{12} - p_{22} = 3 \ \ and \ \ 9p_{12} - 3p_{22} = 9 .$$

Let $p_{12} = 1$ so that $p_{22} = 0$ and $P_2 = \begin{bmatrix} p_{12} \\ p_{22} \end{bmatrix} = \begin{bmatrix} 1 \\ 0 \end{bmatrix}$.

Example 4.7 For a square matrix $A = \begin{bmatrix} -2 & 4 & 0 \\ 0 & -2 & 0 \\ -4 & 4 & -2 \end{bmatrix}$, determine the Eigen values and the associated Eigen vectors.

Solution: The Eigen values are obtained by solving the equation

$$|\lambda I - A| = 0 \qquad or \quad \begin{vmatrix} \lambda+2 & -4 & 0 \\ 0 & \lambda+2 & 0 \\ 4 & -4 & \lambda+2 \end{vmatrix} = 0 \qquad or \quad (\lambda+2)^3 = 0 \; .$$

The Eigen values are $\lambda_1 = -2, \lambda_2 = -2$ and $\lambda_3 = -2$, with repeated Eigen values.

To find the Eigen vector corresponding to $\lambda_1 = -2$,

Cofactors of the first row of $[\lambda I - A]$ is $P_i = \begin{bmatrix} +\{(\lambda+2)^2\} \\ -\{0\} \\ +\{-4((\lambda+2)\} \end{bmatrix} = \begin{bmatrix} (\lambda+2)^2 \\ 0 \\ -4((\lambda+2) \end{bmatrix}$

$$P_1 = \begin{bmatrix} (\lambda+2)^2 \\ 0 \\ -4(\lambda+2) \end{bmatrix}_{\lambda=-2} = \begin{bmatrix} 0 \\ 0 \\ 0 \end{bmatrix}.$$

An Eigen vector with all-zero elements is not acceptable.

Cofactors of the second row of $[\lambda I - A]$ is $P_i = \begin{bmatrix} -\{-4(\lambda+2)\} \\ +\{(\lambda+2)^2\} \\ -\{-4(\lambda+2)+16\} \end{bmatrix} = \begin{bmatrix} 4(\lambda+2) \\ (\lambda+2)^2 \\ 4(\lambda+2)-16 \end{bmatrix}$

$$P_1 = \begin{bmatrix} 4(\lambda+2) \\ (\lambda+2)^2 \\ 4(\lambda+2)-16 \end{bmatrix}_{\lambda=-2} = \begin{bmatrix} 0 \\ 0 \\ -16 \end{bmatrix}.$$

A generalized Eigen vector corresponding to $\lambda_2 = -2$ is

$$P_2 = \frac{d}{d\lambda} \begin{bmatrix} 4(\lambda+2) \\ (\lambda+2)^2 \\ 4(\lambda+2)-16 \end{bmatrix}_{\lambda=-2} = \begin{bmatrix} 4 \\ 2(\lambda+2) \\ 4 \end{bmatrix}_{\lambda=-2} = \begin{bmatrix} 4 \\ 0 \\ 4 \end{bmatrix}.$$

Another generalized Eigen vector corresponding to $\lambda_3 = -2$ is

$$P_3 = \frac{1}{2!}\frac{d^2}{d\lambda^2} \begin{bmatrix} 4(\lambda+2) \\ (\lambda+2)^2 \\ 4(\lambda+2)-16 \end{bmatrix}_{\lambda=-2} = \frac{1}{2}\frac{d}{d\lambda} \begin{bmatrix} 4 \\ 2(\lambda+2) \\ 4 \end{bmatrix}_{\lambda=-2} = \begin{bmatrix} 0 \\ 1 \\ 0 \end{bmatrix}.$$

Example 4.8 Repeat Example 4.7 using the definition of Eigen vectors.

Solution:

$$A = \begin{bmatrix} -2 & 4 & 0 \\ 0 & -2 & 0 \\ -4 & 4 & -2 \end{bmatrix}$$

The Eigen values are obtained by solving the equation

$$|\lambda I - A| = 0 \qquad or \begin{vmatrix} \lambda+2 & -4 & 0 \\ 0 & \lambda+2 & 0 \\ 4 & -4 & \lambda+2 \end{vmatrix} = 0 \qquad or \ (\lambda+2)^3 = 0$$

The Eigen values are $\lambda_1 = -2, \lambda_2 = -2$ *and* $\lambda_3 = -2$, with repeated Eigen values.

To find the Eigen vector $P_1 = \begin{bmatrix} p_{11} \\ p_{21} \\ p_{31} \end{bmatrix}$ corresponding to $\lambda_1 = -2$,

$$\begin{bmatrix} \lambda+2 & -4 & 0 \\ 0 & \lambda+2 & 0 \\ 4 & -4 & \lambda+2 \end{bmatrix}_{\lambda=-2} \begin{bmatrix} p_{11} \\ p_{21} \\ p_{31} \end{bmatrix} = \begin{bmatrix} 0 \end{bmatrix},$$

which gives the following equations:

$$(0)p_{11} - 4p_{21} + (0)p_{31} = 0 \tag{1}$$

$$(0)p_{11} + (0)p_{21} + (0)p_{31} = 0 \tag{2}$$

$$4p_{11} - 4p_{21} + (0)p_{31} = 0 \tag{3}$$

These equations cannot be solved uniquely.
From equation (1), $p_{21} = 0$.
Substituting the value of p_{21} in equation (3), $p_{11} = 0$. Let $p_{31} = 4$.

Hence $P_1 = \begin{bmatrix} 0 \\ 0 \\ 4 \end{bmatrix}$.

An Eigen vector corresponding to $\lambda_2 = -2$ is obtained from solving the equation

$$\begin{bmatrix} \lambda+2 & -4 & 0 \\ 0 & \lambda+2 & 0 \\ 4 & -4 & \lambda+2 \end{bmatrix}_{\lambda=-2} \begin{bmatrix} P_2 \end{bmatrix} = -\begin{bmatrix} P_1 \end{bmatrix}.$$

$$\begin{bmatrix} \lambda+2 & -4 & 0 \\ 0 & \lambda+2 & 0 \\ 4 & -4 & \lambda+2 \end{bmatrix}_{\lambda=-2} \begin{bmatrix} p_{12} \\ p_{22} \\ p_{32} \end{bmatrix} = \begin{bmatrix} 0 \\ 0 \\ -4 \end{bmatrix},$$

which gives the following equations:

$$(0)p_{12} - 4p_{22} + (0)p_{32} = 0 \tag{4}$$

$$(0)p_{12} + (0)p_{22} + (0)p_{32} = 0 \tag{5}$$

$$4p_{12} - 4p_{22} + (0)p_{32} = -4 \tag{6}$$

From equation (4), $p_{22} = 0$.
In equation (6), let $p_{12} = -1$ and $p_{32} = 4$.

$$\text{Hence } P_2 = \begin{bmatrix} -1 \\ 0 \\ 4 \end{bmatrix}.$$

Another Eigen vector corresponding to $\lambda_3 = -2$ is obtained from solving the equation

$$\begin{bmatrix} \lambda+2 & -4 & 0 \\ 0 & \lambda+2 & 0 \\ 4 & -4 & \lambda+2 \end{bmatrix}_{\lambda=-2} [P_3] = -[P_2].$$

$$\begin{bmatrix} \lambda+2 & -4 & 0 \\ 0 & \lambda+2 & 0 \\ 4 & -4 & \lambda+2 \end{bmatrix}_{\lambda=-2} \begin{bmatrix} p_{13} \\ p_{23} \\ p_{33} \end{bmatrix} = \begin{bmatrix} 1 \\ 0 \\ -4 \end{bmatrix},$$

which gives the following equations

$$(0)p_{13} - 4p_{23} + (0)p_{33} = 1 \tag{7}$$

$$(0)p_{13} + (0)p_{23} + (0)p_{33} = 0 \tag{8}$$

$$4p_{13} - 4p_{23} + (0)p_{33} = -4 \tag{9}$$

From equation (7), $p_{23} = -0.25$.
In equation (9), $p_{13} = -1.25$ and let $p_{33} = 0$.

$$\text{Hence } P_3 = \begin{bmatrix} -1.25 \\ -0.25 \\ 0 \end{bmatrix}.$$

4.5 TRANSFORMATION TO DIAGONAL CANONICAL FORM OR DIAGONALIZATION

If the Eigen values of a system matrix are distinct, then the system can be transformed into diagonal canonical form. Consider a state model

$$\dot{X} = AX + Br$$

$$y = CX.$$

Consider that it is transformed into diagonal canonical form with the transformation

$$X = PZ.$$

Then from equations (4.4) and (4.5), the new state model is

$$\dot{Z} = [P^{-1}AP]Z + [P^{-1}B]r,$$

where $[P^{-1}AP]$ is the new system matrix and $[P^{-1}B]$ is the new input matrix.

$$y = [CP]Z$$

The transformation matrix P, composed of Eigen vectors, transforms the given state model into diagonal canonical form. Consider that for a third-order system, the Eigen values are distinct and are λ_1, λ_2 and λ_3. The Eigen vectors associated with each Eigen value can be obtained from $[\lambda_i I - A][P_i] = [0]$, where P_i is an Eigen vector associated with an Eigen value λ_i.

From this equation, $AP_i = \lambda_i P_i$.

Let the Eigen vectors associated with the Eigen values λ_1, λ_2 and λ_3 be

$$P_1 = \begin{bmatrix} p_{11} \\ p_{21} \\ p_{31} \end{bmatrix}, P_2 = \begin{bmatrix} p_{12} \\ p_{22} \\ p_{32} \end{bmatrix} \text{ and } P_3 = \begin{bmatrix} p_{13} \\ p_{23} \\ p_{33} \end{bmatrix}, \text{ respectively.}$$

Now $AP_1 = \lambda_1 P_1$, $AP_2 = \lambda_2 P_2$ and $AP_3 = \lambda_3 P_3$.

$$A \begin{bmatrix} p_{11} \\ p_{21} \\ p_{31} \end{bmatrix} = \begin{bmatrix} \lambda_1 p_{11} \\ \lambda_1 p_{21} \\ \lambda_1 p_{31} \end{bmatrix},$$

$$A \begin{bmatrix} p_{12} \\ p_{22} \\ p_{32} \end{bmatrix} = \begin{bmatrix} \lambda_2 p_{12} \\ \lambda_2 p_{22} \\ \lambda_2 p_{32} \end{bmatrix},$$

$$A \begin{bmatrix} p_{13} \\ p_{23} \\ p_{33} \end{bmatrix} = \begin{bmatrix} \lambda_3 p_{13} \\ \lambda_3 p_{23} \\ \lambda_3 p_{33} \end{bmatrix}.$$

Combining these equations

$$A \begin{bmatrix} p_{11} & p_{12} & p_{13} \\ p_{21} & p_{22} & p_{23} \\ p_{31} & p_{32} & p_{33} \end{bmatrix} = \begin{bmatrix} \lambda_1 p_{11} & \lambda_2 p_{12} & \lambda_3 p_{13} \\ \lambda_1 p_{21} & \lambda_2 p_{22} & \lambda_3 p_{23} \\ \lambda_1 p_{31} & \lambda_2 p_{32} & \lambda_3 p_{33} \end{bmatrix}$$

$$= \begin{bmatrix} p_{11} & p_{12} & p_{13} \\ p_{21} & p_{22} & p_{23} \\ p_{31} & p_{32} & p_{33} \end{bmatrix} \begin{bmatrix} \lambda_1 & 0 & 0 \\ 0 & \lambda_2 & 0 \\ 0 & 0 & \lambda_3 \end{bmatrix}$$

$$= \begin{bmatrix} p_{11} & p_{12} & p_{13} \\ p_{21} & p_{22} & p_{23} \\ p_{31} & p_{32} & p_{33} \end{bmatrix} [\Lambda]. \qquad (4.6)$$

The matrix Λ is a diagonal matrix with the diagonal elements being the Eigen values:

$$\Lambda = \begin{bmatrix} \lambda_1 & 0 & 0 \\ 0 & \lambda_2 & 0 \\ 0 & 0 & \lambda_3 \end{bmatrix}.$$

From equation (4.6),

$$P\Lambda = AP \quad or \quad \Lambda = [P^{-1}AP] \qquad (4.7)$$

This shows that a system matrix A can be transformed into diagonal form using the transformation matrix P

$$P = \begin{bmatrix} p_{11} & p_{12} & p_{13} \\ p_{21} & p_{22} & p_{23} \\ p_{31} & p_{32} & p_{33} \end{bmatrix}.$$

From equation (4.7) it can be observed that using the transformation matrix P, composed of Eigen vectors associated with each Eigen value, a system matrix can be transformed into diagonal form. Each column of P is an Eigen vector. Thus using Eigen vectors, a state model can be transformed into diagonal canonical form. The matrix P which diagonalizes a given system matrix is also known as a *modal matrix*.

4.6 DETERMINATION OF STATE TRANSITION MATRIX USING THE METHOD OF DIAGONALIZATION

For a third-order system, let a state equation be

$$\dot{X} = AX . \tag{4.8}$$

The solution to this state equation is

$$X = e^{At} X(0). \tag{4.9}$$

Let the system matrix A be transformed into a diagonal matrix Λ by the transformation

$$X = PZ \tag{4.10}$$

$$\Lambda = \begin{bmatrix} \lambda_1 & 0 & 0 \\ 0 & \lambda_2 & 0 \\ 0 & 0 & \lambda_3 \end{bmatrix} .$$

P is the transformation matrix composed of Eigen vectors.
Consider that the transformed state equation is

$$\dot{Z} = \Lambda Z . \tag{4.11}$$

The solution of this equation is

$$Z = e^{\Lambda t} Z(0), \tag{4.12}$$

where

$$e^{\Lambda t} = \begin{bmatrix} e^{\lambda_1 t} & 0 & 0 \\ 0 & e^{\lambda_2 t} & 0 \\ 0 & 0 & e^{\lambda_3 t} \end{bmatrix} . \tag{4.13}$$

The method of obtaining $e^{\Lambda t}$ is explained in Section 3.5.1.
From equations (4.9) and (4.10),

$$PZ = e^{At} X(0) = e^{At} PZ(0)$$

Or

$$Z = [P^{-1} e^{At} P] Z(0). \tag{4.14}$$

Comparing equations (4.12) and (4.14),

$$e^{\Lambda t} = [P^{-1} e^{At} P] .$$

Or
$$e^{At} = [Pe^{At}P^{-1}] = P \begin{bmatrix} e^{\lambda_1 t} & 0 & 0 \\ 0 & e^{\lambda_2 t} & 0 \\ 0 & 0 & e^{\lambda_3 t} \end{bmatrix} P^{-1} \tag{4.15}$$

Using equations (4.15), the state transition matrix e^{At} can be determined. For an n-th order system with distinct Eigen values,

$$e^{At} = \begin{bmatrix} e^{\lambda_1 t} & 0 & 0 & 0----0 \\ 0 & e^{\lambda_2 t} & 0 & 0----0 \\ 0 & 0 & e^{\lambda_3 t} & 0----0 \\ -&-&-&-&-&-&- \\ 0 & 0 & 0 & 0---- e^{\lambda_n t} \end{bmatrix}.$$

Example 4.9 A state model for a linear system is

$$\dot{X} = \begin{bmatrix} -2 & 2 \\ 1 & -3 \end{bmatrix} X + \begin{bmatrix} 1 \\ 2 \end{bmatrix} r$$

$$y = [2 \quad -1]X.$$

Transform the system into diagonal canonical form. Also find the state transition matrix.

Solution:

The Eigen values are obtained by solving the equation

$$|\lambda I - A| = 0 \qquad or \begin{vmatrix} \lambda+2 & -2 \\ -1 & \lambda+3 \end{vmatrix} = 0 \qquad or \ \lambda^2 + 5\lambda + 4 = 0.$$

The Eigen values are $\lambda_1 = -1$ and $\lambda_2 = -4$, and are distinct. Hence the system matrix can be transformed to diagonal canonical form.

The Eigen vectors are determined from

$$[\ \lambda I - A] = \begin{bmatrix} +2 & -2 \\ -1 & +3 \end{bmatrix}.$$

The matrix of cofactors of the second row is $P_i = \begin{bmatrix} -(-2) \\ +(\lambda+2) \end{bmatrix} = \begin{bmatrix} 2 \\ (\lambda+2) \end{bmatrix}.$

The Eigen value corresponding to $\lambda_1 = -1$ is $P_1 = \begin{bmatrix} 2 \\ (\lambda+2) \end{bmatrix}_{\lambda=-1} = \begin{bmatrix} 2 \\ 1 \end{bmatrix}.$

The Eigen vector associated with the Eigen value $\lambda_1 = -4$ is

$$P_2 = \begin{bmatrix} 2 \\ \lambda+2 \end{bmatrix}_{\lambda=-4} = \begin{bmatrix} 2 \\ -2 \end{bmatrix} \text{ or } P_2 = \begin{bmatrix} 1 \\ -1 \end{bmatrix}.$$

For the transformation matrix $P = \begin{bmatrix} 2 & 1 \\ 1 & -1 \end{bmatrix}$, $|P| = \begin{vmatrix} 2 & 1 \\ 1 & -1 \end{vmatrix} = -3$

$$P^{-1} = -\frac{1}{3}\begin{bmatrix} -1 & -1 \\ -1 & 2 \end{bmatrix} = \frac{1}{3}\begin{bmatrix} 1 & 1 \\ 1 & -2 \end{bmatrix}$$

Check:

$$P^{-1}P = \frac{1}{3}\begin{bmatrix} 1 & 1 \\ 1 & -2 \end{bmatrix}\begin{bmatrix} 2 & 1 \\ 1 & -1 \end{bmatrix} = \frac{1}{3}\begin{bmatrix} 3 & 0 \\ 0 & 3 \end{bmatrix} = \begin{bmatrix} 1 & 0 \\ 0 & 1 \end{bmatrix} = I$$

$$P^{-1}AP = \frac{1}{3}\begin{bmatrix} 1 & 1 \\ 1 & -2 \end{bmatrix}\begin{bmatrix} -2 & 2 \\ 1 & -3 \end{bmatrix}\begin{bmatrix} 2 & 1 \\ 1 & -1 \end{bmatrix}$$

$$= \frac{1}{3}\begin{bmatrix} 1 & 1 \\ 1 & -2 \end{bmatrix}\begin{bmatrix} -2 & -4 \\ -1 & 4 \end{bmatrix} = \begin{bmatrix} -1 & 0 \\ 0 & -4 \end{bmatrix} = \Lambda.$$

The state model in diagonal canonical form is

$$\dot{Z} = [P^{-1}AP]Z + [P^{-1}B]r$$

and $y = [CP]Z$

$$P^{-1}AP = \begin{bmatrix} -1 & 0 \\ 0 & -4 \end{bmatrix} \qquad P^{-1}B = \frac{1}{3}\begin{bmatrix} 1 & 1 \\ 1 & -2 \end{bmatrix}\begin{bmatrix} 1 \\ 2 \end{bmatrix} = \begin{bmatrix} 1 \\ -1 \end{bmatrix}.$$

$$CP = \begin{bmatrix} 2 & -1 \end{bmatrix}\begin{bmatrix} 2 & 1 \\ 1 & -1 \end{bmatrix} = \begin{bmatrix} 3 & 3 \end{bmatrix}$$

Hence

$$\dot{Z} = \begin{bmatrix} -1 & 0 \\ 0 & -4 \end{bmatrix}Z + \begin{bmatrix} 1 \\ -1 \end{bmatrix}r$$

and $y = \begin{bmatrix} 3 & 3 \end{bmatrix}Z$.

$$e^{\Lambda t} = \begin{bmatrix} e^{-t} & 0 \\ 0 & e^{-4t} \end{bmatrix}$$

The state transition matrix is

$$e^{At} = [Pe^{\Lambda t}P^{-1}] = \frac{1}{3}\begin{bmatrix} 2 & 1 \\ 1 & -1 \end{bmatrix}\begin{bmatrix} e^{-t} & 0 \\ 0 & e^{-4t} \end{bmatrix}\begin{bmatrix} 1 & 1 \\ 1 & -2 \end{bmatrix} = \frac{1}{3}\begin{bmatrix} 2 & 1 \\ 1 & -1 \end{bmatrix}\begin{bmatrix} e^{-t} & e^{-t} \\ e^{-4t} & -2e^{-4t} \end{bmatrix}$$

$$= \begin{bmatrix} \dfrac{2}{3}e^{-t} + \dfrac{1}{3}e^{-4t} & \dfrac{2}{3}e^{-t} - \dfrac{2}{3}e^{-4t} \\ \dfrac{1}{3}e^{-t} - \dfrac{1}{3}e^{-4t} & \dfrac{1}{3}e^{-t} + \dfrac{2}{3}e^{-4t} \end{bmatrix}.$$

Example 4.10 A state model for a system is

$$\dot{X} = \begin{bmatrix} 1 & 6 \\ -1 & -4 \end{bmatrix} X + \begin{bmatrix} 0 \\ 1 \end{bmatrix} r$$

$$y = \begin{bmatrix} 1 & 2 \end{bmatrix} X .$$

Transform the given system into diagonal canonical form. Also find the state transition matrix.

Solution:
The Eigen values are obtained by solving the equation

$$|\lambda I - A| = 0 \quad or \begin{vmatrix} -1 & -6 \\ 1 & 4 \end{vmatrix} = 0 \quad or \ \lambda^2 + 3\lambda + 2 = 0 \ \ or \ (\lambda + 1)(\lambda + 2) = 0 .$$

The Eigen values are $\lambda_1 = -1$ and $\lambda_2 = -2$, and are distinct. Hence, the system matrix can be transformed to diagonal canonical form.
The Eigen vectors are determined from

$$[\lambda I - A] = \begin{bmatrix} \lambda - 1 & -6 \\ 1 & \lambda + 4 \end{bmatrix} .$$

The matrix of cofactors of the first row is $P_i = \begin{bmatrix} +(\lambda + 4) \\ -1 \end{bmatrix} .$

The Eigen value corresponding to $\lambda_1 = -1$ is $P_1 = \begin{bmatrix} (\lambda + 4) \\ -1 \end{bmatrix}_{\lambda = -1} = \begin{bmatrix} 3 \\ -1 \end{bmatrix} .$

The Eigen vector associated with the Eigen value $\lambda_1 = -2$ is

$$P_2 = \begin{bmatrix} (\lambda + 4) \\ -1 \end{bmatrix}_{\lambda = -2} = \begin{bmatrix} 2 \\ -1 \end{bmatrix} .$$

For the transformation matrix $P = \begin{bmatrix} 3 & 2 \\ -1 & -1 \end{bmatrix}, \ |P| = \begin{vmatrix} 3 & 2 \\ -1 & -1 \end{vmatrix} = -1$

$$P^{-1} = -\begin{bmatrix} -1 & -2 \\ 1 & 3 \end{bmatrix} = \begin{bmatrix} 1 & 2 \\ -1 & -3 \end{bmatrix}$$

Check:
$$P^{-1}P = \begin{bmatrix} 1 & 2 \\ -1 & -3 \end{bmatrix}\begin{bmatrix} 3 & 2 \\ -1 & -1 \end{bmatrix} = \begin{bmatrix} 1 & 0 \\ 0 & 1 \end{bmatrix}$$

$$P^{-1}AP = \begin{bmatrix} 1 & 2 \\ -1 & -3 \end{bmatrix}\begin{bmatrix} 1 & 6 \\ -1 & -4 \end{bmatrix}\begin{bmatrix} 3 & 2 \\ -1 & -1 \end{bmatrix} = \begin{bmatrix} 1 & 2 \\ -1 & -3 \end{bmatrix}\begin{bmatrix} -3 & -4 \\ 1 & 2 \end{bmatrix} = \begin{bmatrix} -1 & 0 \\ 0 & -2 \end{bmatrix} = \Lambda .$$

The state model in diagonal canonical form is

$$\dot{Z} = [P^{-1}AP]Z + [P^{-1}B]r$$

$$y = [CP]Z$$

$$P^{-1}AP = \begin{bmatrix} -1 & 0 \\ 0 & -2 \end{bmatrix} \qquad P^{-1}B = \begin{bmatrix} 1 & 2 \\ -1 & -3 \end{bmatrix}\begin{bmatrix} 0 \\ 1 \end{bmatrix} = \begin{bmatrix} 2 \\ -3 \end{bmatrix}$$

$$CP = \begin{bmatrix} 1 & 2 \end{bmatrix}\begin{bmatrix} 3 & 2 \\ -1 & -1 \end{bmatrix} = \begin{bmatrix} 1 & 0 \end{bmatrix}.$$

Hence the state model in diagonal canonical form is

$$\dot{Z} = \begin{bmatrix} -1 & 0 \\ 0 & -2 \end{bmatrix}Z + \begin{bmatrix} 2 \\ -3 \end{bmatrix}r$$

and $y = \begin{bmatrix} 1 & 0 \end{bmatrix}Z$.

For finding the state transition matrix

$$e^{\Lambda t} = \begin{bmatrix} e^{-t} & 0 \\ 0 & e^{-2t} \end{bmatrix}$$

$$e^{At} = Pe^{\Lambda t}P^{-1} = \begin{bmatrix} 3 & 2 \\ -1 & -1 \end{bmatrix}\begin{bmatrix} e^{-t} & 0 \\ 0 & e^{-2t} \end{bmatrix}\begin{bmatrix} 1 & 2 \\ -1 & -3 \end{bmatrix} = \begin{bmatrix} 3 & 2 \\ -1 & -1 \end{bmatrix}\begin{bmatrix} e^{-t} & 2e^{-t} \\ -e^{-2t} & -3e^{-2t} \end{bmatrix}$$

$$= \begin{bmatrix} 3e^{-t} - 2e^{-2t} & 6e^{-t} - 6e^{-2t} \\ -e^{-t} + e^{-2t} & -2e^{-t} + 3e^{-2t} \end{bmatrix}.$$

Example 4.11 A linear time-invariant system is represented by a state model

$$\dot{X} = \begin{bmatrix} -4 & 1 & 0 \\ 0 & -3 & 1 \\ 0 & 0 & -2 \end{bmatrix}X + \begin{bmatrix} 0 \\ 1 \\ 2 \end{bmatrix}r$$

$$y = \begin{bmatrix} 2 & -1 & 1 \end{bmatrix}X.$$

Transform the model into diagonal canonical form. Also find the state transition matrix.

Solution: The Eigen values are obtained by solving the equation

$$|\lambda I - A| = 0 \qquad or \quad \begin{vmatrix} \lambda+4 & -1 & 0 \\ 0 & \lambda+3 & -1 \\ 0 & 0 & \lambda+2 \end{vmatrix} = 0 \qquad or \quad (\lambda+4)(\lambda+3)(\lambda+2) = 0.$$

The Eigen values are $\lambda_1 = -2, \lambda_2 = -3$ *and* $\lambda_3 = -4$ and are distinct. Hence, the system matrix can be transformed to diagonal canonical form.

$$Adj[\lambda I - A] = \begin{bmatrix} +\{(\lambda+3)(\lambda+2)\} & -\{-(\lambda+2)\} & +\{1\} \\ -\{0\} & +\{(\lambda+2)(\lambda+4)\} & -\{-(\lambda+4)\} \\ +\{0\} & -\{0\} & +\{(\lambda+3)(\lambda+4)\} \end{bmatrix}$$

$$= \begin{bmatrix} (\lambda+3)(\lambda+2) & (\lambda+2) & 1 \\ 0 & (\lambda+2)(\lambda+4) & (\lambda+4) \\ 0 & 0 & (\lambda+3)(\lambda+4) \end{bmatrix}.$$

The matrix of cofactors of the third row is $P_i = \begin{bmatrix} 1 \\ (\lambda+4) \\ (\lambda+3)(\lambda+4) \end{bmatrix}$

Eigen vector corresponding to $\lambda_1 = -2$ *is* $P_1 = \begin{bmatrix} 1 \\ (\lambda+4) \\ (\lambda+3)(\lambda+4) \end{bmatrix}_{\lambda_1=-2} = \begin{bmatrix} 1 \\ 2 \\ 2 \end{bmatrix}$

Eigen vector corresponding to $\lambda_2 = -3$ *is* $P_2 = \begin{bmatrix} 1 \\ (\lambda+4) \\ (\lambda+3)(\lambda+4) \end{bmatrix}_{\lambda_2=-3} = \begin{bmatrix} 1 \\ 1 \\ 0 \end{bmatrix}$

Eigen vector corresponding to $\lambda_3 = -4$ *is* $P_3 = \begin{bmatrix} 1 \\ (\lambda+4) \\ (\lambda+3)(\lambda+4) \end{bmatrix}_{\lambda_3=-4} = \begin{bmatrix} 1 \\ 0 \\ 0 \end{bmatrix}.$

Hence $P = [P_1 \ P_2 \ P_3] = \begin{bmatrix} 1 & 1 & 1 \\ 2 & 1 & 0 \\ 2 & 0 & 0 \end{bmatrix}$ $\qquad |P| = \begin{vmatrix} 1 & 1 & 1 \\ 2 & 1 & 0 \\ 2 & 0 & 0 \end{vmatrix} = -2,$

$$P^{-1} = -\frac{1}{2}\begin{bmatrix} +\{0\} & -\{0\} & +\{-1\} \\ -\{0\} & +\{-2\} & -\{-2\} \\ +\{-2\} & -\{-2\} & +\{-1\} \end{bmatrix} = -\frac{1}{2}\begin{bmatrix} 0 & 0 & -1 \\ 0 & -2 & 2 \\ -2 & 2 & -1 \end{bmatrix} = \frac{1}{2}\begin{bmatrix} 0 & 0 & 1 \\ 0 & 2 & -2 \\ 2 & -2 & 1 \end{bmatrix}$$

Check:
$$P^{-1}P = \frac{1}{2}\begin{bmatrix} 0 & 0 & 1 \\ 0 & 2 & -2 \\ 2 & -2 & 1 \end{bmatrix}\begin{bmatrix} 1 & 1 & 1 \\ 2 & 1 & 0 \\ 2 & 0 & 0 \end{bmatrix} = \begin{bmatrix} 1 & 0 & 0 \\ 0 & 1 & 0 \\ 0 & 0 & 1 \end{bmatrix}$$

$$P^{-1}AP = \frac{1}{2}\begin{bmatrix} 0 & 0 & 1 \\ 0 & 2 & -2 \\ 2 & -2 & 1 \end{bmatrix}\begin{bmatrix} -4 & 1 & 0 \\ 0 & -3 & 1 \\ 0 & 0 & -2 \end{bmatrix}\begin{bmatrix} 1 & 1 & 1 \\ 2 & 1 & 0 \\ 2 & 0 & 0 \end{bmatrix}$$

$$= \frac{1}{2}\begin{bmatrix} 0 & 0 & 1 \\ 0 & 2 & -2 \\ 2 & -2 & 1 \end{bmatrix}\begin{bmatrix} -2 & -3 & -4 \\ -4 & -3 & 0 \\ -4 & 0 & 0 \end{bmatrix}$$

$$= \begin{bmatrix} -2 & 0 & 0 \\ 0 & -3 & 0 \\ 0 & 0 & -4 \end{bmatrix} = \Lambda .$$

The state model in diagonal canonical form is

$$\dot{Z} = [P^{-1}AP]Z + [P^{-1}B]r$$

and
$$y = [CP]Z .$$

$$P^{-1}AP = \begin{bmatrix} -2 & 0 & 0 \\ 0 & -3 & 0 \\ 0 & 0 & -4 \end{bmatrix} \quad P^{-1}B = \frac{1}{2}\begin{bmatrix} 0 & 0 & 1 \\ 0 & 2 & -2 \\ 2 & -2 & 1 \end{bmatrix}\begin{bmatrix} 0 \\ 1 \\ 2 \end{bmatrix} = \begin{bmatrix} 1 \\ -1 \\ 0 \end{bmatrix}$$

$$CP = \begin{bmatrix} 2 & -1 & 1 \end{bmatrix}\begin{bmatrix} 1 & 1 & 1 \\ 2 & 1 & 0 \\ 2 & 0 & 0 \end{bmatrix} = \begin{bmatrix} 2 & 1 & 2 \end{bmatrix} .$$

Hence, the state model in diagonal canonical form is

$$\dot{Z} = \begin{bmatrix} -1 & 0 & 0 \\ 0 & -2 & 0 \\ 0 & 0 & -3 \end{bmatrix}Z + \begin{bmatrix} 1 \\ -1 \\ 0 \end{bmatrix}r$$

$$y = \begin{bmatrix} 2 & 1 & 2 \end{bmatrix}Z$$

$$e^{\Lambda t} == \begin{bmatrix} e^{-2t} & 0 & 0 \\ 0 & e^{-3t} & 0 \\ 0 & 0 & e^{-4t} \end{bmatrix} .$$

The state transition matrix is

$$e^{At} = P e^{\Lambda t} P^{-1} = \frac{1}{2} \begin{bmatrix} 1 & 1 & 1 \\ 2 & 1 & 0 \\ 2 & 0 & 0 \end{bmatrix} \begin{bmatrix} e^{-2t} & 0 & 0 \\ 0 & e^{-3t} & 0 \\ 0 & 0 & e^{-4t} \end{bmatrix} \begin{bmatrix} 0 & 0 & 1 \\ 0 & 2 & -2 \\ 2 & -2 & 1 \end{bmatrix}$$

$$= \frac{1}{2} \begin{bmatrix} 1 & 1 & 1 \\ 2 & 1 & 0 \\ 2 & 0 & 0 \end{bmatrix} \begin{bmatrix} 0 & 0 & e^{-2t} \\ 0 & 2e^{-3t} & -2e^{-3t} \\ 2e^{-4t} & -2e^{-4t} & e^{-4t} \end{bmatrix}$$

$$= \begin{bmatrix} e^{-4t} & e^{-3t} - e^{-4t} & \frac{1}{2}e^{-2t} - e^{-3t} + \frac{1}{2}e^{-4t} \\ 0 & e^{-3t} & e^{-2t} - e^{-3t} \\ 0 & 0 & e^{-2t} \end{bmatrix}.$$

4.7 TRANSFORMATION TO JORDAN CANONICAL FORM

If the system matrix has repeated Eigen values, then it can be transformed into Jordan canonical form. The system can be represented as

$$\dot{X} = JX,$$

where J is the system matrix in Jordan canonical form.

The transformation matrix P is composed of Eigen vectors and generalized Eigen vectors. These are found out using the methods discussed in Sections 4.3 and 4.4 respectively. The state transition matrix for the Jordan form is e^{Jt} and can be determined using the method explained in Section 3.5.2. The state transition matrix is given by

$$e^{At} = P e^{Jt} P^{-1}.$$

Example 4.12 For a state model

$$\dot{X} = \begin{bmatrix} -3 & 2 \\ -2 & 1 \end{bmatrix} X + \begin{bmatrix} 1 \\ -2 \end{bmatrix} r$$

$$y = \begin{bmatrix} 1 & 2 \end{bmatrix} X,$$

determine the Eigen values and the associated Eigen vectors. Transform the given state model into Jordan canonical form. Hence, find the state transition matrix.

Solution: The Eigen values are obtained by solving the equation

$$|\lambda I - A| = 0 \quad or \quad \begin{vmatrix} +3 & -2 \\ 2 & 1 \end{vmatrix} = 0 \quad or \quad \lambda^2 + 2\lambda + 1 = 0 \quad or \quad (\lambda + 1)^2 = 0.$$

The Eigen values are $\lambda_1 = -1$ and $\lambda_2 = -1$, and are repeated. Hence, the system matrix can be transformed to Jordan canonical form.

The Eigen vectors are determined from

$$[\lambda I - A] = \begin{bmatrix} \lambda + 3 & -2 \\ 2 & \lambda - 1 \end{bmatrix}.$$

The matrix of cofactors of the first row is $P_i = \begin{bmatrix} +(\lambda - 1) \\ -(2) \end{bmatrix} = \begin{bmatrix} (\lambda - 1) \\ -(2) \end{bmatrix}.$

The Eigen value corresponding to $\lambda_1 = -1$ is $P_1 = \begin{bmatrix} (\lambda - 1) \\ -2 \end{bmatrix}_{\lambda = -1} = \begin{bmatrix} -2 \\ -2 \end{bmatrix}.$

The generalized Eigen vector corresponding to $\lambda_1 = -1$ is

$$[\lambda I - A]_{\lambda = -1} P_2 = -P_1.$$

$$\begin{bmatrix} \lambda + 3 & -2 \\ 2 & \lambda - 1 \end{bmatrix}_{\lambda = -1} \begin{bmatrix} p_{12} \\ p_{22} \end{bmatrix} = \begin{bmatrix} 2 \\ 2 \end{bmatrix}$$

$$2p_{12} - 2p_{22} = 2 \text{ and } 2p_{12} - 2p_{22} = 2.$$

Let $p_{22} = 0$ so that $p_{12} = 1$. Hence $P_2 = \begin{bmatrix} 1 \\ 0 \end{bmatrix}.$

The transformation matrix is

$$P = \begin{bmatrix} -2 & 1 \\ -2 & 0 \end{bmatrix} \qquad |P| = 2 \qquad P^{-1} = \frac{1}{2} \begin{bmatrix} 0 & -1 \\ 2 & -2 \end{bmatrix}.$$

Check: $P^{-1}P = \frac{1}{2} \begin{bmatrix} 0 & -1 \\ 2 & -2 \end{bmatrix} \begin{bmatrix} -2 & 1 \\ -2 & 0 \end{bmatrix} = \begin{bmatrix} 1 & 0 \\ 0 & 1 \end{bmatrix}$

$$J = P^{-1}AP = \frac{1}{2} \begin{bmatrix} 0 & -1 \\ 2 & -2 \end{bmatrix} \begin{bmatrix} -3 & 2 \\ -2 & 1 \end{bmatrix} \begin{bmatrix} -2 & 1 \\ -2 & 0 \end{bmatrix} = \frac{1}{2} \begin{bmatrix} 0 & -1 \\ 2 & -2 \end{bmatrix} \begin{bmatrix} 2 & -3 \\ 2 & -2 \end{bmatrix} = \begin{bmatrix} -1 & 1 \\ 0 & -1 \end{bmatrix}$$

$$P^{-1}AP = \begin{bmatrix} -1 & 1 \\ 0 & -1 \end{bmatrix} \qquad P^{-1}B = \frac{1}{2} \begin{bmatrix} 0 & -1 \\ 2 & -2 \end{bmatrix} \begin{bmatrix} 1 \\ -2 \end{bmatrix} = \begin{bmatrix} 1 \\ 3 \end{bmatrix}$$

$$CP = \begin{bmatrix} 1 & 2 \end{bmatrix} \begin{bmatrix} -2 & 1 \\ -2 & 0 \end{bmatrix} = \begin{bmatrix} -6 & 1 \end{bmatrix}.$$

The state model in diagonal canonical form is

$$\dot{Z} = [P^{-1}AP]Z + [P^{-1}B]r = \begin{bmatrix} -1 & 1 \\ 0 & -1 \end{bmatrix} Z + \begin{bmatrix} 1 \\ 3 \end{bmatrix} r$$

$$y = [CP]Z = [-6 \quad 1]Z$$

$$e^{Jt} = \begin{bmatrix} e^{-t} & te^{-t} \\ 0 & e^{-t} \end{bmatrix}$$

$$e^{At} = Pe^{Jt}P^{-1} = \frac{1}{2}\begin{bmatrix} -2 & 1 \\ -2 & 0 \end{bmatrix}\begin{bmatrix} e^{-t} & te^{-t} \\ 0 & e^{-t} \end{bmatrix}\begin{bmatrix} 0 & -1 \\ 2 & -2 \end{bmatrix} = \frac{1}{2}\begin{bmatrix} -2 & 1 \\ -2 & 0 \end{bmatrix}\begin{bmatrix} 2te^{-t} & -e^{-t}-2te^{-t} \\ 2e^{-t} & -2e^{-t} \end{bmatrix}$$

$$= \begin{bmatrix} e^{-t}-2te^{-t} & 2te^{-t} \\ -2te^{-t} & e^{-t}+2te^{-t} \end{bmatrix}.$$

Example 4.13 For a state model

$$\dot{X} = \begin{bmatrix} -3 & 1 \\ -1 & -1 \end{bmatrix} X + \begin{bmatrix} 1 \\ -2 \end{bmatrix} r$$

$$y = [2 \quad -1]X ,$$

determine the Eigen values and the associated Eigen vectors. Transform the given state model into Jordan canonical form. Also find the state transition matrix.

Solution:
The Eigen values are obtained by solving the equation

$$|\lambda I - A| = 0 \qquad or \begin{vmatrix} \lambda+3 & -1 \\ 1 & \lambda+1 \end{vmatrix} = 0 \qquad or \ \lambda^2 + 4\lambda + 4 = 0 \quad or \ (\lambda+2)^2 = 0.$$

The Eigen values are $\lambda_1 = -2 \ and \ \lambda_2 = -2$ and are repeated. Hence the system matrix can be transformed to Jordan canonical form.
The Eigen vectors are determined from

$$[\lambda I - A] = \begin{bmatrix} \lambda+3 & -1 \\ 1 & \lambda+1 \end{bmatrix}.$$

The cofactors of the first row is $P_i = \begin{bmatrix} +(\lambda+1) \\ -(1) \end{bmatrix} = \begin{bmatrix} (\lambda+1) \\ -(1) \end{bmatrix}.$

The Eigen vector corresponding to $\lambda_1 = -2$ is $P_1 = \begin{bmatrix} (\lambda+1) \\ -1 \end{bmatrix}_{\lambda=-2} = \begin{bmatrix} -1 \\ -1 \end{bmatrix}$.

The generalized Eigen vector corresponding to $\lambda_1 = -2$ is

$$P_2 = \frac{d}{d\lambda} \begin{bmatrix} (\lambda+1) \\ -1 \end{bmatrix}_{\lambda=-2} = \begin{bmatrix} 1 \\ 0 \end{bmatrix}.$$

The transformation matrix is $P = \begin{bmatrix} -1 & 1 \\ -1 & 0 \end{bmatrix}$ $|P| = 1$ $P^{-1} = \begin{bmatrix} 0 & -1 \\ 1 & -1 \end{bmatrix}$.

Check: $PP^{-1} = \begin{bmatrix} -1 & 1 \\ -1 & 0 \end{bmatrix}\begin{bmatrix} 0 & -1 \\ 1 & -1 \end{bmatrix} = \begin{bmatrix} 1 & 0 \\ 0 & 1 \end{bmatrix}$

$$P^{-1}AP = \begin{bmatrix} 0 & -1 \\ 1 & -1 \end{bmatrix}\begin{bmatrix} -3 & 1 \\ -1 & -1 \end{bmatrix}\begin{bmatrix} -1 & 1 \\ -1 & 0 \end{bmatrix} = \begin{bmatrix} 0 & -1 \\ 1 & -1 \end{bmatrix}\begin{bmatrix} 2 & -3 \\ 2 & -1 \end{bmatrix} = \begin{bmatrix} -2 & 1 \\ 0 & -2 \end{bmatrix} = J.$$

The state model in diagonal canonical form is

$$\dot{Z} = [P^{-1}AP]Z + [P^{-1}B]r$$

$$y = [CP]Z$$

$$P^{-1}AP = \begin{bmatrix} -2 & 1 \\ 0 & -2 \end{bmatrix}$$ $$P^{-1}B = \begin{bmatrix} 0 & -1 \\ 1 & -1 \end{bmatrix}\begin{bmatrix} 1 \\ -2 \end{bmatrix} = \begin{bmatrix} 2 \\ 3 \end{bmatrix}$$

$$CP = [2 \quad -1]\begin{bmatrix} -1 & 1 \\ -1 & 0 \end{bmatrix} = [-1 \quad 2]$$

Hence, $\dot{Z} = \begin{bmatrix} -1 & 1 \\ 0 & -1 \end{bmatrix}Z + \begin{bmatrix} 2 \\ 3 \end{bmatrix}r$

$$y = [-1 \quad 2]Z$$

$$e^{Jt} = \begin{bmatrix} e^{-2t} & te^{-2t} \\ 0 & e^{-2t} \end{bmatrix}.$$

The state transition matrix for the given state model is

$$e^{At} = Pe^{Jt}P^{-1} = \begin{bmatrix} -1 & 1 \\ -1 & 0 \end{bmatrix}\begin{bmatrix} e^{-2t} & te^{-2t} \\ 0 & e^{-2t} \end{bmatrix}\begin{bmatrix} 0 & -1 \\ 1 & -1 \end{bmatrix} = \begin{bmatrix} -1 & 1 \\ -1 & 0 \end{bmatrix}\begin{bmatrix} te^{-2t} & -e^{-2t} - te^{-2t} \\ e^{-2t} & -e^{-2t} \end{bmatrix}$$

$$= \begin{bmatrix} e^{-2t} - te^{-2t} & te^{-2t} \\ -te^{-2t} & e^{-2t} + te^{-2t} \end{bmatrix}.$$

Example 4.14   A state model for a linear system is

$$\dot{X} = \begin{bmatrix} -6 & 1 & 0 \\ -12 & 0 & 1 \\ -8 & 0 & 0 \end{bmatrix} X + \begin{bmatrix} 1 \\ 2 \\ 1 \end{bmatrix} r$$

$$y = \begin{bmatrix} 1 & 2 & -1 \end{bmatrix} X .$$

Transform the given state model into Jordan canonical form. Also find the state transition matrix.

Solution:
The Eigen values are obtained by solving the equation

$$|\lambda I - A| = 0 \quad or \begin{vmatrix} \lambda+6 & -1 & 0 \\ 12 & \lambda & -1 \\ 8 & 0 & \lambda \end{vmatrix} = 0 \quad or \quad \lambda^3 + 6\lambda^2 + 12\lambda + 8 = 0 \quad or \quad (\lambda+2)^3 = 0 .$$

The Eigen values are $\lambda_1 = -2, \lambda_2 = -2$ *and* $\lambda_3 = -2$ and are repeated. Hence the system matrix can be transformed to Jordan canonical form.
The Eigen vectors are determined from

$$[\lambda I - A] = \begin{bmatrix} \lambda+6 & -1 & 0 \\ 12 & \lambda & -1 \\ 8 & 0 & \lambda \end{bmatrix} .$$

The matrix of cofactors of the third row is $P_i = \begin{bmatrix} +(1) \\ -\{-(\lambda+6)\} \\ +(\lambda^2+6\lambda+12) \end{bmatrix} = \begin{bmatrix} 1 \\ \lambda+6 \\ \lambda^2+6\lambda+12 \end{bmatrix} .$

The Eigen vector corresponding to $\lambda_1 = -2$ is $P_1 = \begin{bmatrix} 1 \\ \lambda+6 \\ \lambda^2+6\lambda+12 \end{bmatrix}_{\lambda=-2} = \begin{bmatrix} 1 \\ 4 \\ 4 \end{bmatrix} .$

Generalized Eigen vector corresponding to $\lambda_1 = -2$ is

$$P_2 = \frac{d}{d\lambda} \begin{bmatrix} 1 \\ \lambda+6 \\ \lambda^2+6\lambda+12 \end{bmatrix}_{\lambda=-2} = \begin{bmatrix} 0 \\ 1 \\ 2\lambda+6 \end{bmatrix}_{\lambda=-2} = \begin{bmatrix} 0 \\ 1 \\ 2 \end{bmatrix} .$$

Another generalized Eigen vector corresponding to $\lambda_1 = -2$ is $P_3 = \dfrac{1}{2!}\dfrac{d}{d\lambda}\begin{bmatrix} 0 \\ 1 \\ 2\lambda + 6 \end{bmatrix}_{\lambda = -2} = \begin{bmatrix} 0 \\ 0 \\ 1 \end{bmatrix}$.

The transformation matrix is $P = \begin{bmatrix} 1 & 0 & 0 \\ 4 & 1 & 0 \\ 4 & 2 & 1 \end{bmatrix}$.

$$|P| = \begin{vmatrix} 1 & 0 & 0 \\ 4 & 1 & 0 \\ 4 & 2 & 1 \end{vmatrix} = 1 \qquad P^{-1} = \begin{bmatrix} +(1) & -(0) & +(0) \\ -(4) & +(1) & -(0) \\ +(4) & -(2) & +(1) \end{bmatrix} = \begin{bmatrix} 1 & 0 & 0 \\ -4 & 1 & 0 \\ 4 & -2 & 1 \end{bmatrix}.$$

Check: $P^{-1}P = \begin{bmatrix} 1 & 0 & 0 \\ -4 & 1 & 0 \\ 4 & -2 & 1 \end{bmatrix}\begin{bmatrix} 1 & 0 & 0 \\ 4 & 1 & 0 \\ 4 & 2 & 1 \end{bmatrix} = \begin{bmatrix} 1 & 0 & 0 \\ 0 & 1 & 0 \\ 0 & 0 & 1 \end{bmatrix}$

$$J = P^{-1}AP = \begin{bmatrix} 1 & 0 & 0 \\ -4 & 1 & 0 \\ 4 & -2 & 1 \end{bmatrix}\begin{bmatrix} -6 & 1 & 0 \\ -12 & 0 & 1 \\ -8 & 0 & 0 \end{bmatrix}\begin{bmatrix} 1 & 0 & 0 \\ 4 & 1 & 0 \\ 4 & 2 & 1 \end{bmatrix}$$

$$= \begin{bmatrix} 1 & 0 & 0 \\ -4 & 1 & 0 \\ 4 & -2 & 1 \end{bmatrix}\begin{bmatrix} -2 & 1 & 0 \\ -8 & 2 & 1 \\ -8 & 0 & 0 \end{bmatrix} = \begin{bmatrix} -2 & 1 & 0 \\ 0 & -2 & 1 \\ 0 & 0 & -2 \end{bmatrix} = J.$$

The state model in diagonal canonical form is

$$\dot{Z} = [P^{-1}AP]Z + [P^{-1}B]r$$

$$y = [CP]Z.$$

$$P^{-1}AP = \begin{bmatrix} -2 & 1 & 0 \\ 0 & -2 & 1 \\ 0 & 0 & -2 \end{bmatrix} \qquad P^{-1}B = \begin{bmatrix} 1 & 0 & 0 \\ -4 & 1 & 0 \\ 4 & -2 & 1 \end{bmatrix}\begin{bmatrix} 1 \\ 2 \\ 1 \end{bmatrix} = \begin{bmatrix} 1 \\ -2 \\ 1 \end{bmatrix}$$

$$CP = \begin{bmatrix} 1 & 2 & -1 \end{bmatrix}\begin{bmatrix} 1 & 0 & 0 \\ 4 & 1 & 0 \\ 4 & 2 & 1 \end{bmatrix} = \begin{bmatrix} 5 & 0 & -1 \end{bmatrix}$$

Hence $\dot{Z} = \begin{bmatrix} -2 & 1 & 0 \\ 0 & -2 & 1 \\ 0 & 0 & -2 \end{bmatrix} Z + \begin{bmatrix} 1 \\ -2 \\ 1 \end{bmatrix} r$.

$y = \begin{bmatrix} 5 & 0 & -1 \end{bmatrix} Z$

$$e^{Jt} = \begin{bmatrix} e^{-2t} & te^{-2t} & \dfrac{1}{2!}t^2 e^{-2t} \\ 0 & e^{-2t} & te^{-2t} \\ 0 & 0 & e^{-2t} \end{bmatrix}.$$

The state transition matrix is $e^{At} = Pe^{Jt}P^{-1}$

$$= \begin{bmatrix} 1 & 0 & 0 \\ 4 & 1 & 0 \\ 4 & 2 & 1 \end{bmatrix} \begin{bmatrix} e^{-2t} & te^{-2t} & \dfrac{1}{2!}t^2 e^{-2t} \\ 0 & e^{-2t} & te^{-2t} \\ 0 & 0 & e^{-2t} \end{bmatrix} \begin{bmatrix} 1 & 0 & 0 \\ -4 & 1 & 0 \\ 4 & -2 & 1 \end{bmatrix}$$

$$= \begin{bmatrix} 1 & 0 & 0 \\ 4 & 1 & 0 \\ 4 & 2 & 1 \end{bmatrix} \begin{bmatrix} e^{-2t} - 4te^{-2t} + 2t^2 e^{-2t} & te^{-2t} - t^2 e^{-2t} & \dfrac{1}{2}t^2 e^{-2t} \\ -4e^{-2t} + 4te^{-2t} & e^{-2t} - 2te^{-2t} & te^{-2t} \\ 4e^{-2t} & -2e^{-2t} & e^{-2t} \end{bmatrix}$$

$$= \begin{bmatrix} e^{-2t} - 4te^{-2t} + 2t^2 e^{-2t} & te^{-2t} - t^2 e^{-2t} & \dfrac{1}{2}t^2 e^{-2t} \\ -12te^{-2t} + 8t^2 e^{-2t} & e^{-2t} + 2te^{-2t} - 4t^2 e^{-2t} & te^{-2t} + 2t^2 e^{-2t} \\ -8te^{-2t} + 8t^2 e^{-2t} & -4t^2 e^{-2t} & e^{-2t} + 2te^{-2t} + 2t^2 e^{-2t} \end{bmatrix}.$$

4.8 TRANSFORMATION FROM CONTROLLABLE CANONICAL FORM TO DIAGONAL FORM

A state model in controllable canonical form can be directly transformed into diagonal form. If the Eigen values are distinct, the transformation matrix is composed of Eigen vectors associated with each Eigen value. In this case of distinct Eigen values, the given state model can be transformed into diagonal form. If some Eigen values are repeated, the transformation matrix is composed of Eigen vectors and generalized Eigen vectors, in which case the given state model can be transformed into Jordan canonical form. Consider a state model of a system in controllable canonical form

$$\dot{X} = \begin{bmatrix} 0 & 1 & 0 \\ 0 & 0 & 1 \\ -a_0 & -a_1 & -a_2 \end{bmatrix} X + \begin{bmatrix} 0 \\ 0 \\ 1 \end{bmatrix} r .$$

$$[\lambda I - A] = \begin{bmatrix} \lambda & -1 & 0 \\ 0 & \lambda & -1 \\ a_0 & a_1 & \lambda + a_2 \end{bmatrix}$$

The matrix of cofactors of the first row of $[\lambda I - A] = \begin{bmatrix} +\{\lambda^2 + a_2\lambda + a_1\} \\ -\{a_0\} \\ +\{-a_0\lambda\} \end{bmatrix} = \begin{bmatrix} \lambda^2 + a_2\lambda + a_1 \\ -a_0 \\ -a_0\lambda \end{bmatrix} .$

The matrix of cofactors of the second row of $[\lambda I - A] = \begin{bmatrix} -\{-(\lambda + a_2)\} \\ +\{\lambda(\lambda + a_2)\} \\ -\{a_1\lambda + a_0\} \end{bmatrix} = \begin{bmatrix} (\lambda + a_2) \\ \lambda(\lambda + a_2) \\ -(a_1\lambda + a_0) \end{bmatrix} .$

The matrix of cofactors of the third row of $[\lambda I - A] = \begin{bmatrix} +(1) \\ -(-\lambda) \\ +(\lambda^2) \end{bmatrix} = \begin{bmatrix} 1 \\ \lambda \\ \lambda^2 \end{bmatrix} .$

It can be observed that it is convenient to select the matrix of cofactors of the third row of $[\lambda I - A]$ as the Eigen vectors.

Case 1: Distinct Eigen values

When the Eigen values are distinct, the Eigen vectors are given by

$$P_i = \begin{bmatrix} 1 \\ \lambda_i \\ \lambda_i^2 \\ \lambda_i^3 \\ \dots \\ \dots \\ \lambda_i^{n-1} \end{bmatrix} ; i = 1,2,3,\dots,n .$$

For a third-order system with distinct Eigen values λ_1, λ_2 and λ_3, the transformation matrix is

$$P = \begin{bmatrix} 1 & 1 & 1 \\ \lambda_1 & \lambda_2 & \lambda_3 \\ \lambda_1^2 & \lambda_2^2 & \lambda_3^2 \end{bmatrix}.$$

This transformation matrix P is called *Vander Monde matrix*. For an *n*-th order system with distinct Eigen values $\lambda_1, \lambda_2, \lambda_3, ..., \lambda_n$, the transformation matrix, or the Vander Monde matrix, is

$$P = \begin{bmatrix} 1 & 1 & 1 ... & 1 \\ \lambda_1 & \lambda_2 & \lambda_3 ... & \lambda_n \\ \lambda_1^2 & \lambda_2^2 & \lambda_3^2 ... & \lambda_n^2 \\ \lambda_1^3 & \lambda_2^3 & \lambda_3^3 ... & \lambda_n^3 \\ ... \\ ... \\ \lambda_1^{n-1} & \lambda_2^{n-1} & \lambda_3^{n-1} ... & \lambda_n^{n-1} \end{bmatrix}$$

Example 4.15 Show that a state model in controllable canonical form or in phase variable form with distinct Eigen values can be transformed into diagonal canonical form using the Vander Monde matrix.

Solution: Consider a third-order system with distinct Eigen values λ_1, λ_2 and λ_3. If it is represented either in the controllable canonical form or in phase variable form, the system matrix will be of the form

$$A = \begin{bmatrix} 0 & 1 & 0 \\ 0 & 0 & 1 \\ -a_0 & -a_1 & -a_2 \end{bmatrix}.$$

The characteristic equation is

$$|\lambda I - A| = \begin{vmatrix} \lambda & -1 & 0 \\ 0 & \lambda & -1 \\ a_0 & a_1 & \lambda + a_2 \end{vmatrix} = 0 \qquad or \quad \lambda\{(\lambda^2 + a_2\lambda) + a_1\} + a_0 = 0 .$$

Or
$$\lambda^3 + a_2\lambda^2 + a_1\lambda + a_0 = 0 \qquad\qquad (1)$$

The Vander Monde matrix is

$$P = \begin{bmatrix} 1 & 1 & 1 \\ \lambda_1 & \lambda_2 & \lambda_3 \\ \lambda_1^2 & \lambda_2^2 & \lambda_3^2 \end{bmatrix}.$$

$$AP = \begin{bmatrix} 0 & 1 & 0 \\ 0 & 0 & 1 \\ -a_0 & -a_1 & -a_2 \end{bmatrix} \begin{bmatrix} 1 & 1 & 1 \\ \lambda_1 & \lambda_2 & \lambda_3 \\ \lambda_1^2 & \lambda_2^2 & \lambda_3^2 \end{bmatrix}$$

$$= \begin{bmatrix} \lambda_1 & \lambda_2 & \lambda_3 \\ \lambda_1^2 & \lambda_2^2 & \lambda_3^2 \\ -(a_0 + a_1\lambda_1 + a_2\lambda_1^2) & -(a_0 + a_1\lambda_2 + a_2\lambda_2^2) & -(a_0 + a_1\lambda_3 + a_2\lambda_3^2) \end{bmatrix}$$

Using equation (1),

$$AP = \begin{bmatrix} \lambda_1 & \lambda_2 & \lambda_3 \\ \lambda_1^2 & \lambda_2^2 & \lambda_3^2 \\ \lambda_1^3 & \lambda_2^3 & \lambda_3^3 \end{bmatrix} = \begin{bmatrix} 1 & 1 & 1 \\ \lambda_1 & \lambda_2 & \lambda_3 \\ \lambda_1^2 & \lambda_2^2 & \lambda_3^2 \end{bmatrix} \begin{bmatrix} \lambda_1 & 0 & 0 \\ 0 & \lambda_2 & 0 \\ 0 & 0 & \lambda_3 \end{bmatrix} = P\Lambda.$$

$$AP = P\Lambda \qquad or \quad \Lambda = P^{-1}AP$$

Hence, a Vander Monde matrix transforms a state model in controllable canonical form or in phase variable form to the diagonal canonical form.

Case 2: Repeated Eigen values

A system with repeated Eigen values can be transformed into Jordan canonical form. Consider, for example, a fifth-order system with four repeated Eigen values. Let the Eigen values be λ_1, λ_1, λ_1 and λ_5 the Vander Monde matrix is given by $P = [P_1 \quad P_2 \quad P_3 \quad P_4 \quad P_5]$.

Let

$$P_v = \begin{bmatrix} 1 \\ \lambda \\ \lambda^2 \\ \lambda^3 \\ \lambda^4 \end{bmatrix}.$$

$$P_1 = [P_v]_{\lambda=\lambda_1} \; ; P_2 = \frac{d}{d\lambda}[P_v]_{\lambda=\lambda_1} \; ; P_3 = \frac{1}{2!}\frac{d^2}{d\lambda^2}[P_v]_{\lambda=\lambda_1} \; ; P_4 = \frac{1}{3!}\frac{d^3}{d\lambda^3}[P_v]_{\lambda=\lambda_1} \; ; P_5 = [P_v]_{\lambda=\lambda_5}$$

Example 4.16 A state model of a system in controllable canonical form is

$$\dot{X} = \begin{bmatrix} 0 & 1 \\ -2 & -3 \end{bmatrix} X + \begin{bmatrix} 0 \\ 1 \end{bmatrix} r$$

$$y = [3 \quad 1] X .$$

Transform the state model into diagonal canonical form.

Solution: The Eigen values are obtained by solving the equation

$$|\lambda I - A| = 0 \qquad or \begin{vmatrix} \lambda & -1 \\ 2 & \lambda + 3 \end{vmatrix} = 0 \qquad or \ \lambda^2 + 3\lambda + 2 = 0 \quad or \ (\lambda + 1)(\lambda + 2) = 0 .$$

The Eigen values are $\lambda_1 = -1$ *and* $\lambda_2 = -2$ and are distinct. Hence, the system matrix can be transformed to diagonal canonical form. Since the given state model is in controllable canonical form, the transformation matrix is the Vander Monde matrix. It is given by

$$P = \begin{bmatrix} 1 & 1 \\ \lambda_1 & \lambda_2 \end{bmatrix} = \begin{bmatrix} 1 & 1 \\ -1 & -2 \end{bmatrix} \quad |P| = \begin{vmatrix} 1 & 1 \\ -1 & -2 \end{vmatrix} = -1 \qquad P^{-1} = -\begin{bmatrix} -2 & -1 \\ 1 & 1 \end{bmatrix} = \begin{bmatrix} 2 & 1 \\ -1 & -1 \end{bmatrix} .$$

Check: $P^{-1}P = \begin{bmatrix} 2 & 1 \\ -1 & -1 \end{bmatrix} \begin{bmatrix} 1 & 1 \\ -1 & -2 \end{bmatrix} = \begin{bmatrix} 1 & 0 \\ 0 & 1 \end{bmatrix}$

$$P^{-1}AP = \begin{bmatrix} 2 & 1 \\ -1 & -1 \end{bmatrix} \begin{bmatrix} 0 & 1 \\ -2 & -3 \end{bmatrix} \begin{bmatrix} 1 & 1 \\ -1 & -2 \end{bmatrix} = \begin{bmatrix} 2 & 1 \\ -1 & -1 \end{bmatrix} \begin{bmatrix} -1 & -2 \\ 1 & 4 \end{bmatrix} = \begin{bmatrix} -1 & 0 \\ 0 & -2 \end{bmatrix} = \Lambda .$$

$$P^{-1}B = \begin{bmatrix} 2 & 1 \\ -1 & -1 \end{bmatrix} \begin{bmatrix} 0 \\ 1 \end{bmatrix} = \begin{bmatrix} 1 \\ -1 \end{bmatrix} \qquad CP = [3 \quad 1] \begin{bmatrix} 1 & 1 \\ -1 & -2 \end{bmatrix} = [2 \quad 1] .$$

The state model in diagonal canonical form is

$$\dot{Z} = \begin{bmatrix} -1 & 0 \\ 0 & -2 \end{bmatrix} Z + \begin{bmatrix} 1 \\ -1 \end{bmatrix} r \quad y = [2 \quad 1] Z .$$

Example 4.17 A state model of a system in phase variable form is

$$\dot{X} = \begin{bmatrix} 0 & 1 \\ -5 & -6 \end{bmatrix} X + \begin{bmatrix} 1 \\ 3 \end{bmatrix} r \quad y = [3 \quad -1] X .$$

Transform the state model into diagonal canonical form.

Solution: The Eigen values are obtained by solving the equation

$$|\lambda I - A| = 0 \qquad or \begin{vmatrix} \lambda & -1 \\ 5 & \lambda + 6 \end{vmatrix} = 0 \qquad or \ \lambda^2 + 6\lambda + 5 = 0 \quad or \ (\lambda + 1)(\lambda + 5) = 0 .$$

The Eigen values are $\lambda_1 = -1$ *and* $\lambda_2 = -5$ and are distinct. Hence, the system matrix can be transformed to diagonal canonical form. Since the given state model is in phase variable form, the transformation matrix is the Vander Monde matrix. It is given by

$$P = \begin{bmatrix} 1 & 1 \\ \lambda_1 & \lambda_2 \end{bmatrix} = \begin{bmatrix} 1 & 1 \\ -1 & -5 \end{bmatrix} \quad |P| = \begin{vmatrix} 1 & 1 \\ -1 & -5 \end{vmatrix} = -4 \quad P^{-1} = \frac{-1}{4}\begin{bmatrix} -5 & -1 \\ 1 & 1 \end{bmatrix} = \frac{1}{4}\begin{bmatrix} 5 & 1 \\ -1 & -1 \end{bmatrix}.$$

Check: $P^{-1}P = \dfrac{1}{4}\begin{bmatrix} 5 & 1 \\ -1 & -1 \end{bmatrix}\begin{bmatrix} 1 & 1 \\ -1 & -5 \end{bmatrix} = \begin{bmatrix} 1 & 0 \\ 0 & 1 \end{bmatrix}$

$$P^{-1}AP = \frac{1}{4}\begin{bmatrix} 5 & 1 \\ -1 & -1 \end{bmatrix}\begin{bmatrix} 0 & 1 \\ -5 & -6 \end{bmatrix}\begin{bmatrix} 1 & 1 \\ -1 & -5 \end{bmatrix} = \frac{1}{4}\begin{bmatrix} 5 & 1 \\ -1 & -1 \end{bmatrix}\begin{bmatrix} -1 & -5 \\ 1 & 25 \end{bmatrix} = \begin{bmatrix} -1 & 0 \\ 0 & -5 \end{bmatrix} = \Lambda$$

$$P^{-1}B = \frac{1}{4}\begin{bmatrix} 5 & 1 \\ -1 & -1 \end{bmatrix}\begin{bmatrix} 1 \\ 3 \end{bmatrix} = \begin{bmatrix} 2 \\ -1 \end{bmatrix} \qquad CP = \begin{bmatrix} 3 & -1 \end{bmatrix}\begin{bmatrix} 1 & 1 \\ -1 & -5 \end{bmatrix} = \begin{bmatrix} 4 & 8 \end{bmatrix}.$$

The state model in diagonal canonical form is

$$\dot{Z} = \begin{bmatrix} -1 & 0 \\ 0 & -5 \end{bmatrix}Z + \begin{bmatrix} 2 \\ -1 \end{bmatrix}r \quad y = \begin{bmatrix} 4 & 8 \end{bmatrix}Z .$$

Example 4.18 A linear system has a state model in controllable canonical form given by

$$\dot{X} = \begin{bmatrix} 0 & 1 & 0 \\ 0 & 0 & 1 \\ -15 & -23 & -9 \end{bmatrix}X + \begin{bmatrix} 0 \\ 0 \\ 1 \end{bmatrix}r \quad y = \begin{bmatrix} 1 & 2 & 1 \end{bmatrix}X .$$

Transform the state model into diagonal canonical form.

Solution: The Eigen values are obtained by solving the equation

$$|\lambda I - A| = \begin{vmatrix} \lambda & -1 & 0 \\ 0 & \lambda & -1 \\ 15 & 23 & \lambda + 9 \end{vmatrix} = 0$$

$$\lambda(\lambda^2 + 9\lambda + 23) + 15 = 0 \quad or \quad \lambda^3 + 9\lambda^2 + 23\lambda + 15 = 0 .$$

The Eigen values are $\lambda_1 = -1, \lambda_2 = -3$ *and* $\lambda_3 = -5$ and are distinct. Hence, the state model can be transformed into diagonal canonical form. Since the given state model is in controllable canonical form, the transformation matrix is the Vander Monde matrix.

$$P = \begin{bmatrix} 1 & 1 & 1 \\ \lambda_1 & \lambda_2 & \lambda_3 \\ \lambda_1^2 & \lambda_2^2 & \lambda_3^2 \end{bmatrix} = \begin{bmatrix} 1 & 1 & 1 \\ -1 & -3 & -5 \\ 1 & 9 & 25 \end{bmatrix}.$$

$$|P| = \begin{vmatrix} 1 & 1 & 1 \\ -1 & -3 & -5 \\ 1 & 9 & 25 \end{vmatrix} = (-75+45) - (-25+5) + (-9+3) = -16$$

$$P^{-1} = \frac{-1}{16} = \begin{bmatrix} +(-30) & -(16) & +(-2) \\ -(-20) & +(24) & -(-4) \\ +(-6) & -(8) & +(-2) \end{bmatrix} = \frac{1}{16} \begin{bmatrix} 30 & 16 & 2 \\ -20 & -24 & -4 \\ 6 & 8 & 2 \end{bmatrix}.$$

Check: $P^{-1}P = \dfrac{1}{16} \begin{bmatrix} 30 & 16 & 2 \\ -20 & -24 & -4 \\ 6 & 8 & 2 \end{bmatrix} \begin{bmatrix} 1 & 1 & 1 \\ -1 & -3 & -5 \\ 1 & 9 & 25 \end{bmatrix} = \begin{bmatrix} 1 & 0 & 0 \\ 0 & 1 & 0 \\ 0 & 0 & 1 \end{bmatrix}$

$$P^{-1}AP = \frac{1}{16} \begin{bmatrix} 30 & 16 & 2 \\ -20 & -24 & -4 \\ 6 & 8 & 2 \end{bmatrix} \begin{bmatrix} 0 & 1 & 0 \\ 0 & 0 & 1 \\ -15 & -23 & -9 \end{bmatrix} \begin{bmatrix} 1 & 1 & 1 \\ -1 & -3 & -5 \\ 1 & 9 & 25 \end{bmatrix}$$

$$= \frac{1}{16} \begin{bmatrix} 30 & 16 & 2 \\ -20 & -24 & -4 \\ 6 & 8 & 2 \end{bmatrix} \begin{bmatrix} -1 & -3 & -5 \\ 1 & 9 & 25 \\ -1 & -27 & -125 \end{bmatrix} = \begin{bmatrix} -1 & 0 & 0 \\ 0 & -3 & 0 \\ 0 & 0 & -5 \end{bmatrix}$$

$$P^{-1}B = \frac{1}{16} \begin{bmatrix} 30 & 16 & 2 \\ -20 & -24 & -4 \\ 6 & 8 & 2 \end{bmatrix} \begin{bmatrix} 0 \\ 0 \\ 1 \end{bmatrix} = \begin{bmatrix} 1/8 \\ -1/4 \\ 1/8 \end{bmatrix}$$

$$CP = \begin{bmatrix} 1 & 2 & 1 \end{bmatrix} \begin{bmatrix} 1 & 1 & 1 \\ -1 & -3 & -5 \\ 1 & 9 & 25 \end{bmatrix} = \begin{bmatrix} 0 & 4 & 16 \end{bmatrix}.$$

The state model in diagonal canonical form is

$$\dot{Z} = \begin{bmatrix} -1 & 0 & 0 \\ 0 & -3 & 0 \\ 0 & 0 & -5 \end{bmatrix} Z + \begin{bmatrix} 1/8 \\ -1/4 \\ 1/8 \end{bmatrix} r \quad y = \begin{bmatrix} 0 & 4 & 16 \end{bmatrix} Z.$$

Example 4.19 A state model of a system in controllable canonical form is

$$\dot{X} = \begin{bmatrix} 0 & 1 \\ -4 & -4 \end{bmatrix} X + \begin{bmatrix} 0 \\ 1 \end{bmatrix} r \quad y = \begin{bmatrix} 2 & -1 \end{bmatrix} X.$$

Transform the state model into Jordan canonical form.

Solution: The Eigen values are obtained by solving the equation

$$|\lambda I - A| = 0 \qquad or \quad \begin{vmatrix} \lambda & -1 \\ 4 & \lambda + 4 \end{vmatrix} = 0 \qquad or \quad \lambda^2 + 4\lambda + 4 = 0 \quad or \quad (\lambda + 2)^2 = 0.$$

The Eigen values are $\lambda_1 = -2$ *and* $\lambda_2 = -2$ and are repeated. Hence, the system matrix can be transformed to Jordan canonical form. Since the given state model is in controllable canonical form, the transformation matrix is the Vander Monde matrix.

$$P_v = \begin{bmatrix} 1 \\ \lambda \end{bmatrix} \qquad P_1 = [P_v]_{\lambda = -2} = \begin{bmatrix} 1 \\ -2 \end{bmatrix} \qquad P_2 = \frac{d}{d\lambda}[P_v]_{\lambda = -2} = \begin{bmatrix} 0 \\ 1 \end{bmatrix}$$

$$P = \begin{bmatrix} 1 & 0 \\ -2 & 1 \end{bmatrix} \qquad |P| = \begin{vmatrix} 1 & 0 \\ -2 & 1 \end{vmatrix} = 1 \qquad P^{-1} = \begin{bmatrix} 1 & 0 \\ 2 & 1 \end{bmatrix}.$$

Check: $P^{-1}P = \begin{bmatrix} 1 & 0 \\ 2 & 1 \end{bmatrix}\begin{bmatrix} 1 & 0 \\ -2 & 1 \end{bmatrix} = \begin{bmatrix} 1 & 0 \\ 0 & 1 \end{bmatrix}$

$$P^{-1}AP = \begin{bmatrix} 1 & 0 \\ 2 & 1 \end{bmatrix}\begin{bmatrix} 0 & 1 \\ -4 & -4 \end{bmatrix}\begin{bmatrix} 1 & 0 \\ -2 & 1 \end{bmatrix} = \begin{bmatrix} 1 & 0 \\ 2 & 1 \end{bmatrix}\begin{bmatrix} -2 & 1 \\ 4 & -4 \end{bmatrix} = \begin{bmatrix} -2 & 1 \\ 0 & -2 \end{bmatrix} = J$$

$$P^{-1}B = \begin{bmatrix} 1 & 0 \\ 2 & 1 \end{bmatrix}\begin{bmatrix} 0 \\ 1 \end{bmatrix} = \begin{bmatrix} 0 \\ 1 \end{bmatrix} \qquad CP = [2 \quad -1]\begin{bmatrix} 1 & 0 \\ -2 & 1 \end{bmatrix} = [4 \quad -1].$$

The state model in diagonal canonical form is

$$\dot{Z} = \begin{bmatrix} -2 & 1 \\ 0 & -2 \end{bmatrix}Z + \begin{bmatrix} 0 \\ 1 \end{bmatrix}r \quad y = [4 \quad -1]Z.$$

Example 4.20 A state model of a system in phase variable form is

$$\dot{X} = \begin{bmatrix} 0 & 1 \\ -9 & -6 \end{bmatrix}X + \begin{bmatrix} 1 \\ 2 \end{bmatrix}r$$

$$y = [1 \quad 2]X.$$

Transform the state model into Jordan canonical form.

Solution: The Eigen values are obtained by solving the equation

$$|\lambda I - A| = 0 \qquad or \quad \begin{vmatrix} \lambda & -1 \\ 9 & \lambda + 6 \end{vmatrix} = 0 \qquad or \quad \lambda^2 + 6\lambda + 9 = 0 \quad or \quad (\lambda + 3)^2 = 0.$$

The Eigen values are $\lambda_1 = -3$ *and* $\lambda_2 = -3$ and are repeated. Hence, the system matrix can be transformed to Jordan canonical form. Since the given state model is in phase variable form, the transformation matrix is the Vander Monde matrix.

$$P_v = \begin{bmatrix} 1 \\ \lambda \end{bmatrix} \qquad P_1 = [P_v]_{\lambda = -3} = \begin{bmatrix} 1 \\ -3 \end{bmatrix} \qquad P_2 = \frac{d}{d\lambda}[P_v]_{\lambda = -3} = \begin{bmatrix} 0 \\ 1 \end{bmatrix}$$

$$P = \begin{bmatrix} 1 & 0 \\ -3 & 1 \end{bmatrix} \qquad |P| = \begin{vmatrix} 1 & 0 \\ -3 & 1 \end{vmatrix} = 1 \qquad P^{-1} = \begin{bmatrix} 1 & 0 \\ 3 & 1 \end{bmatrix}.$$

Check: $P^{-1}P = \begin{bmatrix} 1 & 0 \\ 3 & 1 \end{bmatrix}\begin{bmatrix} 1 & 0 \\ -3 & 1 \end{bmatrix} = \begin{bmatrix} 1 & 0 \\ 0 & 1 \end{bmatrix}$

$$P^{-1}AP = \begin{bmatrix} 1 & 0 \\ 3 & 1 \end{bmatrix}\begin{bmatrix} 0 & 1 \\ -9 & -6 \end{bmatrix}\begin{bmatrix} 1 & 0 \\ -3 & 1 \end{bmatrix} = \begin{bmatrix} 1 & 0 \\ 3 & 1 \end{bmatrix}\begin{bmatrix} -3 & 1 \\ 9 & -6 \end{bmatrix} = \begin{bmatrix} -3 & 1 \\ 0 & -3 \end{bmatrix} = J$$

$$P^{-1}B = \begin{bmatrix} 1 & 0 \\ 3 & 2 \end{bmatrix}\begin{bmatrix} 1 \\ 2 \end{bmatrix} = \begin{bmatrix} 1 \\ 7 \end{bmatrix} \qquad CP = \begin{bmatrix} 1 & 2 \end{bmatrix}\begin{bmatrix} 1 & 0 \\ -3 & 1 \end{bmatrix} = \begin{bmatrix} -5 & 2 \end{bmatrix}.$$

The state model in diagonal canonical form is

$$\dot{Z} = \begin{bmatrix} -3 & 1 \\ 0 & -3 \end{bmatrix}Z + \begin{bmatrix} 1 \\ 7 \end{bmatrix}r \qquad y = \begin{bmatrix} -5 & 2 \end{bmatrix}Z.$$

Example 4.21 A state model of a linear system is

$$\dot{X} = \begin{bmatrix} 0 & 1 & 0 \\ 0 & 0 & 1 \\ -18 & -21 & -8 \end{bmatrix}X + \begin{bmatrix} 0 \\ 0 \\ 1 \end{bmatrix}r \quad y = \begin{bmatrix} 4 & -2 & 1 \end{bmatrix}X.$$

Transform the state model into Jordan canonical form.

Solution: The Eigen values are obtained by solving the equation

$$|\lambda I - A| = 0 \qquad \begin{vmatrix} \lambda & -1 & 0 \\ 0 & \lambda & -1 \\ 18 & 21 & \lambda+8 \end{vmatrix} = 0$$

$$\lambda^3 + 8\lambda^2 + 21\lambda + 18 = 0 \quad or \quad (\lambda+2)(\lambda+3)^2 = 0.$$

The Eigen values are $\lambda_1 = -2, \lambda_2 = -3$ and $\lambda_3 = -3$ and are repeated. Hence, the system matrix can be transformed to Jordan canonical form. Since the given state model is in controllable canonical form, the transformation matrix is the Vander Monde matrix.

$$P_v = \begin{bmatrix} 1 \\ \lambda \\ \lambda^2 \end{bmatrix} \quad P_1 = [P_v]_{\lambda=-2} = \begin{bmatrix} 1 \\ -2 \\ 4 \end{bmatrix} \quad P_2 = [P_v]_{\lambda=-3} = \begin{bmatrix} 1 \\ -3 \\ 9 \end{bmatrix} \quad P_3 = \frac{d}{d\lambda}[P_v]_{\lambda=-3} = \begin{bmatrix} 0 \\ 1 \\ 2\lambda \end{bmatrix}_{\lambda=-3} = \begin{bmatrix} 0 \\ 1 \\ -6 \end{bmatrix}.$$

$$P = \begin{bmatrix} 1 & 1 & 0 \\ -2 & -3 & 1 \\ 4 & 9 & -6 \end{bmatrix} \qquad |P| = \begin{vmatrix} 1 & 1 & 0 \\ -2 & -3 & 1 \\ 4 & 9 & -6 \end{vmatrix} = 18 - 9 - (12 - 4) = 1$$

$$P^{-1} = \begin{bmatrix} +9 & -(-6) & +1 \\ -(8) & +(-6) & -1 \\ +(-6) & -(5) & +(-1) \end{bmatrix} = \begin{bmatrix} 9 & 6 & 1 \\ -8 & -6 & -1 \\ -6 & -5 & -1 \end{bmatrix}.$$

Check: $P^{-1}P = \begin{bmatrix} 9 & 6 & 1 \\ -8 & -6 & -1 \\ -6 & -5 & -1 \end{bmatrix} \begin{bmatrix} 1 & 1 & 0 \\ -2 & -3 & 1 \\ 4 & 9 & -6 \end{bmatrix} = \begin{bmatrix} 1 & 0 & 0 \\ 0 & 1 & 0 \\ 0 & 0 & 1 \end{bmatrix}$

$$P^{-1}AP = \begin{bmatrix} 9 & 6 & 1 \\ -8 & -6 & -1 \\ -6 & -5 & -1 \end{bmatrix} \begin{bmatrix} 0 & 1 & 0 \\ 0 & 0 & 1 \\ -18 & -21 & -8 \end{bmatrix} \begin{bmatrix} 1 & 1 & 0 \\ -2 & -3 & 1 \\ 4 & 9 & -6 \end{bmatrix}$$

$$= \begin{bmatrix} 9 & 6 & 1 \\ -8 & -6 & -1 \\ -6 & -5 & -1 \end{bmatrix} \begin{bmatrix} -2 & -3 & 1 \\ 4 & 9 & -6 \\ -8 & -27 & 27 \end{bmatrix} = \begin{bmatrix} -2 & 0 & 0 \\ 0 & -3 & 1 \\ 0 & 0 & -3 \end{bmatrix}$$

$$P^{-1}B = \begin{bmatrix} 9 & 6 & 1 \\ -8 & -6 & -1 \\ -6 & -5 & -1 \end{bmatrix} \begin{bmatrix} 0 \\ 0 \\ 1 \end{bmatrix} = \begin{bmatrix} 1 \\ -1 \\ -1 \end{bmatrix} \quad CP = \begin{bmatrix} 4 & -2 & -1 \end{bmatrix} \begin{bmatrix} 1 & 1 & 0 \\ -2 & -3 & 1 \\ 4 & 9 & -6 \end{bmatrix} = \begin{bmatrix} 4 & 1 & 4 \end{bmatrix}.$$

The state model in diagonal canonical form is

$$\dot{Z} = \begin{bmatrix} -2 & 0 & 0 \\ 0 & -3 & 1 \\ 0 & 0 & -3 \end{bmatrix} Z + \begin{bmatrix} 1 \\ -1 \\ -1 \end{bmatrix} r \quad y = \begin{bmatrix} 4 & 1 & 4 \end{bmatrix} Z.$$

4.9 TRANSFORMATION FROM OBSERVABLE CANONICAL FORM TO DIAGONAL FORM

A state model in observable canonical form can be transformed into diagonal form, if the Eigen values are distinct. If some Eigen values are repeated, the given state model can be transformed into Jordan canonical form.

Case 1: Distinct Eigen values

Consider a fifth-order system, the state model of which in observable canonical form can be represented as

$$\dot{X} = \begin{bmatrix} -a_4 & 1 & 0 & 0 & 0 \\ -a_3 & 0 & 1 & 0 & 0 \\ -a_2 & 0 & 0 & 1 & 0 \\ -a_1 & 0 & 0 & 0 & 1 \\ -a_0 & 0 & 0 & 0 & 0 \end{bmatrix} X + \begin{bmatrix} b_4 \\ b_3 \\ b_2 \\ b_1 \\ b_0 \end{bmatrix} r \qquad y = \begin{bmatrix} 1 & 0 & 0 & 0 & 0 \end{bmatrix} X.$$

The characteristic equation is

$$\lambda^5 + a_4\lambda^4 + a_3\lambda^3 + a_2\lambda^2 + a_1\lambda + a_0 = 0.$$

Consider that the Eigen values are distinct. The transformation matrix P that transforms the state model in observable canonical form to diagonal canonical form can be determined as explained below:

$$P_v = \begin{bmatrix} 1 \\ \lambda + a_4 \\ \lambda^2 + a_4\lambda + a_3 \\ \lambda^3 + a_4\lambda^2 + a_3\lambda + a_2 \\ \lambda^4 + a_4\lambda^3 + a_3\lambda^2 + a_2\lambda + a_1 \end{bmatrix}$$

$$P_1 = [P_v]_{\lambda=\lambda_1} \quad P_2 = [P_v]_{\lambda=\lambda_2} \quad P_3 = [P_v]_{\lambda=\lambda_3} \quad P_4 = [P_v]_{\lambda=\lambda_4} \quad P_5 = [P_v]_{\lambda=\lambda_5}$$

$$P = [P_1 \quad P_2 \quad P_3 \quad P_4 \quad P_5].$$

Case 2: Repeated Eigen values

For a fifth-order system, let the characteristic equation be

$$\lambda^5 + a_4\lambda^4 + a_3\lambda^3 + a_2\lambda^2 + a_1\lambda + a_0 = 0.$$

Consider that the Eigen values are $\lambda_1, \lambda_1, \lambda_1, \lambda_1$ and λ_5. Four Eigen values are repeated. The transformation matrix P that transforms the state model in observable canonical form can be determined as explained below:

$$P_v = \begin{bmatrix} 1 \\ \lambda + a_4 \\ \lambda^2 + a_4\lambda + a_3 \\ \lambda^3 + a_4\lambda^2 + a_3\lambda + a_2 \\ \lambda^4 + a_4\lambda^3 + a_3\lambda^2 + a_2\lambda + a_1 \end{bmatrix}$$

$$P_1 = [P_v]_{\lambda=\lambda_1} ; P_2 = \frac{d}{d\lambda}[P_v]_{\lambda=\lambda_1} ; P_3 = \frac{1}{2!}\frac{d^2}{d\lambda^2}[P_v]_{\lambda=\lambda_1} ; P_4 = \frac{1}{3!}\frac{d^3}{d\lambda^3}[P_v]_{\lambda=\lambda_1} ; P_5 = [P_v]_{\lambda=\lambda_5}$$

$$P = [P_1 \quad P_2 \quad P_3 \quad P_4 \quad P_5].$$

Example 4.22 The state model of a system in observable canonical form is

$$\dot{X} = \begin{bmatrix} -3 & 1 \\ -2 & 0 \end{bmatrix} X + \begin{bmatrix} 1 \\ 2 \end{bmatrix} r \quad y = [1 \quad 0]X.$$

Convert this model into diagonal form.

Solution: The given state model is in observable canonical form.

$$A = \begin{bmatrix} -3 & 1 \\ -2 & 0 \end{bmatrix} \qquad [\lambda I - A] = \begin{bmatrix} \lambda+3 & -1 \\ 2 & \lambda \end{bmatrix}.$$

The characteristic equation is

$$\lambda^2 + 3\lambda + 2 = 0 = \lambda^2 + a_1\lambda + a_0.$$

The Eigen values are $\lambda_1 = -1$ *and* $\lambda_2 = -2$.

The Eigen values are distinct so that the given state model can be transformed into diagonal canonical form.

$$P_v = \begin{bmatrix} 1 \\ \lambda+a_1 \end{bmatrix} = \begin{bmatrix} 1 \\ \lambda+3 \end{bmatrix} \qquad P_1 = \begin{bmatrix} 1 \\ \lambda+3 \end{bmatrix}_{\lambda=-1} = \begin{bmatrix} 1 \\ -1+3 \end{bmatrix} = \begin{bmatrix} 1 \\ 2 \end{bmatrix}$$

$$P_2 = \begin{bmatrix} 1 \\ \lambda+3 \end{bmatrix}_{\lambda=-2} = \begin{bmatrix} 1 \\ -2+3 \end{bmatrix} = \begin{bmatrix} 1 \\ 1 \end{bmatrix}.$$

The transformation matrix is

$$P = \begin{bmatrix} 1 & 1 \\ 2 & 1 \end{bmatrix} \quad |P| = -1 \quad P^{-1} = \begin{bmatrix} -1 & 1 \\ 2 & -1 \end{bmatrix}$$

$$P^{-1}AP = \begin{bmatrix} -1 & 1 \\ 2 & -1 \end{bmatrix}\begin{bmatrix} -3 & 1 \\ -2 & 0 \end{bmatrix}\begin{bmatrix} 1 & 1 \\ 2 & 1 \end{bmatrix} = \begin{bmatrix} -1 & 1 \\ 2 & -1 \end{bmatrix}\begin{bmatrix} -1 & -2 \\ -2 & -2 \end{bmatrix} = \begin{bmatrix} -1 & 0 \\ 0 & -2 \end{bmatrix}$$

$$P^{-1}B = \begin{bmatrix} -1 & 1 \\ 2 & -1 \end{bmatrix}\begin{bmatrix} 1 \\ 2 \end{bmatrix} = \begin{bmatrix} 1 \\ 0 \end{bmatrix} \qquad CP = \begin{bmatrix} 1 & 0 \end{bmatrix}\begin{bmatrix} 1 & 1 \\ 2 & 1 \end{bmatrix} = \begin{bmatrix} 1 & 1 \end{bmatrix}.$$

The state model in diagonal canonical form is

$$\dot{X} = \begin{bmatrix} -1 & 0 \\ 0 & -2 \end{bmatrix}X + \begin{bmatrix} 1 \\ 0 \end{bmatrix}r \quad y = \begin{bmatrix} 1 & 1 \end{bmatrix}X.$$

Example 4.23 The state model of a system in observable canonical form is

$$\dot{X} = \begin{bmatrix} -6 & 1 & 0 \\ -11 & 0 & 1 \\ -6 & 0 & 0 \end{bmatrix}X + \begin{bmatrix} 1 \\ 2 \\ 3 \end{bmatrix}r \quad y = \begin{bmatrix} 1 & 0 & 0 \end{bmatrix}X.$$

Convert this model into diagonal form.

Solution:

$$A = \begin{bmatrix} -6 & 1 & 0 \\ -11 & 0 & 1 \\ -6 & 0 & 0 \end{bmatrix} \qquad [\lambda I - A] = \begin{bmatrix} \lambda+6 & -1 & 0 \\ 11 & \lambda & -1 \\ 6 & 0 & \lambda \end{bmatrix}.$$

The characteristic equation is

$$\lambda^3 + 6\lambda^2 + 11\lambda + 6 = 0 = \lambda^3 + a_2\lambda^2 + a_1\lambda + a_0$$

$$a_2 = 6, a_1 = 11 \ and \ a_0 = 6$$

The Eigen values are $\lambda_1 = -1, \lambda_2 = -2$ and $\lambda_3 = -3$.
The Eigen values are distinct so that the given state model can be transformed into diagonal canonical form.

$$P_v = \begin{bmatrix} 1 \\ \lambda + a_2 \\ \lambda^2 + a_2\lambda + a_1 \end{bmatrix} = \begin{bmatrix} 1 \\ \lambda + 6 \\ \lambda^2 + 6\lambda + 11 \end{bmatrix} \qquad P_1 = \begin{bmatrix} 1 \\ \lambda + 6 \\ \lambda^2 + 6\lambda + 11 \end{bmatrix}_{\lambda = -1} = \begin{bmatrix} 1 \\ 5 \\ 6 \end{bmatrix}$$

$$P_2 = \begin{bmatrix} 1 \\ \lambda + 6 \\ \lambda^2 + 6\lambda + 11 \end{bmatrix}_{\lambda = -2} = \begin{bmatrix} 1 \\ 4 \\ 3 \end{bmatrix} \qquad P_3 = \begin{bmatrix} 1 \\ \lambda + 6 \\ \lambda^2 + 6\lambda + 11 \end{bmatrix}_{\lambda = -3} = \begin{bmatrix} 1 \\ 3 \\ 2 \end{bmatrix}.$$

The transformation matrix is

$$P = \begin{bmatrix} 1 & 1 & 1 \\ 5 & 4 & 3 \\ 6 & 3 & 2 \end{bmatrix} \qquad\qquad |P| = -2.$$

$$P^{-1} = \frac{-1}{2}\begin{bmatrix} +\{-1\} & -\{-1\} & +\{-1\} \\ -\{-8\} & +\{-4\} & -\{-2\} \\ +\{-9\} & -\{-3\} & +\{-1\} \end{bmatrix} = \frac{1}{2}\begin{bmatrix} 1 & -1 & 1 \\ -8 & 4 & -2 \\ 9 & -3 & 1 \end{bmatrix}$$

$$P^{-1}AP = \frac{1}{2}\begin{bmatrix} 1 & -1 & 1 \\ -8 & 4 & -2 \\ 9 & -3 & 1 \end{bmatrix}\begin{bmatrix} -6 & 1 & 0 \\ -11 & 0 & 1 \\ -6 & 0 & 0 \end{bmatrix}\begin{bmatrix} 1 & 1 & 1 \\ 5 & 4 & 3 \\ 6 & 3 & 2 \end{bmatrix}$$

$$= \frac{1}{2}\begin{bmatrix} 1 & -1 & 1 \\ -8 & 4 & -2 \\ 9 & -3 & 1 \end{bmatrix}\begin{bmatrix} -1 & -2 & -3 \\ -5 & -8 & -9 \\ -6 & -6 & -6 \end{bmatrix} = \begin{bmatrix} -1 & 0 & 0 \\ 0 & -2 & 0 \\ 0 & 0 & -3 \end{bmatrix}$$

$$P^{-1}B = \frac{1}{2}\begin{bmatrix} 1 & -1 & 1 \\ -8 & 4 & -2 \\ 9 & -3 & 1 \end{bmatrix}\begin{bmatrix} 1 \\ 2 \\ 3 \end{bmatrix} = \begin{bmatrix} 2 \\ -6 \\ 6 \end{bmatrix}$$

$$CP = [1 \quad 0 \quad 0] \begin{bmatrix} 1 & 1 & 1 \\ 5 & 4 & 3 \\ 6 & 3 & 2 \end{bmatrix} = [1 \quad 1 \quad 1]$$

$$\dot{Z} = \begin{bmatrix} -1 & 0 & 0 \\ 0 & -2 & 0 \\ 0 & 0 & -3 \end{bmatrix} Z + \begin{bmatrix} 2 \\ -6 \\ 6 \end{bmatrix} r \qquad y = [1 \quad 1 \quad 1] Z .$$

Example 4.24 The state model of a system in observable canonical form is

$$\dot{X} = \begin{bmatrix} -4 & 1 \\ -4 & 0 \end{bmatrix} X + \begin{bmatrix} 3 \\ 1 \end{bmatrix} r \qquad y = [1 \quad 0] X .$$

Convert this model into diagonal form.

Solution:

$$A = \begin{bmatrix} -4 & 1 \\ -4 & 0 \end{bmatrix} \qquad\qquad [\lambda I - A] = \begin{bmatrix} \lambda + 4 & -1 \\ 4 & \lambda \end{bmatrix} .$$

The characteristic equation is

$$\lambda^2 + 4\lambda + 4 = 0 = \lambda^2 + a_1\lambda + a_0 \qquad a_1 = 4 \qquad a_0 = 4 .$$

The Eigen values are $\lambda_1 = -2$ and $\lambda_2 = -2$.
The Eigen values are repeated so that the given state model can be transformed into Jordan canonical form.

$$P_v = \begin{bmatrix} 1 \\ \lambda + a_1 \end{bmatrix} \qquad P_1 = \begin{bmatrix} 1 \\ \lambda + 4 \end{bmatrix}_{\lambda = -2} = \begin{bmatrix} 1 \\ 2 \end{bmatrix} \qquad P_2 = \frac{d}{d\lambda} \begin{bmatrix} 1 \\ \lambda + 4 \end{bmatrix}_{\lambda = -2} = \begin{bmatrix} 0 \\ 1 \end{bmatrix} .$$

The transformation matrix is

$$P = [P_1 \quad P_2] = \begin{bmatrix} 1 & 0 \\ 2 & 1 \end{bmatrix} \qquad |P| = 1 \qquad P^{-1} = \begin{bmatrix} 1 & 0 \\ -2 & 1 \end{bmatrix}$$

$$P^{-1}AP = \begin{bmatrix} 1 & 0 \\ -2 & 1 \end{bmatrix} \begin{bmatrix} -4 & 1 \\ -4 & 0 \end{bmatrix} \begin{bmatrix} 1 & 0 \\ 2 & 1 \end{bmatrix} = \begin{bmatrix} 1 & 0 \\ -2 & 1 \end{bmatrix} \begin{bmatrix} -2 & 1 \\ -4 & 0 \end{bmatrix} = \begin{bmatrix} -2 & 1 \\ 0 & -2 \end{bmatrix} = J$$

$$P^{-1}B = \begin{bmatrix} 1 & 0 \\ -2 & 1 \end{bmatrix} \begin{bmatrix} 3 \\ 1 \end{bmatrix} = \begin{bmatrix} 3 \\ -5 \end{bmatrix} \qquad CP = [1 \quad 0] \begin{bmatrix} 1 & 0 \\ 2 & 1 \end{bmatrix} = [1 \quad 0] .$$

The state model in Jordan canonical form is

$$\dot{X} = \begin{bmatrix} -2 & 1 \\ 0 & -2 \end{bmatrix} X + \begin{bmatrix} 3 \\ -5 \end{bmatrix} r \qquad y = [1 \quad 0] X .$$

Example 4.25 The state model of a system in observable canonical form is

$$\dot{X} = \begin{bmatrix} -7 & 1 & 0 \\ -15 & 0 & 1 \\ -9 & 0 & 0 \end{bmatrix} X + \begin{bmatrix} 1 \\ 3 \\ 2 \end{bmatrix} r \quad y = \begin{bmatrix} 1 & 0 & 0 \end{bmatrix} X.$$

Convert this model into diagonal form.

Solution:

$$A = \begin{bmatrix} -7 & 1 & 0 \\ -15 & 0 & 1 \\ -9 & 0 & 0 \end{bmatrix} \qquad [\lambda I - A] = \begin{bmatrix} \lambda+7 & -1 & 0 \\ 15 & \lambda & -1 \\ 9 & 0 & \lambda \end{bmatrix}.$$

The characteristic equation is

$$\lambda^3 + 7\lambda^2 + 15\lambda + 9 = 0 = \lambda^3 + a_2\lambda^2 + a_1\lambda + a_0 \qquad\qquad a_2 = 7, a_1 = 15 \text{ and } a_0 = 9$$

The Eigen values are $\lambda_1 = -1, \lambda_2 = -3$ and $\lambda_3 = -3$.

Two Eigen values are repeated so that the given state model can be transformed into Jordan canonical form.

$$P_v = \begin{bmatrix} 1 \\ \lambda+a_2 \\ \lambda^2 + a_2\lambda + a_1 \end{bmatrix} = \begin{bmatrix} 1 \\ \lambda+7 \\ \lambda^2 + 7\lambda + 15 \end{bmatrix}$$

$$P_1 = \begin{bmatrix} 1 \\ \lambda+7 \\ \lambda^2 + 7\lambda + 15 \end{bmatrix}_{\lambda=-1} = \begin{bmatrix} 1 \\ 6 \\ 9 \end{bmatrix} \qquad P_2 = \begin{bmatrix} 1 \\ \lambda+7 \\ \lambda^2 + 7\lambda + 15 \end{bmatrix}_{\lambda=-3} = \begin{bmatrix} 1 \\ 4 \\ 3 \end{bmatrix}$$

$$P_3 = \begin{bmatrix} \dfrac{d}{d\lambda}(1) \\ \dfrac{d}{d\lambda}(\lambda+7) \\ \dfrac{d}{d\lambda}(\lambda^2 + 7\lambda + 15) \end{bmatrix}_{\lambda=-3} = \begin{bmatrix} 0 \\ 1 \\ 2\lambda+7 \end{bmatrix}_{\lambda=-3} = \begin{bmatrix} 0 \\ 1 \\ 1 \end{bmatrix}.$$

The transformation matrix is

$$P = [P_1 \quad P_2 \quad P_3] = \begin{bmatrix} 1 & 1 & 0 \\ 6 & 4 & 1 \\ 9 & 3 & 1 \end{bmatrix} \qquad\qquad |P| = (4-3) - (6-9) = 4$$

$$P^{-1} = \frac{1}{4}\begin{bmatrix} +\{1\} & -\{1\} & +\{1\} \\ -\{-3\} & +\{1\} & -\{1\} \\ +\{-18\} & -\{-6\} & +\{-2\} \end{bmatrix} = \frac{1}{4}\begin{bmatrix} 1 & -1 & 1 \\ 3 & 1 & -1 \\ -18 & 6 & -2 \end{bmatrix}$$

$$P^{-1}AP = \frac{1}{4}\begin{bmatrix} 1 & -1 & 1 \\ 3 & 1 & -1 \\ -18 & 6 & -2 \end{bmatrix}\begin{bmatrix} -7 & 1 & 0 \\ -15 & 0 & 1 \\ -9 & 0 & 0 \end{bmatrix}\begin{bmatrix} 1 & 1 & 0 \\ 6 & 4 & 1 \\ 9 & 3 & 1 \end{bmatrix}$$

$$= \frac{1}{4}\begin{bmatrix} 1 & -1 & 1 \\ 3 & 1 & -1 \\ -18 & 6 & -2 \end{bmatrix}\begin{bmatrix} -1 & -3 & 1 \\ -6 & -12 & 1 \\ -9 & -9 & 0 \end{bmatrix} = \begin{bmatrix} -1 & 0 & 0 \\ 0 & -3 & 1 \\ 0 & 0 & -3 \end{bmatrix} = J$$

$$P^{-1}B = \frac{1}{4}\begin{bmatrix} 1 & -1 & 1 \\ 3 & 1 & -1 \\ -18 & 6 & -2 \end{bmatrix}\begin{bmatrix} 1 \\ 3 \\ 2 \end{bmatrix} = \begin{bmatrix} 0 \\ 1 \\ -1 \end{bmatrix}$$

$$CP = \begin{bmatrix} 1 & 0 & 0 \end{bmatrix}\begin{bmatrix} 1 & 1 & 0 \\ 6 & 4 & 1 \\ 9 & 3 & 1 \end{bmatrix} = \begin{bmatrix} 1 & 1 & 0 \end{bmatrix}.$$

The state model in Jordan canonical form is

$$\dot{X} = \begin{bmatrix} -1 & 0 & 0 \\ 0 & -3 & 1 \\ 0 & 0 & -3 \end{bmatrix}X + \begin{bmatrix} 0 \\ 1 \\ -1 \end{bmatrix}r \qquad y = \begin{bmatrix} 1 & 1 & 0 \end{bmatrix}X.$$

Example 4.26 The state model of a system in observable canonical form is

$$\dot{X} = \begin{bmatrix} -6 & 1 & 0 \\ -12 & 0 & 1 \\ -8 & 0 & 0 \end{bmatrix}X + \begin{bmatrix} 1 \\ 2 \\ 5 \end{bmatrix}r \qquad y = \begin{bmatrix} 1 & 0 & 0 \end{bmatrix}X.$$

Convert this model into diagonal form.

Solution:

$$A = \begin{bmatrix} -6 & 1 & 0 \\ -12 & 0 & 1 \\ -8 & 0 & 0 \end{bmatrix} \qquad [\lambda I - A] = \begin{bmatrix} \lambda+6 & -1 & 0 \\ 12 & \lambda & -1 \\ 8 & 0 & \lambda \end{bmatrix}.$$

The characteristic equation is

$$\lambda^3 + 6\lambda^2 + 12\lambda + 8 = 0 = \lambda^3 + a_2\lambda^2 + a_1\lambda + a_0 \qquad\qquad a_2 = 6, a_1 = 12 \text{ and } a_0 = 8.$$

$(\lambda+2)^3 = 0$. The Eigen values are $\lambda_1 = -2, \lambda_2 = -2$ and $\lambda_3 = -2$.

The three Eigen values are repeated so that the given state model can be transformed into Jordan canonical form.

$$P_v = \begin{bmatrix} 1 \\ \lambda + a_2 \\ \lambda^2 + a_2\lambda + a_1 \end{bmatrix} = \begin{bmatrix} 1 \\ \lambda + 6 \\ \lambda^2 + 6\lambda + 12 \end{bmatrix}$$

$$P_1 = \begin{bmatrix} 1 \\ \lambda + 6 \\ \lambda^2 + 6\lambda + 12 \end{bmatrix}_{\lambda=-2} = \begin{bmatrix} 1 \\ 4 \\ 4 \end{bmatrix}$$

$$P_2 = \frac{d}{d\lambda} \begin{bmatrix} 1 \\ \lambda + 6 \\ \lambda^2 + 6\lambda + 12 \end{bmatrix}_{\lambda=-2} = \begin{bmatrix} 0 \\ 1 \\ 2\lambda + 6 \end{bmatrix}_{\lambda=-2} = \begin{bmatrix} 0 \\ 1 \\ 2 \end{bmatrix}$$

$$P_3 = \frac{1}{2!} \frac{d^2}{d\lambda^2} \begin{bmatrix} 1 \\ \lambda + 6 \\ \lambda^2 + 6\lambda + 12 \end{bmatrix}_{\lambda=-3} = \frac{1}{2!} \frac{d}{d\lambda} \begin{bmatrix} 0 \\ 1 \\ 2\lambda + 6 \end{bmatrix}_{\lambda=-2} = \begin{bmatrix} 0 \\ 0 \\ 1 \end{bmatrix}$$

$$P = [P_1 \quad P_2 \quad P_3] == \begin{bmatrix} 1 & 0 & 0 \\ 4 & 1 & 0 \\ 4 & 2 & 1 \end{bmatrix} \qquad |P| = 1$$

$$P^{-1} = \begin{bmatrix} +\{1\} & -\{0\} & +\{0\} \\ -\{4\} & +\{1\} & -\{0\} \\ +\{4\} & -\{2\} & +\{1\} \end{bmatrix} = \begin{bmatrix} 1 & 0 & 0 \\ -4 & 1 & 0 \\ 4 & -2 & 1 \end{bmatrix}$$

$$P^{-1}AP = \begin{bmatrix} 1 & 0 & 0 \\ -4 & 1 & 0 \\ 4 & -2 & 1 \end{bmatrix} \begin{bmatrix} -6 & 1 & 0 \\ -12 & 0 & 1 \\ -8 & 0 & 0 \end{bmatrix} \begin{bmatrix} 1 & 0 & 0 \\ 4 & 1 & 0 \\ 4 & 2 & 1 \end{bmatrix}$$

$$= \begin{bmatrix} 1 & 0 & 0 \\ -4 & 1 & 0 \\ 4 & -2 & 1 \end{bmatrix} \begin{bmatrix} -2 & 1 & 0 \\ -8 & 2 & 1 \\ -8 & 0 & 0 \end{bmatrix} = \begin{bmatrix} -2 & 1 & 0 \\ 0 & -2 & 1 \\ 0 & 0 & -2 \end{bmatrix} = J.$$

$$P^{-1}B = \begin{bmatrix} 1 & 0 & 0 \\ -4 & 1 & 0 \\ 4 & -2 & 1 \end{bmatrix} \begin{bmatrix} 1 \\ 2 \\ 5 \end{bmatrix} = \begin{bmatrix} 1 \\ -2 \\ 5 \end{bmatrix} \qquad CP = \begin{bmatrix} 1 & 0 & 0 \end{bmatrix} \begin{bmatrix} 1 & 0 & 0 \\ 4 & 1 & 0 \\ 4 & 2 & 1 \end{bmatrix} = \begin{bmatrix} 1 & 0 & 0 \end{bmatrix}.$$

The state model in Jordan canonical form is

$$\dot{X} = \begin{bmatrix} -2 & 1 & 0 \\ 0 & -2 & 1 \\ 0 & 0 & -2 \end{bmatrix} X + \begin{bmatrix} 1 \\ -2 \\ 5 \end{bmatrix} r \qquad y = [1 \quad 0 \quad 0] X.$$

4.10 PROPERTIES OF SIMILARITY TRANSFORMATION

1. Invariance of Eigen values: The Eigen values of a system are not affected by similarity transformation. For a system matrix A, the Eigen values are obtained by solving the equation

$$|\lambda I - A| = 0$$

The system matrix from the transformed state model is $[P^{-1}AP]$ and the Eigen values are calculated by solving the equation

$$|\lambda I - [P^{-1}AP]| = 0.$$

$$|\lambda I - [P^{-1}AP]| = |\lambda P^{-1}P - P^{-1}AP|$$

$$= |P^{-1}\lambda P - P^{-1}AP|$$

$$= |P^{-1}\lambda I P - P^{-1}AP|$$

$$= |P^{-1}(\lambda I - A)P|$$

$$= |P^{-1}||\lambda I - A||P|$$

$$= |\lambda I - A||P^{-1}||P|$$

$$= |\lambda I - A||P^{-1}P|$$

$$= |\lambda I - A||I|$$

$$= |\lambda I - A|$$

This shows the invariance of Eigen values in similarity transformation.

2. Invariance of transfer function: The transfer function remains the same even after similarity transformation. Consider a state model

$$\dot{X} = AX + Br \quad and \quad y = CX.$$

The transfer function is given by

$$\frac{Y(s)}{R(s)} = C[sI - A]^{-1}B .$$

The state model for a transformed system can be represented as

$$\dot{Z} = [P^{-1}AP]Z + P^{-1}Br \text{ and } y = [CP]Z .$$

The transfer function is given by

$$[CP][sI - P^{-1}AP]^{-1}[P^{-1}B]$$

$$= [CP][sP^{-1}P - P^{-1}AP]^{-1}[P^{-1}B]$$

$$= [CP][P^{-1}sP - P^{-1}AP]^{-1}[P^{-1}B]$$

$$= [CP][P^{-1}sIP - P^{-1}AP]^{-1}[P^{-1}B]$$

$$= [CP][P^{-1}(sI - A)P]^{-1}[P^{-1}B]$$

$$= [CP\, P^{-1}(sI - A)^{-1} PP^{-1}B]$$

$$= C[sI - A]^{-1}B .$$

This shows the invariance of transfer function in similarity transformation.

3. $\left|P^{-1}AP\right| = |A|$.

4. The trace of $[P^{-1}AP]$ is equal to the trace of A.

SUMMARY

- State space models are generally transformed into canonical forms for simplicity and convenience of design and analysis of control systems.
- The method of transforming non-canonical state models into canonical forms is known as *similarity transformation*.
- The matrix that transforms a given state model into a canonical state model is called the *transformation matrix*.
- The same system can have different state space models, but these models have the same Eigen values and the same transfer function.
- For a square matrix A with an Eigen value λ, a column vector P that satisfies the relation $AP = \lambda P$ is known as *Eigen vector*.
- For every Eigen value, there is an associated Eigen vector.
- If an $n \times n$ matrix has distinct Eigen values, it is said to have a full set of n linearly independent Eigen vectors.

- If an $n \times n$ matrix has distinct Eigen values, it can be transformed into diagonal form by a similarity transformation.
- If an $n \times n$ matrix has a set of m repeated Eigen values, m<n, it does not possess a full set of n linearly independent Eigen vectors. There are $(n - m + 1)$ Eigen vectors associated with the distinct Eigen values. The remaining $(m - 1)$ Eigen vectors are called the generalized Eigen vectors.
- The transformation matrix that is composed of Eigen vectors transforms a state model into diagonal canonical form.
- A square matrix with repeated Eigen values can be transformed into Jordan canonical form of state model.
- The transformation matrix that transforms a state model in controllable canonical form to diagonal form is the Vander Monde matrix.

PRACTICE PROBLEMS

PP 4.1 Determine the Eigen values and the associated Eigen vectors for the square matrix

$$A = \begin{bmatrix} -1 & 1 \\ -3 & -5 \end{bmatrix} .$$

PP 4.2 Find the Eigen values and the associated Eigen vectors for $A = \begin{bmatrix} 0 & 3 \\ -1 & -4 \end{bmatrix} .$

PP 4.3 Determine the Eigen values and the associated Eigen vectors for

$$A = \begin{bmatrix} 0 & 1 & 0 \\ 3 & 0 & 2 \\ -12 & -7 & -6 \end{bmatrix} .$$

PP 4.4 For a system matrix $A = \begin{bmatrix} -4 & 2 \\ -1 & -1 \end{bmatrix}$, find the state transition matrix.

PP 4.5 For a square matrix $A = \begin{bmatrix} 0 & 1 \\ -2 & -3 \end{bmatrix}$, determine the Eigen values and the associated Eigen vectors. Diagonalize the given matrix, and hence find the state transition matrix.

PP 4.6 Find the Eigen values and the associated Eigen vectors for the system matrix

$$A = \begin{bmatrix} -4 & 1 \\ -4 & 0 \end{bmatrix} .$$

PP 4.7 Find the Eigen values and the associated Eigen vectors for the system matrix

$$A = \begin{bmatrix} -7 & 1 & 0 \\ -16 & 0 & 1 \\ -12 & 0 & 0 \end{bmatrix} .$$

PP 4.8 Find the Eigen values and the associated Eigen vectors for the system matrix

$$A = \begin{bmatrix} -1 & 4 & 0 \\ 0 & -1 & 0 \\ 4 & -4 & -1 \end{bmatrix}.$$

PP 4.9 Transform the given state model of a system into diagonal canonical form and hence determine the state transition matrix.

$$\dot{X} = \begin{bmatrix} -3 & 1 \\ 1 & -3 \end{bmatrix} X + \begin{bmatrix} 2 \\ -1 \end{bmatrix} r \qquad y = [1 \quad 2]X.$$

PP 4.10 Transform the given state model of a system into diagonal canonical form, and hence determine the state transition matrix.

$$\dot{X} = \begin{bmatrix} -2 & 1 \\ 0 & -3 \end{bmatrix} X + \begin{bmatrix} 1 \\ 1 \end{bmatrix} r \qquad y = [2 \quad -1]X.$$

PP 4.11 Transform the given state model of a system into diagonal canonical form, and hence determine the state transition matrix.

$$\dot{X} = \begin{bmatrix} -6 & 2 \\ -8 & 2 \end{bmatrix} X + \begin{bmatrix} 0 \\ 1 \end{bmatrix} r \qquad y = [1 \quad 1]X.$$

PP 4.12 Diagonalize the state model

$$\dot{X} = \begin{bmatrix} -1 & 2 & 0 \\ 0 & -2 & 0 \\ 2 & -2 & -3 \end{bmatrix} X + \begin{bmatrix} 0 \\ 1 \\ 1 \end{bmatrix} r \qquad y = [1 \quad 1 \quad 0]X.$$

PP 4.13 Diagonalize the state model

$$\dot{X} = \begin{bmatrix} -2 & 4 & 0 \\ 0 & -2 & 0 \\ 4 & -4 & -2 \end{bmatrix} X + \begin{bmatrix} 1 \\ 2 \\ 3 \end{bmatrix} r \qquad y = [1 \quad 2 \quad -1]X.$$

PP 4.14 Transform the given state model into diagonal canonical form.

$$\dot{X} = \begin{bmatrix} -5 & 2 & 0 \\ 0 & -1 & 0 \\ -3 & 1 & -1 \end{bmatrix} X + \begin{bmatrix} 1 \\ 3 \\ 0 \end{bmatrix} r \qquad y = [1 \quad 0 \quad -1]X.$$

PP 4.15 Convert the given state model in controllable canonical form to diagonal form. Use Vander Monde matrix.

$$\dot{X} = \begin{bmatrix} 0 & 1 \\ -8 & -6 \end{bmatrix} X + \begin{bmatrix} 0 \\ 1 \end{bmatrix} r \qquad y = [2 \quad -1]X.$$

PP 4.16 Transform the given state model in controllable canonical form to diagonal canonical form. Use Vander Monde matrix.

$$\dot{X} = \begin{bmatrix} 0 & 1 \\ -9 & -6 \end{bmatrix} X + \begin{bmatrix} 0 \\ 1 \end{bmatrix} r \qquad y = [1 \quad 1] X .$$

PP 4.17 Transform the given state model in phase variable form to diagonal canonical form. Use Vander Monde matrix.

$$\dot{X} = \begin{bmatrix} 0 & 1 \\ -25 & -10 \end{bmatrix} X + \begin{bmatrix} 2 \\ 1 \end{bmatrix} r \qquad y = [1 \quad -2] X .$$

PP 4.18 A state model in controllable canonical form is given below. Convert it into diagonal form.

$$\dot{X} = \begin{bmatrix} 0 & 1 & 0 \\ 0 & 0 & 1 \\ -6 & -11 & -6 \end{bmatrix} X + \begin{bmatrix} 0 \\ 0 \\ 1 \end{bmatrix} r \qquad y = [2 \quad 2 \quad 1] X .$$

PP 4.19 A state model in controllable canonical form is given below. Convert it into diagonal form.

$$\dot{X} = \begin{bmatrix} 0 & 1 & 0 \\ 0 & 0 & 1 \\ -8 & -12 & -6 \end{bmatrix} X + \begin{bmatrix} 0 \\ 0 \\ 1 \end{bmatrix} r \qquad y = [1 \quad 2 \quad 0] X .$$

PP 4.20 A state model of a linear system in observable canonical form is given as

$$\dot{X} = \begin{bmatrix} -5 & 1 \\ -4 & 0 \end{bmatrix} X + \begin{bmatrix} 1 \\ 2 \end{bmatrix} r \qquad y = [1 \quad 0] X .$$

Transform this model into diagonal form.

PP 4.21 A state model of a linear system in observable canonical form is given as

$$\dot{X} = \begin{bmatrix} -6 & 1 \\ -9 & 0 \end{bmatrix} X + \begin{bmatrix} 2 \\ 5 \end{bmatrix} r \qquad y = [1 \quad 0] X .$$

Transform this model into diagonal form.

PP 4.22 A state model of a linear system is represented in observable canonical form as

$$\dot{X} = \begin{bmatrix} -9 & 1 & 0 \\ -23 & 0 & 1 \\ -15 & 0 & 0 \end{bmatrix} X + \begin{bmatrix} 2 \\ 1 \\ 3 \end{bmatrix} r \qquad y = [1 \quad 0 \quad 0] X .$$

Transform this model into diagonal canonical form.

PP 4.23 A state model of a linear system is represented in observable canonical form as

$$\dot{X} = \begin{bmatrix} -7 & 1 & 0 \\ -11 & 0 & 1 \\ -5 & 0 & 0 \end{bmatrix} X + \begin{bmatrix} 1 \\ -2 \\ 2 \end{bmatrix} r \qquad y = [1 \quad 0 \quad 0] X .$$

Transform this model into diagonal form.

REVIEW QUESTIONS

4.1 Explain similarity transformation.

4.2 What is a transformation matrix? Explain.

4.3 What is a modal matrix? Explain.

4.4 Define an Eigen vector.

4.5 Define a generalized Eigen vector. Explain by means of an example.

4.6 Describe any two methods of finding Eigen vectors associated with distinct Eigen values of a given square matrix.

4.7 Describe a method to find a generalized Eigen vector associated with repeated Eigen values of a square matrix.

4.8 Consider a state model

$$\dot{X} = AX + Br \qquad y = CX .$$

Assuming that the Eigen values are distinct, show that a transformation matrix that is composed of Eigen vectors transforms the given state model into diagonal canonical form.

4.9 Explain the method of obtaining the state transition matrix for a given system matrix by diagonalization.

4.10 With the help of an example, discuss the Jordan canonical form of state model.

4.11 What is a Vander Monde matrix?

4.12 Discuss the method of finding the Vander Monde matrix for a given square matrix of size (3×3) with distinct Eigen values.

4.13 Discuss the method of finding the Vander Monde matrix for a given square matrix of size (3×3) with repeated Eigen values.

4.14 Explain the method of diagonalization using Vander Monde matrix.

4.15 Discus a method of finding the transformation matrix that transforms a state model in observable canonical form to diagonal form, when the Eigen values are distinct.

4.16 Discus a method of finding the transformation matrix that transforms a state model in observable canonical form to diagonal form, when the Eigen values are repeated.

4.17 Write down the properties of similarity transformation.

5 Controllability and Observability

5.1 CONTROLLABILITY

Consider a linear plant shown in Figure 5.1. The state model for the plant can be written as

$$\dot{X} = AX + Bu$$

$$y = CX + Du.$$

r(t)   u(t)   Linear Plant   y(t)

Figure 5.1 A linear plant

If it is possible to control the state variables via the input signal to the plant, then the system is said to be *controllable*. In other words, if the input signal or the control signal u can control each state variable, then the system is *completely controllable*. A system is said to be *completely controllable* if there exists a control signal u that can transfer any initial state $X(t_0)$ to any chosen final state $X(t_f)$; otherwise, the system is *uncontrollable*.

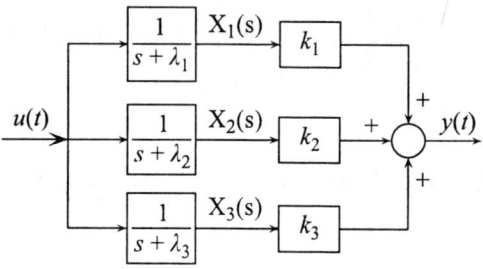

Figure 5.2 Block diagram for diagonal canonical form

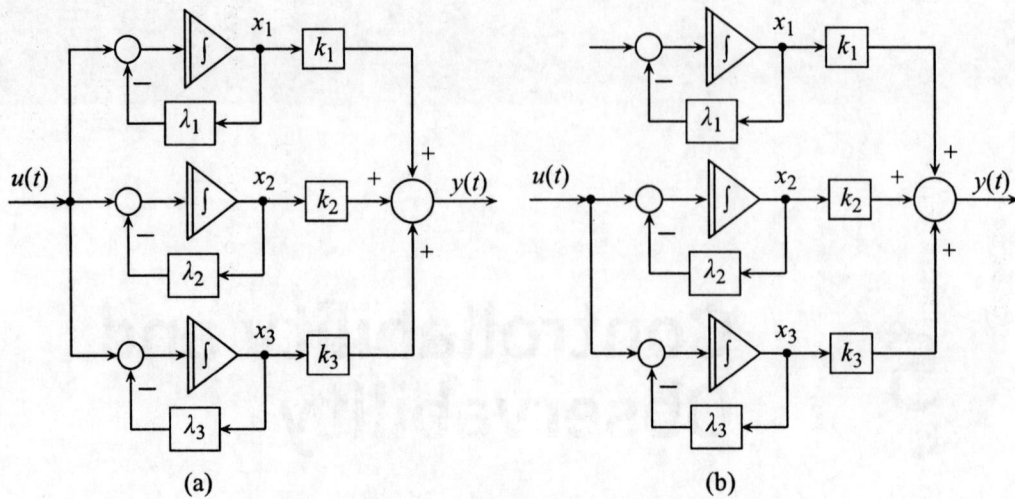

Figure 5.3 Controllable and uncontrollable systems

Consider the state model of the plant in diagonal canonical form, the block diagram of which is shown in Figure 5.2. The state diagram is shown in Figure 5.3. This compares the controllable and uncontrollable systems. In Figure 5.3(a), all the state variables are linked to the input so that they can be controlled by the input and the system is completely controllable. In Figure 5.3(b), the state variable x_1 is not linked to the input so that it cannot be controlled by the input. Controllability is a property of the coupling between the input or control signal u and the state X, and thus it involves the matrices A and B.

5.2 CONTROLLABILITY CRITERIA

There are many tests by which the complete controllability of a system represented by a state model can be assessed. Some of them are discussed here.

5.2.1 Controllability Criterion 1: Gilbert's Test

From Figure 5.3(a), the state model of the system in diagonal canonical form is

$$
\begin{bmatrix} \dot{x}_1 \\ \dot{x}_2 \\ \dot{x}_3 \end{bmatrix} = \begin{bmatrix} -\lambda_1 & 0 & 0 \\ 0 & -\lambda_2 & 0 \\ 0 & 0 & -\lambda_3 \end{bmatrix} \begin{bmatrix} x_1 \\ x_2 \\ x_3 \end{bmatrix} + \begin{bmatrix} 1 \\ 1 \\ 1 \end{bmatrix} u .
\tag{5.1}
$$

This represents a controllable system. From Figure 5.3(b), the state model is

$$\begin{bmatrix} \dot{x}_1 \\ \dot{x}_2 \\ \dot{x}_3 \end{bmatrix} = \begin{bmatrix} -\lambda_1 & 0 & 0 \\ 0 & -\lambda_2 & 0 \\ 0 & 0 & -\lambda_3 \end{bmatrix} \begin{bmatrix} x_1 \\ x_2 \\ x_3 \end{bmatrix} + \begin{bmatrix} 0 \\ 1 \\ 1 \end{bmatrix} u . \qquad (5.2)$$

This represents an uncontrollable system. Comparing equations (5.1) and (5.2), it can be observed that for a system to be completely controllable, the input matrix, in the diagonal canonical form, should not have any zero elements.

A state model represented by

$$\dot{X} = AX + Bu ,$$

if transformed into diagonal canonical form, can be represented as

$$\dot{Z} = [P^{-1}AP]Z + [P^{-1}B]u . \qquad (5.3)$$

The system is completely controllable if and only if any row of the matrix $[P^{-1}B]$ in equation (5.3) is not with all zero elements.

Consider a system with repeated Eigen values. The system can be transformed to Jordan canonical form. A state diagram of a third-order system with two repeated Eigen values is shown in Figure 5.4.

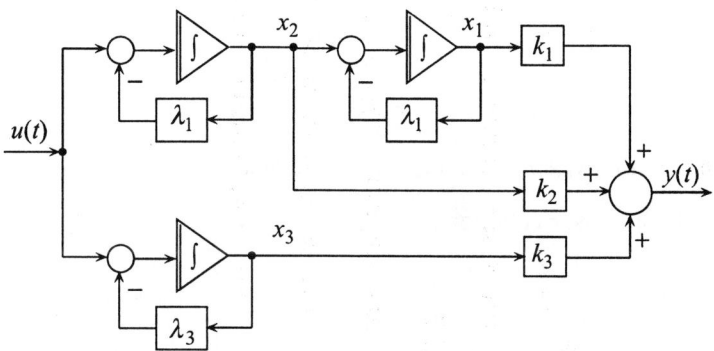

Figure 5.4 State diagram in Jordan canonical form

It can be observed that if x_2 is not linked with the input, x_2 is not controllable and hence the system is not completely controllable. The same is the case with x_3 but not with x_1. In other words, x_2 and x_3 must be linked with the input but not necessarily x_1, for this system to be completely controllable. From the state diagram shown in Figure 5.4, the state equation can be written down as

$$\begin{bmatrix} \dot{x_1} \\ \dot{x_2} \\ \dot{x_3} \end{bmatrix} = \begin{bmatrix} -\lambda_1 & 1 & 0 \\ 0 & -\lambda_1 & 0 \\ 0 & 0 & -\lambda_3 \end{bmatrix} \begin{bmatrix} x_1 \\ x_2 \\ x_3 \end{bmatrix} + \begin{bmatrix} 0 \\ 1 \\ 1 \end{bmatrix} u \qquad (5.4)$$

Observing equation (5.4), it can be concluded that for a system to be completely controllable, the row in the input matrix corresponding to the last row of the Jordan block, should not have all zero elements.

To summarize, a linear time-invariant system is completely controllable if and only if

(a) any row of $[P^{-1}B]$ matrix does not have all zero elements, in the case of distinct Eigen values.

(b) the row of $[P^{-1}B]$ matrix that corresponds to the last row of Jordan block in the $[P^{-1}AP]$ matrix does not have all zero elements and all other rows corresponding to other Eigen values do not have all zero elements, in the case of repeated Eigen values.

The test for controllability using this criterion is called Gilbert's test. For this test to be performed, the given state model is to be transformed into either diagonal or Jordan canonical form, depending on whether the Eigen values are distinct or repeated. An advantage of this test is that it indicates which state variable is uncontrollable.

Example 5.1 A state model of a plant is

$$\dot{X} = \begin{bmatrix} 0 & 1 \\ -2 & -3 \end{bmatrix} X + \begin{bmatrix} 1 \\ 1 \end{bmatrix} u$$

$$y = \begin{bmatrix} 2 & 1 \end{bmatrix}.$$

Check whether the system is completely controllable.

Solution: The characteristic equation is

$$\begin{vmatrix} \lambda & -1 \\ 2 & \lambda+3 \end{vmatrix} = 0 \quad or \quad \lambda^2 + 3\lambda + 2 = 0 \quad \lambda_1 = -1 \ and \ \lambda_2 = -2.$$

The state equation is in the phase variable form. Hence, the transformation matrix that transforms the state model into diagonal canonical form is the Vander Monde matrix.

$$P = \begin{bmatrix} 1 & 1 \\ -1 & -2 \end{bmatrix} \qquad |P| = -1 \qquad P^{-1} = -\begin{bmatrix} -2 & -1 \\ 1 & 1 \end{bmatrix} = \begin{bmatrix} 2 & 1 \\ -1 & -1 \end{bmatrix}.$$

Check:

$$P^{-1}P = \begin{bmatrix} 2 & 1 \\ -1 & -1 \end{bmatrix} \begin{bmatrix} 1 & 1 \\ -1 & -2 \end{bmatrix} = \begin{bmatrix} 1 & 0 \\ 0 & 1 \end{bmatrix}$$

$$P^{-1}AP = \begin{bmatrix} 2 & 1 \\ -1 & -1 \end{bmatrix} \begin{bmatrix} 0 & 1 \\ -2 & -3 \end{bmatrix} \begin{bmatrix} 1 & 1 \\ -1 & -2 \end{bmatrix} = \begin{bmatrix} 2 & 1 \\ -1 & -1 \end{bmatrix} \begin{bmatrix} -1 & -2 \\ 1 & 4 \end{bmatrix} = \begin{bmatrix} -1 & 0 \\ 0 & -2 \end{bmatrix}$$

$$P^{-1}B = \begin{bmatrix} 2 & 1 \\ -1 & -1 \end{bmatrix} \begin{bmatrix} 1 \\ 1 \end{bmatrix} = \begin{bmatrix} 3 \\ -2 \end{bmatrix}.$$

Since no row of $[P^{-1}B]$ contains all zero elements, the system is completely controllable.

Example 5.2 A state equation of a system is

$$\dot{X} = \begin{bmatrix} -3 & -1 \\ 2 & 0 \end{bmatrix} X + \begin{bmatrix} 1 \\ -1 \end{bmatrix} u .$$

Check whether the system is completely controllable.

Solution: The characteristic equation is

$$\begin{vmatrix} \lambda+3 & 1 \\ -2 & \lambda \end{vmatrix} = 0 \quad \lambda^2 + 3\lambda + 2 = 0.$$

The Eigen values are $\lambda_1 = -1$ and $\lambda_2 = -2$.

The Eigen vector associated with each Eigen value is

$$P_i = \begin{bmatrix} \lambda \\ 2 \end{bmatrix}_{\lambda=\lambda_i} \quad P_1 = \begin{bmatrix} \lambda \\ 2 \end{bmatrix}_{\lambda=-1} = \begin{bmatrix} -1 \\ 2 \end{bmatrix} \quad P_2 = \begin{bmatrix} \lambda \\ 2 \end{bmatrix}_{\lambda=-2} = \begin{bmatrix} -2 \\ 2 \end{bmatrix}.$$

The transformation matrix that transforms the state model into diagonal canonical form is

$$P = \begin{bmatrix} -1 & -2 \\ 2 & 2 \end{bmatrix} \quad |P| = \begin{vmatrix} -1 & -2 \\ 2 & 2 \end{vmatrix} = 2 \quad P^{-1} = \frac{1}{2}\begin{bmatrix} 2 & 2 \\ -2 & -1 \end{bmatrix}.$$

Check:

$$P^{-1}P = \frac{1}{2}\begin{bmatrix} 2 & 2 \\ -2 & -1 \end{bmatrix}\begin{bmatrix} -1 & -2 \\ 2 & 2 \end{bmatrix} = \begin{bmatrix} 1 & 0 \\ 0 & 1 \end{bmatrix}$$

$$P^{-1}AP = \frac{1}{2}\begin{bmatrix} 2 & 2 \\ -2 & -1 \end{bmatrix}\begin{bmatrix} -3 & -1 \\ 2 & 0 \end{bmatrix}\begin{bmatrix} -1 & -2 \\ 2 & 2 \end{bmatrix} = \frac{1}{2}\begin{bmatrix} 2 & 2 \\ -2 & -1 \end{bmatrix}\begin{bmatrix} 1 & 4 \\ -2 & -4 \end{bmatrix} = \begin{bmatrix} -1 & 0 \\ 0 & -2 \end{bmatrix}$$

$$P^{-1}B = \frac{1}{2}\begin{bmatrix} 2 & 2 \\ -2 & -1 \end{bmatrix}\begin{bmatrix} 1 \\ -1 \end{bmatrix} = \begin{bmatrix} 0 \\ -0.5 \end{bmatrix}$$

Since one row of [P⁻¹B] has a zero element, the system is not completely controllable.

Example 5.3 A state equation of a system is

$$\dot{X} = \begin{bmatrix} -7 & -2 & 6 \\ 2 & -3 & -2 \\ -2 & -2 & 1 \end{bmatrix} X + \begin{bmatrix} 1 & 1 \\ 1 & -1 \\ 1 & 0 \end{bmatrix} u \ .$$

Check whether the system is completely controllable.

Solution: The characteristic equation is

$$\begin{vmatrix} \lambda+7 & 2 & -6 \\ -2 & \lambda+3 & 2 \\ 2 & 2 & \lambda-1 \end{vmatrix} = 0 \quad \lambda^3 + 9\lambda^2 + 23\lambda + 15 = 0 \ .$$

The Eigen values are $\lambda_1 = -1, \lambda_2 = -3$ and $\lambda_3 = -5$. The Eigen vectors are determined from the cofactors of first row of $[\lambda I - A]$.

$$P_i = \begin{bmatrix} +\{(\lambda+3)(\lambda-1)-4\} \\ -\{-2(\lambda-1)-4\} \\ +\{-4-2(\lambda+3)\} \end{bmatrix}$$

$$P_1 = \begin{bmatrix} (\lambda+3)(\lambda-1)-4 \\ 2(\lambda-1)+4 \\ -4-2(\lambda+3) \end{bmatrix}_{\lambda=-1} = \begin{bmatrix} -8 \\ 0 \\ -8 \end{bmatrix} \quad or \ P_1 = \begin{bmatrix} 1 \\ 0 \\ 1 \end{bmatrix}$$

$$P_2 = \begin{bmatrix} (\lambda+3)(\lambda-1)-4 \\ 2(\lambda-1)+4 \\ -4-2(\lambda+3) \end{bmatrix}_{\lambda=-3} = \begin{bmatrix} -4 \\ -4 \\ -4 \end{bmatrix} \quad or \ P_2 = \begin{bmatrix} 1 \\ 1 \\ 1 \end{bmatrix}$$

$$P_3 = \begin{bmatrix} (\lambda+3)(\lambda-1)-4 \\ 2(\lambda-1)+4 \\ -4-2(\lambda+3) \end{bmatrix}_{\lambda=-5} = \begin{bmatrix} 8 \\ -8 \\ 0 \end{bmatrix} \quad or \ P_3 = \begin{bmatrix} 1 \\ -1 \\ 0 \end{bmatrix} \ .$$

The transformation matrix that transforms the given state model into diagonal canonical form is

$$P = \begin{bmatrix} 1 & 1 & 1 \\ 0 & 1 & -1 \\ 1 & 1 & 0 \end{bmatrix} \quad |P| = -1 \quad P^{-1} = -\begin{bmatrix} +1 & -(-1) & +(-2) \\ -1 & +(-1) & -(-1) \\ +(-1) & -(0) & +1 \end{bmatrix} = \begin{bmatrix} -1 & -1 & 2 \\ 1 & 1 & -1 \\ 1 & 0 & -1 \end{bmatrix}$$

Check:

$$P^{-1}P = \begin{bmatrix} -1 & -1 & 2 \\ 1 & 1 & -1 \\ 1 & 0 & -1 \end{bmatrix}\begin{bmatrix} 1 & 1 & 1 \\ 0 & 1 & -1 \\ 1 & 1 & 0 \end{bmatrix} = \begin{bmatrix} 1 & 0 & 0 \\ 0 & 1 & 0 \\ 0 & 0 & 1 \end{bmatrix}$$

$$P^{-1}AP = \begin{bmatrix} -1 & -1 & 2 \\ 1 & 1 & -1 \\ 1 & 0 & -1 \end{bmatrix}\begin{bmatrix} -7 & -2 & 6 \\ 2 & -3 & -2 \\ -2 & -2 & 1 \end{bmatrix}\begin{bmatrix} 1 & 1 & 1 \\ 0 & 1 & -1 \\ 1 & 1 & 0 \end{bmatrix}$$

$$= \begin{bmatrix} -1 & -1 & 2 \\ 1 & 1 & -1 \\ 1 & 0 & -1 \end{bmatrix}\begin{bmatrix} -1 & -3 & -5 \\ 0 & -3 & 5 \\ -1 & -3 & 0 \end{bmatrix} = \begin{bmatrix} -1 & 0 & 0 \\ 0 & -3 & 0 \\ 0 & 0 & -5 \end{bmatrix}$$

$$P^{-1}B = \begin{bmatrix} -1 & -1 & 2 \\ 1 & 1 & -1 \\ 1 & 0 & -1 \end{bmatrix}\begin{bmatrix} 1 & 1 \\ 1 & -1 \\ 1 & 0 \end{bmatrix} = \begin{bmatrix} 0 & 0 \\ 1 & 0 \\ 0 & 1 \end{bmatrix}.$$

Since the first row of [P⁻¹B] has all zero elements, the system is not completely controllable.

Example 5.4 A state model of a linear plant is

$$\dot{X} = \begin{bmatrix} -4 & 1 \\ -9 & 2 \end{bmatrix}X + \begin{bmatrix} 3 \\ 1 \end{bmatrix}u$$

$$y = [2 \quad 1].$$

Check whether the system is completely controllable.

Solution: The characteristic equation is

$$\begin{vmatrix} \lambda+4 & -1 \\ 9 & \lambda-2 \end{vmatrix} = 0 \qquad \lambda^2 + 2\lambda + 1 = 0 \qquad \lambda_1 = -1, \lambda_2 = -1.$$

The Eigen values are repeated so that the state model can be transformed into Jordan canonical form. The Eigen vector associated with the Eigen values, using the cofactors of the first row of $[\lambda I - A]$ is

$$P_i = \begin{bmatrix} +(\lambda-2) \\ -(9) \end{bmatrix}_{\lambda=\lambda_i} \qquad P_1 = \begin{bmatrix} \lambda-2 \\ -9 \end{bmatrix}_{\lambda=-1} = \begin{bmatrix} -3 \\ -9 \end{bmatrix} \qquad P_2 = \frac{d}{d\lambda}\begin{bmatrix} \lambda-2 \\ -9 \end{bmatrix}_{\lambda=-2} = \begin{bmatrix} 1 \\ 0 \end{bmatrix}$$

$$P = \begin{bmatrix} -3 & 1 \\ -9 & 0 \end{bmatrix} \qquad |P| = 9 \qquad P^{-1} = \frac{1}{9}\begin{bmatrix} 0 & -1 \\ 9 & -3 \end{bmatrix}$$

Check:

$$P^{-1}P = \frac{1}{9}\begin{bmatrix} 0 & -1 \\ 9 & -3 \end{bmatrix}\begin{bmatrix} -3 & 1 \\ -9 & 0 \end{bmatrix} = \begin{bmatrix} 1 & 0 \\ 0 & 1 \end{bmatrix}$$

$$P^{-1}AP = \frac{1}{9}\begin{bmatrix} 0 & -1 \\ 9 & -3 \end{bmatrix}\begin{bmatrix} -4 & 1 \\ -9 & 2 \end{bmatrix}\begin{bmatrix} -3 & 1 \\ -9 & 0 \end{bmatrix} = \frac{1}{9}\begin{bmatrix} 0 & -1 \\ 9 & -3 \end{bmatrix}\begin{bmatrix} 3 & -4 \\ 9 & -9 \end{bmatrix} = \begin{bmatrix} -1 & 1 \\ 0 & -1 \end{bmatrix}$$

$$P^{-1}B = \frac{1}{9}\begin{bmatrix} 0 & -1 \\ 9 & -3 \end{bmatrix}\begin{bmatrix} 3 \\ 1 \end{bmatrix} = \begin{bmatrix} (-1/9) \\ (8/3) \end{bmatrix}.$$

Since all the rows of [P⁻¹B] have all non-zero elements, the system is completely controllable.

Example 5.5 A state model of a linear system is

$$\dot{X} = \begin{bmatrix} 0 & 1 \\ -4 & -4 \end{bmatrix}X + \begin{bmatrix} 1 \\ -2 \end{bmatrix}u .$$

Check whether the system is completely controllable.

Solution: The characteristic equation is

$$|\lambda I - A| = \begin{vmatrix} \lambda & -1 \\ 4 & \lambda+4 \end{vmatrix} = 0 \qquad \lambda^2 + 4\lambda + 4 = 0 \qquad \lambda_1 = -2, \lambda_2 = -2 .$$

The Eigen values are repeated so that the state model can be transformed into Jordan canonical form. The given state model is in phase variable form. Hence, the transformation matrix that transforms the state model to Jordan canonical form is the Vander Monde matrix.

$$P_i = \begin{bmatrix} 1 \\ \lambda \end{bmatrix} \qquad P_1 = \begin{bmatrix} 1 \\ \lambda \end{bmatrix}_{\lambda=-2} = \begin{bmatrix} 1 \\ -2 \end{bmatrix} \qquad P_2 = \frac{d}{d\lambda}\begin{bmatrix} 1 \\ \lambda \end{bmatrix}_{\lambda=-2} = \begin{bmatrix} 0 \\ 1 \end{bmatrix}$$

$$P = \begin{bmatrix} 1 & 0 \\ -2 & 1 \end{bmatrix} \qquad\qquad |P| = 1 \qquad P^{-1} = \begin{bmatrix} 1 & 0 \\ 2 & 1 \end{bmatrix}$$

Check:

$$P^{-1}P = \begin{bmatrix} 1 & 0 \\ 2 & 1 \end{bmatrix}\begin{bmatrix} 1 & 0 \\ -2 & 1 \end{bmatrix} = \begin{bmatrix} 1 & 0 \\ 0 & 1 \end{bmatrix}$$

$$J = P^{-1}AP = \begin{bmatrix} 1 & 0 \\ 2 & 1 \end{bmatrix}\begin{bmatrix} 0 & 1 \\ -4 & -4 \end{bmatrix}\begin{bmatrix} 1 & 0 \\ -2 & 1 \end{bmatrix} = \begin{bmatrix} 1 & 0 \\ 2 & 1 \end{bmatrix}\begin{bmatrix} -2 & 1 \\ 4 & -4 \end{bmatrix} = \begin{bmatrix} -2 & 1 \\ 0 & -2 \end{bmatrix}$$

$$P^{-1}B = \begin{bmatrix} 1 & 0 \\ 2 & 1 \end{bmatrix}\begin{bmatrix} 1 \\ -2 \end{bmatrix} = \begin{bmatrix} 1 \\ 0 \end{bmatrix}.$$

Since the element of [P⁻¹B] that corresponds to the last row of Jordan block of J is zero, the system is not completely controllable.

Example 5.6 Assess the controllability of a linear system, a state equation of which is given as

$$\dot{X} = \begin{bmatrix} 0 & 0 & 1 \\ -2 & -3 & 0 \\ 0 & 2 & -3 \end{bmatrix} X + \begin{bmatrix} 1 \\ -2 \\ 1 \end{bmatrix} u \ .$$

Solution: The characteristic equation is

$$|\lambda I - A| = \begin{vmatrix} \lambda & 0 & -1 \\ 2 & \lambda+3 & 0 \\ 0 & -2 & \lambda+3 \end{vmatrix} = 0 \quad \lambda(\lambda+3)^2 - (-4) = 0 \qquad \lambda^3 + 6\lambda^2 + 9\lambda + 4 = 0$$

$$(\lambda+1)^2(\lambda+4) = 0 \ .$$

The Eigen values are $\lambda_1 = -1, \lambda_2 = -1$ and $\lambda_3 = -4$.

$$[\lambda I - A] = \begin{bmatrix} \lambda & 0 & -1 \\ 2 & \lambda+3 & 0 \\ 0 & -2 & \lambda+3 \end{bmatrix} \ .$$

Matrix of cofactors of the third row of $[\lambda I - A] = \begin{bmatrix} +\{(\lambda+3)\} \\ -\{2\} \\ +\{\lambda(\lambda+3)\} \end{bmatrix} = \begin{bmatrix} (\lambda+3) \\ -2 \\ \lambda(\lambda+3) \end{bmatrix}$

$$P_1 = \begin{bmatrix} (\lambda+3) \\ -2 \\ \lambda(\lambda+3) \end{bmatrix}_{\lambda=-1} = \begin{bmatrix} 2 \\ -2 \\ -2 \end{bmatrix} \qquad P_2 = \frac{d}{d\lambda}\begin{bmatrix} (\lambda+3) \\ -2 \\ \lambda^2+3\lambda \end{bmatrix}_{\lambda=-1} = \begin{bmatrix} 1 \\ 0 \\ 2\lambda+3 \end{bmatrix}_{\lambda=-1} = \begin{bmatrix} 1 \\ 0 \\ 1 \end{bmatrix}$$

$$P_3 = \begin{bmatrix} (\lambda+3) \\ -2 \\ \lambda(\lambda+3) \end{bmatrix}_{\lambda=-4} = \begin{bmatrix} -1 \\ -2 \\ 4 \end{bmatrix} \ .$$

The transformation matrix that transforms the state model into diagonal canonical form is

$$P = \begin{bmatrix} 2 & 1 & -1 \\ -2 & 0 & -2 \\ -2 & 1 & 4 \end{bmatrix} \qquad |P| = (2)(5) + 2(4) = 18$$

$$P^{-1} = \frac{1}{18} \begin{bmatrix} +2 & -5 & +(-2) \\ -(-12) & +6 & -(-6) \\ +(-2) & -4 & +2 \end{bmatrix} = \frac{1}{18} \begin{bmatrix} 2 & -5 & -2 \\ 12 & 6 & 6 \\ -2 & -4 & 2 \end{bmatrix}$$

Check:

$$P^{-1}P = \frac{1}{18} \begin{bmatrix} 2 & -5 & -2 \\ 12 & 6 & 6 \\ -2 & -4 & 2 \end{bmatrix} \begin{bmatrix} 2 & 1 & -1 \\ -2 & 0 & -2 \\ -2 & 1 & 4 \end{bmatrix} = \begin{bmatrix} 1 & 0 & 0 \\ 0 & 1 & 0 \\ 0 & 0 & 1 \end{bmatrix}$$

$$J = P^{-1}AP = \frac{1}{18} \begin{bmatrix} 2 & -5 & -2 \\ 12 & 6 & 6 \\ -2 & -4 & 2 \end{bmatrix} \begin{bmatrix} 0 & 0 & 1 \\ -2 & -3 & 0 \\ 0 & 2 & -3 \end{bmatrix} \begin{bmatrix} 2 & 1 & -1 \\ -2 & 0 & -2 \\ -2 & 1 & 4 \end{bmatrix}$$

$$= \frac{1}{18} \begin{bmatrix} 2 & -5 & -2 \\ 12 & 6 & 6 \\ -2 & -4 & 2 \end{bmatrix} \begin{bmatrix} -2 & 1 & 4 \\ 2 & -2 & 8 \\ 2 & -3 & -16 \end{bmatrix} = \begin{bmatrix} -1 & 1 & 0 \\ 0 & -1 & 0 \\ 0 & 0 & -4 \end{bmatrix}.$$

$$P^{-1}B = \frac{1}{18} \begin{bmatrix} 2 & -5 & -2 \\ 12 & 6 & 6 \\ -2 & -4 & 2 \end{bmatrix} \begin{bmatrix} 1 \\ -2 \\ 1 \end{bmatrix} = \begin{bmatrix} (5/9) \\ (1/3) \\ (4/9) \end{bmatrix}$$

The second row of [P⁻¹B] that corresponds to the last row of Jordan block in J is non-zero. Also the last row [P⁻¹B] is also non-zero. Hence the system is completely controllable.

5.2.2 Controllability Criterion 2: Kalman's Test

A linear time-invariant n-th order system, represented by the state equation,

$$\dot{X} = AX + Bu$$

is completely controllable if and only if the matrix

$$M = [B \quad AB \quad A^2B \quad A^3B \dots A^{n-1}B] \text{ has rank } n.$$

Proof: The state equation of the system is

$$\dot{X} = AX + Bu.$$

The solution of the state equation is

$$X(t) = e^{At}X(0) + \int_0^t e^{A(t-\theta)}Bu(\theta)d\theta. \tag{5.5}$$

By definition, a system is said to be completely controllable if any initial state $X(t_0)$ can be transferred to any chosen final state in the finite time $t_0 < t < t_f$, by the use of certain control signal $u(t)$. In other words, if the state of a system can be changed from one chosen value to another by manipulating the input, then the system can be controlled. Choose the final desired state as $X(t_f) = 0$. Then equation (5.5) becomes

$$X(t_f) = e^{At}X(0) + \int_0^{t_f} e^{A(t-\theta)}Bu(\theta)d\theta = 0$$

$$X(0) = -\int_0^{t_f} e^{-A\theta}Bu(\theta)d\theta. \tag{5.6}$$

The matrix $e^{-A\theta}$ can be expressed in a power series as

$$e^{-A\theta} = \sum_{k=0}^{n-1} \alpha_k(\theta)A^k.$$

Hence equation (5.6) becomes

$$-X(0) = \sum A\ B\int \alpha\ (\theta)u(\theta)d\theta.$$

For each k, the integral will have a finite value and let it be Q_k.

$$-X(0) = \sum_{k=0}^{n-1} A^k B Q_k$$

$$= [B \quad AB \quad A^2B \quad A^3B \, ... \, A^{n-1}B]\begin{bmatrix} Q_0 \\ Q_1 \\ Q_2 \\ ... \\ ... \\ Q_{n-1} \end{bmatrix}$$

This equation will be satisfied only if the vectors B, AB, A²B,..., A^{n-1}B are linearly independent. This is true only if the matrix $M = [B \quad AB \quad A^2B \quad A^3B \, ... \, A^{n-1}B]$ has rank n.

Hence a system represented by the state equation

$$\dot{X} = AX + Bu$$

is completely controllable if and only if the matrix M has full rank.

This matrix M is called the *controllability matrix*. The test for controllability using this criterion is called Kalman's test. For this test, transformation of the system into either diagonal or Jordan canonical form is not required. But it remains unknown which state variable is uncontrollable.

Example 5.7 A state equation for a system is

$$\dot{X} = \begin{bmatrix} 0 & 1 \\ -2 & -3 \end{bmatrix} X + \begin{bmatrix} 0 \\ 2 \end{bmatrix} u \, .$$

Check whether the system represented by this state equation is completely controllable.

Solution:

$$A = \begin{bmatrix} 0 & 1 \\ -2 & -3 \end{bmatrix} \quad B = \begin{bmatrix} 0 \\ 2 \end{bmatrix} \quad AB = \begin{bmatrix} 0 & 1 \\ -2 & -3 \end{bmatrix} \begin{bmatrix} 0 \\ 2 \end{bmatrix} = \begin{bmatrix} 2 \\ -6 \end{bmatrix}$$

The controllability matrix $M = \begin{bmatrix} 0 & 2 \\ 2 & -6 \end{bmatrix}$ $|M| = -4$.

Since the determinant of M is not zero, the rank of M is 2 and the system represented by the state model is completely controllable.

Example 5.8 A state equation for a system is

$$\dot{X} = \begin{bmatrix} -3 & -1 \\ 2 & 0 \end{bmatrix} X + \begin{bmatrix} 1 \\ -1 \end{bmatrix} u \, .$$

Check whether the system represented by this state equation is completely controllable.

Solution:

$$A = \begin{bmatrix} -3 & -1 \\ 2 & 0 \end{bmatrix} \quad\quad B = \begin{bmatrix} 1 \\ -1 \end{bmatrix} \quad\quad AB = \begin{bmatrix} -3 & -1 \\ 2 & 0 \end{bmatrix} \begin{bmatrix} 1 \\ -1 \end{bmatrix} = \begin{bmatrix} -2 \\ 2 \end{bmatrix}$$

The controllability matrix $M = \begin{bmatrix} 1 & -2 \\ -1 & 2 \end{bmatrix}$ $|M| = 0$.

Since the determinant of M is zero, the rank of M is 1 and the system represented by the state model is not completely controllable.

Example 5.9 A state equation for a system is

$$\dot{X} = \begin{bmatrix} -7 & -2 & 6 \\ 2 & -3 & -2 \\ -2 & -2 & 1 \end{bmatrix} X + \begin{bmatrix} 1 & 1 \\ 1 & -1 \\ 1 & 0 \end{bmatrix} u \, .$$

Check whether the system represented by this state equation is completely controllable.

Solution:

$$A = \begin{bmatrix} -7 & -2 & 6 \\ 2 & -3 & -2 \\ -2 & -2 & 1 \end{bmatrix} \qquad B = \begin{bmatrix} 1 & 1 \\ 1 & -1 \\ 1 & 0 \end{bmatrix}.$$

$$AB = \begin{bmatrix} -7 & -2 & 6 \\ 2 & -3 & -2 \\ -2 & -2 & 1 \end{bmatrix} \begin{bmatrix} 1 & 1 \\ 1 & -1 \\ 1 & 0 \end{bmatrix} = \begin{bmatrix} -3 & -5 \\ -3 & 5 \\ -3 & 0 \end{bmatrix}$$

$$A^2 B = \begin{bmatrix} -7 & -2 & 6 \\ 2 & -3 & -2 \\ -2 & -2 & 1 \end{bmatrix} \begin{bmatrix} -3 & -5 \\ -3 & 5 \\ -3 & 0 \end{bmatrix} = \begin{bmatrix} 9 & 25 \\ 9 & -25 \\ 9 & 0 \end{bmatrix}$$

The controllability matrix $M = [B \quad AB \quad A^2 B] = \begin{bmatrix} 1 & 1 & -3 & -5 & 9 & 25 \\ 1 & -1 & -3 & 5 & 9 & -25 \\ 1 & 0 & -3 & 0 & 9 & 0 \end{bmatrix}$

This is not a square matrix. The rank of M is the rank of a square sub matrix.

$$M_1 = \begin{bmatrix} 1 & 1 & -3 \\ 1 & -1 & -3 \\ 1 & 0 & -3 \end{bmatrix} \qquad |M_1| = 0$$

$$M_2 = \begin{bmatrix} 1 & -3 & -5 \\ -1 & -3 & 5 \\ 0 & -3 & 0 \end{bmatrix} \qquad |M_2| = 0$$

$$M_3 = \begin{bmatrix} -3 & -5 & 9 \\ -3 & 5 & 9 \\ -3 & 0 & 9 \end{bmatrix} \qquad |M_3| = 0$$

$$M_4 = \begin{bmatrix} -5 & 9 & 25 \\ 5 & 9 & -25 \\ 0 & 9 & 0 \end{bmatrix} \qquad |M_4| = 0$$

Hence the rank of A is less than 3.

The next submatrix $M_2 = \begin{bmatrix} 1 & 1 \\ 1 & -1 \end{bmatrix}$ $|M_2| = -2$. Hence the rank of M_2 is 2.

The rank of A is 2 and the system represented by the state model is not completely controllable.

5.2.3 Controllability Criterion 3: Factors Cancellation Test

A necessary and sufficient condition that a system defined by

$$\dot{X} = AX + Bu$$

is completely controllable is that the matrix $[sI - A]^{-1}B$ has no cancellation of factors.

Proof: The solution of the state equation in the s-domain is

$$X(s) = [sI - A]^{-1}BU(s).$$

Since $[sI - A]^{-1} = \dfrac{\text{adjoint of } [sI - A]}{|sI - A|}$

$$[sI - A]^{-1}B = \frac{1}{|sI - A|}\begin{bmatrix} \alpha_1(s) \\ \alpha_2(s) \\ \dots \\ \dots \\ \alpha_n(s) \end{bmatrix}.$$

The factors of $|sI - A|$ yields the Eigen values. Consider that a factor $(s + \lambda_1)$ is cancelled out from $[sI - A]^{-1}B$. From Figure 5.2, it can be observed that the state variable x_1 is not linked to the input and it is not controllable. Hence if there is a cancellation between the numerator and denominator factors in the matrix $[sI - A]^{-1}B$, it means that a particular state variable is not linked to the input and the system is not completely controllable.

Example 5.10 A state equation for a system is given as

$$\dot{X} = \begin{bmatrix} -3 & -1 \\ -2 & -2 \end{bmatrix} X + \begin{bmatrix} 1 \\ 1 \end{bmatrix} u .$$

Check whether it is completely controllable.

Solution:

$$[sI - A]^{-1} = \begin{bmatrix} s+3 & 1 \\ 2 & s+2 \end{bmatrix}^{-1} = \frac{1}{s^2 + 5s + 4}\begin{bmatrix} s+2 & -1 \\ -2 & s+3 \end{bmatrix} = \frac{1}{(s+1)(s+4)}\begin{bmatrix} s+2 & -1 \\ -2 & s+3 \end{bmatrix}$$

$$[sI - A]^{-1}B = \frac{1}{(s+1)(s+4)}\begin{bmatrix} s+2 & -1 \\ -2 & s+3 \end{bmatrix}\begin{bmatrix} 1 \\ 1 \end{bmatrix} = \frac{1}{(s+1)(s+4)}\begin{bmatrix} s+1 \\ s+1 \end{bmatrix}.$$

It can be observed that there is the cancellation of the factor $(s + 1)$ and so the system is not completely controllable.

Example 5.11 A state equation for a system is given as

$$\dot{X} = \begin{bmatrix} 1 & 0 & 0 \\ 0 & 1 & 0 \\ -1 & 0 & 2 \end{bmatrix} X + \begin{bmatrix} 1 \\ 1 \\ 1 \end{bmatrix} u.$$

Check whether the system is completely controllable.

Solution:

$$[sI - A]^{-1} = \begin{bmatrix} (s-1) & 0 & 0 \\ 0 & (s-1) & 0 \\ 1 & 0 & (s-2) \end{bmatrix}^{-1}$$

$$= \frac{1}{(s-1)^2(s-2)} \begin{bmatrix} +\{(s-1)(s-2)\} & -\{0\} & +\{0\} \\ -\{0\} & +\{(s-1)(s-2)\} & -\{0\} \\ +\{-(s-1)\} & -\{0\} & +\{(s-1)^2\} \end{bmatrix}$$

$$= \frac{1}{(s-1)^2(s-2)} \begin{bmatrix} (s-1)(s-2) & 0 & 0 \\ 0 & (s-1)(s-2) & 0 \\ -(s-1) & 0 & (s-1)^2 \end{bmatrix}$$

$$[sI - A]^{-1} B = \frac{1}{(s-1)^2(s-2)} \begin{bmatrix} (s-1)(s-2) & 0 & 0 \\ 0 & (s-1)(s-2) & 0 \\ -(s-1) & 0 & (s-1)^2 \end{bmatrix} \begin{bmatrix} 1 \\ 1 \\ 1 \end{bmatrix}$$

$$= \frac{1}{(s-1)^2(s-2)} \begin{bmatrix} (s-1)(s-2) \\ (s-1)(s-2) \\ (s-1)(s-2) \end{bmatrix}.$$

There exists cancellation of factors and hence the system is not completely controllable.

5.2.4 Controllability Criterion 4: PBH Test

A linear system represented by the state equation

$$\dot{X} = AX + Bu$$

is completely controllable if the matrix pencil $[(\lambda I - A) \quad B]$ has full rank n, for every Eigen value λ. The test for controllability using this criterion is known as **Popov–Belevitch–Hautus** test. A matrix pencil is a matrix whose elements are a linear function of a parameter.

Example 5.12 A state equation for a linear system is

$$\dot{X} = \begin{bmatrix} -3 & -2 \\ 1 & 0 \end{bmatrix} X + \begin{bmatrix} 1 \\ 0 \end{bmatrix} u .$$

Assess the controllability of the system.

Solution: The Eigen values are given by

$$|\lambda I - A| = \begin{vmatrix} \lambda+3 & 2 \\ -1 & \lambda \end{vmatrix} = 0 \quad \lambda^2 + 3\lambda + 2 = 0 \quad \lambda = -1, -2$$

$$[\lambda I - A] = \begin{bmatrix} \lambda+3 & 2 \\ -1 & \lambda \end{bmatrix} \qquad [\lambda I - A \quad B] = \begin{bmatrix} \lambda+3 & 2 & 1 \\ -1 & \lambda & 0 \end{bmatrix}$$

$$\text{rank}[\lambda I - A \quad B]_{\lambda=-1} = \text{rank} \begin{bmatrix} \lambda+3 & 2 & 1 \\ -1 & \lambda & 0 \end{bmatrix}_{\lambda=-1} = \text{rank} \begin{bmatrix} 2 & 2 & 1 \\ -1 & -1 & 0 \end{bmatrix} = 2$$

$$\text{rank}[\lambda I - A \quad B]_{\lambda=-2} = \text{rank} \begin{bmatrix} \lambda+3 & 2 & 1 \\ -1 & \lambda & 0 \end{bmatrix}_{\lambda=-2} = \text{rank} \begin{bmatrix} 1 & 2 & 1 \\ -1 & -2 & 0 \end{bmatrix} = 2 .$$

Hence the system is completely controllable.

Example 5.13 A state equation for a linear system is

$$\dot{X} = \begin{bmatrix} -2 & 0 & 0 \\ 0 & -3 & 0 \\ 1 & 1 & -1 \end{bmatrix} X + \begin{bmatrix} -1 \\ 2 \\ 0 \end{bmatrix} u .$$

Assess the controllability of the system.

Solution: The Eigen values are given by

$$|\lambda I - A| = \begin{vmatrix} \lambda+2 & 0 & 0 \\ 0 & \lambda+3 & 0 \\ -1 & -1 & \lambda+1 \end{vmatrix} = 0 \quad (\lambda+2)(\lambda+3)(\lambda+1) = 0 \quad \lambda = -1, -2, -3$$

$$\text{rank}[\lambda I - A \quad B]_{\lambda=-1} = \text{rank} \begin{bmatrix} \lambda+2 & 0 & 0 & -1 \\ 0 & \lambda+3 & 0 & 2 \\ -1 & -1 & \lambda+1 & 0 \end{bmatrix}_{\lambda=-1} = \text{rank} \begin{bmatrix} 1 & 0 & 0 & -1 \\ 0 & 2 & 0 & 2 \\ -1 & -1 & 0 & 0 \end{bmatrix} = 2$$

$$\text{rank}\begin{bmatrix} \lambda I - A & B \end{bmatrix}_{\lambda=-2} = \text{rank}\begin{bmatrix} \lambda+2 & 0 & 0 & -1 \\ 0 & \lambda+3 & 0 & 2 \\ -1 & -1 & \lambda+1 & 0 \end{bmatrix}_{\lambda=-2} = \text{rank}\begin{bmatrix} 0 & 0 & 0 & -1 \\ 0 & 1 & 0 & 2 \\ -1 & -1 & -1 & 0 \end{bmatrix} = 3$$

$$\text{rank}\begin{bmatrix} (\lambda I - A) & B \end{bmatrix}_{\lambda=-3} = \text{rank}\begin{bmatrix} \lambda+2 & 0 & 0 & -1 \\ 0 & \lambda+3 & 0 & 2 \\ -1 & -1 & \lambda+1 & 0 \end{bmatrix}_{\lambda=-3} = \text{rank}\begin{bmatrix} -1 & 0 & 0 & -1 \\ 0 & 0 & 0 & 2 \\ -1 & -1 & -2 & 0 \end{bmatrix} = 3$$

The mode corresponding to $\lambda = -1$ is not controllable. Hence the system represented by the given state model is not completely controllable.

5.3 OBSERVABILITY

The ability to estimate a state variable from the known output and input is known as the *observability* of a system. If any state variable has no effect on the output, this particular state variable cannot be evaluated from the output. Consider a system represented by the state model in diagonal canonical form, the block diagram of which is shown in Figure 5.4. The concept of observability can be best understood from this figure which compares the observable and unobservable systems. In Figure 5.4(a), all state variables are connected to the output and each of these state variables can be observed at the output. Hence, the system is said to be completely observable. In Figure 5.4(b), one state variable is not connected to the output and it cannot be estimated from a measurement of the output.

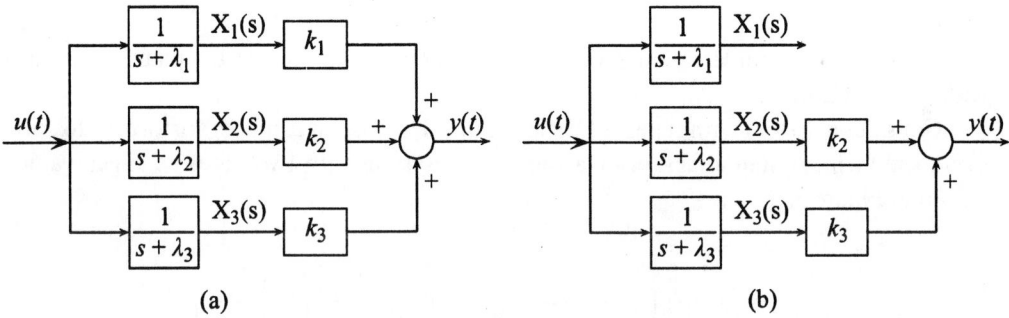

(a)   (b)

Figure 5.4 Observable and unobservable systems

A linear system represented by a state model

$$\dot{X} = AX + Bu$$

$$y = CX + Du$$

is *completely observable* if and only if all the state variables can be uniquely determined using the input $u(t)$ and output $y(t)$; otherwise the system is *unobservable*.

Observability is a property of the coupling between the state $X(t)$ and the output $y(t)$ and thus it involves the matrices A and C.

5.4 OBSERVABILITY CRITERIA

There are many tests by which the complete observability of a system represented by a state model can be assessed. Some of them are discussed here.

5.4.1 Observability Criterion 1: Gilbert's Test

Referring to Figure 5.4(a), each state variable is connected to the output. The output equation is

$$y = [k_1 \quad k_2 \quad k_3]X . \tag{5.7}$$

This represents an observable system. Referring to Figure 5.4(b), the state variable x_1 is not connected to the output. The output equation is

$$y = [0 \quad k_2 \quad k_3]X . \tag{5.8}$$

This represents an unobservable system.

Comparing equations (5.7) and (5.8), it can be observed that for a system, in the diagonal canonical form, to be completely observable, the output matrix should not have any zero elements.

The output equation $y = CX$, if transformed into diagonal canonical form, can be represented as

$$y = [CP]Z .$$

The system is completely observable if and only if any column of the matrix $[CP]$ is not with all zero elements.

Consider a system with repeated Eigen values. The system can be transformed to Jordan canonical form. A state diagram of a third-order system with two repeated Eigen values is shown in Figure 5.5.

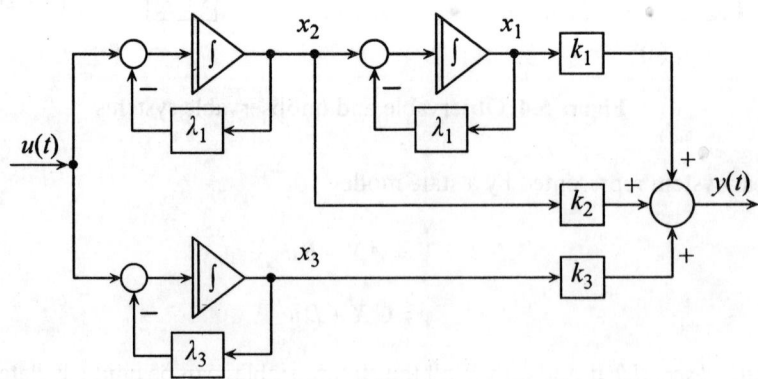

Figure 5.5 State diagram in Jordan canonical form

It can be observed that if x_1 is not linked with the output, x_1 is not observable and hence the system is not completely observable. If k_2 is zero but not k_1, both x_1 and x_2 can be determined. Hence, if k_1 and k_3 are not zeros, the system is completely observable. From the state diagram, the output equation can be written down as

$$y = [k_1 \quad k_2 \quad k_3]X = [CP]X .$$

For the system to be completely observable, it is necessary that both k_1 and k_3 should not be zero, whereas and k_2 can be zero.

Thus it can be concluded that, if a state model is transformed into diagonal form or Jordan canonical form, the system is completely observable if
   (a) no columns of CP that correspond to distinct Eigen value consists of all zero elements.
   (b) no columns of CP that corresponds to the first row of each Jordan block consists of all zero elements and all other columns corresponding to other Eigen values do not have all zero elements.

Example 5.14 A linear system is represented by a state model

$$\dot{X} = \begin{bmatrix} 0 & 1 \\ 8 & -2 \end{bmatrix} X + \begin{bmatrix} 1 \\ 1 \end{bmatrix} u$$

$$y = [4 \quad 1]X .$$

Check whether the system is completely observable.

Solution: The characteristic equation is

$$|\lambda I - A| = \begin{vmatrix} \lambda & -1 \\ -8 & \lambda+2 \end{vmatrix} = 0 \quad \lambda^2 + 2\lambda - 8 = 0 \quad (\lambda+4)(\lambda-2) = 0$$

The Eigen values are $\lambda_1 = -4$, $\lambda_2 = 2$ and are distinct. The given state model is in phase variable form. The transformation matrix that transforms the state model into diagonal form is the Vander Monde matrix.

$$P = \begin{bmatrix} 1 & 1 \\ \lambda_1 & \lambda_2 \end{bmatrix} = \begin{bmatrix} 1 & 1 \\ -4 & 2 \end{bmatrix} \quad |P| = 6 \qquad P^{-1} = \frac{1}{6}\begin{bmatrix} 2 & -1 \\ 4 & 1 \end{bmatrix}$$

Check:

$$P^{-1}P = \frac{1}{6}\begin{bmatrix} 2 & -1 \\ 4 & 1 \end{bmatrix}\begin{bmatrix} 1 & 1 \\ -4 & 2 \end{bmatrix} = \begin{bmatrix} 1 & 0 \\ 0 & 1 \end{bmatrix} = I$$

$$P^{-1}AP = \frac{1}{6}\begin{bmatrix} 2 & -1 \\ 4 & 1 \end{bmatrix}\begin{bmatrix} 0 & 1 \\ 8 & -2 \end{bmatrix}\begin{bmatrix} 1 & 1 \\ -4 & 2 \end{bmatrix} = \frac{1}{6}\begin{bmatrix} 2 & -1 \\ 4 & 1 \end{bmatrix}\begin{bmatrix} -4 & 2 \\ 16 & 4 \end{bmatrix} = \begin{bmatrix} -4 & 0 \\ 0 & 2 \end{bmatrix}$$

$$CP = \begin{bmatrix} 4 & 1 \end{bmatrix} \begin{bmatrix} 1 & 1 \\ -4 & 2 \end{bmatrix} = \begin{bmatrix} 0 & 6 \end{bmatrix}$$

One column of CP is zero and so the system represented by the state model is not completely observable.

Example 5.15 A linear system is represented by a state model

$$\dot{X} = \begin{bmatrix} -3 & -1 \\ 2 & 0 \end{bmatrix} X + \begin{bmatrix} 1 \\ 0 \end{bmatrix} u$$

$$y = \begin{bmatrix} 1 & 2 \end{bmatrix} X .$$

Check whether the system is completely observable.

Solution: The characteristic equation is

$$|\lambda I - A| = \begin{vmatrix} \lambda+3 & 1 \\ -2 & \lambda \end{vmatrix} = 0 \quad \lambda^2 + 3\lambda + 2 = 0 \quad (\lambda+1)(\lambda+2) = 0 .$$

The Eigen values are $\lambda_1 = -1$, $\lambda_2 = -2$ and are distinct. The Eigen vectors are

$$P_i = \begin{bmatrix} \lambda \\ 2 \end{bmatrix} \quad P_1 = \begin{bmatrix} \lambda \\ 2 \end{bmatrix}_{\lambda=-1} = \begin{bmatrix} -1 \\ 2 \end{bmatrix} \quad P_2 = \begin{bmatrix} \lambda \\ 2 \end{bmatrix}_{\lambda=-2} = \begin{bmatrix} -2 \\ 2 \end{bmatrix}$$

$$P = \begin{bmatrix} -1 & -2 \\ 2 & 2 \end{bmatrix} \quad |P| = 2 \quad P^{-1} = \frac{1}{2} \begin{bmatrix} 2 & 2 \\ -2 & -1 \end{bmatrix}.$$

Check:

$$P^{-1}P = \frac{1}{2} \begin{bmatrix} 2 & 2 \\ -2 & -1 \end{bmatrix} \begin{bmatrix} -1 & -2 \\ 2 & 2 \end{bmatrix} = \begin{bmatrix} 1 & 0 \\ 0 & 1 \end{bmatrix}$$

$$P^{-1}AP = \frac{1}{2} \begin{bmatrix} 2 & 2 \\ -2 & -1 \end{bmatrix} \begin{bmatrix} -3 & -1 \\ 2 & 0 \end{bmatrix} \begin{bmatrix} -1 & -2 \\ 2 & 2 \end{bmatrix} = \frac{1}{2} \begin{bmatrix} 2 & 2 \\ -2 & -1 \end{bmatrix} \begin{bmatrix} 1 & 4 \\ -2 & -4 \end{bmatrix} = \begin{bmatrix} -1 & 0 \\ 0 & -2 \end{bmatrix}$$

$$CP = \begin{bmatrix} 1 & 2 \end{bmatrix} \begin{bmatrix} -1 & -2 \\ 2 & 2 \end{bmatrix} = \begin{bmatrix} 3 & 2 \end{bmatrix}.$$

Since the elements of any column of CP are not all zeros, the system represented by the given state model is completely observable.

Example 5.16 A linear system is represented by a state model

$$\dot{X} = \begin{bmatrix} -2 & -2 & 0 \\ 0 & 0 & 1 \\ 0 & -3 & -4 \end{bmatrix} X + \begin{bmatrix} 1 & 0 \\ 0 & 1 \\ 1 & 2 \end{bmatrix} u$$

$$y = \begin{bmatrix} 1 & 3 & 1 \\ 1 & 2 & 0 \end{bmatrix} X .$$

Check whether the system is completely observable.

Solution: The characteristic equation is

$$|\lambda I - A| = \begin{vmatrix} \lambda+2 & 2 & 0 \\ 0 & \lambda & -1 \\ 0 & 3 & \lambda+4 \end{vmatrix} = 0 \quad (\lambda+2)(\lambda^2+4\lambda+3) = 0 \quad (\lambda+2)(\lambda+1)(\lambda+3) = 0.$$

The Eigen values are $\lambda_1 = -1$, $\lambda_2 = -2$ and $\lambda_3 = -3$ and are distinct. Taking the cofactors of the first row of $[\lambda I - A]$

$$P_i = \begin{bmatrix} +\{\lambda(\lambda+4)+3\} \\ -\{0\} \\ +\{0\} \end{bmatrix} = \begin{bmatrix} \lambda(\lambda+4)+3 \\ 0 \\ 0 \end{bmatrix} \qquad P_1 = \begin{bmatrix} \lambda(\lambda+4)+3 \\ 0 \\ 0 \end{bmatrix}_{\lambda=-1} = \begin{bmatrix} 0 \\ 0 \\ 0 \end{bmatrix}.$$

Selecting the cofactors of the third row of $[\lambda I - A]$,

$$P_i = \begin{bmatrix} +\{-2\} \\ -\{-(\lambda+2)\} \\ +\{\lambda(\lambda+2)\} \end{bmatrix} = \begin{bmatrix} -2 \\ \lambda+2 \\ \lambda(\lambda+2) \end{bmatrix} \qquad P_1 = \begin{bmatrix} -2 \\ \lambda+2 \\ \lambda(\lambda+2) \end{bmatrix}_{\lambda=-1} = \begin{bmatrix} -2 \\ 1 \\ -1 \end{bmatrix}$$

$$P_2 = \begin{bmatrix} -2 \\ \lambda+2 \\ \lambda(\lambda+2) \end{bmatrix}_{\lambda=-2} = \begin{bmatrix} -2 \\ 0 \\ 0 \end{bmatrix} \quad Or \; P_2 = \begin{bmatrix} 1 \\ 0 \\ 0 \end{bmatrix} \qquad P_3 = \begin{bmatrix} -2 \\ \lambda+2 \\ \lambda(\lambda+2) \end{bmatrix}_{\lambda=-3} = \begin{bmatrix} -2 \\ -1 \\ 3 \end{bmatrix}$$

$$P = \begin{bmatrix} -2 & 1 & -2 \\ 1 & 0 & -1 \\ -1 & 0 & 3 \end{bmatrix} \qquad |P| = -2(0) - 1(3-1) - 2(0) = -2$$

$$P^{-1} = \frac{-1}{2} \begin{bmatrix} +\{0\} & -\{3\} & +\{-1\} \\ -\{2\} & +\{-8\} & -\{4\} \\ +\{0\} & -\{1\} & +\{-1\} \end{bmatrix} = \frac{1}{2} \begin{bmatrix} 0 & 3 & 1 \\ 2 & 8 & 4 \\ 0 & 1 & 1 \end{bmatrix}$$

Check:

$$P^{-1}P = \frac{1}{2}\begin{bmatrix} 0 & 3 & 1 \\ 2 & 8 & 4 \\ 0 & 1 & 1 \end{bmatrix}\begin{bmatrix} -2 & 1 & -2 \\ 1 & 0 & -1 \\ -1 & 0 & 3 \end{bmatrix} = \begin{bmatrix} 1 & 0 & 0 \\ 0 & 1 & 0 \\ 0 & 0 & 1 \end{bmatrix}$$

$$P^{-1}AP = \frac{1}{2}\begin{bmatrix} 0 & 3 & 1 \\ 2 & 8 & 4 \\ 0 & 1 & 1 \end{bmatrix}\begin{bmatrix} -2 & -2 & 0 \\ 0 & 0 & 1 \\ 0 & -3 & -4 \end{bmatrix}\begin{bmatrix} -2 & 1 & -2 \\ 1 & 0 & -1 \\ -1 & 0 & 3 \end{bmatrix}$$

$$= \frac{1}{2}\begin{bmatrix} 0 & 3 & 1 \\ 2 & 8 & 4 \\ 0 & 1 & 1 \end{bmatrix}\begin{bmatrix} 2 & -2 & 6 \\ -1 & 0 & 3 \\ 1 & 0 & -9 \end{bmatrix} = \begin{bmatrix} -1 & 0 & 0 \\ 0 & -2 & 0 \\ 0 & 0 & -3 \end{bmatrix}$$

$$CP = \begin{bmatrix} 1 & 3 & 1 \\ 1 & 2 & 0 \end{bmatrix}\begin{bmatrix} -2 & 1 & -2 \\ 1 & 0 & -1 \\ -1 & 0 & 3 \end{bmatrix} = \begin{bmatrix} 0 & 1 & -2 \\ 0 & 1 & -4 \end{bmatrix}.$$

Since the first column of [CP] has all zero elements, the system represented by the state model is not completely observable.

5.4.2 Observability Criterion 2: Kalman's Test

A linear time-invariant system represented by a state model

$$\dot{X} = AX + Bu$$

$$y = CX$$

is completely observable if and only if the matrix

$$Q = \begin{bmatrix} C \\ CA \\ CA^2 \\ \dots \\ \dots \\ CA^{n-1} \end{bmatrix} \text{ has full rank } n.$$

Proof:

$$\dot{X} = AX + Bu.$$

The solution of this state equation is

$$X = e^{At}X(0) + \int_0^t e^{A(t-\theta)} B u(\theta) d\theta.$$

The output is

$$y = CX = Ce^{At}X(0) + \int_0^t Ce^{A(t-\theta)} B u(\theta) d\theta.$$

Since the matrices A, B and C and the input $u(t)$ are known, the zero state response (ZSR) can be completely determined and can be subtracted from $y(t)$ to get the zero input response (ZIR). The ZIR is

$$y_0(t) = Ce^{At}X(0).$$

Hence only the state model

$$\dot{X} = AX \qquad \text{and} \quad y = CX$$

is to be considered for deriving the condition for observability.

The matrix exponential e^{At} can be expressed as

$$e^{At} = \alpha_0(t)I + \alpha_1(t)A + \alpha_2(t)A^2 + \dots + \alpha_{n-1}(t)A^{n-1}.$$

The ZIR is

$$y_0(t) = Ce^{At}X(0) = \alpha_0(t)CX(0) + \alpha_1(t)CAX(0) + \alpha_2(t)CA^2X(0) + \alpha_3(t)CA^3X(0)$$
$$+ \dots + \alpha_{n-1}(t)CA^{n-1}X(0)$$

$$y_0(t) = [\alpha_0(t) \quad \alpha_1(t) \quad \alpha_2(t) \dots \alpha_{n-1}(t)] \begin{bmatrix} C \\ CA \\ CA^2 \\ \dots \\ \dots \\ CA^{n-1} \end{bmatrix} X(0)$$

If the system is completely observable, then the output $y_0(t)$ can be uniquely determined from the above equation. This requires that the vectors $C, \ CA, \ CA^2, \ -----, \dots, CA^{n-1}$ are linearly independent. This means that the matrix

$$Q = \begin{bmatrix} C \\ CA \\ CA^2 \\ \dots \\ \dots \\ CA^{n-1} \end{bmatrix}$$

should have full rank n. This matrix is called the *observability matrix*. Hence the system represented by a state model is completely observable if and only if the observability matrix has full rank.

Example 5.17 A linear system is represented by a state model

$$\dot{X} = \begin{bmatrix} -3 & -1 \\ 2 & 0 \end{bmatrix} X + \begin{bmatrix} 1 \\ 0 \end{bmatrix} u \quad y = [1 \quad -1] X .$$

Check whether the system is completely observable.

Solution:

$$A = \begin{bmatrix} -3 & -1 \\ 2 & 0 \end{bmatrix} \qquad C = [1 \quad -1] \qquad CA = [1 \quad -1] \begin{bmatrix} -3 & -1 \\ 2 & 0 \end{bmatrix} = [-5 \quad -1] .$$

The observability matrix is

$$Q = \begin{bmatrix} C \\ CA \end{bmatrix} = \begin{bmatrix} 1 & -1 \\ -5 & -1 \end{bmatrix} \quad |Q| = -6 .$$

The rank of Q is 2 and hence the system represented by the state model is completely observable.

Example 5.18 A linear system is represented by a state model

$$\dot{X} = \begin{bmatrix} 2 & -2 & 3 \\ 1 & 1 & 1 \\ 1 & 3 & -1 \end{bmatrix} X + \begin{bmatrix} 1 \\ 0 \\ 1 \end{bmatrix} u \qquad\qquad y = [-3 \quad 5 \quad -2] X .$$

Check whether the system is completely observable.

Solution:

$$A = \begin{bmatrix} 2 & -2 & 3 \\ 1 & 1 & 1 \\ 1 & 3 & -1 \end{bmatrix} \qquad\qquad C = [-3 \quad 5 \quad -2]$$

$$CA = [-3 \quad 5 \quad -2] \begin{bmatrix} 2 & 2 & 3 \\ 1 & 1 & 1 \\ 1 & 3 & 1 \end{bmatrix} = [-3 \quad 5 \quad -2]$$

$$CA^2 = [-3 \quad 5 \quad -2] \begin{bmatrix} 2 & -2 & 3 \\ 1 & 1 & 1 \\ 1 & 3 & -1 \end{bmatrix} = [-3 \quad 5 \quad -2] .$$

The observability matrix

$$Q = \begin{bmatrix} C \\ CA \\ CA^2 \end{bmatrix} = \begin{bmatrix} -3 & 5 & -2 \\ -3 & 5 & -2 \\ -3 & 5 & -2 \end{bmatrix}.$$

Its rank is 1. Hence only one mode is observable and the system represented by the state model is not completely observable.

Example 5.19 A linear system is represented by a state model

$$\dot{X} = \begin{bmatrix} -2 & -2 & 0 \\ 0 & 0 & 1 \\ 0 & -3 & -4 \end{bmatrix} X + \begin{bmatrix} 1 & 0 \\ 0 & 1 \\ 1 & 2 \end{bmatrix} u \qquad y = \begin{bmatrix} 1 & 3 & 1 \\ 1 & 2 & 0 \end{bmatrix} X.$$

Check whether the system is completely observable.

Solution:

$$A = \begin{bmatrix} -2 & -2 & 0 \\ 0 & 0 & 1 \\ 0 & -3 & -4 \end{bmatrix} \qquad C = \begin{bmatrix} 1 & 3 & 1 \\ 1 & 2 & 0 \end{bmatrix}$$

$$CA = \begin{bmatrix} 1 & 3 & 1 \\ 1 & 2 & 0 \end{bmatrix} \begin{bmatrix} -2 & -2 & 0 \\ 0 & 0 & 1 \\ 0 & -3 & -4 \end{bmatrix} = \begin{bmatrix} -2 & -5 & -1 \\ -2 & -2 & 2 \end{bmatrix}$$

$$CA^2 = \begin{bmatrix} -2 & -5 & -1 \\ -2 & -2 & 2 \end{bmatrix} \begin{bmatrix} -2 & -2 & 0 \\ 0 & 0 & 1 \\ 0 & -3 & -4 \end{bmatrix} = \begin{bmatrix} 4 & 7 & -1 \\ 4 & -2 & -10 \end{bmatrix}.$$

The observability matrix

$$Q = \begin{bmatrix} C \\ CA \\ CA^2 \end{bmatrix} = \begin{bmatrix} 1 & 3 & 1 \\ 1 & 2 & 0 \\ -2 & -5 & -1 \\ -2 & -2 & 2 \\ 4 & 7 & -1 \\ 4 & -2 & -10 \end{bmatrix}$$

$$|Q_1| = \begin{vmatrix} 1 & 3 & 1 \\ 1 & 2 & 0 \\ -2 & -5 & -1 \end{vmatrix} = 0 \qquad\qquad |Q_2| = \begin{vmatrix} 1 & 2 & 0 \\ -2 & -5 & -1 \\ -2 & -2 & 2 \end{vmatrix} = 0$$

$$|Q_3| = \begin{vmatrix} -2 & -5 & -1 \\ -2 & -2 & 2 \\ 4 & 7 & -1 \end{vmatrix} = 0 \qquad |Q_4| = \begin{vmatrix} -2 & -2 & 2 \\ 4 & 7 & -1 \\ 4 & -2 & -10 \end{vmatrix} = 0$$

$$|Q_5| = \begin{vmatrix} 1 & 3 \\ 1 & 2 \end{vmatrix} = -1 .$$

The rank of Q is 2 and hence the system represented by the state model is not completely observable.

5.4.3 Observability Criterion 3: Factors Cancellation Test

A necessary and sufficient condition that the system

$$\dot{X} = AX + Bu$$

$$y = CX$$

is completely observable is that the matrix $\{C[sI - A]^{-1}\}$ has no cancellation of factors.

Example 5.20 For a linear system, a state model is given as

$$\dot{X} = \begin{bmatrix} -5 & 1 \\ -8 & 1 \end{bmatrix} X + \begin{bmatrix} 0 \\ 1 \end{bmatrix} u \qquad\qquad y = [2 \quad 0] X .$$

Assess the observability of the system.

Solution:

$$[sI - A]^{-1} = \begin{bmatrix} s+5 & -1 \\ 8 & s-1 \end{bmatrix}^{-1} = \frac{1}{s^2 + 4s + 3} \begin{bmatrix} s-1 & 1 \\ -8 & s+5 \end{bmatrix}$$

$$= \frac{1}{(s+1)(s+3)} \begin{bmatrix} s-1 & 1 \\ -8 & s+5 \end{bmatrix} .$$

$$C[sI - A]^{-1} = \frac{1}{(s+1)(s+3)} [2 \quad 0] \begin{bmatrix} s-1 & 1 \\ -8 & s+5 \end{bmatrix} = \frac{1}{(s+1)(s+3)} [2(s-1) \quad 2] .$$

There is no cancellation of factors and hence the system represented by the given state model is completely observable.

Example 5.21 For a linear system, a state model is given as

$$\dot{X} = \begin{bmatrix} -1 & 1 \\ -1 & -3 \end{bmatrix} X + \begin{bmatrix} 1 \\ 1 \end{bmatrix} u$$

$$y = [1 \quad 1]X .$$

Assess the observability of the system.

Solution:

$$[sI - A]^{-1} = \begin{bmatrix} s+1 & -1 \\ 1 & s+3 \end{bmatrix}^{-1} = \frac{1}{s^2 + 4s + 4} \begin{bmatrix} s+3 & 1 \\ -1 & s+1 \end{bmatrix}$$

$$= \frac{1}{(s+2)^2} \begin{bmatrix} s+3 & 1 \\ -1 & s+1 \end{bmatrix}.$$

$$C[sI - A]^{-1} = \frac{1}{(s+2)^2} [1 \quad 1] \begin{bmatrix} s+3 & 1 \\ -1 & s+1 \end{bmatrix} = \frac{1}{(s+2)^2} [(s+2) \quad (s+2)] .$$

There is cancellation of factors and hence the system represented by the given state model is not completely observable.

Example 5.22 A linear system is represented by a state model

$$\dot{X} = \begin{bmatrix} 0 & 1 & 0 \\ 0 & 0 & 1 \\ -6 & -11 & -6 \end{bmatrix} X + \begin{bmatrix} 0 \\ 0 \\ 1 \end{bmatrix} u \qquad\qquad y = [1 \quad 2 \quad 1]X .$$

Check whether the system is completely observable.

Solution:

$$A = \begin{bmatrix} 0 & 1 & 0 \\ 0 & 0 & 1 \\ -6 & -11 & -6 \end{bmatrix} \qquad\qquad C = [1 \quad 2 \quad 1].$$

$$[sI - A]^{-1} = \begin{bmatrix} s & -1 & 0 \\ 0 & s & -1 \\ 6 & 11 & s+6 \end{bmatrix}^{-1}$$

$$= \frac{1}{s(s^2 + 6s + 11) + 6} \begin{bmatrix} +\{s^2 + 6s + 11\} & -\{-(s+6)\} & +\{1\} \\ -\{6\} & +\{s(s+6)\} & -\{-s\} \\ +\{-6s\} & -\{11s+6\} & +\{s^2\} \end{bmatrix}$$

$$= \frac{1}{(s+1)(s+2)(s+3)} \begin{bmatrix} s^2 + 6s + 11 & (s+6) & 1 \\ -6 & s(s+6) & s \\ -6s & -(11s+6) & s^2 \end{bmatrix}.$$

$$C[sI - A]^{-1} = \frac{1}{(s+1)(s+2)(s+3)}[1 \quad 2 \quad 1]\begin{bmatrix} s^2+6s+11 & (s+6) & 1 \\ -6 & s(s+6) & s \\ -6s & -(11s+6) & s^2 \end{bmatrix}$$

$$= \frac{1}{(s+1)(s+2)(s+3)}\left[(s+1)(s-1) \quad 2s(s+1) \quad (s+1)^2\right].$$

The factor $(s+1)$ cancels out and hence the mode corresponding to the Eigen value $s=-1$ is unobservable. The system represented by the given state model is not completely observable.

5.4.4 OBSERVABILITY CRITERION 4: PBH TEST

A linear system represented by the state equation

$$\dot{X} = AX + Bu \qquad\qquad y = CX$$

is completely observable if the matrix pencil $\begin{bmatrix} C \\ \lambda I - A \end{bmatrix}$ has full rank n, for every Eigen value λ.

The test for observability using this criterion is known as Popov–Belevitch–Hautus test.

Example 5.23 For a linear system, a state model is given as

$$\dot{X} = \begin{bmatrix} -5 & 1 \\ -8 & 1 \end{bmatrix}X + \begin{bmatrix} 0 \\ 1 \end{bmatrix}u \qquad\qquad y = [2 \quad 0]X .$$

Assess the observability of the system, using PBH test.

Solution:

$$[\lambda I - A] = \begin{bmatrix} \lambda+5 & -1 \\ 8 & \lambda-1 \end{bmatrix}.$$

The Eigen values are obtained by solving

$$\begin{vmatrix} \lambda+5 & -1 \\ 8 & \lambda-1 \end{vmatrix} = 0 \qquad\qquad \lambda^2 + 4\lambda + 3 = 0 \quad \lambda_1 = -1 \text{ and } \lambda_2 = -3$$

$$\begin{bmatrix} C \\ \lambda I - A \end{bmatrix} = \begin{bmatrix} 2 & 0 \\ \lambda+5 & -1 \\ 8 & \lambda-1 \end{bmatrix}.$$

$$\text{Rank}\begin{bmatrix} C \\ \lambda I - A \end{bmatrix}_{\lambda=-1} = \text{Rank}\begin{bmatrix} 2 & 0 \\ \lambda+5 & -1 \\ 8 & \lambda-1 \end{bmatrix}_{\lambda=-1} = \text{Rank}\begin{bmatrix} 2 & 0 \\ 4 & -1 \\ 8 & -2 \end{bmatrix} = 2$$

$$\text{Rank}\begin{bmatrix} C \\ \lambda I - A \end{bmatrix}_{\lambda=-3} = \text{Rank}\begin{bmatrix} 2 & 0 \\ \lambda+5 & -1 \\ 8 & \lambda-1 \end{bmatrix}_{\lambda=-3} = \text{Rank}\begin{bmatrix} 2 & 0 \\ 2 & -1 \\ 8 & -4 \end{bmatrix} = 2.$$

The rank of $\begin{bmatrix} C \\ \lambda I - A \end{bmatrix}$ is 2 for both the Eigen values. Hence, the system represented by the state model is completely observable.

Example 5.24 For a linear system, a state model is given as

$$\dot{X} = \begin{bmatrix} -1 & 1 \\ -1 & -3 \end{bmatrix} X + \begin{bmatrix} 1 \\ 1 \end{bmatrix} u \qquad\qquad y = \begin{bmatrix} 1 & 1 \end{bmatrix} X .$$

Assess the observability of the system, using PBH test.

Solution:

$$[\lambda I - A] = \begin{bmatrix} \lambda+1 & -1 \\ 1 & \lambda+3 \end{bmatrix}.$$

The Eigen values are obtained by solving

$$\begin{vmatrix} \lambda+1 & -1 \\ 1 & \lambda+3 \end{vmatrix} = 0 \qquad\qquad \lambda^2 + 4\lambda + 4 = 0 \quad \lambda_1 = -2 \text{ and } \lambda_2 = -2.$$

$$\begin{bmatrix} C \\ \lambda I - A \end{bmatrix} = \begin{bmatrix} 1 & 1 \\ \lambda+1 & -1 \\ 1 & \lambda+3 \end{bmatrix}$$

$$\text{Rank}\begin{bmatrix} C \\ \lambda I - A \end{bmatrix}_{\lambda=-2} = \text{Rank}\begin{bmatrix} 1 & 1 \\ \lambda+1 & -1 \\ 1 & \lambda+3 \end{bmatrix}_{\lambda=-2} = \text{Rank}\begin{bmatrix} 1 & 1 \\ -1 & -1 \\ 1 & 1 \end{bmatrix} = 1.$$

The matrix $\begin{bmatrix} C \\ \lambda I - A \end{bmatrix}$ does not have full rank for the repeated Eigen values. Hence, the system represented by the state model is not completely observable.

Example 5.25 A linear system is represented by a state model

$$\dot{X} = \begin{bmatrix} 0 & 1 & 0 \\ 0 & 0 & 1 \\ -6 & -11 & -6 \end{bmatrix} X + \begin{bmatrix} 0 \\ 0 \\ 1 \end{bmatrix} u$$

$$y = \begin{bmatrix} 1 & 2 & 1 \end{bmatrix} X .$$

Check whether the system is completely observable, using PBH test.

Solution:

$$A = \begin{bmatrix} 0 & 1 & 0 \\ 0 & 0 & 1 \\ -6 & -11 & -6 \end{bmatrix} \qquad C = \begin{bmatrix} 1 & 2 & 1 \end{bmatrix}.$$

$$|\lambda I - A| = \begin{vmatrix} \lambda & -1 & 0 \\ 0 & \lambda & -1 \\ 6 & 11 & \lambda+6 \end{vmatrix} = 0 \qquad \lambda^3 + 6\lambda^2 + 11\lambda + 6 = 0 \quad \lambda = -1, -2 \, and -3$$

$$\begin{bmatrix} C \\ \lambda I - A \end{bmatrix} = \begin{bmatrix} 1 & 2 & 1 \\ \lambda & -1 & 0 \\ 0 & \lambda & -1 \\ 6 & 11 & \lambda+6 \end{bmatrix}.$$

$$\text{Rank} \begin{bmatrix} C \\ \lambda I - A \end{bmatrix}_{\lambda=-1} = \text{Rank} \begin{bmatrix} 1 & 2 & 1 \\ \lambda & -1 & 0 \\ 0 & \lambda & -1 \\ 6 & 11 & \lambda+6 \end{bmatrix}_{\lambda=-1} = \text{Rank} \begin{bmatrix} 1 & 2 & 1 \\ -1 & -1 & 0 \\ 0 & -1 & -1 \\ 6 & 11 & 5 \end{bmatrix} = 2$$

$$\text{Rank} \begin{bmatrix} C \\ \lambda I - A \end{bmatrix}_{\lambda=-2} = \text{Rank} \begin{bmatrix} 1 & 2 & 1 \\ \lambda & -1 & 0 \\ 0 & \lambda & -1 \\ 6 & 11 & \lambda+6 \end{bmatrix}_{\lambda=-2} = \text{Rank} \begin{bmatrix} 1 & 2 & 1 \\ -2 & -1 & 0 \\ 0 & -2 & -1 \\ 6 & 11 & 4 \end{bmatrix} = 3$$

$$\text{Rank} \begin{bmatrix} C \\ \lambda I - A \end{bmatrix}_{\lambda=-3} = \text{Rank} \begin{bmatrix} 1 & 2 & 1 \\ \lambda & -1 & 0 \\ 0 & \lambda & -1 \\ 6 & 11 & \lambda+6 \end{bmatrix}_{\lambda=-3} = \text{Rank} \begin{bmatrix} 1 & 2 & 1 \\ -3 & -1 & 0 \\ 0 & -3 & -1 \\ 6 & 11 & 3 \end{bmatrix} = 3.$$

Hence the mode corresponding to ($\lambda = -1$) is unobservable, and the system represented by the state model is completely observable.

5.5 TRANSFORMATION TO CONTROLLABLE CANONICAL FORM

In the design of a feedback controller for a system, it is often necessary to transfer the given state model of the system into controllable canonical form. Consider a state model for a system which is not in controllable canonical form:

$$\dot{X} = AX + Bu .$$

Consider, for convenience, a system matrix A of size (3x3) and B of size (3x1). The controllability matrix is

$$M = [B \quad AB \quad A^2 B] .$$

Suppose that the state model is transferred into controllable canonical form by the transformation

$$X = PZ$$

$$\dot{X} = P\dot{Z}$$

$$\dot{Z} = P^{-1} \dot{X} = P^{-1}[AX + Bu] = [P^{-1}AP]Z + [P^{-1}B]u = A_z Z + B_z u .$$

The controllability matrix corresponding to the state model in controllable canonical form is

$$M_Z = [B_Z \quad A_z B_Z \quad A_z^2 B_z]$$

$$= [(P^{-1} B) \quad (P^{-1} AP)(P^{-1} B) \quad (P^{-1} AP)(P^{-1} AP)(P^{-1}B)]$$

$$= [(P^{-1} B) \quad (P^{-1} AB) \quad (P^{-1} A^2 B)]$$

$$= P^{-1}[B \quad AB \quad A^2 B]$$

$$= P^{-1} M .$$

From this $PM_Z = M$

$$P = M M_Z^{-1} .$$

This is the transformation matrix P that transforms the given state model into controllable canonical form. The new system matrix in the controllable canonical form is

$$A_z = [P^{-1}AP] = [M M_Z^{-1}]^{-1} AP = [M_Z M^{-1}]AP .$$

To determine this matrix, M^{-1} should exist. For this $|M|$ should not be zero and M should have full rank. Hence, for a system represented by a state model to be transformed into controllable canonical form, it must be completely controllable.

Example 5.26 A state model for a linear second-order system is

$$\dot{X} = AX + Bu \qquad\qquad y = CX .$$

The characteristic equation is

$$\lambda^2 + a_1\lambda + a_0 = 0 .$$

Using this, obtain the controllability matrix and its inverse when the system is transformed into controllable canonical form.

Solution: When a state model is transformed into controllable canonical form, the state model is of the form

$$\dot{Z} = [P^{-1}AP]Z + [P^{-1}B]u$$

$$= \begin{bmatrix} 0 & 1 \\ -a_0 & -a_1 \end{bmatrix} Z + \begin{bmatrix} 0 \\ 1 \end{bmatrix} u .$$

The controllability matrix is

$$M_Z = [B_Z \quad (A_Z B_Z)]$$

$$= [(P^{-1}B) \quad (P^{-1}AP)(P^{-1}B)]$$

$$(P^{-1}AP)(P^{-1}B) = \begin{bmatrix} 0 & 1 \\ -a_0 & -a_1 \end{bmatrix}\begin{bmatrix} 0 \\ 1 \end{bmatrix} = \begin{bmatrix} 1 \\ -a_1 \end{bmatrix}$$

$$M_Z = \begin{bmatrix} 0 & 1 \\ 1 & -a_1 \end{bmatrix} \qquad M_Z^{-1} = -\begin{bmatrix} -a_1 & -1 \\ -1 & 0 \end{bmatrix} = \begin{bmatrix} a_1 & 1 \\ 1 & 0 \end{bmatrix} .$$

Example 5.27 A state model for a third order linear system is

$$\dot{X} = AX + Bu \qquad\qquad y = CX .$$

The characteristic equation is

$$s^3 + a_2 s^2 + a_1 s + a_0 = 0 .$$

Using this, obtain the controllability matrix and its inverse when the state model is transformed into controllable canonical form.

Solution: When a state model is transformed into controllable canonical form, the state model is of the form

$$\dot{Z} = [P^{-1}AP]Z + [P^{-1}B]u$$

$$= \begin{bmatrix} 0 & 1 & 0 \\ 0 & 0 & 1 \\ -a_0 & -a_1 & -a_2 \end{bmatrix} Z + \begin{bmatrix} 0 \\ 0 \\ 1 \end{bmatrix} u ,$$

where a_0, a_1 and a_2 are the coefficients of the characteristic equation

$$s^3 + a_2 s^2 + a_1 s + a_0 = 0.$$

The controllability matrix is

$$M_Z = [B_Z \quad A_Z B_Z \quad A_Z^2 B_Z]$$

$$= [(P^{-1}B) \quad (P^{-1}AP)(P^{-1}B) \quad (P^{-1}AP)^2(P^{-1}B)].$$

$$(P^{-1}AP)(P^{-1}B) = \begin{bmatrix} 0 & 1 & 0 \\ 0 & 0 & 1 \\ -a_0 & -a_1 & -a_2 \end{bmatrix} \begin{bmatrix} 0 \\ 0 \\ 1 \end{bmatrix} = \begin{bmatrix} 0 \\ 1 \\ -a_2 \end{bmatrix}$$

$$(P^{-1}AP)^2(P^{-1}B) = \begin{bmatrix} 0 & 1 & 0 \\ 0 & 0 & 1 \\ -a_0 & -a_1 & -a_2 \end{bmatrix} \begin{bmatrix} 0 \\ 1 \\ -a_2 \end{bmatrix} = \begin{bmatrix} 1 \\ -a_2 \\ -a_1 + a_2^2 \end{bmatrix}.$$

The controllability matrix

$$M_Z = \begin{bmatrix} 0 & 0 & 1 \\ 0 & 1 & -a_2 \\ 1 & -a_2 & -a_1 + a_2^2 \end{bmatrix} \qquad |M_Z| = -1$$

$$M_Z^{-1} = -\begin{bmatrix} +\{-a_1\} & -\{a_2\} & +\{-1\} \\ -\{a_2\} & +\{-1\} & -\{0\} \\ +\{-1\} & -\{0\} & +\{0\} \end{bmatrix} = \begin{bmatrix} a_1 & a_2 & 1 \\ a_2 & 1 & 0 \\ 1 & 0 & 0 \end{bmatrix}$$

Example 5.28

A third-order linear system is represented by the state equation

$$\dot{X} = AX + Bu$$

and has the characteristic equation

$$s^3 + a_2 s^2 + a_1 s + a_0 = 0.$$

If this is transformed into controllable canonical form

$$\dot{Z} = [P^{-1}AP]Z + [P^{-1}B]u$$

with the transformation

$$X = PZ,$$

show that the output matrix is $[P^{-1}B] = \begin{bmatrix} 0 & 0 & 1 \end{bmatrix}^T$.

Solution:
The characteristic equation of the system is given as

$$s^3 + a_2 s^2 + a_1 s + a_0 = 0.$$

The controllability matrix of the given state model is

$$M = [B \quad AB \quad A^2 B].$$

It is to be proved that the output matrix in the controllable canonical form is

$$[P^{-1}B] = \begin{bmatrix} 0 \\ 0 \\ 1 \end{bmatrix}.$$

From this $B = P \begin{bmatrix} 0 \\ 0 \\ 1 \end{bmatrix}$.

Hence it is sufficient to prove $P \begin{bmatrix} 0 \\ 0 \\ 1 \end{bmatrix} = B$

The matrix P is the transformation matrix and is given by

$$P = M M_Z^{-1}.$$

$$M_Z = \begin{bmatrix} a_1 & a_2 & 1 \\ a_2 & 1 & 0 \\ 1 & 0 & 0 \end{bmatrix}^{-1}$$

$$P = M M_Z^{-1} = [B \quad AB \quad A^2 B] \begin{bmatrix} a_1 & a_2 & 1 \\ a_2 & 1 & 0 \\ 1 & 0 & 0 \end{bmatrix}$$

$$P \begin{bmatrix} 0 \\ 0 \\ 1 \end{bmatrix} = [B \quad AB \quad A^2 B] \begin{bmatrix} a_1 & a_2 & 1 \\ a_2 & 1 & 0 \\ 1 & 0 & 0 \end{bmatrix} \begin{bmatrix} 0 \\ 0 \\ 1 \end{bmatrix}$$

$$= [B \quad AB \quad A^2 B] \begin{bmatrix} 1 \\ 0 \\ 0 \end{bmatrix} = B$$

Hence $[P^{-1}B] = \begin{bmatrix} 0 \\ 0 \\ 1 \end{bmatrix}$.

Example 5.29

The state equation for a linear system is

$$\dot{X} = \begin{bmatrix} -2 & 1 \\ 0 & -1 \end{bmatrix} X + \begin{bmatrix} 2 \\ 1 \end{bmatrix} u \qquad\qquad y = [2 \quad -1]X .$$

Convert the state model into controllable canonical form.

Solution:

$$\dot{X} = \begin{bmatrix} -2 & 1 \\ 0 & -1 \end{bmatrix} X + \begin{bmatrix} 2 \\ 1 \end{bmatrix} u .$$

$$A = \begin{bmatrix} -2 & 1 \\ 0 & -1 \end{bmatrix} \qquad\qquad B = \begin{bmatrix} 2 \\ 1 \end{bmatrix}.$$

The Eigen values are obtained by solving the equation

$$\begin{vmatrix} \lambda+2 & -1 \\ 0 & \lambda+1 \end{vmatrix} = 0 \qquad\qquad (\lambda+1)(\lambda+2) = 0$$

$$\lambda^2 + 3\lambda + 2 = 0 = \lambda^2 + a_1\lambda + a_0 \qquad\qquad a_1 = 3 \text{ and } a_0 = 2 .$$

$$AB = \begin{bmatrix} -2 & 1 \\ 0 & -1 \end{bmatrix} \begin{bmatrix} 2 \\ 1 \end{bmatrix} = \begin{bmatrix} -3 \\ -1 \end{bmatrix}.$$

The controllability matrix is

$$M = [B \quad AB] = \begin{bmatrix} 2 & -3 \\ 1 & -1 \end{bmatrix}.$$

The rank of M is 2, and hence, the system representation is completely controllable. The state model can be transformed into controllable canonical form.

$$A_{\mathcal{Z}} = \begin{bmatrix} 0 & 1 \\ -a_0 & -a_1 \end{bmatrix} = \begin{bmatrix} 0 & 1 \\ -2 & -3 \end{bmatrix} \qquad\qquad B_{\mathcal{Z}} = \begin{bmatrix} 0 \\ 1 \end{bmatrix}$$

$$A_\chi B_\chi = \begin{bmatrix} 0 & 1 \\ -2 & -3 \end{bmatrix} \begin{bmatrix} 0 \\ 1 \end{bmatrix} = \begin{bmatrix} 1 \\ -3 \end{bmatrix}.$$

The controllability matrix of the system in controllable canonical form is

$$M_\chi = [B_\chi \quad A_\chi B_\chi] = \begin{bmatrix} 0 & 1 \\ 1 & -3 \end{bmatrix} \qquad M_Z^{-1} = -\begin{bmatrix} -3 & -1 \\ -1 & 0 \end{bmatrix} = \begin{bmatrix} 3 & 1 \\ 1 & 0 \end{bmatrix}.$$

The transformation matrix that transforms the state model into controllable canonical form is

$$P = M M_\chi^{-1} = \begin{bmatrix} 2 & -3 \\ 1 & -1 \end{bmatrix} \begin{bmatrix} 3 & 1 \\ 1 & 0 \end{bmatrix} = \begin{bmatrix} 3 & 2 \\ 2 & 1 \end{bmatrix}$$

$$|P| = -1 \quad P^{-1} = -\begin{bmatrix} 1 & -2 \\ -2 & 3 \end{bmatrix} = \begin{bmatrix} -1 & 2 \\ 2 & -3 \end{bmatrix}$$

Check:

$$A_\chi = P^{-1}AP = \begin{bmatrix} -1 & 2 \\ 2 & -3 \end{bmatrix} \begin{bmatrix} -2 & 1 \\ 0 & -1 \end{bmatrix} \begin{bmatrix} 3 & 2 \\ 2 & 1 \end{bmatrix} = \begin{bmatrix} -1 & 2 \\ 2 & -3 \end{bmatrix} \begin{bmatrix} -4 & -3 \\ -2 & -1 \end{bmatrix} = \begin{bmatrix} 0 & 1 \\ -2 & -3 \end{bmatrix}$$

$$B_\chi = P^{-1}B = \begin{bmatrix} -1 & 2 \\ 2 & -3 \end{bmatrix} \begin{bmatrix} 2 \\ 1 \end{bmatrix} = \begin{bmatrix} 0 \\ 1 \end{bmatrix} \qquad C_\chi = CP = [2 \quad -1] \begin{bmatrix} 3 & 2 \\ 2 & 1 \end{bmatrix} = [4 \quad 3].$$

The state model in controllable canonical form is

$$\dot{Z} = \begin{bmatrix} 0 & 1 \\ -2 & -3 \end{bmatrix} Z + \begin{bmatrix} 0 \\ 1 \end{bmatrix} u \qquad y = [4 \quad 3]Z.$$

Example 5.30

The state equation for a linear system is

$$\dot{X} = \begin{bmatrix} -7 & 1 & 0 \\ 0 & -8 & 1 \\ 0 & 0 & -9 \end{bmatrix} X + \begin{bmatrix} 0 \\ 0 \\ 1 \end{bmatrix} u \qquad y = [-2 \quad 2 \quad 1]X.$$

Convert the state model into controllable canonical form.

Solution:

$$A = \begin{bmatrix} -7 & 1 & 0 \\ 0 & -8 & 1 \\ 0 & 0 & -9 \end{bmatrix} \qquad B = \begin{bmatrix} 0 \\ 0 \\ 1 \end{bmatrix} \qquad C = [-2 \quad 2 \quad 1].$$

The characteristic equation is

$$|\lambda I - A| = \begin{vmatrix} \lambda+7 & -1 & 0 \\ 0 & \lambda+8 & -1 \\ 0 & 0 & \lambda+9 \end{vmatrix} = 0 \qquad (\lambda+7)(\lambda+8)(\lambda+9) = 0$$

$$\lambda^3 + 24\lambda^2 + 191\lambda + 504 = 0 = \lambda^3 + a_2\lambda^2 + a_1\lambda + a_0 \qquad a_2 = 24 \quad a_1 = 191 \quad a_0 = 504.$$

$$AB = \begin{bmatrix} -7 & 1 & 0 \\ 0 & -8 & 1 \\ 0 & 0 & -9 \end{bmatrix} \begin{bmatrix} 0 \\ 0 \\ 1 \end{bmatrix} = \begin{bmatrix} 0 \\ 1 \\ -9 \end{bmatrix} \qquad A^2B = \begin{bmatrix} -7 & 1 & 0 \\ 0 & -8 & 1 \\ 0 & 0 & -9 \end{bmatrix} \begin{bmatrix} 0 \\ 1 \\ -9 \end{bmatrix} = \begin{bmatrix} 1 \\ -17 \\ 81 \end{bmatrix}.$$

The controllability matrix is

$$M = [B \quad AB \quad A^2B] = \begin{bmatrix} 0 & 0 & 1 \\ 0 & 1 & -17 \\ 1 & -9 & 81 \end{bmatrix}.$$

The rank of M is 3 and hence the system representation is completely controllable. The state model can be transformed into controllable canonical form.

$$M_Z^{-1} = \begin{bmatrix} a_1 & a_2 & 1 \\ a_2 & 1 & 0 \\ 1 & 0 & 0 \end{bmatrix} = \begin{bmatrix} 191 & 24 & 1 \\ 24 & 1 & 0 \\ 1 & 0 & 0 \end{bmatrix}.$$

The transformation matrix that transforms the given state model into controllable canonical form is

$$P = M M_Z^{-1} = \begin{bmatrix} 0 & 0 & 1 \\ 0 & 1 & -17 \\ 1 & -9 & 81 \end{bmatrix} \begin{bmatrix} 191 & 24 & 1 \\ 24 & 1 & 0 \\ 1 & 0 & 0 \end{bmatrix} = \begin{bmatrix} 1 & 0 & 0 \\ 7 & 1 & 0 \\ 56 & 15 & 1 \end{bmatrix}.$$

$$|P| = 1 \qquad P^{-1} = \begin{bmatrix} +\{1\} & -\{0\} & +\{0\} \\ -\{7\} & +\{1\} & -\{0\} \\ +\{49\} & -\{15\} & +\{1\} \end{bmatrix} = \begin{bmatrix} 1 & 0 & 0 \\ -7 & 1 & 0 \\ 49 & -15 & 1 \end{bmatrix}.$$

Check:

$$A_Z = P^{-1}AP = \begin{bmatrix} 1 & 0 & 0 \\ -7 & 1 & 0 \\ 49 & -15 & 1 \end{bmatrix} \begin{bmatrix} -7 & 1 & 0 \\ 0 & -8 & 1 \\ 0 & 0 & -9 \end{bmatrix} \begin{bmatrix} 1 & 0 & 0 \\ 7 & 1 & 0 \\ 56 & 15 & 1 \end{bmatrix}$$

$$= \begin{bmatrix} 1 & 0 & 0 \\ -7 & 1 & 0 \\ 49 & -15 & 1 \end{bmatrix} \begin{bmatrix} 0 & 1 & 0 \\ 0 & 7 & 1 \\ -504 & -135 & -9 \end{bmatrix} = \begin{bmatrix} 0 & 1 & 0 \\ 0 & 0 & 1 \\ -504 & -191 & -24 \end{bmatrix}$$

$$B_{\chi} = P^{-1}B = \begin{bmatrix} 1 & 0 & 0 \\ -7 & 1 & 0 \\ 49 & -15 & 1 \end{bmatrix}\begin{bmatrix} 0 \\ 0 \\ 1 \end{bmatrix} = \begin{bmatrix} 0 \\ 0 \\ 1 \end{bmatrix}$$

$$C_{\chi} = CP = \begin{bmatrix} -2 & 2 & 1 \end{bmatrix}\begin{bmatrix} 1 & 0 & 0 \\ 7 & 1 & 0 \\ 56 & 15 & 1 \end{bmatrix} = \begin{bmatrix} 68 & 17 & 1 \end{bmatrix}.$$

The state model in controllable canonical form is

$$\dot{Z} = \begin{bmatrix} 0 & 1 & 0 \\ 0 & 0 & 1 \\ -504 & -191 & -24 \end{bmatrix} Z + \begin{bmatrix} 0 \\ 0 \\ 1 \end{bmatrix} u \qquad y = \begin{bmatrix} 68 & 17 & 1 \end{bmatrix} Z.$$

5.6 TRANSFORMATION TO OBSERVABLE CANONICAL FORM

It is often necessary to transfer the given state model of the system into observable canonical form. Consider a third-order system represented by a state model

$$\dot{X} = AX + Bu \qquad\qquad y = CX.$$

Assume that this is to be transferred to observable canonical form. The observability matrix of the given state model is

$$Q = \begin{bmatrix} C \\ CA \\ CA^2 \end{bmatrix}.$$

Assume that the state model can be transformed into observable canonical form using the transformation

$$X = PZ,$$

so that $\dot{X} = P\dot{Z}$.

$$\dot{Z} = P^{-1}\dot{X} = P^{-1}AX + P^{-1}Bu = [P^{-1}AP]Z + [P^{-1}B]u = A_{\chi} Z + B_{\chi} u$$

$$y = CX = [CP] Z = C_Z Z.$$

The observability matrix for this state model is

$$Q_{\chi} = \begin{bmatrix} CP \\ (CP)(P^{-1}AP) \\ CP(P^{-1}AP)^2 \end{bmatrix} = \begin{bmatrix} CP \\ (CP)(P^{-1}AP) \\ CP(P^{-1}AP)(P^{-1}AP) \end{bmatrix}$$

$$= \begin{bmatrix} CP \\ CAP \\ CAAP \end{bmatrix} = \begin{bmatrix} C \\ CA \\ CA^2 \end{bmatrix} P = QP.$$

Hence $Q_\chi = QP$ and $P = Q^{-1}Q_\chi$.

The matrix P transforms the given state model into observable canonical form. To determine this matrix, Q^{-1} should exist. In other words, Q should have full rank and for this $|Q|$ should not be zero. Hence, if the given state space model of a system is to be transformed into observable canonical form, it should be completely observable.

Example 5.31

A state model for a linear system is

$$\dot{X} = \begin{bmatrix} -2 & 1 \\ 0 & -1 \end{bmatrix} X + \begin{bmatrix} -1 \\ 1 \end{bmatrix} u \qquad y = \begin{bmatrix} 1 & 2 \end{bmatrix} X.$$

Convert this into observable canonical form.

Solution:

$$A = \begin{bmatrix} -2 & 1 \\ 0 & -1 \end{bmatrix} \qquad C_X = \begin{bmatrix} 1 & 2 \end{bmatrix}$$

$$[\lambda I - A] = \begin{bmatrix} \lambda+2 & -1 \\ 0 & \lambda+1 \end{bmatrix}.$$

The characteristic equation is

$$(\lambda+2)(\lambda+1) = 0 \qquad Or \quad \lambda^2 +3\lambda+2 = 0 = \lambda^2 + a_1\lambda + a_0 \qquad\qquad a_1 = 3;\ a_0 = 2.$$

The observability matrix for the given state model is

$$Q = \begin{bmatrix} C \\ C A \end{bmatrix}$$

$$C A = \begin{bmatrix} 1 & 2 \end{bmatrix} \begin{bmatrix} -2 & 1 \\ 0 & -1 \end{bmatrix} = \begin{bmatrix} -2 & -1 \end{bmatrix}$$

$$Q = \begin{bmatrix} 1 & 2 \\ -2 & -1 \end{bmatrix}.$$

The rank of Q is 2 and hence the system representation is completely observable. Therefore, the state model can be transformed into observable canonical form.

$$Q^{-1} = \frac{1}{3}\begin{bmatrix} -1 & -2 \\ 2 & 1 \end{bmatrix}$$

$$A_{\not{z}} = \begin{bmatrix} -a_1 & 1 \\ -a_0 & 0 \end{bmatrix} = \begin{bmatrix} -3 & 1 \\ -2 & 0 \end{bmatrix} \qquad\qquad C_{\not{z}} = \begin{bmatrix} 1 & 0 \end{bmatrix}$$

$$C_{\not{z}} A_{\not{z}} = \begin{bmatrix} 1 & 0 \end{bmatrix} \begin{bmatrix} -3 & 1 \\ -2 & 0 \end{bmatrix} = \begin{bmatrix} -3 & 1 \end{bmatrix}$$

$$Q_Z = \begin{bmatrix} 1 & 0 \\ -3 & 1 \end{bmatrix}$$

$$P = Q^{-1} Q_Z = \frac{1}{3} \begin{bmatrix} -1 & -2 \\ 2 & 1 \end{bmatrix} \begin{bmatrix} 1 & 0 \\ -3 & 1 \end{bmatrix} = \frac{1}{3} \begin{bmatrix} 5 & -2 \\ -1 & 1 \end{bmatrix} \qquad P^{-1} = \begin{bmatrix} 1 & 2 \\ 1 & 5 \end{bmatrix}$$

Check:

$$A_{\not{z}} = P^{-1} A P = \frac{1}{3} \begin{bmatrix} 1 & 2 \\ 1 & 5 \end{bmatrix} \begin{bmatrix} -2 & 1 \\ 0 & -1 \end{bmatrix} \begin{bmatrix} 5 & -2 \\ -1 & 1 \end{bmatrix} = \frac{1}{3} \begin{bmatrix} 1 & 2 \\ 1 & 5 \end{bmatrix} \begin{bmatrix} -11 & 5 \\ 1 & -1 \end{bmatrix} = \begin{bmatrix} -3 & 1 \\ -2 & 0 \end{bmatrix}$$

$$B_{\not{z}} = P^{-1} B = \begin{bmatrix} 1 & 2 \\ 1 & 5 \end{bmatrix} \begin{bmatrix} -1 \\ 1 \end{bmatrix} = \begin{bmatrix} 1 \\ 4 \end{bmatrix}$$

$$C_{\not{z}} = C P = \frac{1}{3} \begin{bmatrix} 1 & 2 \end{bmatrix} \begin{bmatrix} 5 & -2 \\ -1 & 1 \end{bmatrix} = \begin{bmatrix} 1 & 0 \end{bmatrix}.$$

The state model in observable canonical form is

$$\dot{Z} = \begin{bmatrix} -3 & 1 \\ -2 & 0 \end{bmatrix} Z + \begin{bmatrix} 1 \\ 4 \end{bmatrix} u \qquad\qquad y = \begin{bmatrix} 1 & 0 \end{bmatrix} Z.$$

Example 5.32

The state equation for a linear system is

$$\dot{X} = \begin{bmatrix} 0 & 1 & 0 \\ 3 & 0 & 2 \\ -12 & -7 & -6 \end{bmatrix} X + \begin{bmatrix} 0 \\ 0 \\ 1 \end{bmatrix} u \qquad\qquad y = \begin{bmatrix} 2 & -2 & -1 \end{bmatrix} X.$$

Convert the state model into observable canonical form.

Solution:

$$A = \begin{bmatrix} 0 & 1 & 0 \\ 3 & 0 & 2 \\ -12 & -7 & -6 \end{bmatrix} \qquad B = \begin{bmatrix} 0 \\ 0 \\ 1 \end{bmatrix} \qquad C = \begin{bmatrix} 2 & -2 & -1 \end{bmatrix}.$$

The characteristic equation is

$$|\lambda I - A| = \begin{vmatrix} \lambda & -1 & 0 \\ -3 & \lambda & -2 \\ 12 & 7 & \lambda+6 \end{vmatrix} = 0$$

$$\lambda^3 + 6\lambda^2 + 11\lambda + 6 = 0 = \lambda^3 + a_2\lambda^2 + a_1\lambda + a_0 \qquad\qquad a_2 = 6 \quad a_1 = 11 \quad a_0 = 6 .$$

$$CA = [2 \quad -2 \quad -1]\begin{bmatrix} 0 & 1 & 0 \\ 3 & 0 & 2 \\ -12 & -7 & -6 \end{bmatrix} = [6 \quad 9 \quad 2]$$

$$CA^2 = [6 \quad 9 \quad 2]\begin{bmatrix} 0 & 1 & 0 \\ 3 & 0 & 2 \\ -12 & -7 & -6 \end{bmatrix} = [3 \quad -8 \quad 6] .$$

The controllability matrix is

$$Q = \begin{bmatrix} C \\ CA \\ CA^2 \end{bmatrix} = \begin{bmatrix} 2 & -2 & -1 \\ 6 & 9 & 2 \\ 3 & -8 & 6 \end{bmatrix}$$

$$|Q| = 2(54+16) + 2(36-6) - (-48-27) = 275 .$$

The rank of Q is 3 and hence the system representation is completely observable. The state model can be transformed into observable canonical form.

$$Q^{-1} = \frac{1}{275}\begin{bmatrix} +\{70\} & -\{-20\} & +\{5\} \\ -\{30\} & +\{15\} & -\{10\} \\ +\{-75\} & -\{-10\} & +\{30\} \end{bmatrix} = \frac{1}{275}\begin{bmatrix} 70 & 20 & 5 \\ -30 & 15 & -10 \\ -75 & 10 & 30 \end{bmatrix}$$

$$A_{\chi} = \begin{bmatrix} -a_2 & 1 & 0 \\ -a_1 & 0 & 1 \\ -a_0 & 0 & 0 \end{bmatrix} = \begin{bmatrix} -6 & 1 & 0 \\ -11 & 0 & 1 \\ -6 & 0 & 0 \end{bmatrix} \qquad\qquad C_{\chi} = [1 \quad 0 \quad 0]$$

$$C_{\chi}A_{\chi} = [1 \quad 0 \quad 0]\begin{bmatrix} -6 & 1 & 0 \\ -11 & 0 & 1 \\ -6 & 0 & 0 \end{bmatrix} = [-6 \quad 1 \quad 0]$$

$$C_{\chi}A_{\chi}^2 = [-6 \quad 1 \quad 0]\begin{bmatrix} -6 & 1 & 0 \\ -11 & 0 & 1 \\ -6 & 0 & 0 \end{bmatrix} = [25 \quad -6 \quad 1] .$$

The observability matrix of the observable canonical form is

$$Q_z = \begin{bmatrix} 1 & 0 & 0 \\ -6 & 1 & 0 \\ 25 & -6 & 1 \end{bmatrix}.$$

The transformation matrix that transforms the given state model into observable canonical form is given by

$$P = Q^{-1} Q_z = \frac{1}{275} \begin{bmatrix} 70 & 20 & 5 \\ -30 & 15 & -10 \\ -75 & 10 & 30 \end{bmatrix} \begin{bmatrix} 1 & 0 & 0 \\ -6 & 1 & 0 \\ 25 & -6 & 1 \end{bmatrix} = \frac{1}{275} \begin{bmatrix} 75 & -10 & 5 \\ -370 & 75 & -10 \\ 615 & -170 & 30 \end{bmatrix}$$

$$|P| = \frac{1}{275}\{75(550) + 10(-4950) + 5(16775) = 275$$

$$P^{-1} = \frac{1}{275} \begin{bmatrix} +\{550\} & -\{550\} & +\{275\} \\ -\{-4950\} & +\{825\} & -\{1100\} \\ +\{16775\} & -\{6600\} & +\{1925\} \end{bmatrix} = \begin{bmatrix} 2 & -2 & -1 \\ 18 & -3 & -4 \\ 61 & 24 & 7 \end{bmatrix}$$

Check:

$$A_z = P^{-1}AP = \frac{1}{275} \begin{bmatrix} 2 & -2 & -1 \\ 18 & -3 & -4 \\ 61 & 24 & 7 \end{bmatrix} \begin{bmatrix} 0 & 1 & 0 \\ 3 & 0 & 2 \\ -12 & -7 & -6 \end{bmatrix} \begin{bmatrix} 75 & -10 & 5 \\ -370 & 75 & -10 \\ 615 & -170 & 30 \end{bmatrix}$$

$$= \frac{1}{275} \begin{bmatrix} 2 & -2 & -1 \\ 18 & -3 & -4 \\ 61 & 24 & 7 \end{bmatrix} \begin{bmatrix} -370 & 75 & -10 \\ 1455 & -370 & 75 \\ -2000 & 615 & -170 \end{bmatrix} = \begin{bmatrix} -6 & 1 & 0 \\ -11 & 0 & 1 \\ -6 & 0 & 0 \end{bmatrix}$$

$$B_z = P^{-1}B = \begin{bmatrix} 2 & -2 & -1 \\ 18 & -3 & -4 \\ 61 & 24 & 7 \end{bmatrix} \begin{bmatrix} 0 \\ 0 \\ 1 \end{bmatrix} = \begin{bmatrix} -1 \\ -4 \\ 7 \end{bmatrix}$$

$$C_z = CP = \frac{1}{275} \begin{bmatrix} 2 & -2 & -1 \end{bmatrix} \begin{bmatrix} 75 & -10 & 5 \\ -370 & 75 & -10 \\ 615 & -170 & 30 \end{bmatrix} = \begin{bmatrix} 1 & 0 & 0 \end{bmatrix}.$$

The state model in observable canonical form is

$$\dot{Z} = \begin{bmatrix} -6 & 1 & 0 \\ -11 & 0 & 1 \\ -6 & 0 & 0 \end{bmatrix} Z + \begin{bmatrix} -1 \\ -4 \\ 7 \end{bmatrix} u \qquad\qquad y = \begin{bmatrix} 1 & 0 & 0 \end{bmatrix} Z.$$

SUMMARY

- If it is possible to control the state variables via the system input signal, then the system is said to be controllable.
- If any state variable is not linked to the input, it cannot be controlled by the input and so the system is not completely controllable.
- Controllability is a property of the coupling between input or control signal and the state X, and thus it involves the matrices A and B.
- The ability to estimate a state variable from the known output and input is known as observability of a system.
- If any state variable has no effect on the output, this particular state variable cannot be evaluated from the output.
- Observability is a property of the coupling between the state vector **X** and the output $y(t)$ and thus involves the matrices A and C.
- In the design of a feedback controller for a system, it is often necessary to transfer the given state model of the system into controllable canonical form.
- To carry out the design of observers or estimators, it is often necessary to transfer the given state model of the system into observable canonical form.
- A given state model can be transformed into controllable canonical form using the controllability matrices of the two state models.
- A given state model can be transformed into observable canonical form using the observability matrices of the two state models.

PRACTICE PROBLEMS

PP 5.1 Check by using Gilbert's test the complete controllability and complete observability of the state model

$$\dot{X} = \begin{bmatrix} 0 & 1 \\ 2.5 & -1.5 \end{bmatrix} X + \begin{bmatrix} 1 \\ 1 \end{bmatrix} u \qquad\qquad y = [2 \quad -1]X .$$

PP 5.2 Check by using Gilbert's test the complete controllability and complete observability of the state model

$$\dot{X} = \begin{bmatrix} -1 & 0 \\ 1 & 1 \end{bmatrix} X + \begin{bmatrix} -2 \\ 1 \end{bmatrix} u \qquad\qquad y = [1 \quad 2]X .$$

PP 5.3 Repeat PP5.1 using Kalman's test.
PP 5.4 Repeat PP5.2 using Kalman's test.
PP 5.5 Determine the condition for complete controllability and complete observability for the system represented by the state model

$$\dot{X} = \begin{bmatrix} -\alpha & 0 \\ 1 & -\beta \end{bmatrix} X + \begin{bmatrix} k_1 \\ k_2 \end{bmatrix} u \qquad\qquad y = [c_1 \quad c_2]X .$$

PP 5.6 A state model of a system is given as

$$\dot{X} = \begin{bmatrix} 0 & 1 & 0 \\ 0 & 0 & 1 \\ -1 & -3 & -3 \end{bmatrix} X + \begin{bmatrix} 2 \\ 0 \\ -1 \end{bmatrix} u \qquad y = [1 \quad 2 \quad -1]X \ .$$

Check whether the system is completely controllable and completely observable using Gilbert's test.

PP 5.7 Repeat PP5.6 using Kalman's test.

PP 5.8 A state equation of a system is given as

$$\dot{X} = \begin{bmatrix} 0 & 1 & 0 \\ 0 & 0 & 1 \\ 0 & -2 & -3 \end{bmatrix} X + \begin{bmatrix} 0 \\ 0 \\ 1 \end{bmatrix} u \qquad y = [2 \quad 2 \quad 1]X \ .$$

Check whether the system is completely controllable and completely observable, using Gilbert's test.

PP 5.9 Repeat PP5.8 using Kalman's test.

PP 5.10 A state equation of a system is given as

$$\dot{X} = \begin{bmatrix} -7 & -2 & 6 \\ 2 & -3 & -2 \\ -2 & -2 & 1 \end{bmatrix} X + \begin{bmatrix} 1 & 1 \\ 1 & -1 \\ 1 & 0 \end{bmatrix} u$$

$$y = \begin{bmatrix} -1 & -1 & 2 \\ 1 & 1 & -1 \end{bmatrix}.$$

Check whether the system is completely controllable and completely observable.

PP 5.11 A state equation of a system is given as

$$\dot{X} = \begin{bmatrix} 0 & 1 & 0 \\ 0 & 0 & 1 \\ -18 & -21 & -8 \end{bmatrix} X + \begin{bmatrix} 1 \\ 0 \\ 1 \end{bmatrix} u$$

$$y = [1 \quad 0.5 \quad 0] \ .$$

Check whether the system is completely controllable and completely observable.

PP 5.12 Check whether the system represented by the state model of PP5.1 is completely controllable and completely observable using the factor-cancellation test.

PP 5.13 Check whether the system represented by the state model of problem 5.2 is completely controllable and completely observable using the factor-cancellation test.

PP 5.14 A state equation of a system is given as

$$\dot{X} = \begin{bmatrix} 0 & 1 & 0 \\ 3 & 0 & 2 \\ -12 & -7 & -6 \end{bmatrix} X + \begin{bmatrix} 0 \\ 0 \\ 1 \end{bmatrix} u \qquad y = [4 \quad 2 \quad 0]X \ .$$

Check the complete controllability and complete observability of the system represented by the state model, using **PBH** test.

PP 5.15 A state space representation for a linear system is

$$\dot{X} = \begin{bmatrix} 0 & 1 \\ -8 & -6 \end{bmatrix} X + \begin{bmatrix} 1 \\ 2 \end{bmatrix} u \qquad\qquad y = [2 \quad 1]X .$$

Convert this into controllable canonical form.

PP 5.16 A state space representation for a linear system is

$$\dot{X} = \begin{bmatrix} -3 & 1 \\ 1 & -3 \end{bmatrix} X + \begin{bmatrix} 1 \\ 2 \end{bmatrix} u \qquad\qquad y = [1 \quad -2]X .$$

Convert this into controllable canonical form.

PP 5.17 A state space representation for a linear system is

$$\dot{X} = \begin{bmatrix} -5 & 1 \\ -6 & 0 \end{bmatrix} X + \begin{bmatrix} 0 \\ 1 \end{bmatrix} u \qquad\qquad y = [3 \quad -2]X .$$

Convert this into controllable canonical form.

PP 5.18 A state space representation for a linear system is

$$\dot{X} = \begin{bmatrix} -3 & 1 \\ 0 & -3 \end{bmatrix} X + \begin{bmatrix} 1 \\ 2 \end{bmatrix} u \qquad\qquad y = [1 \quad -2]X .$$

Convert this into observable canonical form.

PP 5.19 A state space representation for a linear system is

$$\dot{X} = \begin{bmatrix} 0 & 1 \\ -4 & -5 \end{bmatrix} X + \begin{bmatrix} 1 \\ 0 \end{bmatrix} u \qquad\qquad y = [1 \quad 2]X .$$

Convert this into observable canonical form.

PP 5.20 A state space representation for a linear system is

$$\dot{X} = \begin{bmatrix} -3 & 1 \\ 1 & -3 \end{bmatrix} X + \begin{bmatrix} 1 \\ 2 \end{bmatrix} u \qquad\qquad y = [1 \quad -2]X .$$

Convert this into observable canonical form.

PP 5.21 A state equation of a system is given as

$$\dot{X} = \begin{bmatrix} 2 & -2 & 3 \\ 1 & 1 & 1 \\ 1 & 3 & -1 \end{bmatrix} X + \begin{bmatrix} 11 \\ 1 \\ -14 \end{bmatrix} u \qquad\qquad y = [-3 \quad 5 \quad -2]X .$$

Check the complete controllability and complete observability of the system represented by the state model, using **Kalman's** test.

PP 5.22 Repeat PP5.21 using factor-cancellation method.
PP 5.23 Repeat PP5.21 using PBH test.
PP 5.24 Repeat PP5.6 using PBH test.
PP 5.25 Repeat PP5.8 using PBH test.

REVIEW QUESTIONS

5.1   State and explain the controllability of a system.
5.2   Explain the Gilbert's test for assessing the controllability of a system.
5.3   Discuss the Kalman's test for assessing the controllability of a system.
5.4   A linear system is represented by the state equation $\dot{X} = AX + Bu$. Obtain the controllability matrix for determining the controllability of the system.
5.5   Discuss the 'factors-cancellation test' for determining the controllability of the system.
5.6   Explain the PBH test for determining the controllability of the system.
5.7   State and explain the observability of a system.
5.8   Explain the Gilbert's test for assessing the observability of a system.
5.9   Discuss the Kalman's test for assessing the observability of a system.
5.10 A linear system is represented by the state equation

$$\dot{X} = AX + Bu \qquad\qquad y = CX .$$

Obtain the observability matrix for determining the observability of the system.
5.11 Discuss the 'factors-cancellation test' for determining the observability of the system.
5.12 Explain the PBH test for determining the observability of the system.
5.13 A state model for a linear third-order system is

$$\dot{X} = AX + Bu \qquad\qquad y = CX .$$

Using this, obtain the transformation matrix that transforms the given state model into controllable canonical form.
5.14 A state model for a linear third order system is

$$\dot{X} = AX + Bu \qquad\qquad y = CX .$$

Using this, obtain the transformation matrix that transforms the given state model into observable canonical form.

6 Pole Placement Design

6.1 CONCEPT OF POLE PLACEMENT DESIGN

The transient performance of a closed loop control system depends on the location of the closed loop poles in the s-plane. The performance indices like rise time, peak time, settling time, peak overshoot, damping ratio and natural frequency of oscillations of a system response depend on the location of the dominant pole. If the location of a dominant pole on the left half of the s-plane is far away from the imaginary axis, then the system response will be faster but the frequency of oscillations will be higher. On the other hand, if the dominant poles are located close to the imaginary axis, the system response will be slow and sluggish. To effect a change in the system response, it is sufficient to change the locations of closed loop poles to some desired locations in the s-plane. Using control signals, it is possible to shift the dominant closed loop poles from the existing locations to some other desired locations. This is called *pole placement technique*. By pole placement technique, it is also possible to stabilize an otherwise unstable system.

Using the state space method, it is possible to design a system with desired closed loop poles. Such a design that enables the engineer to place the closed loop poles at the desired locations, using state feedback, is known as *pole placement design*. It is assumed that all state variables can be used as feedback signals. One of the design objectives is to achieve desired pole locations in order to ensure satisfactory transient response of the system.

As an illustration, consider a system with the damping ratio $\delta = 0.5$ and the settling time $t_S = 2s$.

$$t_S = \frac{4}{\delta \omega_n} = 2 \qquad \text{or} \quad \omega_n = \frac{4}{2\delta} = 4 \ rad \ / \ s$$

The natural frequency is 4 rad/s. The characteristic equation is

$$s^2 + 2\delta\omega_n s + \omega_n^2 = s^2 + 4s + 16 = 0 \ .$$

The closed loop poles are $s = -2 \pm j\sqrt{12}$.

Now assume that it is desired to alter the transient response by reducing the settling time to 1 s, with the same damping ratio. The new natural frequency of oscillations of the response is

$$t_{S2} = \frac{4}{\delta \omega_{n2}} = 1 \qquad \text{or} \quad \omega_{n2} = \frac{4}{\delta} = 8 \ rad \, / \, s \, .$$

The new characteristic equation is

$$s^2 + 2\delta \omega_n s + \omega_n^2 = s^2 + 8s + 64 = 0 \, .$$

The closed loop poles are $s = -4 \pm j\sqrt{48}$.

Hence, to change the transient response of the system by reducing the settling time from 2 s to 1 s, the closed loop poles at $s = -2 \pm j\sqrt{12}$ are to be shifted to $s = -4 \pm j\sqrt{48}$ and this is the essence of pole placement design. Pole placement design is carried out by feeding back the state variables to the input signal. The controller with constant gains that is used to feed back state variables is called a *state feedback controller.*

6.2 PRINCIPLE OF FEEDBACK CONTROLLER DESIGN

Consider a plant represented by a state equation

$$\dot{X} = AX + Bu \qquad\qquad y = CX + Du \, .$$

The Eigen values or the closed loop poles are obtained by solving the equation

$$|sI - A| = 0 \, .$$

Consider that each state variable is fed back to the input signal. This is shown in Figure 6.1. The state feedback signal is

$$u_C = KX = [k_1 \quad k_2 \quad k_3 \dots k_n][x_1 \ x_2 \ x_3 \dots x_n]^T \, .$$

The equation that describes the state feedback is

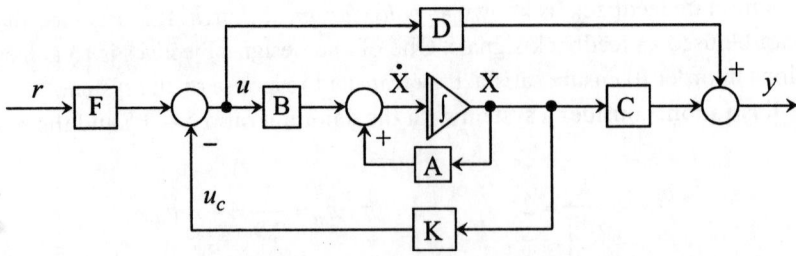

Figure 6.1 State feedback controller

$$u = F r - K X \qquad (6.1)$$

The state equation now becomes

$$\dot{X} = A X + B[F r - K X] = [A - BK]X + [BF]r . \qquad (6.2)$$

The new characteristic equation with state feedback is

$$\left|sI - (A - BK)\right| = 0 . \qquad (6.3)$$

With state feedback, the new system matrix is $[A - BK]$. The new closed poles are the new desired locations of closed loop poles and are given by equation (6.3).

Each state variable is fed back to the control signal u, through a gain $K = [k_1 \quad k_2 \quad k_3 \, ... k_n]$ that could be adjusted to yield the required closed loop poles. Hence, by introducing the matrix K, that is, by introducing each state feedback through a pre-designed gain, all the poles of the closed loop system can be placed at desired locations. The desired characteristic equation is obtained from the design specifications, which are the performance indices. The characteristic equation with state feedback, given in equation (6.3), is compared with the desired characteristic equation to evaluate the feedback matrix K. This method is called the *method of matching coefficients*. In order to synthesize the system with real hardware, all elements of K must be real.

In this chapter, the state space design technique applicable only to a linear time-invariant system is discussed. If any one of the state variables cannot be controlled by the control signal u, then, it is not possible to place the system poles at the desired locations. Hence, in order to carry out the pole placement design, it is necessary that the system represented by the state model is completely state controllable.

6.3 METHOD OF MATCHING COEFFICIENTS

The design of a feedback controller by the method of matching coefficients is carried out as follows.

Step 1: Check the complete controllability of the given state model.

Step 2: Find the desired closed loop poles from the design specifications and hence find the desired characteristic equation.

Step 3: Find the characteristic equation with the state feedback gain matrix K.

Step 4: Evaluate K by comparing the coefficients of like powers of the two characteristic equations of step 2 and step 3.

Example 6.1 A linear continuous time system is represented by the state model

$$\dot{X} = \begin{bmatrix} -3 & 1 \\ 0 & -2 \end{bmatrix} X + \begin{bmatrix} 0 \\ 1 \end{bmatrix} u \qquad y = [2 \quad 1]X .$$

Design a feedback controller that places the closed loop poles at $s = -3 \pm j3$.

Solution:

Step 1: Check for controllability

$$A = \begin{bmatrix} -3 & 1 \\ 0 & -2 \end{bmatrix} \qquad B = \begin{bmatrix} 0 \\ 1 \end{bmatrix} \qquad AB = \begin{bmatrix} -3 & 1 \\ 0 & -2 \end{bmatrix}\begin{bmatrix} 0 \\ 1 \end{bmatrix} = \begin{bmatrix} 1 \\ -2 \end{bmatrix}.$$

The controllability matrix

$$M = [B \quad AB] = \begin{bmatrix} 0 & 1 \\ 1 & -2 \end{bmatrix} \qquad |M| = -1.$$

M has rank 2 and the system is completely controllable.

Step 2: Desired characteristic equation

The required pole locations are $s = -3 \pm j3$. The desired characteristic equation is

$$(s+3)^2 + (3)^2 = 0 \text{ or } s^2 + 6s + 18 = 0. \tag{1}$$

Step 3: The characteristic equation with state feedback

Let the feedback gain matrix be $K = [k_0 \quad k_1]$.

The system matrix with state feedback

$$[A - BK] = \begin{bmatrix} -3 & 1 \\ 0 & -2 \end{bmatrix} - \begin{bmatrix} 0 \\ 1 \end{bmatrix}[k_0 \quad k_1] = \begin{bmatrix} -3 & 1 \\ -k_0 & -(2+k_1) \end{bmatrix}.$$

The characteristic equation with state feedback is

$$|sI - (A - BK)| = 0 \qquad Or \qquad \begin{vmatrix} s+3 & -1 \\ k_0 & s+(2+k_1) \end{vmatrix} = 0.$$

$$s^2 + (5 + k_1)s + (6 + 3k_1 + k_0) = 0. \tag{2}$$

Step 4: Evaluation of gain matrix K

By comparing equations (1) and (2)

$$5 + k_1 = 6 \qquad\qquad 6 + 3k_1 + k_0 = 18.$$

Solving these equations $k_1 = 1 \qquad k_0 = 9$.

Hence the feedback gain matrix $K = [9 \quad 1]$

The feedback signal is $-u_C = -[9 \quad 1]X$

Example 6.2 A linear continuous time system is represented by the state model

$$\dot{X} = \begin{bmatrix} -3 & 1 \\ 0 & -2 \end{bmatrix} X + \begin{bmatrix} 1 \\ 2 \end{bmatrix} u \qquad\qquad y = [2 \quad 1]X.$$

Design a feedback controller that places the closed loop poles at $s = -3 \pm j4$.

Solution:
Step 1: Check for controllability

$$A = \begin{bmatrix} -3 & 1 \\ 0 & -2 \end{bmatrix} \qquad B = \begin{bmatrix} 1 \\ 2 \end{bmatrix} \qquad AB = \begin{bmatrix} -3 & 1 \\ 0 & -2 \end{bmatrix}\begin{bmatrix} 1 \\ 2 \end{bmatrix} = \begin{bmatrix} -1 \\ -4 \end{bmatrix}.$$

The controllability matrix

$$M = [B \quad AB] = \begin{bmatrix} 1 & -1 \\ 2 & -4 \end{bmatrix} \qquad |M| = -2.$$

M has rank 2 and the system is completely controllable.
Step 2: Desired characteristic equation
The required pole locations are $s = -3 \pm j4$. The desired characteristic equation is

$$(s+3)^2 + (4)^2 = 0 \; or \; s^2 + 6s + 25 = 0. \tag{1}$$

Step 3: The characteristic equation with state feedback
Let the feedback gain matrix be $K = [k_0 \quad k_1]$.
The system matrix with state feedback

$$[A - BK] = \begin{bmatrix} -3 & 1 \\ 0 & -2 \end{bmatrix} - \begin{bmatrix} 1 \\ 2 \end{bmatrix}[k_0 \quad k_1]$$

$$= \begin{bmatrix} -3 & 1 \\ 0 & -2 \end{bmatrix} - \begin{bmatrix} k_0 & k_1 \\ 2k_0 & 2k_1 \end{bmatrix} = \begin{bmatrix} -(3+k_0) & 1-k_1 \\ -2k_0 & -(2+2k_1) \end{bmatrix}.$$

The characteristic equation with state feedback is

$$|sI - (A - BK)| = 0 \quad Or \quad \begin{vmatrix} s+(3+k_0) & -1+k_1 \\ 2k_0 & s+(2+2k_1) \end{vmatrix} = 0$$

$$s^2 + (5 + 2k_1 + k_0)s + (6 + 6k_1 + 4k_0) = 0. \tag{2}$$

Step 4: Evaluation of gain matrix K
Comparing equations (1) and (2)

$$5 + 2k_1 + k_0 = 6 \qquad\qquad 6 + 6k_1 + 4k_0 = 25.$$

Solving these equations $k_1 = -7.5$ $\qquad\qquad k_0 = 16.$
Hence the feedback gain matrix $K = [16 \quad -7.5]$.
The feedback signal is $-u_C = -[16 \quad -7.5]X$.

Example 6.3 A linear continuous time system is represented by the state model

$$\dot{X} = \begin{bmatrix} -4 & 1 \\ 2 & -3 \end{bmatrix}X + \begin{bmatrix} 2 \\ 3 \end{bmatrix}u \qquad y = [3 \quad -2]X.$$

Design a feedback controller to have a system response with a damping ratio of 0.707 and a settling time of 1 s.

Solution:

Step 1: Check for controllability

$$A = \begin{bmatrix} -4 & 1 \\ 2 & -3 \end{bmatrix} \qquad B = \begin{bmatrix} 2 \\ 3 \end{bmatrix} \qquad AB = \begin{bmatrix} -4 & 1 \\ 2 & -3 \end{bmatrix}\begin{bmatrix} 2 \\ 3 \end{bmatrix} = \begin{bmatrix} -5 \\ -5 \end{bmatrix}.$$

The controllability matrix

$$M = [B \quad AB] = \begin{bmatrix} 2 & 5 \\ 1 & 5 \end{bmatrix} \qquad |M| = -5 .$$

M has rank 2 and the system is completely controllable.

Step 2: Desired characteristic equation

$$\text{Damping ratio } \delta = .707 \qquad \text{Settling time} = \frac{4}{\delta \, \omega_n} = 1s .$$

$$\delta \omega_n = 4 \qquad \text{Natural frequency } \omega_n = \frac{4}{\delta} = \frac{4}{0.707} = 5.658 \; rad/s .$$

The desired pole locations are $s = -\delta \omega_n \pm j\omega_n \sqrt{1-\delta^2} = -4 \pm j4$. The desired characteristic equation is

$$(s+4)^2 + 4^2 = 0 \quad or \quad s^2 + 8s + 32 = 0 . \tag{1}$$

Step 3: The characteristic equation with state feedback

Let the feedback gain matrix be $K = [k_0 \quad k_1]$.

The system matrix with state feedback

$$[A - BK] = \begin{bmatrix} -4 & 1 \\ 2 & -3 \end{bmatrix} - \begin{bmatrix} 2 \\ 3 \end{bmatrix}[k_0 \quad k_1]$$

$$= \begin{bmatrix} -4 & 1 \\ 2 & -3 \end{bmatrix} - \begin{bmatrix} 2k_0 & 2k_1 \\ 3k_0 & 3k_1 \end{bmatrix} = \begin{bmatrix} -(4+2k_0) & 1-2k_1 \\ (2-3k_0) & -(3+3k_1) \end{bmatrix} .$$

The characteristic equation with state feedback is

$$|sI - (A - BK)| = 0 \qquad Or \qquad \begin{vmatrix} s+(4+2k_0) & -(1-2k_1) \\ -(2-3k_0) & s+(3+3k_1) \end{vmatrix} = 0 .$$

$$s^2 + (7 + 2k_0 + 3k_1)s + (10 + 9k_0 + 16k_1) = 0 . \tag{2}$$

Step 4: Evaluation of gain matrix K
    Comparing equations (1) and (2)

$$7 + 2k_0 + 3k_1 = 8 \qquad\qquad 10 + 9k_0 + 16k_1 = 32 .$$

Solving these equations $k_0 = -10$ $\qquad\qquad k_1 = 7 .$
Hence the feedback gain matrix $K = [-10 \quad 7]$.

6.4 DESIGNS BY TRANSFORMATION TO CONTROLLABLE CANONICAL FORM

The method of matching coefficients may be tedious with higher-order systems as the forma-
tion of equations and solving them can be much more intricate. In such cases, it is convenient
to transform the given state model into controllable canonical form to evaluate the feedback
gain matrix. This gain matrix is then transferred back to the original representation by proper
transformation.

    Consider the state space model for a third-order closed loop system

$$\dot{X} = AX + Bu \qquad\qquad y = CX .$$

    Let this be transformed into controllable canonical form by using the transformation
$X = PZ$. The transformation matrix is

$$P = M M_Z^{-1},$$

where M is the controllability matrix of the given state space model and M_Z is that of the state
model of the system in controllable canonical form. Let the characteristic equation be

$$s^3 + a_2 s^2 + a_1 s + a_0 = 0 . \qquad\qquad (6.4)$$

$$M_Z^{-1} = \begin{bmatrix} a_1 & a_2 & 1 \\ a_2 & 1 & 0 \\ 1 & 0 & 0 \end{bmatrix}$$

$$\dot{Z} = A_Z Z + B_Z u \qquad\qquad y = C_Z Z$$

$$\dot{Z} = \begin{bmatrix} 0 & 1 & 0 \\ 0 & 0 & 1 \\ -a_0 & -a_1 & -a_2 \end{bmatrix} Z + \begin{bmatrix} 0 \\ 0 \\ 1 \end{bmatrix} u \qquad\qquad y = [c_0 \quad c_1 \quad c_2] Z .$$

    Let the feedback matrix be $K_Z = [k_0 \quad k_1 \quad k_2]$.
    With state feedback, the new system matrix is

$$[A_Z - B_Z K_Z] = \begin{bmatrix} 0 & 1 & 0 \\ 0 & 0 & 1 \\ -a_0 & -a_1 & -a_2 \end{bmatrix} - \begin{bmatrix} 0 \\ 0 \\ 1 \end{bmatrix} \begin{bmatrix} k_0 & k_1 & k_2 \end{bmatrix}$$

$$= \begin{bmatrix} 0 & 1 & 0 \\ 0 & 0 & 1 \\ -(a_0 + k_0) & -(a_1 + k_1) & -(a_2 + k_2) \end{bmatrix}.$$

The characteristic equation with state feedback is

$$|sI - (A_Z - B_Z K_Z)| = \begin{vmatrix} s & -1 & 0 \\ 0 & s & -1 \\ (a_0 + k_0) & (a_1 + k_1) & s + (a_2 + k_2) \end{vmatrix} = 0$$

Or $s^3 + (a_2 + k_2)s^2 + (a_1 + k_1)s + (a_0 + k_0) = 0$.   (6.5)

From the given design specifications, which are the performance parameters, the desired characteristic equation can be obtained and can be represented as

$$s^3 + \alpha_2 s^2 + \alpha_1 s + \alpha_0 = 0.$$ (6.6)

Equating coefficients of like powers in equations (6.5) and (6.6)

$(a_2 + k_2) = \alpha_2$ $(a_1 + k_1) = \alpha_1$ $(a_0 + k_0) = \alpha_0$

$k_2 = \alpha_2 - a_2$ $k_1 = \alpha_1 - a_1$ $k_0 = \alpha_0 - a_0$.

Hence the feedback gain matrix is $K_Z = [(\alpha_0 - a_0) \quad (\alpha_1 - a_1) \quad (\alpha_2 - a_2)]$.   (6.7)

Thus, if the system is represented in controllable canonical form, the feedback gain matrix can be written down by inspection of the characteristic equation (6.4) of the given system and the desired characteristic equation (6.6) of the system. This is the advantage of transforming the given state model into controllable canonical form in designing the feedback controller. Now equation (6.7) is to be modified with respect to the given state model.

From equation (6.2), the given state model with state feedback is

$$\dot{X} = [A - BK]X + [BF]r.$$ (6.2)

The state model with state feedback in controllable canonical form is

$$\dot{Z} = [A_Z - B_Z K_Z]Z + [B_Z F]r.$$ (6.8)

Since $Z = P^{-1} X$ $A_Z = P^{-1}AP$ and $B_Z = P^{-1}B$, equation (6.8) becomes

$$P^{-1}\dot{X} = [P^{-1}AP - P^{-1}BK_Z]P^{-1}X + [P^{-1}BF]r$$

$$\dot{X} = [AP - BK_Z]P^{-1}X + [BF]r$$

$$\dot{X} = [A - BK_Z P^{-1}]X + [BF]r .$$ (6.9)

Comparing equations (6.2) and (6.9)

$$K = K_Z P^{-1} .$$ (6.10)

The matrix given by equation (6.10) is the state feedback gain matrix for the given system.

6.5 FEEDBACK CONTROLLER DESIGN PROCEDURE

Step 1: Obtain the controllability matrix M of the given system and check for complete controllability of the given system.

Step 2: Find the characteristic equation of the given system

$$s^n + a_{n-1}s^{n-1} + a_{n-2}s^{n-2} + \ldots + a_1 s + a_0 = 0$$

Step 3: From the state model in controllable canonical form, find the controllability matrix M_Z as

$$M_Z = \begin{bmatrix} a_1 & a_2 & a_3 & \ldots & & a_{n-1} & 1 \\ a_2 & a_3 & a_4 & \ldots & a_{n-1} & 1 & 0 \\ a_3 & a_4 & a_5 & \ldots & 1 & 0 & 0 \\ \ldots & & & & & & \\ a_{n-1} & 1 & 0 & \ldots & 0 & 0 & 0 \\ 1 & 0 & 0 & \ldots & 0 & 0 & 0 \end{bmatrix}^{-1} .$$

Step 4: Determine the transformation matrix P that transforms the given state model into controllable canonical form and find its inverse P^{-1} .

$$P = M M_Z^{-1}$$

Step 5: Obtain the desired characteristic equation from the given design specifications.

$$s^n + \alpha_{n-1}s^{n-1} + \alpha_{n-2}s^{n-2} + \ldots + \alpha_1 s + \alpha_0 = 0$$

Step 6: Feedback controller gain matrix

$$K_Z = [(\alpha_0 - a_0) \quad (\alpha_1 - a_1) \quad (\alpha_2 - a_2) \ldots (\alpha_{n-1} - a_{n-1})] .$$

Step 7: Feedback controller gain matrix for the given system

$$K = K_\chi P^{-1}.$$

Example 6.4 A linear continuous time system is represented by the state model

$$\dot{X} = \begin{bmatrix} 0 & 1 \\ -6 & -5 \end{bmatrix} X + \begin{bmatrix} 0 \\ 1 \end{bmatrix} u \qquad y = [2 \quad 1] X.$$

Design a feedback controller that places the closed loop poles at $s = -3 \pm j4$.

Solution:

Step 1: Check for controllability

The given system is already in controllable canonical form and hence it is completely controllable.

Step 2: Characteristic equation of the given system

$$\begin{vmatrix} s & -1 \\ 6 & s+5 \end{vmatrix} = 0 \quad s^2 + 5s + 6 = 0 = s^2 + a_1 s + a_0.$$

Step 3: Desired characteristic equation

Desired pole locations are $s = -3 \pm j4$.

Desired characteristic equation is

$$(s+3)^2 + 4^2 = 0 \quad \text{Or} \quad s^2 + 6s + 25 = 0 = s^2 + \alpha_1 s + \alpha_0.$$

Step 4: Evaluation of feedback gain matrix K

$$K = [(\alpha_0 - a_0) \quad (\alpha_1 - a_1)] = [(25 - 6) \quad (6 - 5)] = [19 \quad 1].$$

Example 6.5 A linear continuous time system is represented by the state model

$$\dot{X} = \begin{bmatrix} 0 & 1 & 0 \\ 0 & 0 & 1 \\ 0 & -4 & -5 \end{bmatrix} X + \begin{bmatrix} 0 \\ 0 \\ 1 \end{bmatrix} u \qquad y = [1 \quad 2 \quad -2] X.$$

Design a feedback controller to yield a peak overshoot of less than 10% and a settling time less than 1 s.

Solution:

Step 1: Check for controllability

The given system is already in controllable canonical form and hence it is completely controllable.

Step 2: Characteristic equation of the given system

$$\begin{vmatrix} s & -1 & 0 \\ 0 & s & -1 \\ 0 & 4 & s+5 \end{vmatrix} = 0 \qquad s^3 + 5s^2 + 4s = 0 = s^3 + a_2 s^2 + a_1 s + a_0. \qquad (1)$$

Step 3: Desired characteristic equation
Peak overshoot $M_p < 0.10$.

The damping ratio $\delta = \dfrac{-ln\, M_p}{\sqrt{\pi^2 + (ln\, M_p)^2}} = \dfrac{-ln\,(0.1)}{\sqrt{\pi^2 + (ln\, 0.1)^2}} = 0.59$.

Choose $\delta = 0.6$.

Settling time $t_s = \dfrac{4}{\delta\omega_n} < 1 \qquad \omega_n > \dfrac{4}{0.6 \times 1} = 6.67$ rad/s.

Choose $\omega_n = 8$ rad/s.

Desired pole locations are $s = -\delta\omega_n \pm j\omega_n \sqrt{1 - \delta^2} = -4.8 \pm j6.4$.
The third pole can be assumed to be on the negative real axis far away from the dominant pole. If it is placed five times farther to the left than the dominant poles, its effect on the system response is negligibly small. Let it be at $s = -25$.
Desired characteristic equation is

$$(s + 25)\{(s + 4.8)^2 + (6.4)^2\} = 0 \qquad Or \quad (s + 25)(s^2 + 9.6s + 64) = 0$$

$$s^3 + 34.6s^2 + 304s + 1600 = 0 = s^3 + \alpha_2 s^2 + \alpha_1 s + \alpha_0. \tag{2}$$

Step 4: Evaluation of feedback gain matrix $K = [k_0 \quad k_1 \quad k_2]$.
Using equations (1) and (2)

$$K = [(\alpha_0 - a_0) \quad (\alpha_1 - a_1) \quad (\alpha_2 - a_2)]$$

$$= [(1600 - 0) \quad (304 - 4) \quad (34.6 - 5)]$$

$$= [1600 \quad 300 \quad 29.6].$$

Example 6.6 A linear continuous time system is represented by the state model

$$\dot{X} = \begin{bmatrix} -2 & 1 \\ 0 & -1 \end{bmatrix} X + \begin{bmatrix} 0 \\ 1 \end{bmatrix} u \qquad\qquad y = [1 \quad 3]X.$$

Design a feedback controller to yield a peak overshoot of less than 17% and a settling time less than 0.5 s.

Solution:
Step 1: Check for controllability

$$A = \begin{bmatrix} -2 & 1 \\ 0 & -1 \end{bmatrix} \qquad B = \begin{bmatrix} 0 \\ 1 \end{bmatrix} \qquad AB = \begin{bmatrix} -2 & 1 \\ 0 & -1 \end{bmatrix}\begin{bmatrix} 0 \\ 1 \end{bmatrix} = \begin{bmatrix} 1 \\ -1 \end{bmatrix}.$$

The given system is not in controllable canonical form.
The controllability matrix

$$M = [B \quad AB] = \begin{bmatrix} 0 & 1 \\ 1 & -1 \end{bmatrix} \qquad |M| = -1.$$

The rank of M is 2 and so the system is completely controllable.

Step 2: Characteristic equation of the given system

$$|sI - A| = \begin{bmatrix} s+2 & -1 \\ 0 & s+1 \end{bmatrix} \qquad s^2 + 3s + 2 = 0 = s^2 + a_1 s + a_0. \tag{1}$$

Step 3: State model in controllable canonical form is

$$\dot{Z} = \begin{bmatrix} 0 & 1 \\ -2 & -3 \end{bmatrix} Z + \begin{bmatrix} 0 \\ 1 \end{bmatrix} Z$$

$$A_Z = \begin{bmatrix} 0 & 1 \\ -2 & -3 \end{bmatrix} \qquad B_Z = \begin{bmatrix} 0 \\ 1 \end{bmatrix} \qquad A_Z B_Z = \begin{bmatrix} 0 & 1 \\ -2 & -3 \end{bmatrix} \begin{bmatrix} 0 \\ 1 \end{bmatrix} = \begin{bmatrix} 1 \\ -3 \end{bmatrix}.$$

The controllability matrix

$$M_Z = [B_Z \quad A_Z B_Z] = \begin{bmatrix} 0 & 1 \\ 1 & -3 \end{bmatrix} \qquad |M_Z| = -1 \qquad M_Z^{-1} = -\begin{bmatrix} -3 & -1 \\ -1 & 0 \end{bmatrix} = \begin{bmatrix} 3 & 1 \\ 1 & 0 \end{bmatrix}.$$

This can also be obtained as

$$M_Z^{-1} = \begin{bmatrix} a_1 & 1 \\ 1 & 0 \end{bmatrix} = \begin{bmatrix} 3 & 1 \\ 1 & 0 \end{bmatrix}.$$

Step 4: The transformation matrix that transforms the given state model into controllable canonical form is

$$P = M M_Z^{-1} = \begin{bmatrix} 0 & 1 \\ 1 & -1 \end{bmatrix} \begin{bmatrix} 3 & 1 \\ 1 & 0 \end{bmatrix} = \begin{bmatrix} 1 & 0 \\ 2 & 1 \end{bmatrix}.$$

$$P^{-1} = \begin{bmatrix} 1 & 0 \\ -2 & 1 \end{bmatrix}$$

Step 5: Desired characteristic equation
Peak overshoot $M_P < 0.17$.

The damping ratio $\delta = \dfrac{-\ln M_P}{\sqrt{\pi^2 + (\ln M_P)^2}} = \dfrac{-\ln(0.17)}{\sqrt{\pi^2 + (\ln 0.17)^2}} = 0.49$.

Choose $\delta = 0.5$.

Settling time $t_S = \dfrac{4}{\delta \omega_n} < 0.5$ $\qquad \omega_n > \dfrac{4}{0.5 \times 0.5} = 16$ rad/s .

Choose $\omega_n = 17$ rad/s .

Desired pole locations are $s = -\delta\omega_n \pm j\omega_n\sqrt{1-\delta^2} = -8.5 \pm j14.72$.
The desired characteristic equation is

$$(s+8.5)^2 + (14.72)^2 = 0 \quad \text{or} \quad s^2 + 17s + 289 = 0 = s^2 + \alpha_1 s + \alpha_0 . \tag{2}$$

Step 6: Feedback controller gain matrix
From equations (1) and (2)

$$K_Z = [(\alpha_0 - a_0) \qquad (\alpha_1 - a_1)] = [(289 - 2) \qquad (17 - 3)] = [287 \qquad 14] .$$

Step 7: Feedback controller gain matrix for the given system

$$K = K_Z P^{-1} = [287 \qquad 14] \begin{bmatrix} 1 & 0 \\ -2 & 1 \end{bmatrix} = [259 \qquad 14] .$$

Example 6.7 A linear continuous time system is represented by the state model

$$\dot{X} = \begin{bmatrix} -2 & 1 \\ 0 & -3 \end{bmatrix} X + \begin{bmatrix} 3 \\ 2 \end{bmatrix} u \qquad\qquad y = [2 \qquad 3] X .$$

Design a feedback controller to yield a peak overshoot of less than 17% and a settling time less than 1 s.

Solution:
Step 1: Check for controllability

$$A = \begin{bmatrix} -2 & 1 \\ 0 & -3 \end{bmatrix} \qquad B = \begin{bmatrix} 3 \\ 2 \end{bmatrix} \qquad AB = \begin{bmatrix} -2 & 1 \\ 0 & -3 \end{bmatrix}\begin{bmatrix} 3 \\ 2 \end{bmatrix} = \begin{bmatrix} -4 \\ -6 \end{bmatrix} .$$

The given system is not in controllable canonical form.
The controllability matrix

$$M = [B \qquad AB] = \begin{bmatrix} 3 & -4 \\ 2 & -6 \end{bmatrix} \qquad\qquad |M_X| = -10 .$$

The rank of M is 2 and so the system is completely controllable.
Step 2: Characteristic equation of the given system

$$|sI - A| = \begin{bmatrix} s+2 & -1 \\ 0 & s+3 \end{bmatrix} \qquad s^2 + 5s + 6 = 0 = s^2 + a_1 s + a_0 . \tag{1}$$

Step 3: State model in controllable canonical form is

$$\dot{Z} = \begin{bmatrix} 0 & 1 \\ -6 & -5 \end{bmatrix} Z + \begin{bmatrix} 0 \\ 1 \end{bmatrix} Z$$

$$A_\chi = \begin{bmatrix} 0 & 1 \\ -6 & -5 \end{bmatrix} \qquad B_\chi = \begin{bmatrix} 0 \\ 1 \end{bmatrix} \qquad A_\chi B_\chi = \begin{bmatrix} 1 \\ -5 \end{bmatrix}.$$

The controllability matrix

$$M_Z = [B_Z \quad A_Z B_Z] = \begin{bmatrix} 0 & 1 \\ 1 & -5 \end{bmatrix} \qquad |M_Z| = -1 \qquad M_Z^{-1} = -\begin{bmatrix} -5 & -1 \\ -1 & 0 \end{bmatrix} = \begin{bmatrix} 5 & 1 \\ 1 & 0 \end{bmatrix}.$$

This can also be obtained as

$$M_Z^{-1} = \begin{bmatrix} a_1 & 1 \\ 1 & 0 \end{bmatrix} = \begin{bmatrix} 5 & 1 \\ 1 & 0 \end{bmatrix}.$$

Step 4: The transformation matrix that transforms the given state model into controllable canonical form is

$$P = M \, M_Z^{-1} = \begin{bmatrix} 3 & -4 \\ 2 & -6 \end{bmatrix} \begin{bmatrix} 5 & 1 \\ 1 & 0 \end{bmatrix} = \begin{bmatrix} 11 & 3 \\ 4 & 2 \end{bmatrix}$$

$$P^{-1} = \frac{1}{10} \begin{bmatrix} 2 & -3 \\ -4 & 11 \end{bmatrix}.$$

Step 5: Desired characteristic equation

Peak overshoot $M_p < 0.17$.

The damping ratio $\delta = \dfrac{-\ln M_p}{\sqrt{\pi^2 + (\ln M_p)^2}} = \dfrac{-\ln(0.17)}{\sqrt{\pi^2 + (\ln 0.17)^2}} = 0.49$.

Choose $\delta = 0.5$.

Settling time $t_S = \dfrac{4}{\delta \omega_n} < 1 \qquad \omega_n > \dfrac{4}{0.5 \times 1} = 8$ rad/s.

Choose $\omega_n = 9$ rad/s.

Desired pole locations are $s = -\delta \omega_n \pm j\omega_n \sqrt{1 - \delta^2} = -4.5 \pm j7.8$.

The desired characteristic equation is

$$(s + 4.5)^2 + (7.8)^2 = 0 \qquad \text{or} \quad s^2 + 9s + 81 = 0 = s^2 + \alpha_1 s + \alpha_0. \qquad (2)$$

Step 6: Feedback controller gain matrix with respect to controllable canonical form

From equations (1) and (2)

$$K_\chi = [(\alpha_0 - a_0) \quad (\alpha_1 - a_1)] = [(81 - 6) \quad (9 - 5)] = [75 \quad 4].$$

Step 7: Feedback controller gain matrix for the given system

$$K = K_\chi P^{-1} = \frac{1}{10} [75 \quad 4] \begin{bmatrix} 2 & -3 \\ -4 & 11 \end{bmatrix} = [13.4 \quad -18.1].$$

Example 6.8 Design a linear state feedback controller to yield a peak overshoot of about 20% and a settling time of less than 2 s for the unit step response of a closed loop system represented by the state space model

$$\dot{X} = \begin{bmatrix} -7 & 1 & 0 \\ 0 & -8 & 1 \\ 0 & 0 & -9 \end{bmatrix} X + \begin{bmatrix} 0 \\ 0 \\ 1 \end{bmatrix} u \qquad y = [-2 \quad 2 \quad 0]X .$$

Solution:

Step 1: Check for controllability

$$A = \begin{bmatrix} -7 & 1 & 0 \\ 0 & -8 & 1 \\ 0 & 0 & -9 \end{bmatrix} \qquad B = \begin{bmatrix} 0 \\ 0 \\ 1 \end{bmatrix}.$$

The given system is not in controllable canonical form.

$$AB = \begin{bmatrix} -7 & 1 & 0 \\ 0 & -8 & 1 \\ 0 & 0 & -9 \end{bmatrix}\begin{bmatrix} 0 \\ 0 \\ 1 \end{bmatrix} = \begin{bmatrix} 0 \\ 1 \\ -9 \end{bmatrix} \qquad A^2B = \begin{bmatrix} -7 & 1 & 0 \\ 0 & -8 & 1 \\ 0 & 0 & -9 \end{bmatrix}\begin{bmatrix} 0 \\ 1 \\ -9 \end{bmatrix} = \begin{bmatrix} 1 \\ -17 \\ 81 \end{bmatrix}.$$

The controllability matrix

$$M = [B \quad AB \quad A^2 B] = \begin{bmatrix} 0 & 0 & 1 \\ 0 & 1 & -17 \\ 1 & -9 & 81 \end{bmatrix} \qquad |M| = -1.$$

The rank of M is 3 and so the system is completely controllable.

Step 2: Characteristic equation of the given system

$$|sI - A| = \begin{bmatrix} s+7 & -1 & 0 \\ 0 & s+8 & -1 \\ 0 & 0 & s+9 \end{bmatrix} \qquad (s+7)(s+8)(s+9) = 0$$

$$s^3 + 24s^2 + 191s + 504 = 0 = s^3 + a_2 s^2 + a_1 s + a_0 . \tag{1}$$

Step 3: State model in controllable canonical form is

$$\dot{Z} = \begin{bmatrix} 0 & 1 & 0 \\ 0 & 0 & 1 \\ -a_0 & -a_1 & -a_2 \end{bmatrix} Z + \begin{bmatrix} 0 \\ 0 \\ 1 \end{bmatrix} u = \begin{bmatrix} 0 & 1 & 0 \\ 0 & 0 & 1 \\ -504 & -191 & -24 \end{bmatrix} Z + \begin{bmatrix} 0 \\ 0 \\ 1 \end{bmatrix} u .$$

The controllability matrix

$$M_\chi = \begin{bmatrix} a_1 & a_2 & 1 \\ a_2 & 1 & 0 \\ 1 & 0 & 0 \end{bmatrix}^{-1} \qquad M_Z^{-1} = \begin{bmatrix} a_1 & a_2 & 1 \\ a_2 & 1 & 0 \\ 1 & 0 & 0 \end{bmatrix} = \begin{bmatrix} 191 & 24 & 1 \\ 24 & 1 & 0 \\ 1 & 0 & 0 \end{bmatrix}.$$

Step 4: The transformation matrix that transforms the given state model into controllable canonical form

$$P = M M_Z^{-1} = \begin{bmatrix} 0 & 0 & 1 \\ 0 & 1 & -17 \\ 1 & -9 & 81 \end{bmatrix} \begin{bmatrix} 191 & 24 & 1 \\ 24 & 1 & 0 \\ 1 & 0 & 0 \end{bmatrix} = \begin{bmatrix} 1 & 0 & 0 \\ 7 & 1 & 0 \\ 56 & 15 & 1 \end{bmatrix}.$$

$$|P| = 1 \qquad P^{-1} = \begin{bmatrix} 1 & 0 & 0 \\ -7 & 1 & 0 \\ 49 & -15 & 1 \end{bmatrix}.$$

Step 5: Desired characteristic equation

Peak overshoot $M_P < 0.2 \qquad t_S < 2$ s.

The damping ratio $\delta = \dfrac{-\ln M_P}{\sqrt{\pi^2 + (\ln M_P)^2}} = \dfrac{-\ln(0.2)}{\sqrt{\pi^2 + (\ln 0.2)^2}} = 0.456$.

Choose $\delta = 0.5$.

Settling time $t_S = \dfrac{4}{\delta \omega_n} < 2 \qquad \omega_n > \dfrac{4}{0.5 \times 2} = 4$ rad/s . Choose $\omega_n = 5$ rad/s .

Desired pole locations are $s = -\delta \omega_n \pm j \omega_n \sqrt{1 - \delta^2} = -2.5 \pm j4.33$.

Assume that the third pole is far away from the dominant pole at $s = -15$.

The desired characteristic equation is

$$(s+15)\{(s+2.5)^2 + (4.33)^2\} = 0$$

$$(s+15)(s^2 + 5s + 25) = 0$$

$$s^3 + 20s^2 + 100s + 375 = 0 = s^3 + \alpha_2 s^2 + \alpha_1 s + \alpha_0 . \tag{2}$$

Step 6: Feedback controller gain matrix

Using equations (1) and (2),

$$K_\chi = [(\alpha_0 - a_0) \quad (\alpha_1 - a_1) \quad (\alpha_2 - a_2)]$$

$$= [(375 - 504) \quad (100 - 191) \quad (20 - 24)] = [-129 \quad -91 \quad -4].$$

Step 7: Feedback controller gain matrix for the given system

$$K_X = K_Z P^{-1} = [-129 \quad -91 \quad -4] \begin{bmatrix} 1 & 0 & 0 \\ -7 & 1 & 0 \\ 49 & -15 & 1 \end{bmatrix} = [312 \quad -31 \quad -4].$$

Example 6.9 Design a linear state feedback controller to yield a peak overshoot of about 20% and a settling time of less than 4 s for a closed loop system represented by the state space model

$$\dot{X} = \begin{bmatrix} -5 & 1 & 0 \\ 0 & -2 & 1 \\ 0 & 0 & -1 \end{bmatrix} X + \begin{bmatrix} 0 \\ 0 \\ 1 \end{bmatrix} u \qquad y = [-5 \quad 5 \quad 0]X.$$

The closed loop system has a zero at $s = -6$.

Solution:

Step 1: Check for controllability

$$A = \begin{bmatrix} -5 & 1 & 0 \\ 0 & -2 & 1 \\ 0 & 0 & -1 \end{bmatrix} \qquad B = \begin{bmatrix} 0 \\ 0 \\ 1 \end{bmatrix}.$$

The given system is not in controllable canonical form.

$$AB = \begin{bmatrix} -5 & 1 & 0 \\ 0 & -2 & 1 \\ 0 & 0 & -1 \end{bmatrix}\begin{bmatrix} 0 \\ 0 \\ 1 \end{bmatrix} = \begin{bmatrix} 0 \\ 1 \\ -1 \end{bmatrix} \qquad A^2 B = \begin{bmatrix} -5 & 1 & 0 \\ 0 & -2 & 1 \\ 0 & 0 & -1 \end{bmatrix}\begin{bmatrix} 0 \\ 1 \\ -1 \end{bmatrix} = \begin{bmatrix} 1 \\ -3 \\ 1 \end{bmatrix}.$$

The controllability matrix

$$M = \begin{bmatrix} 0 & 0 & 1 \\ 0 & 1 & -3 \\ 1 & -1 & 1 \end{bmatrix} \qquad |M| = \begin{bmatrix} 0 & 0 & 1 \\ 0 & 1 & -3 \\ 1 & -1 & 1 \end{bmatrix} = -1.$$

The rank of M is 3 and hence the system is completely controllable.

Step 2: Characteristic equation of the given system

$$|sI - A| = \begin{bmatrix} s+5 & -1 & 0 \\ 0 & s+2 & -1 \\ 0 & 0 & s+1 \end{bmatrix} \qquad (s+5)(s+2)(s+1) = 0$$

$$s^3 + 8s^2 + 17s + 10 = 0 = s^3 + a_2 s^2 + a_1 s + a_0. \tag{1}$$

Step 3: State model in controllable canonical form is

$$\dot{Z} = \begin{bmatrix} 0 & 1 & 0 \\ 0 & 0 & 1 \\ -a_0 & -a_1 & -a_2 \end{bmatrix} Z + \begin{bmatrix} 0 \\ 0 \\ 1 \end{bmatrix} u = \begin{bmatrix} 0 & 1 & 0 \\ 0 & 0 & 1 \\ -10 & -17 & -8 \end{bmatrix} Z + \begin{bmatrix} 0 \\ 0 \\ 1 \end{bmatrix} u .$$

The controllability matrix

$$M_{\chi} = \begin{bmatrix} a_1 & a_2 & 1 \\ a_2 & 1 & 0 \\ 1 & 0 & 0 \end{bmatrix}^{-1} \qquad M_Z^{-1} = \begin{bmatrix} a_1 & a_2 & 1 \\ a_2 & 1 & 0 \\ 1 & 0 & 0 \end{bmatrix} = \begin{bmatrix} 17 & 8 & 1 \\ 8 & 1 & 0 \\ 1 & 0 & 0 \end{bmatrix} .$$

The transformation matrix that transforms the given state model into controllable canonical form

$$P = M M_Z^{-1} = \begin{bmatrix} 0 & 0 & 1 \\ 0 & 1 & -3 \\ 1 & -1 & 1 \end{bmatrix} \begin{bmatrix} 17 & 8 & 1 \\ 8 & 1 & 0 \\ 1 & 0 & 0 \end{bmatrix} = \begin{bmatrix} 1 & 0 & 0 \\ 5 & 1 & 0 \\ 10 & 7 & 1 \end{bmatrix}$$

$$P^{-1} = \begin{bmatrix} 1 & 0 & 0 \\ -5 & 1 & 0 \\ 25 & -7 & 1 \end{bmatrix} .$$

Step 4: Desired characteristic equation

Peak overshoot $M_P < 0.2$ $t_s < 4$ s.

The damping ratio $\delta = \dfrac{-\ln M_P}{\sqrt{\pi^2 + (\ln M_P)^2}} = \dfrac{-\ln(0.2)}{\sqrt{\pi^2 + (\ln 0.2)^2}} = 0.456$.

Choose $\delta = 0.5$.

Settling time $t_s = \dfrac{4}{\delta \omega_n} < 4$ $\omega_n > \dfrac{4}{0.5 \times 4} = 2 \text{ rad/s}$.

Choose $\omega_n = 3$ rad/s .

Desired pole locations are $s = -\delta \omega_n \pm j\omega_n \sqrt{1 - \delta^2} = -1.5 \pm j2.6$.
Assume that the third pole is at $s = -6$, which is also the location of the closed loop zero.
The desired characteristic equation is

$$(s+6)\{(s+1.5)^2 + (2.6)^2\} = (s+6)(s^2 + 3s + 9) = 0$$

$$s^3 + 9s^2 + 27s + 54 = 0 = s^3 + \alpha_2 s^2 + \alpha_1 s + \alpha_0 . \qquad (2)$$

Step 5: Feedback controller gain matrix

Comparing equations (1) and (2),

$$K_\chi = [(\alpha_0 - a_0) \quad (\alpha_1 - a_1) \quad (\alpha_2 - a_2)]$$

$$= [(54-10) \quad (27-17) \quad (9-8)] = [44 \quad 10 \quad 1].$$

Feedback controller gain matrix for the given system

$$K = K_\chi P^{-1} = [44 \quad 10 \quad 1] \begin{bmatrix} 1 & 0 & 0 \\ -5 & 1 & 0 \\ 25 & -7 & 1 \end{bmatrix} = [19 \quad 3 \quad 1].$$

6.6 ACKERMANN'S FORMULA FOR CONTROLLER DESIGN

The feedback controller gain matrix can be obtained directly by using Ackermann's formula. Consider, for simplicity, a third-order system represented by the state model

$$\dot{X} = AX + Bu \qquad\qquad y = CX.$$

The controllability matrix

$$M = [B \quad AB \quad A^2 B].$$

Assume that M has full rank so that the system is completely controllable. With state feedback, let the state equation be

$$\dot{X} = [A - BK]X + (BF)r. \tag{6.2}$$

Let the feedback controller gain matrix be $K = [k_0 \quad k_1 \quad k_2]$, and the system matrix be $A_F = [A - BK]$. The characteristic equation with state feedback is

$$|sI - A_F| = [sI - (A - BK)] = 0.$$

Let this characteristic equation be

$$s^3 + \alpha_2 s^2 + \alpha_1 s + \alpha_0 = 0.$$

Cayley–Hamilton theorem states that every square matrix satisfies its own characteristic equation. Hence

$$\phi(A_F) = A_F{}^3 + \alpha_2 A_F{}^2 + \alpha_1 A_F + \alpha_0 I = 0.$$

$$A_F = [A - BK]$$

$$A_F{}^2 = [A - BK][A - BK] = A^2 - ABK - BK[A - BK] = A^2 - ABK - BKA_F.$$

$$A_F{}^3 = [A - BK][A^2 - ABK - BKA_F]$$

$$= A^3 - A^2 BK - ABKA_F - BK[A^2 - ABK - BKA_F]$$

$$= A^3 - A^2 BK - ABKA_F - BKA_F^2$$

$$\phi(A_F) = A_F^3 + \alpha_2 A_F^2 + \alpha_1 A_F + \alpha_0 I$$

$$= [A^3 - A^2 BK - ABKA_F - BKA_F^2] + \alpha_2[A^2 - ABK - BKA_F]$$
$$+ \alpha_1[A - BK] + \alpha_0 I = 0.$$

Rearranging this equation

$$A^3 + \alpha_2 A^2 + \alpha_1 A + \alpha_0 I = [A^2 BK + ABKA_F + BKA_F^2] + \alpha_2[ABK + BKA_F] + \alpha_1 BK .$$

This can be written as

$$\phi(A) = [A^2 BK + ABKA_F + BKA_F^2] + \alpha_2[ABK + BKA_F] + \alpha_1 BK$$

$$= B[\alpha_2 KA_F + \alpha_1 K + KA_F^2] + AB[\alpha_2 K + KA_F] + A^2 BK$$

$$= [B \quad AB \quad A^2 B] \begin{bmatrix} \alpha_2 KA_F + \alpha_1 K + KA_F^2 \\ \alpha_2 K + KA_F \\ K \end{bmatrix}$$

$$= M \begin{bmatrix} \alpha_2 KA_F + \alpha_1 K + KA_F^2 \\ \alpha_2 K + KA_F \\ K \end{bmatrix}.$$

Or $\quad M^{-1}\phi(A) = \begin{bmatrix} \alpha_2 KA_F + \alpha_1 K + KA_F^2 \\ \alpha_2 K + KA_F \\ K \end{bmatrix}.$

Pre-multiplying both sides by $[0 \quad 0 \quad 1]$,

$$[0 \quad 0 \quad 1]M^{-1}\phi(A) = [0 \quad 0 \quad 1] \begin{bmatrix} \alpha_2 KA_F + \alpha_1 K + KA_F^2 \\ \alpha_2 K + KA_F \\ K \end{bmatrix} = K .$$

Hence $K = [0 \quad 0 \quad 1]M^{-1}\phi(A)$.

In general, for an n-th order system,

$$K = [0 \quad 0 \quad 0 \ldots 0 \quad 1]M^{-1}\phi(A). \tag{6.11}$$

The matrix defined by equation (6.11) is the Ackermann's formula for the state feedback gain matrix.

Example 6.10 Repeat Example 6.1 using Ackermann's formula.

Solution:

$$\dot{X} = \begin{bmatrix} -3 & 1 \\ 0 & -2 \end{bmatrix} X + \begin{bmatrix} 0 \\ 1 \end{bmatrix} u \qquad\qquad y = [2 \quad 1]X .$$

Step 1: Check for controllability

$$A = \begin{bmatrix} -3 & 1 \\ 0 & -2 \end{bmatrix} \qquad\qquad B = \begin{bmatrix} 0 \\ 1 \end{bmatrix} u$$

$$M = [B \quad AB] = \begin{bmatrix} 0 & 1 \\ 1 & -2 \end{bmatrix} \qquad\qquad |M| = -1 .$$

The controllability matrix M has rank 2 and hence the system is completely controllable.

$$M^{-1} = -\begin{bmatrix} -2 & -1 \\ -1 & 0 \end{bmatrix} = \begin{bmatrix} 2 & 1 \\ 1 & 0 \end{bmatrix}.$$

Step 2: Desired characteristic equation
 Desired pole locations are at $s = -3 \pm j3$. The desired characteristic equation is

$$(s+3)^2 + 3^2 = 0 \quad\text{or}\quad s^2 + 6s + 18 = 0 = s^2 + \alpha_1 s + \alpha_0 .$$

Step 3: Evaluation of $\phi(A)$

$$\phi(A) = A^2 + \alpha_1 A + \alpha_0 I \qquad\qquad A^2 = \begin{bmatrix} -3 & 1 \\ 0 & -2 \end{bmatrix}\begin{bmatrix} -3 & 1 \\ 0 & -2 \end{bmatrix} = \begin{bmatrix} 9 & -5 \\ 0 & 4 \end{bmatrix}$$

$$\phi(A) = \begin{bmatrix} 9 & -5 \\ 0 & 4 \end{bmatrix} + 6\begin{bmatrix} -3 & 1 \\ 0 & -2 \end{bmatrix} + 18\begin{bmatrix} 1 & 0 \\ 0 & 1 \end{bmatrix} = \begin{bmatrix} 9 & 1 \\ 0 & 10 \end{bmatrix}.$$

Step 4: Feedback controller gain matrix

$$K = [0 \quad 1]M^{-1}\phi(A)$$

$$= [0 \quad 1]\begin{bmatrix} 2 & 1 \\ 1 & 0 \end{bmatrix}\begin{bmatrix} 9 & 1 \\ 0 & 10 \end{bmatrix} = [0 \quad 1]\begin{bmatrix} 18 & 12 \\ 9 & 1 \end{bmatrix} = [9 \quad 1] .$$

Example 6.11 Repeat Example 6.6 using Ackermann's formula.

Solution: Given state model is

$$\dot{X} = \begin{bmatrix} -2 & 1 \\ 0 & -1 \end{bmatrix} X + \begin{bmatrix} 0 \\ 1 \end{bmatrix} u \qquad\qquad y = [1 \quad 3]X .$$

Step 1: Check for controllability

$$A = \begin{bmatrix} -2 & 1 \\ 0 & -1 \end{bmatrix} \qquad B = \begin{bmatrix} 0 \\ 1 \end{bmatrix}.$$

The given system is not in controllable canonical form.
The controllability matrix

$$M = [B \quad AB] = \begin{bmatrix} 0 & 1 \\ 1 & -1 \end{bmatrix} \qquad |M| = -1.$$

The rank of M is 2 and so the system is completely controllable.

$$M^{-1} = -\begin{bmatrix} -1 & -1 \\ -1 & 0 \end{bmatrix} = \begin{bmatrix} 1 & 1 \\ 1 & 0 \end{bmatrix}.$$

Step 2: Desired characteristic equation

Peak overshoot $M_P < 0.17$.

The damping ratio $\delta = \dfrac{-\ln M_P}{\sqrt{\pi^2 + (\ln M_P)^2}} = \dfrac{-\ln(0.17)}{\sqrt{\pi^2 + (\ln 0.17)^2}} = 0.49$.

Choose $\delta = 0.5$.

Settling time $t_S = \dfrac{4}{\delta \omega_n} < 0.5 \qquad \omega_n > \dfrac{4}{0.5 \times 0.5} = 16$ rad/s.

Choose $\omega_n = 17$ rad/s.

Desired pole locations are $s = -\delta \omega_n \pm j\omega_n \sqrt{1 - \delta^2} = -8.5 \pm j14.72$.
The desired characteristic equation is

$$(s + 8.5)^2 + (14.72)^2 = 0 \qquad or \quad s^2 + 17s + 289 = 0 = s^2 + \alpha_1 s + \alpha_0.$$

Step 3: Evaluation of $\phi(A)$

$$\phi(A) = A^2 + \alpha_1 A + \alpha_0 I \qquad A^2 = \begin{bmatrix} -2 & 1 \\ 0 & -1 \end{bmatrix}\begin{bmatrix} -2 & 1 \\ 0 & -1 \end{bmatrix} = \begin{bmatrix} 4 & -3 \\ 0 & 1 \end{bmatrix}$$

$$\phi(A) = \begin{bmatrix} 4 & -3 \\ 0 & 1 \end{bmatrix} + 17\begin{bmatrix} -2 & 1 \\ 0 & -1 \end{bmatrix} + 289\begin{bmatrix} 1 & 0 \\ 0 & 1 \end{bmatrix} = \begin{bmatrix} 259 & 14 \\ 0 & 273 \end{bmatrix}.$$

Step 4: Feedback controller gain matrix

$$K = [0 \quad 1]M^{-1}\phi(A)$$

$$= [0 \quad 1]\begin{bmatrix} 1 & 1 \\ 1 & 0 \end{bmatrix}\begin{bmatrix} 259 & 14 \\ 0 & 273 \end{bmatrix} = [0 \quad 1]\begin{bmatrix} 259 & 287 \\ 259 & 14 \end{bmatrix} = [259 \quad 14].$$

Example 6.12 Repeat Example 6.9 using Ackermann's formula.

$$\dot{X} = \begin{bmatrix} -5 & 1 & 0 \\ 0 & -2 & 1 \\ 0 & 0 & -1 \end{bmatrix} X + \begin{bmatrix} 0 \\ 0 \\ 1 \end{bmatrix} u \qquad y = [-5 \quad 5 \quad 0]X.$$

The closed loop system has a zero at $s = -6$.

Solution:

Step 1: Check for controllability

$$A = \begin{bmatrix} -5 & 1 & 0 \\ 0 & -2 & 1 \\ 0 & 0 & -1 \end{bmatrix} \qquad B = \begin{bmatrix} 0 \\ 0 \\ 1 \end{bmatrix}$$

$$AB = \begin{bmatrix} -5 & 1 & 0 \\ 0 & -2 & 1 \\ 0 & 0 & -1 \end{bmatrix}\begin{bmatrix} 0 \\ 0 \\ 1 \end{bmatrix} = \begin{bmatrix} 0 \\ 1 \\ -1 \end{bmatrix} \qquad A^2 B = \begin{bmatrix} -5 & 1 & 0 \\ 0 & -2 & 1 \\ 0 & 0 & -1 \end{bmatrix}\begin{bmatrix} 0 \\ 1 \\ -1 \end{bmatrix} = \begin{bmatrix} 1 \\ -3 \\ 1 \end{bmatrix}.$$

The controllability matrix

$$M = \begin{bmatrix} 0 & 0 & 1 \\ 0 & 1 & -3 \\ 1 & -1 & 1 \end{bmatrix} \qquad |M| = \begin{bmatrix} 0 & 0 & 1 \\ 0 & 1 & -3 \\ 1 & -1 & 1 \end{bmatrix} = -1.$$

The rank of M is 3 and hence the system is completely controllable.

$$M^{-1} = \begin{bmatrix} 2 & 1 & 1 \\ 3 & 1 & 0 \\ 1 & 0 & 0 \end{bmatrix}.$$

Step 2: Desired characteristic equation

Peak overshoot $M_P < 0.2$.

The damping ratio $\delta = \dfrac{-\ln M_P}{\sqrt{\pi^2 + (\ln M_P)^2}} = \dfrac{-\ln(0.2)}{\sqrt{\pi^2 + (\ln 0.2)^2}} = 0.456$.

Choose $\delta = 0.5$.

Settling time $t_S = \dfrac{4}{\delta\omega_n} < 4 \qquad \omega_n > \dfrac{4}{0.5 \times 4} = 2$ rad/s.

Choose $\omega_n = 3$ rad/s.

Desired pole locations are $s = -\delta\omega_n \pm j\omega_n\sqrt{1-\delta^2} = -1.5 \pm j2.6$.

Assume that the third pole is at $s = -6$, which is the location of the closed loop zero. The desired characteristic equation is

$$(s+6)\{(s+1.5)^2 + (2.6)^2\} = (s+6)(s^2 + 3s + 9) = 0$$

$$s^3 + 9s^2 + 27s + 54 = 0 = s^3 + \alpha_2 s^2 + \alpha_1 s + \alpha_0 .$$

Step 3: Evaluation of $\phi(A)$

$$A^2 = \begin{bmatrix} -5 & 1 & 0 \\ 0 & -2 & 1 \\ 0 & 0 & -1 \end{bmatrix}\begin{bmatrix} -5 & 1 & 0 \\ 0 & -2 & 1 \\ 0 & 0 & -1 \end{bmatrix} = \begin{bmatrix} 25 & -7 & 1 \\ 0 & 4 & -3 \\ 0 & 0 & 1 \end{bmatrix}$$

$$A^3 = \begin{bmatrix} -5 & 1 & 0 \\ 0 & -2 & 1 \\ 0 & 0 & -1 \end{bmatrix}\begin{bmatrix} 25 & -7 & 1 \\ 0 & 4 & -3 \\ 0 & 0 & 1 \end{bmatrix} = \begin{bmatrix} -125 & 39 & -8 \\ 0 & -8 & 7 \\ 0 & 0 & -1 \end{bmatrix}.$$

$$\phi(A) = A^3 + \alpha_2 A^2 + \alpha_1 A + \alpha_0 I$$

$$\phi(A) = \begin{bmatrix} -125 & 39 & -8 \\ 0 & -8 & 7 \\ 0 & 0 & -1 \end{bmatrix} + 9\begin{bmatrix} 25 & -7 & 1 \\ 0 & 4 & -3 \\ 0 & 0 & 1 \end{bmatrix} + 27\begin{bmatrix} -5 & 1 & 0 \\ 0 & -2 & 1 \\ 0 & 0 & -1 \end{bmatrix} + 54\begin{bmatrix} 1 & 0 & 0 \\ 0 & 1 & 0 \\ 0 & 0 & 1 \end{bmatrix}$$

$$= \begin{bmatrix} 19 & 3 & 1 \\ 0 & 28 & 7 \\ 0 & 0 & 35 \end{bmatrix}$$

Step 4: Feedback controller gain matrix

$$K = \begin{bmatrix} 0 & 0 & 1 \end{bmatrix} M^{-1} \phi(A)$$

$$= \begin{bmatrix} 0 & 0 & 1 \end{bmatrix}\begin{bmatrix} 2 & 1 & 1 \\ 3 & 1 & 0 \\ 1 & 0 & 0 \end{bmatrix}\begin{bmatrix} 19 & 3 & 1 \\ 0 & 28 & 7 \\ 0 & 0 & 35 \end{bmatrix}$$

$$= \begin{bmatrix} 0 & 0 & 1 \end{bmatrix}\begin{bmatrix} 38 & 34 & 44 \\ 57 & 37 & 10 \\ 19 & 3 & 1 \end{bmatrix} = \begin{bmatrix} 19 & 3 & 1 \end{bmatrix}$$

6.7 STATE OBSERVERS

In the pole placement technique for designing a feedback controller, it was assumed that all state variables are available for feedback. However, all state variables may not be available for feedback or it is impractical to measure the state variables for reasons of complexity, cost and accuracy. In such cases, it is necessary to estimate the states which are otherwise unavailable or non-measurable. Estimated states, rather than actual states, are then fed to the controller.

Estimation of un-measurable state variables is called *observation*. Generally, a digital computer is used to estimate these un-measurable state variables and is called a *state observer*. The observer or the estimator is a subsystem that reconstructs the states of the plant. If a state observer observes or estimates all state variables even if some state variables are available for direct measurement, it is called a *full-order state observer*. A state observer which estimates only the non-measurable state variables is called a *reduced-order observer*.

A state observer estimates the state variables using the output and the input. Hence, a state observer can be designed only if the system is completely observable. It is easy to design a state observer for a given system if the state space model of the system is in observable canonical form.

6.8 PRINCIPLE OF OBSERVER DESIGN

The block diagram shown in Figure 6.2 illustrates the functioning of an observer.

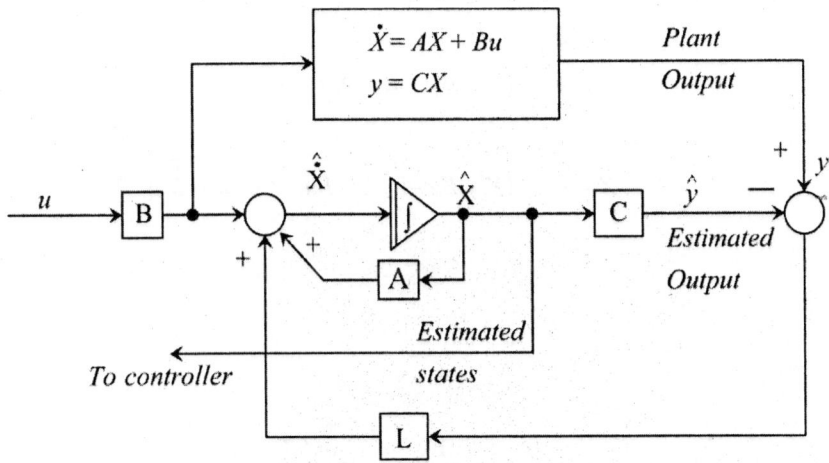

Figure 6.2 State observer

The state model for the plant is

$$\dot{X} = AX + Bu \qquad\qquad y = CX . \qquad\qquad (6.12)$$

The state equation for the observer is

$$\dot{\hat{X}} = A\hat{X} + Bu \qquad\qquad \hat{y} = C\hat{X}. \qquad\qquad (6.13)$$

From these equations (6.12) and (6.13)

$$\dot{X} - \dot{\hat{X}} = A(X - \hat{X}) \qquad\qquad y - \hat{y} = C(X - \hat{X}). \qquad\qquad (6.14)$$

The actual difference between the actual state and the estimated state is independent of the input u. If the plant is stable, that is, if the closed loop poles are located on the left half of s-plane, the error between the actual state and the estimated state $(X - \hat{X})$ approaches zero. The speed of convergence of the error to zero value is the same as that of the transient response of the plant and it may be too slow. To speed up the rate of convergence, the transient response of the observer is made faster than that of the plant so that the controller will receive the estimated states almost instantly. To make the response of the observer faster than that of the given closed loop system, the difference between the actual plant output y and the estimated output $\hat{y}$ is fed back through a gain vector L. This yields a rapidly updated estimate of the state vector.

The state equation for the observer with feedback of error $(y - \hat{y})$ is

$$\dot{\hat{X}} = A\hat{X} + Bu + L(y - \hat{y}).$$

Using equation (6.14)

$$\dot{\hat{X}} = A\hat{X} + Bu + LC(X - \hat{X}).$$

$$\dot{X} - \dot{\hat{X}} = AX + Bu - [A\hat{X} + Bu + LC(X - \hat{X})]$$

$$= [A - LC](X - \hat{X}).$$

Let $E_X = (X - \hat{X})$ so that

$$\dot{E}_X = (\dot{X} - \dot{\hat{X}}) = [A - LC]E_X \qquad and \quad (y - \hat{y}) = C E_X. \qquad\qquad (6.15)$$

Thus, the observer design consists of solving the vector L to yield a characteristic equation that corresponds to a faster response for the observer. The characteristic equation of the system with observer is

$$\left| sI - (A - LC) \right| = 0.$$

The desired characteristic equation, for faster response, is obtained from the design specification for the observer. Generally, the observer response is to be made ten times faster than the response of the closed loop system. For this, observer poles are to be ten times the closed loop poles of the system.

6.9 DESIGN OF FULL ORDER OBSERVER – METHOD OF MATCHING COEFFICIENTS

Consider a third-order closed loop system.

Step 1: Check for complete observability. Design of observer is possible only if the system is completely observable.

Step 2: Obtain the desired characteristic equation from the design specifications. Assume that the observer response is to be made ten times faster than the response of the closed loop system. If the dominant poles of the closed loop system are $s = -a \pm jb$, then the observer poles can be placed at $s = -10a \pm j10b$, for faster response. Select a third pole at a location on the negative real axis five times the distance of the dominant pole of the observer from the imaginary axis. Hence, the characteristic equation is

$$(s+50a)\{(s+10a)^2 + (10b)^2\} = 0 .$$ (6.16)

Step 3: Find the characteristic equation with the addition of the observer. Let the error feedback vector L be $L = \begin{bmatrix} l_2 \\ l_1 \\ l_0 \end{bmatrix}$.

With the observer, the system matrix is

$$[A - LC].$$

The characteristic equation with error feedback is

$$|sI - (A - LC)| = 0 .$$

The characteristic equation can be written down as

$$s^3 + a_2 s^2 + a_1 s + a_0 = 0 .$$ (6.17)

Comparing equations (6.16) and (6.17), the elements of the vector L can be determined.

Example 6.13 A linear continuous time system is represented by the state model

$$\dot{X} = \begin{bmatrix} -3 & 1 \\ 0 & -2 \end{bmatrix} X + \begin{bmatrix} 0 \\ 1 \end{bmatrix} u \qquad y = [2 \quad 1]X .$$

Design an observer for the system with closed loop poles located at $s = -1 \pm j2$.

Solution:

Step 1: Check for observability

$$A = \begin{bmatrix} -3 & 1 \\ 0 & -2 \end{bmatrix} \quad B = \begin{bmatrix} 0 \\ 1 \end{bmatrix} \quad C = \begin{bmatrix} 2 & 1 \end{bmatrix}$$

$$CA = \begin{bmatrix} 2 & 1 \end{bmatrix} \begin{bmatrix} -3 & 1 \\ 0 & -2 \end{bmatrix} = \begin{bmatrix} -6 & 0 \end{bmatrix}.$$

The observability matrix

$$Q = \begin{bmatrix} C \\ CA \end{bmatrix} = \begin{bmatrix} 2 & 1 \\ -6 & 0 \end{bmatrix} \qquad |Q| = 6.$$

Q has rank 2 and the system is completely observable.

Step 2: Desired characteristic equation

The closed loop poles' locations are $s = -1 \pm j2$.

Assume that the observer response is ten times faster than the system response. Hence, the dominant pole for the observer is to be placed at $s = -10 \pm j20$. The desired characteristic equation for the observer is

$$(s+10)^2 + (20)^2 = 0 \qquad or \quad s^2 + 20s + 500 = 0. \tag{1}$$

Step 3: The characteristic equation with state error feedback

Let the error feedback gain matrix be $L = \begin{bmatrix} l_1 \\ l_0 \end{bmatrix}$.

The observer matrix with state error feedback

$$[A - LC] = \begin{bmatrix} -3 & 1 \\ 0 & -2 \end{bmatrix} - \begin{bmatrix} l_1 \\ l_0 \end{bmatrix} \begin{bmatrix} 2 & 1 \end{bmatrix} = \begin{bmatrix} -3 & 1 \\ 0 & -2 \end{bmatrix} - \begin{bmatrix} 2l_1 & l_1 \\ 2l_0 & l_0 \end{bmatrix} = \begin{bmatrix} -3-2l_1 & 1-l_1 \\ -2l_0 & -2-l_0 \end{bmatrix}.$$

The characteristic equation with state error feedback is

$$|sI - (A - LC)| = 0 \qquad Or \quad \begin{vmatrix} s+(3+2l_1) & -1+l_1 \\ 2l_0 & s+(2+l_0) \end{vmatrix} = 0$$

$$s^2 + (5 + 2l_1 + l_0)s + (6 + 5l_0 + 4l_1) = 0. \tag{2}$$

Step 4: Evaluation of vector L

Comparing equations (1) and (2),

$$5 + 2l_1 + l_0 = 20 \qquad\qquad 6 + 5l_0 + 4l_1 = 500.$$

Solving these equations $l_1 = -(419/6)$ $\qquad\qquad l_0 = (464/3)$.

$$L = \begin{bmatrix} l_1 \\ l_0 \end{bmatrix} = \begin{bmatrix} -(419/6) \\ (464/3) \end{bmatrix}.$$

Example 6.14 A linear continuous time system is represented by the state model

$$\dot{X} = \begin{bmatrix} 0 & 2 \\ 0 & 2 \end{bmatrix} X + \begin{bmatrix} 0 \\ 2 \end{bmatrix} u \qquad y = \begin{bmatrix} 1 & 0 \end{bmatrix} X.$$

Design an observer with its poles at $s = -8, -8$.

Solution:

Step 1: Check for observability

$$A = \begin{bmatrix} 0 & 2 \\ 0 & 2 \end{bmatrix} \qquad B = \begin{bmatrix} 0 \\ 2 \end{bmatrix} \qquad C = \begin{bmatrix} 1 & 0 \end{bmatrix}$$

$$CA = \begin{bmatrix} 1 & 0 \end{bmatrix} \begin{bmatrix} 0 & 2 \\ 0 & 2 \end{bmatrix} = \begin{bmatrix} 0 & 2 \end{bmatrix}.$$

The observability matrix

$$Q = \begin{bmatrix} C \\ CA \end{bmatrix} = \begin{bmatrix} 1 & 0 \\ 0 & 2 \end{bmatrix} \qquad |Q| = 2.$$

Q has rank 2 and the system is completely observable.

Step 2: Desired characteristic equation

The observer poles are at $s = -8, -8$. The desired characteristic equation for the observer is

$$(s+8)^2 = 0 \qquad \text{or } s^2 + 16s + 64 = 0. \qquad (1)$$

Step 3: The characteristic equation with state error feedback

Let the error feedback gain matrix be $L = \begin{bmatrix} l_1 \\ l_0 \end{bmatrix}$.

The observer matrix with state error feedback

$$[A - LC] = \begin{bmatrix} 0 & 2 \\ 0 & 2 \end{bmatrix} - \begin{bmatrix} l_1 \\ l_0 \end{bmatrix} \begin{bmatrix} 1 & 0 \end{bmatrix} = \begin{bmatrix} -l_1 & 2 \\ -l_0 & 2 \end{bmatrix}.$$

The characteristic equation with state error feedback is

$$|sI - (A - LC)| = 0 \qquad \text{Or} \qquad \begin{vmatrix} s + l_1 & -2 \\ l_0 & s-2 \end{vmatrix} = 0$$

$$s^2 + (l_1 - 2)s + (2l_0 - 2l_1) = 0. \qquad (2)$$

Step 4: Evaluation of vector L

Comparing equations (1) and (2),

$$l_1 - 2 = 16 \qquad\qquad 2l_0 - 2l_1 = 64 .$$

Solving these equations $l_1 = 18$ $\qquad\qquad l_0 = 50 .$

$$L = \begin{bmatrix} l_1 \\ l_0 \end{bmatrix} = \begin{bmatrix} 18 \\ 50 \end{bmatrix} .$$

6.10 OBSERVER DESIGN BY TRANSFORMATION TO OBSERVABLE CANONICAL FORM

The solution of equations to evaluate the observer gain vector by the method of matching coefficients can be quite intricate with higher-order systems. In such cases, it is very convenient to carry out the observer design by transforming the given state model into observable canonical form. It is easy to design a state observer if the system model is in observable canonical form. The evaluated observer gain vector is then transferred back to the original representation by proper transformation.

Consider a third-order system that is not represented in observable canonical form.

$$\dot{X} = AX + Bu \qquad\qquad y = CX$$

The observability matrix is

$$Q = \begin{bmatrix} C \\ CA \\ CA^2 \end{bmatrix} .$$

Assume that Q has full rank so that system model is completely observable. The system model is now transferred to observable canonical form by the transformation

$$X = PZ ,$$

where Z is the new state vector and P is the transformation matrix.

$$\dot{Z} = P^{-1} \dot{X} = P^{-1}[AX + Bu]$$

$$= [P^{-1} AP]Z + [P^{-1}B]u = A_Z \ Z + B_Z \ u$$

$$y = CX = (CP)Z = C_Z Z$$

Consider the characteristic equation of the given system given by

$$s^3 + a_2 s^2 + a_1 s + a_0 = 0$$

The system matrix in controllable canonical form is

$$A_Z = \begin{bmatrix} -a_2 & 1 & 0 \\ -a_1 & 0 & 1 \\ -a_0 & 0 & 0 \end{bmatrix} \quad \text{and} \quad C_Z = \begin{bmatrix} 1 & 0 & 0 \end{bmatrix}$$

Let L_Z be the error feedback gain matrix for the system in observable canonical form.

Let $$L_Z = \begin{bmatrix} l_2 \\ l_1 \\ l_0 \end{bmatrix}$$

Then the system matrix with error feedback is

$$A_Z - L_Z C_Z = \begin{bmatrix} -a_2 & 1 & 0 \\ -a_1 & 0 & 1 \\ -a_0 & 0 & 0 \end{bmatrix} - \begin{bmatrix} l_2 \\ l_1 \\ l_0 \end{bmatrix} \begin{bmatrix} 1 & 0 & 0 \end{bmatrix}$$

$$= \begin{bmatrix} -(a_2 + l_2) & 1 & 0 \\ -(a_1 + l_1) & 0 & 1 \\ -(a_0 + l_0) & 0 & 0 \end{bmatrix}$$

The characteristic equation with error feedback is

$$s^3 + (a_2 + l_2)s^2 + (a_1 + l_1)s + (a_0 + l_0) = 0$$

Let the desired characteristic equation be

$$s^3 + \alpha_2 s^2 + \alpha_1 s + \alpha_0 = 0$$

Comparing the characteristic equation of the given system with error feedback with the desired characteristic equation

$$a_2 + l_2 = \alpha_2 \qquad\qquad a_1 + l_1 = \alpha_1 \qquad\qquad a_0 + l_0 = \alpha_0$$

So that $\qquad l_2 = \alpha_2 - a_2 \qquad\qquad l_1 = \alpha_1 - a_1 \qquad\qquad l_0 = \alpha_0 - a_0$

and $$L_Z = \begin{bmatrix} l_2 \\ l_1 \\ l_0 \end{bmatrix} = \begin{bmatrix} \alpha_2 - a_2 \\ \alpha_1 - a_1 \\ \alpha_0 - a_0 \end{bmatrix}$$

This is to be transferred back to the given representation. The state error equation is

$$\dot{E}_X = [A - LC]E_X .$$
(6.18)

With respect to the observable canonical form, the error equation is

$$\dot{E}_Z = [P \ AP - L_Z \ CP]E_Z$$
(6.19)

$$E_X = [X - \hat{X}] = P[Z - \hat{Z}] = P E_Z .$$
(6.20)

Differentiating equation (6.20)

$$\dot{E}_X = P\dot{E}_Z .$$
(6.21)

From equations (6.19) and (6.21)

$$\dot{E}_X = P[P^{-1}AP - L_Z \ CP]E_Z = [AP - PL_Z \ CP]E_Z = [A - PL_Z \ C]PE_Z .$$
(6.22)

From equations (6.20), (6.22)

$$\dot{E}_X = [A - PL_Z C]E_X .$$
(6.23)

Comparing equations (6.18) and (6.23)

$$L = P L_Z .$$
(6.24)

Equation (6.24) is the transformation that transforms the observer gain vector L_Z in observable canonical form to L, which is the observer gain vector in the original representation.

$$P = Q^{-1} Q_Z$$

6.11 OBSERVER DESIGN PROCEDURE IN OBSERVABLE CANONICAL FORM

Step 1: Obtain the observability matrix Q of the given system and check for complete observability of the given system. Also determine Q^{-1}.

Step 2: Find the characteristic equation of the given system

$$s^n + a_{n-1}s^{n-1} + a_{n-2}s^{n-2} + ... + a_1 s + a_0 = 0 .$$

Step 3: Obtain the desired characteristic equation from the given design specifications.

$$s^n + \alpha_{n-1}s^{n-1} + \alpha_{n-2}s^{n-2} + ... + \alpha_1 s + \alpha_0 = 0 .$$

Step 4: Observer gain vector

$$L_\chi = \begin{bmatrix} (\alpha_{n-1} - a_{n-1}) \\ (\alpha_{n-2} - a_{n-2}) \\ \cdots \\ (\alpha_2 - a_2) \\ (\alpha_1 - a_1) \\ (\alpha_0 - a_0) \end{bmatrix}.$$

Step 5: From the state model in observable canonical form, find the controllability matrix Q_z.
State model in observable canonical form is

$$\dot{Z} = \begin{bmatrix} -a_{n-1} & 1 & 0 & \cdots & & 0 & 0 \\ -a_{n-2} & 0 & 1 & \cdots & & 0 & 0 \\ & & \cdots & & & & \\ -a_2 & 0 & 0 & \cdots & & 1 & 0 \\ -a_1 & 0 & 0 & \cdots & & 0 & 1 \\ -a_0 & 0 & 0 & \cdots & & 0 & 0 \end{bmatrix} Z + [B]u \qquad y = [1 \quad 0 \quad 0 \quad \cdots \quad 0 \quad 0]X$$

Step 6: Determine the observability matrix Q_z and the transformation matrix P that transforms the given state model into observable canonical form.

$$P = Q^{-1} Q_\chi.$$

Step 7: Observer gain vector for the given system

$$L = PL_\chi.$$

Example 6.15 A linear continuous time system is represented by the state model

$$\dot{X} = \begin{bmatrix} -3 & 1 \\ -2 & 0 \end{bmatrix} X + \begin{bmatrix} 1 \\ 2 \end{bmatrix} u \qquad y = [1 \quad 0]X.$$

The unit step response of the plant has to have a peak overshoot of less than 21% and a settling time of 2 s. Design an observer.

Solution:
Step 1: Check for observability
    The representation is already in observable canonical form. Hence transformation is not required.
Step 2: Characteristic equation of the given system

$$|sI - A| = \begin{vmatrix} s+3 & -1 \\ 2 & s \end{vmatrix} = 0 \quad \text{Or} \quad s^2 + 3s + 2 = 0 = s^2 + a_1 s + a_0.$$

Step 3: Desired characteristic equation from the given design specifications

Damping ratio $\delta = \dfrac{-\ln M_P}{\sqrt{\pi^2 + (\ln M_P)^2}} = \dfrac{-\ln(0.21)}{\sqrt{\pi^2 + (\ln 0.21)^2}} = 0.45$.

Choose $\delta = 0.5$.

Settling time $t_S = \dfrac{4}{\delta \omega_n} = 2$ $\qquad \omega_n = \dfrac{4}{0.5 \times 2} = 4 \,\text{rad/s}$.

Frequency of damped oscillations $\omega_d = \omega_n \sqrt{1-\delta^2} = 4\sqrt{1-(0.5)^2} = 3.464$ rad/s .

Second-order dominant closed loop poles $= -\delta \omega_n \pm j\omega_d = -2 \pm j3.464$.

For fast convergence, let the observer response speed be ten times the speed of response of the closed loop system. For this, let the observer poles be at $= -20 \pm j34.64$. The desired characteristic equation is

$$(s+20)^2 + (34.64)^2 = 0 \qquad \text{Or} \quad s^2 + 40s + 1600 = 0 = s^2 + \alpha_1 s + \alpha_0 .$$

Step 4: The observer gain vector

$$L = \begin{bmatrix} \alpha_1 - a_1 \\ \alpha_0 - a_0 \end{bmatrix} = \begin{bmatrix} 37 \\ 1598 \end{bmatrix} .$$

Example 6.16 A state space model for a linear system has

$$A = \begin{bmatrix} -10 & 1 & 0 \\ -31 & 0 & 1 \\ -30 & 0 & 0 \end{bmatrix} \qquad B = \begin{bmatrix} 2 \\ 1 \\ 0 \end{bmatrix} \qquad C = \begin{bmatrix} 1 & 0 & 0 \end{bmatrix} .$$

The closed loop poles are at $s = -1 \pm j3$. Design an observer for the system.

Solution:
Step 1: Check the observability of the system.

The representation is already in observable canonical form. Hence transformation is not required.

Step 2: Characteristic equation of the given system

$$|sI - A| = \begin{vmatrix} s+10 & -1 & 0 \\ 31 & s & -1 \\ 30 & 0 & s \end{vmatrix} = 0$$

$$s^3 + 10s^2 + 31s + 30 = 0 = s^3 + a_2 s^2 + a_1 s + a_0 . \qquad (1)$$

Step 3: Desired characteristic equation from the given design specifications

The closed loop poles are at $s = -1 \pm j3$. The observer poles are kept at $s = -10 \pm j30$, for faster response. Choose the third pole of the observer at $s = -50$. The desired characteristic equation is

$$(s+50)\{(s+10)^2 + (30)^2\} = 0$$

$$\text{or } s^3 + 70s^2 + 2000s + 50000 = 0 = s^3 + \alpha_2 s^2 + \alpha_1 s + \alpha_0 . \tag{2}$$

Step 4: The observer gain vector
    Comparing equations (1) and (2),

$$L = \begin{bmatrix} \alpha_2 - a_2 \\ \alpha_1 - a_1 \\ \alpha_0 - a_0 \end{bmatrix} = \begin{bmatrix} 60 \\ 1969 \\ 49970 \end{bmatrix}.$$

Example 6.17 Repeat Example 6.13 using transformation to observable canonical form.

Solution:

$$\dot{X} = \begin{bmatrix} -3 & 1 \\ 0 & -2 \end{bmatrix} X + \begin{bmatrix} 0 \\ 1 \end{bmatrix} u \qquad y = \begin{bmatrix} 2 & 1 \end{bmatrix} X .$$

Step 1: Check for complete observability.

$$CA = \begin{bmatrix} 2 & 1 \end{bmatrix} \begin{bmatrix} -3 & 1 \\ 0 & -2 \end{bmatrix} = \begin{bmatrix} -6 & 0 \end{bmatrix}.$$

The observability matrix is

$$Q = \begin{bmatrix} 2 & 1 \\ -6 & 0 \end{bmatrix} \qquad |Q| = 6$$

The rank of this is 2 and so the system is completely observable.

$$Q^{-1} = \frac{1}{6} \begin{bmatrix} 0 & -1 \\ 6 & 2 \end{bmatrix}.$$

Step 2: Find the characteristic equation of the given system

$$\begin{vmatrix} s+3 & -1 \\ 0 & s+2 \end{vmatrix} = 0 \qquad \text{Or } s^2 + 5s + 6 = 0 = s^2 + a_1 s + a_0 .$$

Step 3: Desired characteristic equation
    The closed loop poles are at $s = -1 \pm j2$.
    Assume that the observer response is ten times faster than the system response. Hence, the dominant pole for the observer is to be placed at $s = -10 \pm j20$. The desired characteristic equation for the observer is

$$(s+10)^2 + (20)^2 = 0 \qquad or \quad s^2 + 20s + 500 = 0 = s^2 + \alpha_1 s + \alpha_0 .$$

Step 4: Observer gain vector

$$L_z = \begin{bmatrix} (\alpha_1 - a_1) \\ (\alpha_0 - a_0) \end{bmatrix} = \begin{bmatrix} 20 - 5 \\ 500 - 6 \end{bmatrix} = \begin{bmatrix} 15 \\ 494 \end{bmatrix}.$$

Step 5: From the state model in observable canonical form, find the controllability matrix Q_z. The system matrix in observable canonical form is

$$A_Z = \begin{bmatrix} -a_1 & 1 \\ -a_0 & 0 \end{bmatrix} = \begin{bmatrix} -5 & 1 \\ -6 & 0 \end{bmatrix} \quad and \quad C_Z = \begin{bmatrix} 1 & 0 \end{bmatrix}.$$

$$C_z A_z = \begin{bmatrix} 1 & 0 \end{bmatrix} \begin{bmatrix} -5 & 1 \\ -6 & 0 \end{bmatrix} = \begin{bmatrix} -5 & 1 \end{bmatrix}.$$

The observability matrix is

$$Q_z = \begin{bmatrix} C_z \\ C_z A_z \end{bmatrix} = \begin{bmatrix} 1 & 0 \\ -5 & 1 \end{bmatrix}.$$

Step 6: Determine the transformation matrix P that transforms the given state model into observable canonical form.

$$P = Q^{-1} Q_z = \frac{1}{6} \begin{bmatrix} 0 & -1 \\ 6 & 2 \end{bmatrix} \begin{bmatrix} 1 & 0 \\ -5 & 1 \end{bmatrix} = \frac{1}{6} \begin{bmatrix} 5 & -1 \\ -4 & 2 \end{bmatrix}.$$

Step 7: Observer gain vector for the given system

$$L = P L_z = \frac{1}{6} \begin{bmatrix} 5 & -1 \\ -4 & 2 \end{bmatrix} \begin{bmatrix} 15 \\ 494 \end{bmatrix} = \begin{bmatrix} -(419/6) \\ (464/3) \end{bmatrix}.$$

Example 6.18 A state model of a system has

$$A = \begin{bmatrix} 0 & 1 \\ 0 & 2 \end{bmatrix} \qquad C = \begin{bmatrix} 2 & 1 \end{bmatrix}.$$

Design an observer for the system with its poles at $s = -5 \pm j10$.

Solution:

Step 1: Check for complete observability.

$$CA = \begin{bmatrix} 2 & 1 \end{bmatrix} \begin{bmatrix} 0 & 1 \\ 0 & 2 \end{bmatrix} = \begin{bmatrix} 0 & 4 \end{bmatrix}.$$

The observability matrix is

$$Q = \begin{bmatrix} 2 & 1 \\ 0 & 4 \end{bmatrix} \qquad |Q| = 8.$$

The rank of this is 2 and so the system is completely observable.

$$Q^{-1} = \frac{1}{8} \begin{bmatrix} 4 & -1 \\ 0 & 2 \end{bmatrix}$$

Step 2: Find the characteristic equation of the given system

$$\begin{vmatrix} s & -1 \\ 0 & s-2 \end{vmatrix} = 0 \qquad \text{Or } s^2 - 2s + 0 = 0 = s^2 + a_1 s + a_0 . \qquad (1)$$

Step 3: Desired characteristic equation

The observer poles are $s = -5 \pm j10$. The desired characteristic equation for the observer is

$$(s+5)^2 + (10)^2 = 0 \qquad \text{or } s^2 + 10s + 125 = 0 = s^2 + \alpha_1 s + \alpha_0 . \qquad (2)$$

Step 4: Observer gain vector

Comparing equations (1) and (2),

$$L_\chi = \begin{bmatrix} (\alpha_1 - a_1) \\ (\alpha_0 - a_0) \end{bmatrix} = \begin{bmatrix} 10 - (-2) \\ 125 - (0) \end{bmatrix} = \begin{bmatrix} 12 \\ 125 \end{bmatrix}.$$

Step 5: From the state model in observable canonical form, find the controllability matrix Q_z.
The system matrix in observable canonical form is

$$A_\chi = \begin{bmatrix} -a_1 & 1 \\ -a_0 & 0 \end{bmatrix} = \begin{bmatrix} 2 & 1 \\ 0 & 0 \end{bmatrix} \text{ and } C_\chi = \begin{bmatrix} 1 & 0 \end{bmatrix}$$

$$C_\chi A_\chi = \begin{bmatrix} 1 & 0 \end{bmatrix} \begin{bmatrix} 2 & 1 \\ 0 & 0 \end{bmatrix} = \begin{bmatrix} 2 & 1 \end{bmatrix}.$$

The observability matrix is

$$Q_\chi = \begin{bmatrix} C_\chi \\ C_\chi A_\chi \end{bmatrix} = \begin{bmatrix} 1 & 0 \\ 2 & 1 \end{bmatrix}.$$

Step 6: Determine the transformation matrix P that transforms the given state model into observable canonical form.

$$P = Q^{-1} Q_\chi = \frac{1}{8} \begin{bmatrix} 4 & -1 \\ 0 & 2 \end{bmatrix} \begin{bmatrix} 1 & 0 \\ 2 & 1 \end{bmatrix} = \frac{1}{8} \begin{bmatrix} 2 & -1 \\ 4 & 2 \end{bmatrix}.$$

Step 7: Observer gain vector for the given system

$$L = PL = -\begin{bmatrix} 2 & 1 \\ 4 & 2 \end{bmatrix}\begin{bmatrix} 12 \\ 125 \end{bmatrix} = \begin{bmatrix} \dfrac{101}{149} \end{bmatrix}.$$

Example 6.19 A closed loop system is represented by a state model

$$\dot{X} = \begin{bmatrix} -3 & 1 & 0 \\ 0 & -2 & 1 \\ 0 & 0 & -1 \end{bmatrix}X + \begin{bmatrix} 2 \\ 1 \\ 0 \end{bmatrix}u \qquad C = [1 \quad 1 \quad 1]X.$$

The closed loop system has the dominant poles at $s = -1 \pm j2$. Design an observer the response of which is ten times faster than that of the closed loop system.

Solution:

Step 1: Check for complete observability.

$$\dot{X} = \begin{bmatrix} -3 & 1 & 0 \\ 0 & -2 & 1 \\ 0 & 0 & -1 \end{bmatrix}X + \begin{bmatrix} 2 \\ 1 \\ 0 \end{bmatrix}u \qquad C = [1 \quad 1 \quad 1]X$$

$$CA = [1 \quad 1 \quad 1]\begin{bmatrix} -3 & 1 & 0 \\ 0 & -2 & 1 \\ 0 & 0 & -1 \end{bmatrix} = [-3 \quad -1 \quad 0]$$

$$CA^2 = [-3 \quad -1 \quad 0]\begin{bmatrix} -3 & 1 & 0 \\ 0 & -2 & 1 \\ 0 & 0 & -1 \end{bmatrix} = [9 \quad -1 \quad -1].$$

The observability matrix

$$Q = \begin{bmatrix} C \\ CA \\ CA^2 \end{bmatrix} = \begin{bmatrix} 1 & 1 & 1 \\ -3 & -1 & 0 \\ 9 & -1 & -1 \end{bmatrix} \qquad |Q| = \begin{vmatrix} 1 & 1 & 1 \\ -3 & -1 & 0 \\ 9 & -1 & -1 \end{vmatrix} = 1-3+12 = 10.$$

The rank of Q is 3 and so the system is completely observable.

$$Q^{-1} = \frac{1}{10}\begin{bmatrix} +\{1\} & -\{0\} & +\{1\} \\ -\{3\} & +\{-10\} & -\{3\} \\ +\{12\} & -\{-10\} & +\{2\} \end{bmatrix} = \frac{1}{10}\begin{bmatrix} 1 & 0 & 1 \\ -3 & -10 & -3 \\ 12 & 10 & 2 \end{bmatrix}.$$

Step 2: Find the characteristic equation of the given system

$$\begin{vmatrix} s+3 & -1 & 0 \\ 0 & s+2 & -1 \\ 0 & 0 & s+1 \end{vmatrix} = 0$$

$$s^3 + 6s^2 + 11s + 6 = 0 = s^3 + a_2 s^2 + a_1 s + a_0 . \tag{1}$$

Step 3: Desired characteristic equation

Closed loop poles are at $s = -1 \pm j2$.

Observer response is to be ten times faster than the closed loop system. Hence the observer poles are to be placed at $s = -10 \pm j20$. The third pole is assumed to be at $s = -50$. The desired characteristic equation for the observer is

$$(s+50)\{(s+10)^2 + (20)^2 = 0 \qquad or \ (s+50)(s^2 + 20s + 500) = 0$$

$$s^3 + 70s^2 + 1500s + 25000 = 0 = s^3 + \alpha_2 s^2 + \alpha_1 s + \alpha_0 . \tag{2}$$

Step 4: Observer gain vector

Comparing equations (1) and (2),

$$L_Z = \begin{bmatrix} (\alpha_2 - a_2) \\ (\alpha_1 - a_1) \\ (\alpha_0 - a_0) \end{bmatrix} = \begin{bmatrix} 70 - 6 \\ 1500 - 11 \\ 25000 - 6 \end{bmatrix} = \begin{bmatrix} 64 \\ 1489 \\ 24994 \end{bmatrix} .$$

Step 5: From the state model in observable canonical form, find the observability matrix Q_z.

The system matrix in observable canonical form is

$$A_Z = \begin{bmatrix} -a_2 & 1 & 0 \\ -a_1 & 0 & 1 \\ -a_0 & 0 & 0 \end{bmatrix} = \begin{bmatrix} -6 & 1 & 0 \\ -11 & 0 & 1 \\ -6 & 0 & 0 \end{bmatrix} \ and \ C_Z = \begin{bmatrix} 1 & 0 & 0 \end{bmatrix}$$

$$C_Z A_Z = \begin{bmatrix} 1 & 0 & 0 \end{bmatrix} \begin{bmatrix} -6 & 1 & 0 \\ -11 & 0 & 1 \\ -6 & 0 & 0 \end{bmatrix} = \begin{bmatrix} -6 & 1 & 0 \end{bmatrix}$$

$$C_Z A_Z^2 = \begin{bmatrix} -6 & 1 & 0 \end{bmatrix} \begin{bmatrix} -6 & 1 & 0 \\ -11 & 0 & 1 \\ -6 & 0 & 0 \end{bmatrix} = \begin{bmatrix} 25 & -6 & 1 \end{bmatrix} .$$

The observability matrix is

$$Q_z = \begin{bmatrix} C_z \\ C_z A_z \\ C_z A_z^2 \end{bmatrix} = \begin{bmatrix} 1 & 0 & 0 \\ -6 & 1 & 0 \\ 25 & -6 & 1 \end{bmatrix}$$

Step 6: Determine the transformation matrix P that transforms the given state model into observable canonical form.

$$P = Q^{-1}Q_z = \frac{1}{10}\begin{bmatrix} 1 & 0 & 1 \\ -3 & -10 & -3 \\ 12 & 10 & 2 \end{bmatrix}\begin{bmatrix} 1 & 0 & 0 \\ -6 & 1 & 0 \\ 25 & -6 & 1 \end{bmatrix} = \begin{bmatrix} 26 & -6 & 1 \\ -18 & 8 & -3 \\ 2 & -2 & 2 \end{bmatrix}.$$

Step 7: Observer gain vector for the given system

$$L = PL_z = \frac{1}{10}\begin{bmatrix} 26 & -6 & 1 \\ -18 & 8 & -3 \\ 2 & -2 & 2 \end{bmatrix}\begin{bmatrix} 64 \\ 1489 \\ 24994 \end{bmatrix} = \begin{bmatrix} 1772.4 \\ -6422.2 \\ 4713.8 \end{bmatrix}.$$

6.12 ACKERMANN'S FORMULA FOR OBSERVER DESIGN

Consider a third-order system represented by the state model

$$\dot{X} = AX + Bu \qquad\qquad y = CX .$$

Let the desired characteristic equation for the observer be

$$s^3 + \alpha_2 s^2 + \alpha_1 s + \alpha_0 = 0 .$$

With the inclusion of an observer, the system matrix for the observer is

$$A_F = [A - LC] . \tag{6.25}$$

By Cayley–Hamilton theorem, every square matrix satisfies its own characteristic equation.

$$\phi(A_F) = A_F^3 + \alpha_2 A_F^2 + \alpha_1 A_F + \alpha_0 I = 0 \tag{6.26}$$

$$A_F^2 = [A - LC][A - LC] = [A - LC]A - [A - LC]LC$$

$$= A^2 - LCA - A_F LC . \tag{6.27}$$

$$A_F^3 = [A - LC][A - LC]^2 = [A - LC][A^2 - LCA - A_F LC]$$

$$= [A - LC]A^2 - [A - LC]LCA - [A - LC]A_F LC$$

$$= A^3 - LCA^2 - A_F LCA - A_F^2 LC . \tag{6.28}$$

Substituting equations (6.25), (6.27) and (6.28) in equation (6.26),

$$[A^3 - LCA^2 - A_F LCA - A_F^2 LC] + \alpha_2[A^2 - LCA - A_F LC] + \alpha_1[A - LC] + \alpha_0 I = 0.$$

Rearranging the terms,

$$A^3 + \alpha_2 A^2 + \alpha_1 A + \alpha_0 I = [LCA^2 + A_F LCA + A_F^2 LC] + \alpha_2 LCA + \alpha_2 A_F LC + \alpha_1 LC.$$

This can be written as

$$\phi(A) = [LCA^2 + A_F LCA + A_F^2 LC] + \alpha_2 LCA + \alpha_2 A_F LC + \alpha_1 LC$$

$$= [\alpha_2 A_F L + \alpha_1 L + A_F^2 L]C + [\alpha_2 L + A_F L]CA + LCA^2$$

$$= \left[(\alpha_2 A_F L + \alpha_1 L + A_F^2 L) \quad (\alpha_2 L + A_F L) \quad L \right] \begin{bmatrix} C \\ CA \\ CA^2 \end{bmatrix}$$

$$= \left[(\alpha_2 A_F L + \alpha_1 L + A_F^2 L) \quad (\alpha_2 L + A_F L) \quad L \right] Q,$$

where Q is the observability matrix determined from the given state model.

$$\phi(A)Q^{-1} = \left[(\alpha_2 A_F L + \alpha_1 L + A_F^2 L) \quad (\alpha_2 L + A_F L) \quad L \right].$$

From this equation,

$$\phi(A)Q^{-1} \begin{bmatrix} 0 \\ 0 \\ 1 \end{bmatrix} = \left[(\alpha_2 A_F L + \alpha_1 L + A_F^2 L) \quad (\alpha_2 L + A_F L) \quad L \right] \begin{bmatrix} 0 \\ 0 \\ 1 \end{bmatrix} = L.$$

Hence the observer gain vector

$$L = \phi(A)Q^{-1} \begin{bmatrix} 0 \\ 0 \\ 1 \end{bmatrix}.$$

This is Ackermann's formula for the observer gain vector.

Example 6.20 Repeat Example 6.13, using Ackermann's formula.

Solution:

Step 1: Check for complete observability

$$\dot{X} = \begin{bmatrix} -3 & 1 \\ 0 & -2 \end{bmatrix} X + \begin{bmatrix} 0 \\ 1 \end{bmatrix} u \qquad y = [2 \quad 1]X.$$

The observability matrix

$$Q = \begin{bmatrix} 2 & 1 \\ -6 & 0 \end{bmatrix} \qquad\qquad Q^{-1} = \frac{1}{6}\begin{bmatrix} 0 & -1 \\ 6 & 2 \end{bmatrix}.$$

The rank of Q is 2 and so the system is completely observable.

Step 2: Desired characteristic equation

The closed loop poles are at $s = -1 \pm j2$. The observer poles are placed at $s = -10 \pm j20$ for faster response. Hence, for the observer the desired characteristic equation is

$$(s+10)^2 + (20)^2 = 0 \qquad\qquad Or\ s^2 + 20s + 500 = 0 = s^2 + \alpha_1 s + \alpha_0 .$$

Step 3: Evaluation of $\phi(A)$

$$\phi(A) = A^2 + \alpha_1 A + \alpha_0 I$$

$$A^2 = \begin{bmatrix} -3 & 1 \\ 0 & -2 \end{bmatrix}\begin{bmatrix} -3 & 1 \\ 0 & -2 \end{bmatrix} = \begin{bmatrix} 9 & -5 \\ 0 & 4 \end{bmatrix}$$

$$\phi(A) = \begin{bmatrix} 9 & -5 \\ 0 & 4 \end{bmatrix} + 20\begin{bmatrix} -3 & 1 \\ 0 & -2 \end{bmatrix} + 500\begin{bmatrix} 1 & 0 \\ 0 & 1 \end{bmatrix} = \begin{bmatrix} 449 & 15 \\ 0 & 464 \end{bmatrix}.$$

Step 4: Evaluation of observer gain vector

$$L = \phi(A)Q^{-1}\begin{bmatrix} 0 \\ 1 \end{bmatrix} = \frac{1}{6}\begin{bmatrix} 449 & 15 \\ 0 & 464 \end{bmatrix}\begin{bmatrix} 0 & -1 \\ 6 & 2 \end{bmatrix}\begin{bmatrix} 0 \\ 1 \end{bmatrix} = \begin{bmatrix} -(419/6) \\ (464/3) \end{bmatrix}.$$

Example 6.21 Repeat Example 6.14, using Ackermann's formula.

Solution:

Step 1: Check for complete observability

$$\dot{X} = \begin{bmatrix} 0 & 2 \\ 0 & 2 \end{bmatrix}X + \begin{bmatrix} 0 \\ 2 \end{bmatrix}u \qquad\qquad y = [1 \quad 0]X .$$

The observability matrix

$$Q = \begin{bmatrix} 1 & 0 \\ 0 & 2 \end{bmatrix} \qquad\qquad Q^{-1} = \frac{1}{2}\begin{bmatrix} 2 & 0 \\ 0 & 1 \end{bmatrix}.$$

The rank of Q is 2 and so the system is completely observable.

Step 2: Desired characteristic equation

The observer poles are placed at $= -8, -8$. Hence for the observer the desired characteristic equation is

$$(s+8)^2 = 0 \qquad\qquad Or\ s^2 + 16s + 64 = 0 = s^2 + \alpha_1 s + \alpha_0 .$$

Step 3: Evaluation of $\phi(A)$

$$\phi(A) = A^2 + \alpha_1 A + \alpha_0 I$$

$$A^2 = \begin{bmatrix} 0 & 2 \\ 0 & 2 \end{bmatrix}\begin{bmatrix} 0 & 2 \\ 0 & 2 \end{bmatrix} = \begin{bmatrix} 0 & 4 \\ 0 & 4 \end{bmatrix}$$

$$\phi(A) = \begin{bmatrix} 0 & 4 \\ 0 & 4 \end{bmatrix} + 16\begin{bmatrix} 0 & 2 \\ 0 & 2 \end{bmatrix} + 64\begin{bmatrix} 1 & 0 \\ 0 & 1 \end{bmatrix} = \begin{bmatrix} 64 & 36 \\ 0 & 100 \end{bmatrix}.$$

Step 4: Evaluation of observer gain vector

$$L = \phi(A)Q^{-1}\begin{bmatrix} 0 \\ 1 \end{bmatrix} = \frac{1}{2}\begin{bmatrix} 64 & 36 \\ 0 & 100 \end{bmatrix}\begin{bmatrix} 2 & 0 \\ 0 & 1 \end{bmatrix}\begin{bmatrix} 0 \\ 1 \end{bmatrix} = \frac{1}{2}\begin{bmatrix} 64 & 36 \\ 0 & 100 \end{bmatrix}\begin{bmatrix} 0 \\ 1 \end{bmatrix}\begin{bmatrix} 18 \\ 50 \end{bmatrix}.$$

Example 6.21 Repeat Example 6.19, using Ackermann's formula.

Solution:

Step 1: Check for complete observability

$$\dot{X} = \begin{bmatrix} -3 & 1 & 0 \\ 0 & -2 & 1 \\ 0 & 0 & -1 \end{bmatrix}X + \begin{bmatrix} 2 \\ 1 \\ 0 \end{bmatrix}u \qquad\qquad C = \begin{bmatrix} 1 & 1 & 1 \end{bmatrix}X$$

$$CA = \begin{bmatrix} 1 & 1 & 1 \end{bmatrix}\begin{bmatrix} -3 & 1 & 0 \\ 0 & -2 & 1 \\ 0 & 0 & -1 \end{bmatrix} = \begin{bmatrix} -3 & -1 & 0 \end{bmatrix}$$

$$CA^2 = \begin{bmatrix} -3 & -1 & 0 \end{bmatrix}\begin{bmatrix} -3 & 1 & 0 \\ 0 & -2 & 1 \\ 0 & 0 & -1 \end{bmatrix} = \begin{bmatrix} 9 & -1 & -1 \end{bmatrix}.$$

The observability matrix

$$Q = \begin{bmatrix} C \\ CA \\ CA^2 \end{bmatrix} = \begin{bmatrix} 1 & 1 & 1 \\ -3 & -1 & 0 \\ 9 & -1 & -1 \end{bmatrix} \qquad |Q| = \begin{vmatrix} 1 & 1 & 1 \\ -3 & -1 & 0 \\ 9 & -1 & -1 \end{vmatrix} = 1 - 3 + 12 = 10.$$

The rank of Q is 3 and so the system is completely observable.

$$Q^{-1} = \frac{1}{10}\begin{bmatrix} +\{1\} & -\{0\} & +\{1\} \\ -\{3\} & +\{-10\} & -\{3\} \\ +\{12\} & -\{-10\} & +\{2\} \end{bmatrix} = \frac{1}{10}\begin{bmatrix} 1 & 0 & 1 \\ -3 & -10 & -3 \\ 12 & 10 & 2 \end{bmatrix}.$$

Step 2: Desired characteristic equation

Closed loop poles are at $s = -1 \pm j2$.

Observer response is to be ten times faster than the closed loop system. Hence, the observer poles are to be placed at $s = -10 \pm j20$. The third pole is assumed to be at $s = -50$. The desired characteristic equation for the observer is

$$(s+50)\{(s+10)^2 + (20)^2 = 0$$

$$s^3 + 70s^2 + 1500s + 25000 = 0 = s^3 + \alpha_2 s^2 + \alpha_1 s + \alpha_0.$$

Step 3: Evaluation of $\phi(A)$

$$\phi(A) = A^3 + \alpha_2 A^2 + \alpha_1 A + \alpha_0 I$$

$$A^2 = \begin{bmatrix} -3 & 1 & 0 \\ 0 & -2 & 1 \\ 0 & 0 & -1 \end{bmatrix} \begin{bmatrix} -3 & 1 & 0 \\ 0 & -2 & 1 \\ 0 & 0 & -1 \end{bmatrix} = \begin{bmatrix} 9 & -5 & 1 \\ 0 & 4 & -3 \\ 0 & 0 & 1 \end{bmatrix}$$

$$A^3 = \begin{bmatrix} -3 & 1 & 0 \\ 0 & -2 & 1 \\ 0 & 0 & -1 \end{bmatrix} \begin{bmatrix} 9 & -5 & 1 \\ 0 & 4 & -3 \\ 0 & 0 & 1 \end{bmatrix} = \begin{bmatrix} -27 & 19 & -6 \\ 0 & -8 & 7 \\ 0 & 0 & -1 \end{bmatrix}.$$

$$\phi(A) = \begin{bmatrix} -27 & 19 & -6 \\ 0 & -8 & 7 \\ 0 & 0 & -1 \end{bmatrix} + 70 \begin{bmatrix} 9 & -5 & 1 \\ 0 & 4 & -3 \\ 0 & 0 & 1 \end{bmatrix} + 1500 \begin{bmatrix} -3 & 1 & 0 \\ 0 & -2 & 1 \\ 0 & 0 & -1 \end{bmatrix} + 25000 \begin{bmatrix} 1 & 0 & 0 \\ 0 & 1 & 0 \\ 0 & 0 & 1 \end{bmatrix}$$

$$= \begin{bmatrix} 21103 & 1169 & 64 \\ 0 & 22272 & 1297 \\ 0 & 0 & 23569 \end{bmatrix}.$$

Step 4: Evaluation of observer gain vector

$$L = \phi(A)Q^{-1} \begin{bmatrix} 0 \\ 0 \\ 1 \end{bmatrix} = \frac{1}{10} \begin{bmatrix} 21103 & 1169 & 64 \\ 0 & 22272 & 1297 \\ 0 & 0 & 23569 \end{bmatrix} \begin{bmatrix} 1 & 0 & 1 \\ -3 & -10 & -3 \\ 12 & 10 & 2 \end{bmatrix} \begin{bmatrix} 0 \\ 0 \\ 1 \end{bmatrix}$$

$$= \frac{1}{10} \begin{bmatrix} 21103 & 1169 & 64 \\ 0 & 22272 & 1297 \\ 0 & 0 & 23569 \end{bmatrix} \begin{bmatrix} 1 \\ -3 \\ 2 \end{bmatrix} = \begin{bmatrix} 1772.4 \\ -6422.2 \\ 4713.8 \end{bmatrix}.$$

6.13 REDUCED-ORDER OBSERVER DESIGN

In cases where some state variables are available for direct measurement, a full-order observer is not required. A *reduced-order observer* estimates only the non-measurable state variables. The design of reduced-order observer is developed as outlined in the following lines.

Consider a system represented by the a model

$$\dot{X} = AX + Bu \qquad\qquad y = CX.$$

The state vector $X(t)$ is partitioned into two parts $X_1(t)$ *and* $X_2(t)$. The state vector $X_1(t)$ is the portion of the state vector that can be directly measured. The state vector $X_2(t)$ contains the state variables which are to be estimated. Hence, an observer is necessary for estimating the state variables of $X_2(t)$. This observer is thus a reduced order observer.

The partitioned state equation can be represented as

$$\begin{bmatrix} \dot{X_1}(t) \\ \cdots \\ \dot{X_2}(t) \end{bmatrix} = \begin{bmatrix} A_{11} & A_{12} \\ \cdots & \\ A_{21} & A_{22} \end{bmatrix}\begin{bmatrix} X_1(t) \\ \cdots \\ X_2(t) \end{bmatrix} + \begin{bmatrix} B_1 \\ \cdots \\ B_2 \end{bmatrix} u(t).$$

Expanding this matrix equation

$$\dot{X_1}(t) = A_{11}X_1(t) + A_{12}X_2(t) + B_1 u(t) \tag{6.29}$$

And
$$\dot{X_2}(t) = A_{21}X_1(t) + A_{22}X_2(t) + B_2 u(t). \tag{6.30}$$

Equation (6.29) can be written as

$$\dot{X_1}(t) - A_{11}X_1(t) - B_1 u(t) = A_{12}X_2(t). \tag{6.31}$$

The right side of this equation is of the form CX and can be considered as the "output equation" of the reduced-order observer.

$$y_r = A_{12} X_2 = C_r X_2. \tag{6.32}$$

Equation (6.30) can be considered as the "state equation" for the reduced-order observer. The state equation for a full-order observer with feedback error $(y - \hat{y})$ is

$$\dot{\hat{X}} = A\hat{X} + Bu + L(y - \hat{y})$$

$$= A\hat{X} + Bu + Ly - LC\hat{X}$$

$$= [A - LC]\hat{X} + Bu + Ly. \tag{6.33}$$

In line with equations (6.30) and (6.33), an equation for the reduced-order observer can be written as

$$\dot{\hat{X}}_2 = [A_{22} - L_r C_r]\hat{X}_2 + A_{21} X_1 + B_2 u + L_r y_r.$$

Using equation (6.32), this equation becomes

$$\dot{\hat{X}}_2 = [A_{22} - L_r A_{12}]\hat{X}_2 + A_{21} X_1 + B_2 u + L_r A_{12} X_2. \tag{6.34}$$

Subtracting equation (6.34) from equation (6.30), the observer error equation is

$$\dot{X}_2 - \dot{\hat{X}}_2 = [A_{21}X_1 + A_{22}X_2 + B_2 u] - \{[A_{22} - L_r A_{12}]\hat{X}_2 + A_{21} X_1 + B_2 u + L_r A_{12} X_2\}$$

$$= A_{22}X_2 - [A_{22} - L_r A_{12}]\hat{X}_2 - L_r A_{12} X_2$$

$$= A_{22}X_2 - A_{22}\hat{X}_2 + L_r A_{12}\hat{X}_2 - L_r A_{12} X_2$$

$$= A_{22}[X_2 - \hat{X}_2] - L_r A_{12}[X_2 - \hat{X}_2]$$

$$= [A_{22} - L_r A_{12}][X_2 - \hat{X}_2] \tag{6.35}$$

If $E_{Xr} = X_2 - \hat{X}_2$ is the observer error, then Equation (6.35) becomes

$$\dot{E}_{Xr} = [A_{22} - L_r A_{12}]E_{Xr}. \tag{6.36}$$

From this error equation for the reduced-order observer, the characteristic equation can be written as

$$|sI - [A_{22} - L_r A_{12}]| = 0. \tag{6.37}$$

Comparing the equation (6.37) with the desired characteristic equation, the reduced-order observer gain vector L_r can be determined.

6.14 DESIGN PROCEDURE FOR REDUCED-ORDER OBSERVER

Step 1: Check whether the system is completely observable.

Step 2: From the given desired Eigen values, obtain the characteristic equation.

Step 3: Partition the state vector $X(t)$ into measurable states X_1 and un-measurable states X_2.

$$\begin{bmatrix} \dot{X}_1(t) \\ \cdots \\ \dot{X}_2(t) \end{bmatrix} = \begin{bmatrix} A_{11} & A_{12} \\ \cdots & \\ A_{21} & A_{22} \end{bmatrix} \begin{bmatrix} X_1(t) \\ \cdots \\ X_2(t) \end{bmatrix} + \begin{bmatrix} B_1 \\ B_2 \end{bmatrix} u(t).$$

Determine the matrices A_{12} and A_{22} and the characteristic equation

$$\left| sI - (A_{22} - L_r A_{12}) \right| = 0.$$

Step 4: Compare the coefficients of like powers of s in characteristic equations in steps 2 and 3 to evaluate the elements of the reduced-order observer vector L_r.

Example 6.22 A linear system is represented by a state equation

$$\dot{X} = \begin{bmatrix} 0 & 1 & 0 \\ 0 & 0 & 1 \\ -6 & -11 & -6 \end{bmatrix} X + \begin{bmatrix} 0 \\ 0 \\ 1 \end{bmatrix} u \qquad\qquad y = \begin{bmatrix} 1 & 0 & 0 \end{bmatrix} X.$$

Assume that the output can be accurately measured. Design a reduced order observer with its Eigen values at $s = -4 \pm j5$.

Solution: It is given that the output is accurately measurable. Since $y = x_1$, the state variable x_1 is not to be estimated. Partition the given state equation to separate x_1.

$$\begin{bmatrix} \dot{X}_1 \\ -- \\ \dot{X}_2 \end{bmatrix} = \begin{bmatrix} 0 & \vdots & 1 & 0 \\ --&\vdots& -- & -- \\ 0 & \vdots & 0 & 1 \\ -6 & \vdots & -11 & -6 \end{bmatrix} \begin{bmatrix} X_1 \\ -- \\ X_2 \end{bmatrix} + \begin{bmatrix} 0 \\ - \\ 0 \\ 1 \end{bmatrix} u.$$

$$A_{11} = 0 \qquad\qquad A_{12} = \begin{bmatrix} 1 & 0 \end{bmatrix} \qquad\qquad A_{21} = \begin{bmatrix} 0 \\ -6 \end{bmatrix} \qquad\qquad A_{22} = \begin{bmatrix} 0 & 1 \\ -11 & -6 \end{bmatrix}$$

$$[A_{22} - L_r A_{12}] = \begin{bmatrix} 0 & 1 \\ -11 & -6 \end{bmatrix} - \begin{bmatrix} l_1 \\ l_0 \end{bmatrix} \begin{bmatrix} 1 & 0 \end{bmatrix} = \begin{bmatrix} -l_1 & 1 \\ -(11+l_0) & -6 \end{bmatrix}.$$

The characteristic equation for the reduced-order observer is

$$\left| sI - (A_{22} - L_r A_{12}) \right| = 0 \qquad\qquad \begin{vmatrix} s+l_1 & -1 \\ (11+l_0) & s+6 \end{vmatrix} = 0$$

$$\text{Or } s^2 + (6+l_1)s + (6l_1 + l_0 + 11) = 0. \tag{1}$$

The desired Eigen values for the reduced observer is at $s = -4 \pm j5$.

The desired characteristic equation is

$$(s+4)^2 + (5)^2 = 0 \quad \text{Or} \quad s^2 + 8s + 41 = 0 . \tag{2}$$

Comparing equations (1) and (2),

$$6 + l_1 = 8 \qquad\qquad 6l_1 + l_0 + 11 = 41.$$

Solving these equations, $l_1 = 2$ and $l_0 = 18$.

Hence the reduced order observer gain vector is

$$L_r = \begin{bmatrix} 2 \\ 18 \end{bmatrix} .$$

Example 6.23 A linear system is represented by a state equation

$$\dot{X} = \begin{bmatrix} -1 & 1 & 0 \\ 0 & -2 & 1 \\ 0 & 0 & -2 \end{bmatrix} X + \begin{bmatrix} 1 \\ 0 \\ 1 \end{bmatrix} u \qquad\qquad y = \begin{bmatrix} 0 & 1 & 0 \end{bmatrix} X .$$

If the state variable x_2 is measurable, design a reduced-order observer so that the observer poles are placed at $s = -5 \pm j10$.

Solution: The given state equations are

$$\dot{x}_1 = -x_1 + x_2 + (0)x_3 + u$$

$$\dot{x}_2 = (0)x_1 - 2x_2 + x_3 + (0)u$$

$$\dot{x}_3 = (0)x_1 + (0)x_2 - 2x_3 + u .$$

The state variable x_2 is measurable. Rearranging these equations, by writing the equation for $\dot{x}_2$ first

$$\dot{x}_2 = (0)x_1 - 2x_2 + x_3 + (0)u$$

$$\dot{x}_1 = -x_1 + x_2 + (0)x_3 + u$$

$$\dot{x}_3 = (0)x_1 + (0)x_2 - 2x_3 + u .$$

Keeping these equations in matrix form

$$\begin{bmatrix} \dot{x_2} \\ -- \\ \dot{x_1} \\ \dot{x_3} \end{bmatrix} = \left[\begin{array}{c:cc} 0 & -2 & 1 \\ \hdashline -1 & 1 & 0 \\ 0 & 0 & -2 \end{array} \right] \begin{bmatrix} x_2 \\ -- \\ x_1 \\ x_3 \end{bmatrix} + \begin{bmatrix} 0 \\ -- \\ 1 \\ 1 \end{bmatrix} u$$

$$A_{11} = 0 \qquad A_{12} = [-2 \quad 1] \qquad A_{21} = \begin{bmatrix} -1 \\ 0 \end{bmatrix} \qquad A_{22} = \begin{bmatrix} 1 & 0 \\ 0 & -2 \end{bmatrix}$$

$$[A_{22} - L_r A_{12}] = \begin{bmatrix} 1 & 0 \\ 0 & -2 \end{bmatrix} - \begin{bmatrix} l_1 \\ l_0 \end{bmatrix}[-2 \quad 1] = \begin{bmatrix} 1 + 2l_1 & -l_1 \\ 2l_0 & -(2 + l_0) \end{bmatrix}.$$

The characteristic equation for the reduced-order observer is

$$\left| sI - (A_{22} - L_r A_{12}) \right| = 0 \qquad \begin{vmatrix} s - (1 + 2l_1) & l_1 \\ -2l_0 & s + 2 + l_0 \end{vmatrix} = 0$$

Or $s^2 + (1 + l_0 - 2l_1)s - (2 + l_0 + 4l_1) = 0.$ (1)

The desired Eigen values for the reduced observer is at $s = -5 \pm j10$.
The desired characteristic equation is

$$(s+5)^2 + (10)^2 = 0 \qquad\qquad Or \quad s^2 + 10s + 125 = 0. \tag{2}$$

Comparing equations (1) and (2),

$$1 + l_0 - 2l_1 = 10 \qquad\qquad 2 + l_0 + 4l_1 = -125.$$

Solving these equations, $l_1 = -(68/3)$ and $l_0 = (-109/3)$.
Hence, the reduced-order observer gain vector is

$$L_r = \begin{bmatrix} -(68/3) \\ -(109/3) \end{bmatrix}.$$

6.15 ACKERMANN'S FORMULA FOR REDUCED-ORDER OBSERVER DESIGN

For a full-order observer, the error equation is

$$\dot{E}_X = (\dot{X} - \dot{\hat{X}}) = [A - LC]E_X \qquad\qquad \text{and } (y - \hat{y}) = CE_X.$$

The Ackermann's formula for the full-order observer gain matrix is

$$L = \phi(A)Q^{-1}\begin{bmatrix} 0 \\ 0 \\ \dots \\ 0 \\ 1 \end{bmatrix},$$

where $\phi(A) = A^n + \alpha_{n-1}A^{n-1} + \alpha_{n-2}A^{n-2} + \dots + \alpha_1 A + \alpha_0 I.$

And $Q = \begin{bmatrix} C \\ CA \\ CA^2 \\ \dots \\ CA^{n-1} \end{bmatrix}.$

For a reduced-order observer, the error equation is

$$\dot{E}_{Xr} = [A_{22} - L_r\ A_{12}]E_{Xr}.$$

Hence the Ackermann's formula for a reduced-order observer gain matrix can be written as

$$L_r = \phi(A_{22})Q_r^{-1}\begin{bmatrix} 0 \\ 0 \\ \dots \\ 0 \\ 1 \end{bmatrix}, \tag{6.38}$$

where $\phi(A_{22}) = A_{22}^n + \alpha_{n-1}A_{22}^{n-1} + \alpha_{n-2}A_{22}^{n-2} + \dots + \alpha_1 A_{22} + \alpha_0 I.$ \hfill (6.39)

$$Q_r = \begin{bmatrix} A_{12} \\ A_{12}A_{22} \\ A_{12}A_{22}^2 \\ \dots \\ A_{12}A_{22}^{n-2} \end{bmatrix}. \tag{6.40}$$

Example 6.24 Repeat Example 6.22, using Ackermann's formula.

Solution:

$$\dot{X} = \begin{bmatrix} 0 & 1 & 0 \\ 0 & 0 & 1 \\ -6 & -11 & -6 \end{bmatrix}X + \begin{bmatrix} 0 \\ 0 \\ 1 \end{bmatrix}u \qquad y = [1 \quad 0 \quad 0]X.$$

Solution: It is given that the output is accurately measurable. Since $y = x_1$, the state variable x_1 is not to be estimated. Partition the given state equation to separate x_1.

$$\left[\begin{array}{c} \dot{X}_1 \\ -- \\ \dot{X}_2 \end{array}\right] = \left[\begin{array}{ccc} 0 & 1 & 0 \\ ---+------- \\ 0 & 0 & 1 \\ -6 & -11 & -6 \end{array}\right]\left[\begin{array}{c} X_1 \\ -- \\ X_2 \end{array}\right] + \left[\begin{array}{c} 0 \\ - \\ 0 \\ 1 \end{array}\right]u$$

$$A_{11} = 0 \quad A_{12} = [1 \quad 0] \quad A_{21} = \left[\begin{array}{c} 0 \\ -6 \end{array}\right] \quad A_{22} = \left[\begin{array}{cc} 0 & 1 \\ -11 & -6 \end{array}\right]$$

$$A_{12}A_{22} = [1 \quad 0]\left[\begin{array}{cc} 0 & 1 \\ -11 & -6 \end{array}\right] = [0 \quad 1] \; .$$

$$Q_r = \left[\begin{array}{cc} 1 & 0 \\ 0 & 1 \end{array}\right] \qquad Q_r^{-1} = \left[\begin{array}{cc} 1 & 0 \\ 0 & 1 \end{array}\right] \; .$$

The reduced-order observer has its Eigen values at $s = -4 \pm j5$ and the desired characteristic equation is

$$(s+4)^2 + (5)^2 = 0 \qquad Or \quad s^2 + 8s + 41 = 0 = s^2 + \alpha_1 s + \alpha_0 \; .$$

$$A_{22}^2 = \left[\begin{array}{cc} 0 & 1 \\ -11 & -6 \end{array}\right]\left[\begin{array}{cc} 0 & 1 \\ -11 & -6 \end{array}\right] = \left[\begin{array}{cc} -11 & -6 \\ 66 & 25 \end{array}\right]$$

$$\phi(A_{22}) = A_{22}^2 + \alpha_1 A_{22} + \alpha_0 I$$

$$= \left[\begin{array}{cc} -11 & -6 \\ 66 & 25 \end{array}\right] + 8\left[\begin{array}{cc} 0 & 1 \\ -11 & -6 \end{array}\right] + 41\left[\begin{array}{cc} 1 & 0 \\ 0 & 1 \end{array}\right] = \left[\begin{array}{cc} 30 & 2 \\ -22 & 18 \end{array}\right]$$

$$L_r = \phi(A_{22})Q_r^{-1}\left[\begin{array}{c} 0 \\ 1 \end{array}\right] = \left[\begin{array}{cc} 30 & 2 \\ -22 & 18 \end{array}\right]\left[\begin{array}{cc} 1 & 0 \\ 0 & 1 \end{array}\right]\left[\begin{array}{c} 0 \\ 1 \end{array}\right] = \left[\begin{array}{c} 2 \\ 18 \end{array}\right] \; .$$

Example 6.25 Repeat Example 6.23, using Ackermann's formula.

Solution: From example 6.23

$$A_{11} = 0 \qquad A_{12} = [-2 \quad 1] \qquad A_{21} = \left[\begin{array}{c} -1 \\ 0 \end{array}\right] \qquad A_{22} = \left[\begin{array}{cc} 1 & 0 \\ 0 & -2 \end{array}\right]$$

$$A_{12}A_{22} = [-2 \quad 1]\begin{bmatrix} 1 & 0 \\ 0 & -2 \end{bmatrix} = [-2 \quad -2]$$

$$Q_r = \begin{bmatrix} -2 & 1 \\ -2 & -2 \end{bmatrix} \qquad Q_r^{-1} = \frac{1}{6}\begin{bmatrix} -2 & -1 \\ 2 & -2 \end{bmatrix}.$$

The reduced-order observer has its Eigen values at $s = -4 \pm j5$ and the desired characteristic equation is $(s+5)^2 + (10)^2 = 0$ Or $s^2 + 10s + 125 = 0 = s^2 + \alpha_1 s + \alpha_0$.

$$A_{22}^2 = \begin{bmatrix} 1 & 0 \\ 0 & -2 \end{bmatrix}\begin{bmatrix} 1 & 0 \\ 0 & -2 \end{bmatrix} = \begin{bmatrix} 1 & 0 \\ 0 & 4 \end{bmatrix}.$$

$$\phi(A_{22}) = A_{22}^2 + \alpha_1 A_{22} + \alpha_0 I$$

$$= \begin{bmatrix} 1 & 0 \\ 0 & 4 \end{bmatrix} + 10\begin{bmatrix} 1 & 0 \\ 0 & -2 \end{bmatrix} + 125\begin{bmatrix} 1 & 0 \\ 0 & 1 \end{bmatrix} = \begin{bmatrix} 136 & 0 \\ 0 & 109 \end{bmatrix}.$$

$$L_r = \phi(A_{22})Q_r^{-1}\begin{bmatrix} 0 \\ 1 \end{bmatrix} = \frac{1}{6}\begin{bmatrix} 136 & 0 \\ 0 & 109 \end{bmatrix}\begin{bmatrix} -2 & -1 \\ 2 & -2 \end{bmatrix}\begin{bmatrix} 0 \\ 1 \end{bmatrix} = \begin{bmatrix} -(68/3) \\ -(109/3) \end{bmatrix}.$$

SUMMARY

- Pole placement design of a closed loop control system is based on shifting the closed loop poles of the system from the present location in the s-plane to another, using state variable feedback.
- One of the design objectives is to achieve a desired transient response for the system.
- For pole placement design, it is necessary that the system represented by a given state model is completely controllable.
- The state variable feedback gain matrix is obtained by comparing the coefficients of like powers in the characteristic equation of the system with state feedback, with the characteristic equation formed out of the desired closed loop poles.
- A convenient way of pole placement design is to transform the given state model into controllable canonical form.
- The feedback gain matrix can also be determined by using Ackermann's formula.
- A state observer is a system, usually a digital computer, that estimates the values of state variables from the knowledge about the output and the input.
- State observers are generally used to estimate the values of the state variables when they are not available for measurement or not economical for making the measurement.
- If an observer estimates all the state variables, including the ones which are measurable, it is called a full-order observer.

- If an observer estimates only those state variables which are not available for measurement, then it is called a reduced-order observer.
- For an observer design, it is necessary that the system represented by the state model is completely observable.
- For the sake of convenience of observer design, the given state model is transformed into observable canonical form.
- The observer response is to be much faster than the system response. For this, the dominant poles of the observer are placed in the left half of the s-plane nearly 10 times farther away from the imaginary axis than the closed loop poles of the given system.

PRACTICE PROBLEMS

PP 6.1 A linear plant is represented by a state model

$$\dot{X} = \begin{bmatrix} 0 & 1 \\ -8 & -6 \end{bmatrix} X + \begin{bmatrix} 1 \\ -1 \end{bmatrix} u \qquad y = [3 \quad -1]X.$$

Design a feedback controller gain matrix that places the Eigen values at $s = -3 \pm j3$. Use the method of matching coefficients

PP 6.2 Repeat PP6.1 by the method applicable to state model in controllable canonical form.

PP 6.3 Repeat PP6.1 by using Ackermann's formula.

PP 6.4 A linear plant is represented by the state model

$$\dot{X} = \begin{bmatrix} 0 & 1 \\ -2 & -3 \end{bmatrix} X + \begin{bmatrix} 0 \\ 2 \end{bmatrix} u \qquad y = [1 \quad 0]X.$$

Design a feedback controller gain matrix that places the closed loop poles at $s = -3$ and $s = -5$. Use the method of matching coefficients.

PP 6.5 Repeat PP6.4 using the method applicable to state model in controllable canonical form.

PP 6.6 Repeat PP6.4 by using Ackermann's formula.

PP 6.7 A linear closed loop system is represented by the state model

$$\dot{X} = \begin{bmatrix} 0 & 1 \\ -4 & -4 \end{bmatrix} X + \begin{bmatrix} 0 \\ 1 \end{bmatrix} u \qquad y = [2 \quad 1]X.$$

Design a feedback controller gain matrix that places the closed loop poles at $s = -3 \pm j5$. Use the method of matching coefficients.

PP 6.8 Repeat PP6.7 using the method applicable to controllable canonical form of state model.

PP 6.9 Repeat PP6.7 by using Ackermann's formula.

PP 6.10 Design a linear state feedback controller to yield unit step response with a peak overshoot less than 20% and a settling time of 4 s, for a system with characteristic equation

$$s^3 + 8s^2 + 17s + 10 = 0.$$

Use the controllable canonical form of state model.

PP 6.11 Verify the result of PP6.10 by using Ackermann's formula.

PP 6.12 Design a feedback controller for a plant, the closed loop transfer function of which is

$$\frac{Y(s)}{U(s)} = \frac{5(s+10)}{s(s+3)(s+12)}.$$

The peak overshoot is less than 5% and the peak time is 0.32 s. The third pole is at $s = -50$. Use the controllable canonical form of the state model.

PP 6.13 Repeat PP6.12 using Ackermann's formula.

PP 6.14 A linear plant is represented by a state model

$$\dot{X} = \begin{bmatrix} -3 & 2 \\ 4 & -5 \end{bmatrix} X + \begin{bmatrix} 1 \\ 0 \end{bmatrix} u \qquad\qquad y = [1 \qquad 1]X.$$

Design a control law $u = -KX$ that places the closed loop poles at $s = -4$ and $s = -7$. Use the method of matching coefficients.

PP 6.15 Repeat PP6.14 using transformation to controllable canonical form.

PP 6.16 Repeat PP6.14 using Ackermann's formula.

PP 6.17 A linear plant is represented by the state model

$$\dot{X} = \begin{bmatrix} -5 & 1 & 0 \\ 0 & -2 & 1 \\ 0 & 0 & -1 \end{bmatrix} X + \begin{bmatrix} 0 \\ 0 \\ 1 \end{bmatrix} u \qquad\qquad y = [-5 \qquad 5 \qquad 0]X.$$

Design a state feedback controller that places the dominant poles at $s = -1 \pm j2$. Choose the third pole at $s = -5$. Use the method of matching coefficients.

PP 6.18 Repeat PP6.17, by using transformation to controllable canonical form.

PP 6.19 Repeat PP6.17 using Ackermann's formula.

PP 6.20 The transfer function of a linear system is

$$\frac{Y(s)}{U(s)} = \frac{5s^2 + 3s + 10}{(s+2)(s+3)(s+6)}.$$

Represent the system in controllable canonical form and hence design a feedback controller to place the system poles at $s = -2 \pm j6$. Choose the third pole at $s = -10$.

PP 6.21 A linear plant is represented by a state model

$$\dot{X} = \begin{bmatrix} -2 & 1 \\ 0 & -3 \end{bmatrix} X + \begin{bmatrix} 1 \\ 2 \end{bmatrix} u \qquad\qquad y = [1 \qquad 0]X.$$

(a) Design a feedback controller gain matrix that places the Eigen values at $s = -1 \pm j2$.

(b) Design a full-order observer with its poles placed at $s = -5$ and $s = -10$.

PP 6.22 Repeat PP6.21 using Ackermann's formula.

PP 6.23 A state model of a system is

$$\dot{X} = \begin{bmatrix} -5 & 1 \\ -6 & 0 \end{bmatrix} X + \begin{bmatrix} 1 \\ 2 \end{bmatrix} u \qquad\qquad y = [1 \qquad 0] X.$$

Design a full-order observer with its poles at $s = -5 \pm j5$. Verify the result using Ackermann's formula.

PP 6.24 A closed loop system is represented by a state model

$$\dot{X} = \begin{bmatrix} 0 & 1 & 0 \\ 0 & 0 & 1 \\ -15 & -23 & -9 \end{bmatrix} X + \begin{bmatrix} 0 \\ 0 \\ 2 \end{bmatrix} u \qquad\qquad y = [1 \qquad 0 \qquad 0] X.$$

Design a state feedback controller that places the Eigen values at $s = -1 \pm j3$ and at $s = -5$.

PP 6.25 Repeat PP6.24 using Ackermann's formula.

PP 6.26 For the system of PP6.24, design a full-order observer with its response ten times faster than that of the closed loop system.

PP 6.27 For the system of PP6.24, design a reduced-order observer assuming that the output is accurately measurable. The observer poles are to be at $s = -10 \pm j30$.

PP 6.28 A state space model of a system is

$$A = \begin{bmatrix} -4 & 1 \\ -4 & 0 \end{bmatrix} \qquad\qquad B = \begin{bmatrix} 2 \\ 1 \end{bmatrix} \quad C = [1 \qquad 0].$$

The system response has a damping ratio of 0.5 and a natural frequency of oscillation of 4 rad/s. Design a state feedback controller with its speed of response five times faster than that of the closed loop system. Verify the result using Ackermann's formula.

PP 6.29 A linear system is represented by a state model

$$\dot{X} = \begin{bmatrix} -3 & 0 & 1 \\ 1 & -2 & 0 \\ 0 & 0 & -1 \end{bmatrix} X + \begin{bmatrix} 1 \\ 1 \\ 1 \end{bmatrix} u \qquad\qquad y = [0 \qquad 1 \qquad 0] X.$$

Design a feedback controller that places the system poles at $s = -1 \pm j2$ and at $s = -5$. Use the method of transformation to controllable canonical form.

PP 6.30 Repeat PP6.29 using Ackermann's formula.

PP 6.31 For PP6.29, design a full-order observer with its poles at $s = -10 \pm j20$ and at $s = -50$.

PP 6.32 For the system of PP6.29, design a reduced-order observer. The observer poles are to be at $s = -10 \pm j20$. Assume that the state variable x_2 is accurately measurable.

PP 6.33 For a closed loop system represented by the state equation

$$\dot{X} = \begin{bmatrix} 0 & 1 \\ -2 & -3 \end{bmatrix} X + \begin{bmatrix} 0 \\ 1 \end{bmatrix} u,$$

the feedback control law is $u = -[11 \qquad 1]$. Determine the closed loop poles.

PP 6.34 For a closed loop system represented by the state equation

$$\dot{X} = \begin{bmatrix} 0 & 1 & 0 \\ 0 & 0 & 1 \\ -6 & -11 & -6 \end{bmatrix} X + \begin{bmatrix} 0 \\ 0 \\ 1 \end{bmatrix} u,$$

the feedback control law is $u = -[19 \quad 4 \quad 1]$. Determine the closed loop poles.

PP 6.35 For a closed loop system represented by the state model

$$\dot{X} = \begin{bmatrix} 0 & 2 \\ 0 & 2 \end{bmatrix} X + \begin{bmatrix} 0 \\ 1 \end{bmatrix} u \qquad\qquad y = [1 \quad -2]X,$$

find the feedback control law that places the closed loop poles at $s = -3$ and $s = -4$. Verify the result using Ackermann's formula.

PP 6.36 For a closed loop system represented by the state model

$$\dot{X} = \begin{bmatrix} -2 & 1 & 0 \\ 0 & -2 & 1 \\ 0 & 0 & -2 \end{bmatrix} X + \begin{bmatrix} 0 \\ 0 \\ 1 \end{bmatrix} u \qquad\qquad y = [1 \quad 2 \quad 1]X,$$

find the feedback control law that places the closed loop poles at $s = -1, -1$ and -1.

PP 6.37 Repeat PP6.36, if the closed loop poles are to be placed at $s = -2 \pm j1$ and $s = -10$.

PP 6.38 For the system of PP6.35, design a full-order observer with its poles at $s = -8, -8$.

PP 6.39 A linear closed loop system is represented by a state equation

$$\dot{X} = \begin{bmatrix} -3 & 1 \\ -2 & 0 \end{bmatrix} X + \begin{bmatrix} 1 \\ 2 \end{bmatrix} u \qquad\qquad y = [1 \quad 0]X.$$

The system response has a peak overshoot of about 20% and a settling time of 2 s. Design a full-order observer for the system. The observer response is to be five times faster than that of the closed loop system.

PP 6.40 A closed loop system is represented by a state model

$$\dot{X} = \begin{bmatrix} -10 & 1 & 0 \\ -31 & 0 & 1 \\ -30 & 0 & 0 \end{bmatrix} X + \begin{bmatrix} 2 \\ 1 \\ 0 \end{bmatrix} u \qquad\qquad y = [1 \quad 0 \quad 0]X.$$

Design a full-order observer with its poles at $s = -10 \pm j10$ and $s = -10s$.

PP 6.41 For the system of PP6.40, design a reduced-order observer with its poles at $s = -10 \pm j10$. Assume that the output is accurately measurable.

REVIEW QUESTIONS

6.1 Explain what is meant by pole placement design.

6.2 Discuss the principle of pole placement design using the state variable feedback.

6.3 State the necessary condition for carrying out the pole placement design.

6.4 Mention the steps involved in the method of matching coefficients in pole placement design.

6.5 Discuss the steps involved in pole placement design by transforming the given state model to controllable canonical form.

6.6 Explain the method of pole placement design using Ackermann's formula.

6.7 What is a state observer? Explain.

6.8 Bring out the difference between a full-order observer and a reduced-order observer.

6.9 Discuss the principle of full-order observer design and derive the design equation in terms of error.

6.10 Mention the steps involved in the method of matching coefficients for a full-order observer design.

6.11 Discuss the method of full-order observer design by transforming the given state model into observable canonical form.

6.12 Discuss the method of full-order observer design by using Ackermann's formula.

6.13 Derive the design equation for a reduced-order observer.

6.14 Discuss the design procedure for a reduced-order observer.

6.15 Discuss the method of reduced-order observer design using Ackermann's formula.

6.16 For a third order system, derive the Ackermann's formula for the design of a state feedback controller.

6.17 For a third order system, derive the Ackermann's formula for the design of a state observer.

7 Describing Function Analysis

7.1 NONLINEAR SYSTEMS

Systems in nature are generally nonlinear. They have nonlinear characteristics. In a physical system, relationships among physical quantities are not linear. The voltage–current characteristics of electronic devices are nonlinear. The displacement of a mechanical spring when subjected to a force is nonlinear, but can be considered to be linear up to a certain limit. A physical system which is considered as linear is truly linear only within certain limited operating ranges. Many physical systems are represented by nonlinear models. The study of a system based on a nonlinear model is sometimes extremely difficult. Procedures for finding solution to problems based on mathematical nonlinear models can be extremely complicated and difficult to be solved. An exact solution may be seldom achieved. Even if a result is obtained, it may be applicable only to that particular case and may not be generalized.

In most control systems, the normal operation of the system is around an operating point. In such cases, a nonlinear system can be approximated to a linear system around that operating point for the purpose of study and analysis. Nonlinearities, usually encountered in physical systems, are saturation, dead zone, hysteresis, Coulomb friction, nonlinear spring, backlash in gears, etc.

7.2 PROPERTIES OF NONLINEAR SYSTEMS

Most of the theories and methods of analysis applicable to linear systems and their results do not apply to nonlinear systems.

1. Nonlinear systems do not obey the principle of superposition. The response due to an input $k_1 u(t)$ is not just k_1 times the response due to $u(t)$. Similarly response due to an initial condition $k_2 x(t_0)$ is not just k_2 times the response due to the initial condition $x(t_0)$. If the response of a nonlinear system due to an individual input or due to an initial condition is known, then based on this, the response of the system due to many other initial conditions and due to several simultaneous inputs cannot be predicted. The natures of responses due to

different magnitudes of the same input or due to different initial conditions are entirely different and they do not have any similarity from one another.

2. In a linear system, the nature of the system response depends on the locations of the
poles and zeros of the transfer function in the s-plane. Such a correlation is invalid in the case
of a nonlinear system.

3. In a linear system, the output response is oscillatory when there is a pair of poles of the
transfer function on the $j\omega$ axis. The amplitude of the oscillation depends on the value of initial conditions. Some nonlinear systems also exhibit oscillatory response even in the absence
of external input. The amplitude of the oscillation can be independent of the initial conditions. The frequency of oscillation depends on the type of nonlinear element or the system
parameters. Such oscillations are called *self-excited oscillations* or *limit cycles*. Note that the
limit cycle is independent of any external input. A limit cycle can be assumed to be periodic
with constant magnitude and constant frequency. A limit cycle is the stability boundary for
linear and nonlinear systems. A system with stable limit cycle can exhibit sustained oscillations in its output.

4. In a linear system, a sinusoidal input produces a sinusoidal output of the same frequency as the input frequency and with a phase shift. But in a nonlinear system, a sinusoidal
input can produce not only a sinusoidal output of the same frequency but also several higher
harmonics and sub-harmonics.

5. Nonlinear systems sometimes exhibit the phenomenon of *jump resonance*. The frequency response may show a jump at a certain frequency, when the frequency is gradually
increased. When the frequency is gradually decreased, the response once again may show
a jump but at another frequency. This is shown in Figure 7.1. When the frequency is
increased, the response takes the path A, F, B, C, D with a jump at frequency ω_2. When
the frequency is gradually reduced, the response follows the path D, C, E, F, A with a
jump at frequency ω_1.

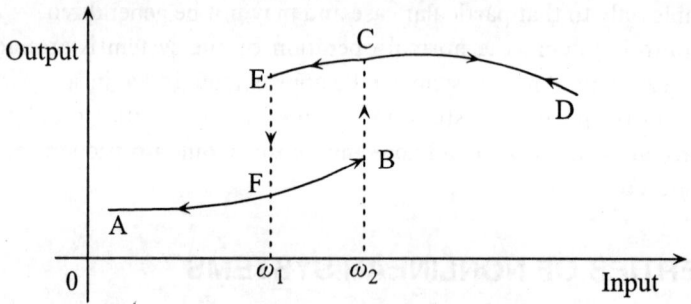

Figure 7.1 Jump resonance

6. Nonlinear systems often exhibit frequency entrainment. When a nonlinear system is
excited with a sinusoidal input of constant frequency and when the magnitude is gradually
increased from low values, at a certain point, the output frequency exactly matches the input
frequency. This is called *frequency entrainment*.

7. A linear system has only one equilibrium point, whereas nonlinear systems may have many equilibrium points or even an equilibrium zone. The system can settle down to any equilibrium point in this zone. The convergence of the system response to an equilibrium point depends on a set of initial conditions. The system response may converge to one equilibrium point for one set of initial conditions. At the same time, it can become unstable for another set of initial conditions.

7.3 CLASSIFICATION OF NONLINEAR SYSTEMS

Nonlinearities in physical systems can be classified into *inherent* and *intentional*. The nonlinearities mentioned earlier are inherent to the system and are unavoidable. They are called *inherent nonlinearities*. In certain cases, nonlinearities are intentionally or deliberately introduced in a system for the sake of convenience and simple design. They are called *intentional nonlinearities*. The insertion of nonlinearity in a system may be necessary to improve the system performance and to reduce the constructional complexities of a system. One of the simplest examples of intentionally introduced nonlinearity is a relay which has the ON–OFF operation. A very common example is a water-level control system. When the water in an overhead storage tank reaches a certain low level, a pump is switched ON to pump water into the tank and is switched OFF when the water reaches a certain higher level, and overflow is avoided.

Another classification is the multi-valued nonlinearity and single-valued nonlinearity. In multi valued nonlinearity, there can be more than one value of the output corresponding to a single input. Such a system is called a system with memory. Examples are magnetic hysteresis and backlash in gears. In single-valued nonlinearity, there is only one output corresponding to a certain value of input. This is also called a memory-less system. Saturation and ideal relays are examples of this type of nonlinearity.

7.4 PHYSICAL NONLINEARITIES

1. **Saturation:** In a system that suffers from saturation, the output initially increases linearly as the input is gradually increased. The increase is proportional to the input up to a certain point and remains constant or increases only slightly with any increase in the input. The output–input characteristic curve for both positive and negative inputs is shown in Figure 7.2. The curve is linear up to a point and droops thereafter. The output is said to reach a level of *saturation*.

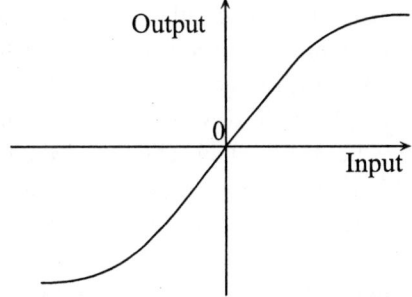

Figure 7.2 Saturation nonlinearity

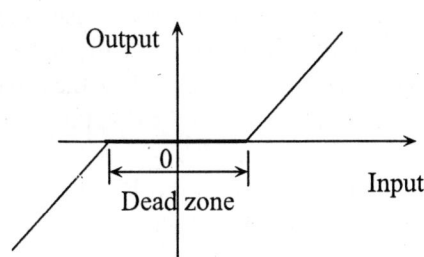

Figure 7.3 Dead-zone nonlinearity

2. Dead zone: A system for which the output remains zero for small values of input is said to have a *dead zone*. Only when the input is greater than a base value, the output starts increasing from zero value. This particular base value of the input up to which the output remains zero is called the *threshold value*. This is shown in Figure 7.3 for both positive and negative inputs. Dead zone in a system will cause steady-state error.

3. Hysteresis: In magnetic circuits, hysteresis is the property by which the flux density lags behind the magnetizing force. In general, in physical systems, if the output lags behind the input, it is said to have hysteresis. When the input is increased from negative values to positive values, the output traces one path, and when the input is increased from positive values to negative values, the output traces another path. This is shown in Figure 7.4.

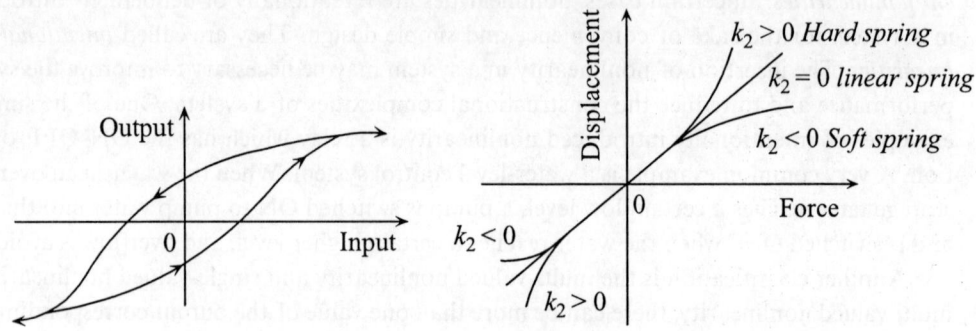

Figure 7.4 Hysteresis   **Figure 7.5** Linear and nonlinear spring

4. Nonlinear spring: A spring is a nonlinear element. Its characteristic is nonlinear. If linearized, its displacement can be accepted to be proportional to the applied force. The force–displacement relationship can be expressed as $f = k\,x$, where f is the applied force, x is the displacement and k is the spring constant. For a nonlinear spring, the force–displacement relationship can be written as

$$f = k_1 x + k_2 x^3 .$$

If k_2 is positive, the spring is called a *hard spring* and if it is negative, the spring is called a *soft spring*. The characteristics are shown in Figure 7.5.

5. Backlash in gears: In mechanical rotational systems, gears are used either to amplify the output torque with a reduction in speed or to decrease the torque with an increase in speed. The teeth on the driving gear wheel mesh with the teeth on the driven gear wheel for torque transmission. As there is a small gap between the teeth on the two gear wheels, the driven gear wheel does not truthfully follow the driving gear wheel. When the driving gear rotates through a small angle less than the gap, the driven wheel does not rotate at all. This is called *backlash in gears* and this causes nonlinear characteristics. After contact has been established between these two gear wheels, the driven gear wheel follows the rotation of the driving gear wheel in a linear fashion. Backlash in gears may cause instability in the system.

6. Viscous friction and Coulomb friction: Whenever there is a relative motion between two surfaces that are in contact, friction takes place. Coulomb friction creates frictional force acting opposite to the direction of motion. Its magnitude is constant and is independent of the relative velocity. Viscous friction also produces a force opposite to the direction of motion, but its magnitude is directly proportional to the relative velocity. The characteristics are shown in Figure 7.6(a). The combined effect of viscous and Coulomb friction is shown in figure 7(b).

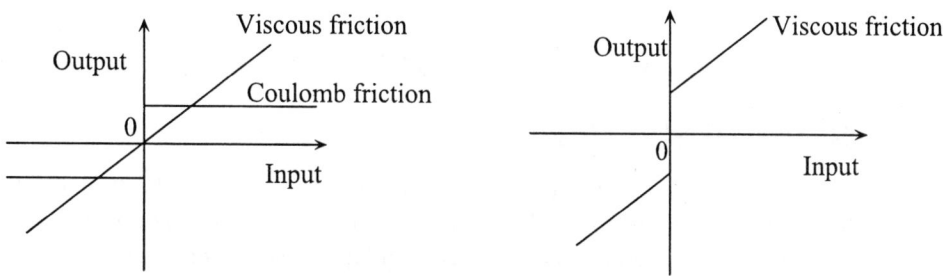

Figure 7.6 (a) Viscous and Coulomb friction.   (b) Combined viscous and Coulomb friction

7. Ideal relay: In an ideal relay, the output has only two states namely ON and OFF. Hence it is also called an ON–OFF relay. When the input is positive, the output is positive at a constant value, and when the input is negative, the output remains negative at a constant value. The characteristics for both positive and negative inputs are shown in Figure 7.7.

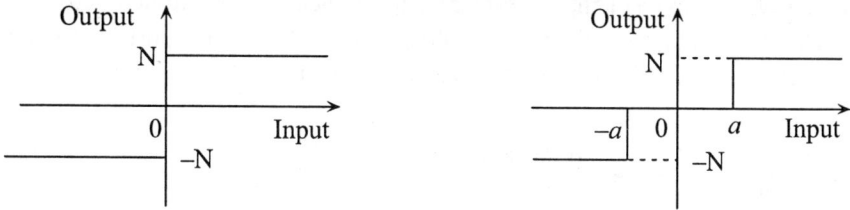

Figure 7.7 Ideal relay   **Figure 7.8** Ideal relay with dead zone

8. Ideal relay with Dead-zone: The relay switches to a constant positive output when the input is greater than the threshold value 'a'. Similarly, the relay switches a constant negative output when the negative input is greater than a negative threshold value '$-a$'. The characteristic is shown in Figure 7.8.

9. Ideal relay with hysteresis: The input–output characteristic curve is shown in Figure 7.9. Referring to this figure, when the input is greater than 'a', the output remains constant at a positive value N. When the negative input is greater than '$-a$', the output remains constant at a negative value '$-N$'. When the input increases from '$-a$' to 'a', the output remains negative and when the input decreases from 'a' to '$-a$', the output is positive. It can be seen that if the input is increased from negative values, the output traces one curve, and if the input is decreased from positive values, the output traces a different curve.

10. Ideal relay with dead-zone and hysteresis The input–output characteristic of an ideal relay with dead zone and hysteresis is shown in figure 7.10. The threshold value for positive input is 'b' and for negative input it is '$-b$'.

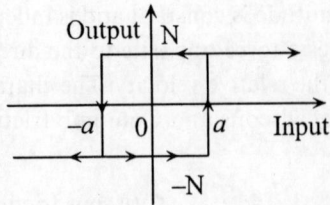

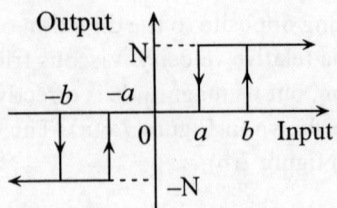

Figure 7.9 Ideal relay with hysteresis **Figure 7.10** Ideal relay with dead zone and hysteresis

7.5 DESCRIBING FUNCTION

For a nonlinear system, a sinusoidal input generally produces a non-sinusoidal output. If the output is periodic with the same period as that of the input, the output can be considered to be composed of many harmonics in addition to the fundamental. The *describing function* of a nonlinear element is defined as the ratio of the fundamental component of the output to the sinusoidal input.

Assumptions

1. Only the magnitude of the fundamental component of the output is significant and the harmonics are negligibly small.

2. The plant that follows a nonlinear element behaves as a low pass filter, filtering out the higher harmonic components.

3. The average value of the fundamental component of the output is zero.

4. The system may contain many nonlinear elements, but it is possible to combine them into a single element and can be kept preceding the plant as shown in Figure 7.11.

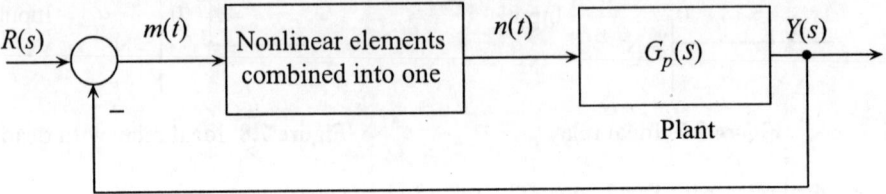

Figure 7.11 Nonlinear control system

Let the sinusoidal input to the nonlinear element be

$$m(t) = M \sin \omega t .$$

This can be expressed in the phasor form as

$$\hat{M} = M \angle 0° .$$

The output $n(t)$ is non-sinusoidal and can be expressed as

$$n(t) = \sum_{n=0}^{\infty} \{A_n \cos n\omega t + B_n \sin n\omega t\} .$$

Considering only the fundamental component and neglecting all higher harmonics,

$$n(t) = A_1 \cos \omega t + B_1 \sin \omega t,$$

where A_1 and B_1 are Fourier coefficients and can be obtained as

$$A_1 = \frac{1}{\pi} \int_0^{2\pi} n(t)\cos \omega t \, d(\omega t) \text{ and } B_1 = \frac{1}{\pi} \int_0^{2\pi} n(t)\sin \omega t \, d(\omega t).$$

Let $A_1 = N \sin \phi$ and $B_1 = N \cos \phi$.

Hence $n(t) = N(\sin \phi \cos \omega t + \cos \phi \sin \omega t)$

$$= N \sin(\omega t + \phi).$$

This can be written in the phasor form as

$$\hat{N} = N \angle \phi = N \cos \phi + jN \sin \phi = B_1 + jA_1.$$

By definition, the describing function is

$$G_N = \frac{N \angle \phi}{M \angle 0°} = \frac{B_1 + jA_1}{M}.$$

7.6 DESCRIBING FUNCTION OF CERTAIN NONLINEAR ELEMENTS

The input $m(t)$ to a nonlinear element and its output $n(t)$ are shown in Figure 7. 12.

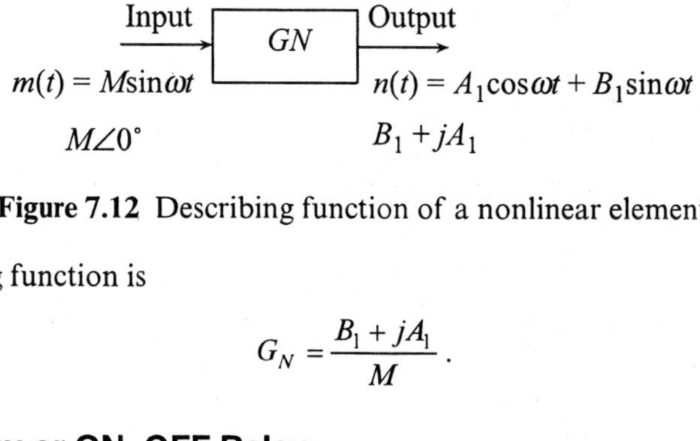

Input | GN | Output
$m(t) = M\sin\omega t$ | | $n(t) = A_1\cos\omega t + B_1\sin\omega t$
$M\angle 0°$ | | $B_1 + jA_1$

Figure 7.12 Describing function of a nonlinear element

The describing function is

$$G_N = \frac{B_1 + jA_1}{M}.$$

7.6.1 Ideal Relay or ON–OFF Relay

The characteristic of an ideal relay is shown in Figure 7.13. The sinusoidal input to it and the output wave form are also shown in the same figure. The input to the nonlinear element is

$$m(t) = M \sin \omega t.$$

The sinusoidal input is positive from $\omega t = 0$ to $\omega t = \pi$. Accordingly, the relay output is positive and constant at 'N' for the same period. From $\omega t = \pi$ to $\omega t = 2\pi$, the sinusoidal input is negative so that the relay output is constant at '$-N$'. The output is periodic.

$$n(t) = \begin{cases} N; 0 < \omega t < \pi \\ -N; \pi < \omega t < 2\pi \end{cases}.$$

By inspection of the waveform of $n(t)$, it can be seen that the average value of the output is zero. Thus $A_0 = 0$. The waveform has odd symmetry so that cosine terms are absent in the Fourier series expansion. Hence $A_1 = 0$.

$$B_1 = \frac{1}{\pi} \int_0^{2\pi} n(t) \sin \omega t \, d(\omega t)$$

$$= \frac{1}{\pi} \left[\int_0^{\pi} N \sin \theta \, d\theta + \int_{\pi}^{2\pi} -N \sin \theta \, d\theta \right]$$

$$= \frac{N}{\pi} \left[\{-\cos \theta\}_0^{\pi} - \{-\cos \theta\}_{\pi}^{2\pi} \right]$$

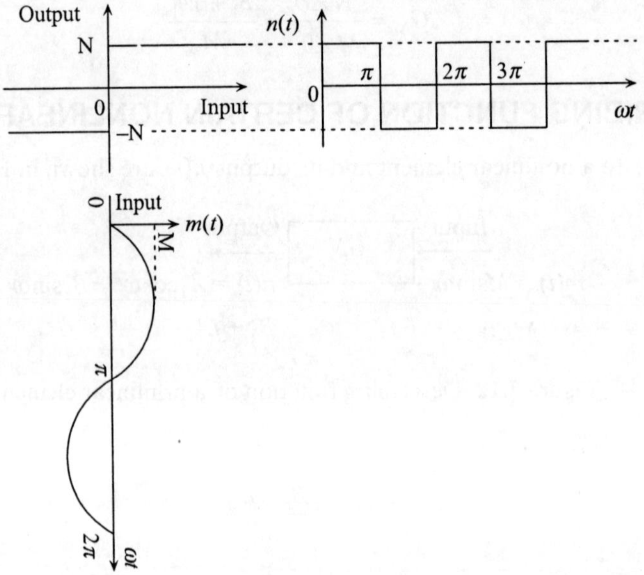

Figure 7.13 Input-output waveforms for ideal relay with sinusoidal input

$$= \frac{N}{\pi} \left[1 - \cos \pi - \{\cos \pi - \cos 2\pi\} \right] = \frac{4N}{\pi}$$

Describing function is

$$G_N = \frac{B_1 + jA_1}{M} = \frac{B_1 + 0}{M} = \frac{4N}{\pi M}$$

7.6.2 Ideal Relay with Dead Zone

The characteristics of an ideal relay with dead zone are shown in Figure 7.14. The dead zone is $2a$. The output is periodic. The input to the sinusoidal element is

$$m(t) = M \sin \omega t .$$

For $0 < \text{input} < a$, the output is zero. From the figure it can be observed that the input

$$a = M \sin \beta = M \sin(\pi - \beta); \; M > a .$$

When the input is less than the magnitude $M \sin \beta$, the output is zero. When the input is greater than a, the output is constant at N. Accordingly, the output equation for one period is

$$n(t) = \begin{cases} 0: 0 < \omega t < \beta \\ N; \beta < \omega t < \pi - \beta \\ 0; \pi - \beta < \omega t < \pi + \beta \\ -N; \pi + \beta < \omega t < 2\pi - \beta \\ 0; 2\pi - \beta < \omega t < 2\pi \end{cases} .$$

By inspection of the waveform of $n(t)$, it can be seen that the average value of the output is zero. Thus $A_0 = 0$. The waveform has odd symmetry so that cosine terms are absent in the Fourier series expansion. Hence $A_1 = 0$.

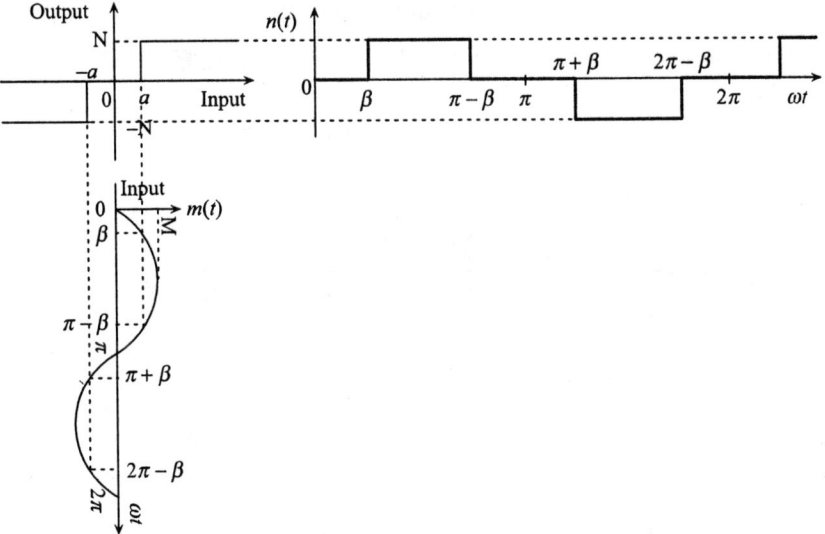

Figure 7.14 Input–output waveforms for ideal relay with dead zone and with sinusoidal input

$$B_1 = \frac{1}{\pi} \int_0^{2\pi} n(t) \sin \omega t \, d(\omega t)$$

$$= \frac{2}{\pi} \int_0^\pi n(t) \sin \omega t \, d(\omega t)$$

$$= \frac{2}{\pi} \int_\beta^{\pi-\beta} N \sin \theta \, d\theta$$

$$= \frac{2N}{\pi} [-\cos \theta]_\beta^{\pi-\beta} = \frac{4N}{\pi} \cos \beta$$

Describing function is

$$G_N = \frac{B_1 + jA_1}{M} = \frac{B_1 + 0}{M} = \frac{4N}{\pi M} \cos \beta \; ; \; M \geq a$$

$$G_N = 0 \; ; \; M < a.$$

7.6.3 Saturation

The input to the nonlinear element is

$$m(t) = M \sin \omega t.$$

For $-a <$ input $< a$, the output is linear and is given by

$$\text{Output} = k \, (\text{Input}),$$

where k is the slope of the linear portion of the nonlinear characteristics.

Hence the output in this region is

$$\text{Output} = k \, M \sin \omega t.$$

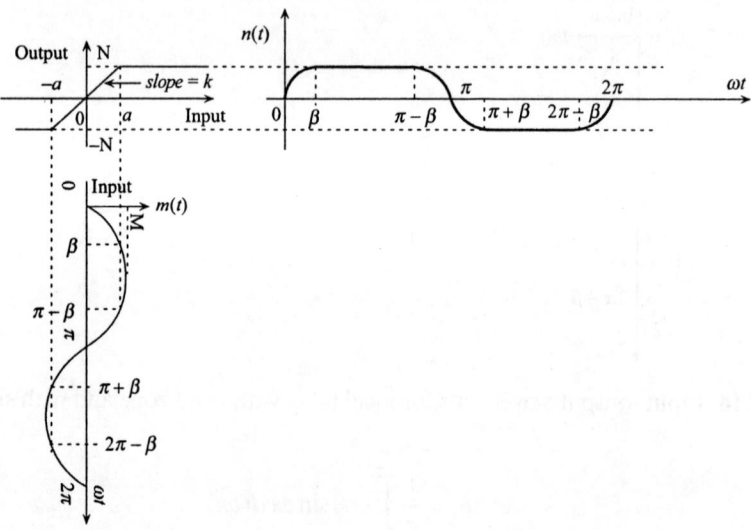

Figure 7.15 Input-output waveforms for saturation nonlinearity with sinusoidal input

For input $> a$, the output is constant at N. The output is periodic. The input–output characteristics are shown in Figure 7.15. The equation for output for half a period is

$$n(t) = \begin{cases} kM \sin \omega t; 0 < \omega t < \beta \\ N; \beta < \omega t < \pi - \beta \\ kM \sin \omega t; \pi - \beta < \omega t < \pi \end{cases}$$

$$N = ka \quad and \quad a = M \sin \beta; M > a.$$

By inspection of the waveform of $n(t)$, it can be observed that the average value of the output is zero. Thus $A_0 = 0$. The waveform has odd symmetry so that cosine terms are absent in the Fourier series expansion. Hence $A_1 = 0$.

$$B_1 = \frac{1}{\pi} \int_0^{2\pi} n(t) \sin \omega t \, d(\omega t)$$

$$= \frac{4}{\pi} \int_0^{\pi/2} n(t) \sin \omega t \, d(\omega t)$$

$$= \frac{4}{\pi} \left[\int_0^\beta (kM \sin \omega t) \sin \omega t \, d\omega t + \int_\beta^{\pi/2} N \sin \omega t \, d\omega t \right]$$

$$= \frac{4}{\pi} \left[\int_0^\beta kM \sin^2 \theta \, d\theta + \int_\beta^{\pi/2} ka \sin \theta \, d\theta \right]$$

$$= \frac{4kM}{\pi} \left[\int_0^\beta \frac{1 - \cos 2\theta}{2} d\theta + \frac{a}{M} \int_\beta^{\pi/2} \sin \theta \, d\theta \right]$$

$$= \frac{4kM}{\pi} \left[\left\{ \frac{\theta}{2} - \frac{\sin 2\theta}{4} \right\}_0^\beta + \frac{a}{M} \{ -\cos \theta \}_\beta^{\pi/2} \right]$$

$$= \frac{4kM}{\pi} \left[\frac{\beta}{2} - \frac{\sin 2\beta}{4} + \sin \beta \cos \beta \right]; \quad since \quad a = M \sin \beta$$

$$= \frac{4kM}{\pi} \left[\frac{\beta}{2} - \frac{\sin 2\beta}{4} + \frac{1}{2} \sin 2\beta \right]$$

$$= \frac{kM}{\pi} [2\beta - \sin 2\beta + 2 \sin 2\beta]$$

$$= \frac{kM}{\pi} [2\beta + \sin 2\beta]$$

Describing function is

$$G_N = \frac{B_1 + jA_1}{M} = \frac{B_1 + 0}{M} = \frac{k}{\pi}[2\beta + \sin 2\beta] \; ; \; M \geq a$$

$$G_N = k \; ; \; M < a .$$

7.6.4 Dead Zone

A dead zone nonlinearity and its characteristic input-output waveform for a sinusoidal input are shown in Figure 7.16. The dead zone is from $-a$ to a.

$$\text{Output} = 0 ; \text{for } -a < \text{input} < a$$

When the input is greater than a, the output linearly increases and is given by

$$\text{Output} = k(\text{Input} - a); \text{ for Input} > a .$$

The input to the nonlinear element is sinusoidal.
$$m(t) = M \sin \omega t \; ; \; M > a$$

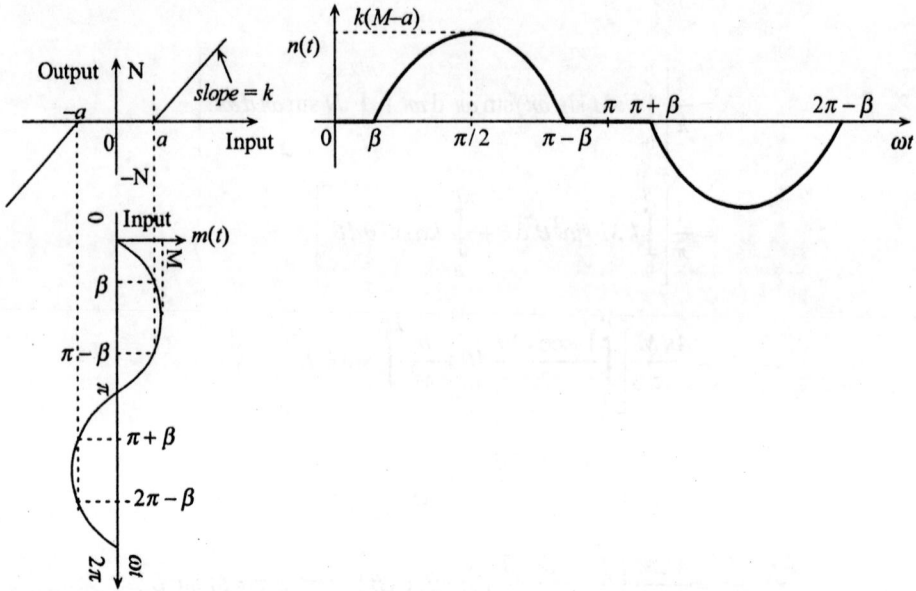

Figure 7.16 Input–output waveforms for dead zone nonlinearity with sinusoidal input

The output waveform is periodic, and for half a period, it can be expressed as

$$n(t) = \begin{cases} 0; 0 < \omega t < \beta \\ k(M \sin \omega t - a); \beta < \omega t < \pi - \beta \\ 0; \pi - \beta < \omega t < \pi \end{cases}$$

$$a = M \sin \beta .$$

By inspection of the waveform of $n(t)$, it can be seen that the average value of the output is zero. Thus $A_0 = 0$. The waveform has odd symmetry so that cosine terms are absent in the Fourier series expansion. Hence $A_1 = 0$.

$$B_1 = \frac{1}{\pi} \int_0^{2\pi} n(t) \sin \omega t \, d(\omega t)$$

$$= \frac{4}{\pi} \int_0^{\pi/2} n(t) \sin \omega t \, d(\omega t)$$

$$= \frac{4}{\pi} \int_\beta^{\pi/2} k(M \sin \omega t - a) \sin \omega t \, d\omega t$$

$$= \frac{4}{\pi} \int_\beta^{\pi/2} k(M \sin \theta - a) \sin \theta d\theta$$

$$= \frac{4}{\pi} \int_\beta^{\pi/2} k(M \sin \theta - M \sin \beta) \sin \theta d\theta$$

$$= \frac{4kM}{\pi} \int_\beta^{\pi/2} (\sin^2 \theta - \sin \beta \sin \theta) d\theta$$

$$= \frac{4kM}{\pi} \int_\beta^{\pi/2} \left[\frac{(1 - \cos 2\theta)}{2} - \sin \beta \sin \theta \right] d\theta$$

$$= \frac{4kM}{\pi} \left[\left\{ \frac{\theta}{2} - \frac{\sin 2\theta}{4} \right\}_\beta^{\pi/2} - \sin \beta \{ -\cos \theta \}_\beta^{\pi/2} \right]$$

$$= \frac{4kM}{\pi} \left[\frac{\pi}{4} - \frac{\sin \pi}{4} - \frac{\beta}{2} + \frac{\sin 2\beta}{4} - \sin \beta \cos \beta \right]$$

$$= \frac{4kM}{\pi} \left[\frac{\pi}{4} - \frac{\beta}{2} + \frac{\sin 2\beta}{4} - \frac{1}{2} \sin 2\beta \right]$$

$$= \frac{kM}{\pi} [\pi - 2\beta + \sin 2\beta - 2 \sin 2\beta]$$

$$= \frac{kM}{\pi} [\pi - 2\beta - \sin 2\beta]$$

Describing function is

$$G_N = \frac{B_1 + jA_1}{M} = \frac{B_1 + 0}{M} = \frac{k}{\pi} [\pi - 2\beta - \sin 2\beta] \; ; \; M \geq a$$

$$G_N = 0 \; ; \; M < a.$$

7.6.5 Saturation with Dead Zone

The input–output characteristics of saturation nonlinearity with dead zone, for a sinusoidal input, are shown in Figure 7.17. For the nonlinear element,

$$\text{Output} = \begin{cases} 0; & 0 < a < \text{input} \\ k(\text{input} - a); & a < \text{input} < b \,. \\ N; & \text{input} > b \end{cases}$$

The input to the nonlinear element is sinusoidal.

$$m(t) = M \sin \omega t \; ; \; M > b.$$

The output of the nonlinear element is periodic and for half a period, it can be represented as

$$n(t) = \begin{cases} 0; 0 < \omega t < \alpha \\ k(M \sin \omega t - a); \alpha < \omega t < \beta \\ N; \beta < \omega t < \pi - \beta \\ k(M \sin \omega t - a); (\pi - \beta) < \omega t < (\pi - \alpha) \\ 0; (\pi - \alpha) < \omega t < \pi \end{cases}$$

$$a = M \sin \alpha \text{ and } b = M \sin \beta \qquad\qquad k = \frac{N}{b-a} \qquad \text{or} \quad N = k(b-a).$$

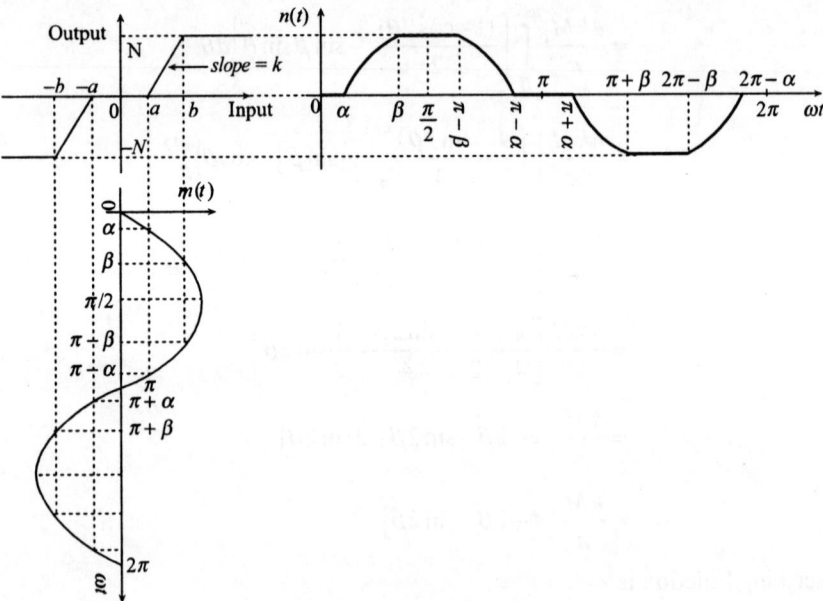

Figure 7.17 Input–output waveforms for saturation and dead zone with sinusoidal input

By inspection of the waveform of $n(t)$, it can be seen that the average value of the output is zero. Thus $A_0 = 0$. The waveform has odd symmetry so that cosine terms are absent in the Fourier series expansion. Hence $A_1 = 0$.

$$B_1 = \frac{1}{\pi} \int_0^{2\pi} n(t) \sin \omega t \, d(\omega t)$$

$$= \frac{4}{\pi} \int_0^{\pi/2} n(t) \sin \omega t \, d(\omega t)$$

$$= \frac{4}{\pi} \left[\int_\alpha^\beta k(M \sin \omega t - a) \sin \omega t \, d(\omega t) + \int_\beta^{\pi/2} N \sin \omega t \, d(\omega t) \right]$$

$$= \frac{4}{\pi} \left[\int_\alpha^\beta k(M \sin \theta - a) \sin \theta \, d(\theta) + \int_\beta^{\pi/2} N \sin \theta \, d(\theta) \right]$$

$$= \frac{4}{\pi} \left[k \int_\alpha^\beta (M \sin^2 \theta - a \sin \theta) d\theta + \int_\beta^{\pi/2} k(b - a) \sin \theta \, d\theta \right]$$

$$= \frac{4}{\pi} \left[k \int_\alpha^\beta \{ M \frac{(1 - \cos 2\theta)}{2} - M \sin \alpha \sin \theta \} d\theta + \int_\beta^{\pi/2} kM (\sin \beta - \sin \alpha) \sin \theta \, d\theta \right]$$

$$= \frac{4kM}{\pi} \left[\left\{ \frac{\theta}{2} - \frac{\sin 2\theta}{4} - \sin \alpha(-\cos \theta) \right\}_\alpha^\beta + (\sin \beta - \sin \alpha) \{ -\cos \theta \}_\beta^{\pi/2} \right]$$

$$= \frac{4kM}{\pi} \left[\frac{\beta}{2} - \frac{\sin 2\beta}{4} - \frac{\alpha}{2} + \frac{\sin 2\alpha}{4} - \sin \alpha(\cos \alpha - \cos \beta) + (\sin \beta - \sin \alpha) \cos \beta \right]$$

$$= \frac{kM}{\pi} [2(\beta - \alpha) + \sin 2\alpha - \sin 2\beta - 4 \sin \alpha \cos \alpha + 4 \sin \beta \cos \beta]$$

$$= \frac{kM}{\pi} [2(\beta - \alpha) + \sin 2\alpha - \sin 2\beta - 2 \sin 2\alpha + 2 \sin 2\beta]$$

$$= \frac{kM}{\pi} [2(\beta - \alpha) + \sin 2\beta - \sin 2\alpha]$$

Describing function is

$$G_N = \frac{B_1 + jA_1}{M} = \frac{B_1 + 0}{M} = \frac{k}{\pi} [2(\beta - \alpha) + \sin 2\beta - \sin 2\alpha] ; \, M \geq b$$

$$G_N = k \; ; \; a \le M < b$$
$$G_N = 0 \; ; \; M < a.$$

7.6.6 Ideal Relay with Hysteresis

The input–output characteristics of an ideal relay with hysteresis, for a sinusoidal input, are shown in Figure 7.18. When the input exceeds a, the output is positive and constant at 'N'. When the input is gradually reduced, the output becomes '$-N$' when the input is '$-a$' and remains at '$-N$'. The sinusoidal input to the nonlinear element is

   $m(t) = M \sin \omega t \; ; \; M > a.$

   The output of the nonlinear element is periodic and for half a period it can be represented as

$$n(t) = \begin{cases} -N; 0 < \omega t < \beta \\ N; \beta < \omega t < (\pi + \beta) \\ -N; (\pi + \beta) < (2\pi + \beta) \end{cases}.$$

$$a = M \sin \beta$$

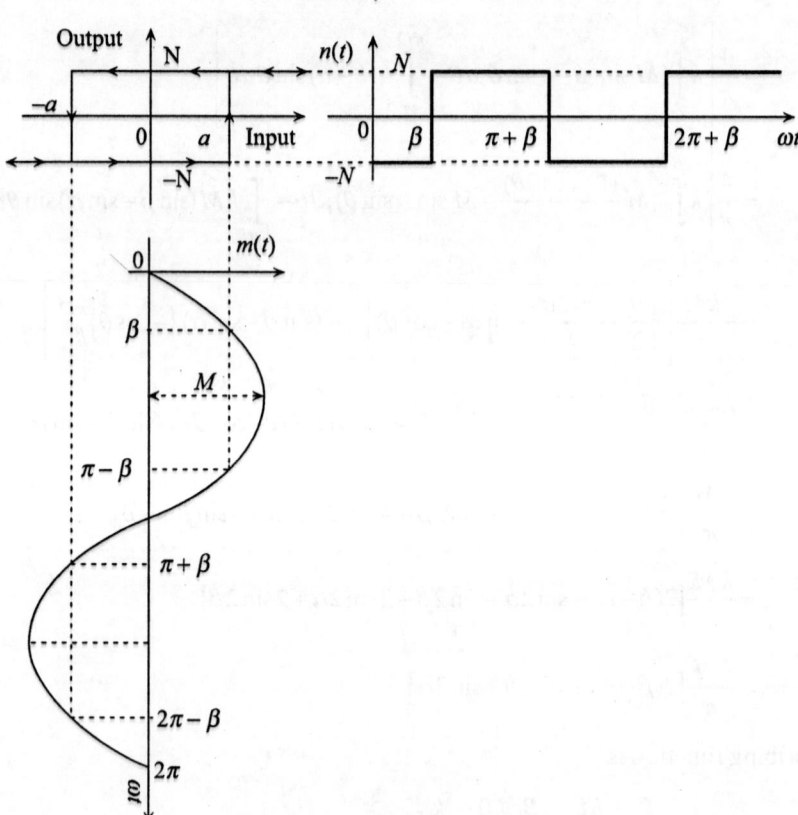

Figure 7.18 Input–output waveforms for ideal relay with hysteresis for sinusoidal input

The output waveform has no symmetry so that its Fourier series contains both sine and cosine terms. The average value of the output is zero and hence $A_0 = 0$.

$$A_1 = \frac{2}{\pi} \int_0^\pi n(t) \cos \omega t \, d(\omega t)$$

$$= \frac{2}{\pi} \int_\beta^{\pi+\beta} N \cos \theta \, d\theta = \frac{-4N}{\pi} \sin \beta.$$

$$B_1 = \frac{2}{\pi} \int_0^\pi n(t) \sin \omega t \, d(\omega t)$$

$$= \frac{2}{\pi} \int_\beta^{\pi+\beta} N \sin \theta \, d\theta = \frac{4N}{\pi} \cos \beta.$$

Describing function is

$$G_N = \frac{B_1 + jA_1}{M} = \frac{4N}{\pi M}(\cos \beta - j \sin \beta) = \frac{4N}{\pi M} e^{-j\beta}$$

$$= \frac{4N}{\pi M} \angle -\beta \, ; \, M \geq a.$$

$$\beta = \sin^{-1}(a/M)$$

The describing function is a complex quantity.

7.6.7 Ideal Relay with Dead Zone and Hysteresis

The input–output characteristics of an ideal relay with dead zone and hysteresis for a sinusoidal input are shown in Figure 7.19. When the input exceeds b, the output is positive and constant at 'N'. When the input is gradually reduced, the output remains at 'N' and becomes zero only when the input is a. A similar characteristic exists for negative inputs.

The sinusoidal input to the nonlinear element is

$m(t) = M \sin \omega t \, ; \, M > b.$

The output of the nonlinear element is periodic and for half a period it can be represented as

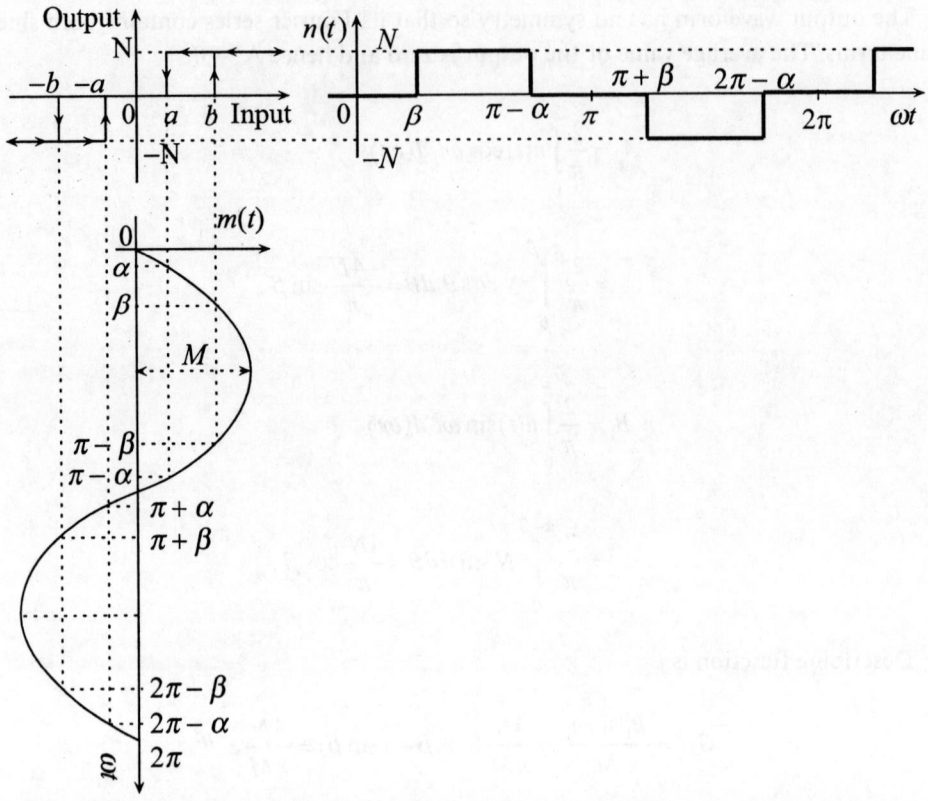

Figure 7.19 Input-output waveforms for Ideal relay with dead zone and hysteresis for sinusoidal input

$$n(t) = \begin{cases} 0; 0 < \omega t < \beta \\ N; \beta < \omega t < (\pi - \alpha) \\ 0; (\pi - \alpha) < \omega t < \pi \end{cases} \qquad a = M \sin \alpha \qquad b = M \sin \beta \qquad M > b.$$

By inspection, it can be observed that the average value of $n(t)$ is zero. $A_0 = 0$.

$$A_1 = \frac{2}{\pi} \int_0^\pi n(t) \cos \omega t \ d(\omega t)$$

$$= \frac{2}{\pi} \int_\beta^{\pi - \alpha} N \cos \theta \ d\theta = \frac{2N}{\pi} (\sin \alpha - \sin \beta).$$

$$B_1 = \frac{2}{\pi} \int_0^\pi n(t) \sin \omega t \ d(\omega t)$$

$$= \frac{2}{\pi} \int_{\beta}^{\pi-\alpha} N \sin\theta \, d\theta = \frac{2N}{\pi}(\cos\beta + \cos\alpha).$$

Describing function is

$$G_N = \frac{B_1 + jA_1}{M} = \frac{2N}{\pi M}[(\cos\alpha + \cos\beta) + j(\sin\alpha - \sin\beta)] \; ; \; M \geq b$$

$$|G_N| = \frac{2N}{\pi M}\sqrt{(\cos\alpha + \cos\beta)^2 + (\sin\alpha - \sin\beta)^2}$$

$$= \frac{2N}{\pi M}\sqrt{\cos^2\alpha + \cos^2\beta + 2\cos\alpha\,\cos\beta + \sin^2\alpha + \sin^2\beta - 2\sin\alpha\,\sin\beta}$$

$$= \frac{2N}{\pi M}\sqrt{2 + 2(\cos\alpha\,\cos\beta - \sin\alpha\,\sin\beta)}$$

$$= \frac{2N}{\pi M}\sqrt{2 + 2\cos(\alpha + \beta)}$$

$$= \frac{2N}{\pi M}\sqrt{2(1 + \cos(\alpha + \beta))}$$

$$= \frac{2N}{\pi M}\sqrt{4\cos^2\frac{(\alpha+\beta)}{2}}$$

$$= \frac{4N}{\pi M}\cos\frac{(\alpha+\beta)}{2}.$$

$$\angle G_N = \tan^{-1}\left(\frac{\sin\alpha - \sin\beta}{\cos\alpha + \cos\beta}\right) = \tan^{-1}\left[\frac{2\cos\left(\frac{\alpha+\beta}{2}\right)\sin\left(\frac{\alpha-\beta}{2}\right)}{2\cos\left(\frac{\alpha+\beta}{2}\right)\cos\left(\frac{\alpha-\beta}{2}\right)}\right]$$

$$= \tan^{-1}\left(\tan\frac{\alpha-\beta}{2}\right) = \frac{\alpha-\beta}{2}.$$

Hence $G_N = \dfrac{4N}{\pi M}\cos\dfrac{(\alpha+\beta)}{2}\angle\left(\dfrac{(\alpha-\beta)}{2}\right).$

The describing function is a complex quantity.

7.6.8 Combined Coulomb and Viscous Friction

The characteristics of a nonlinear element with combined Coulomb and viscous friction for a sinusoidal input are shown in Figure 7.20. When the input is positive,

$$\text{Output} = N + k(\text{input}).$$

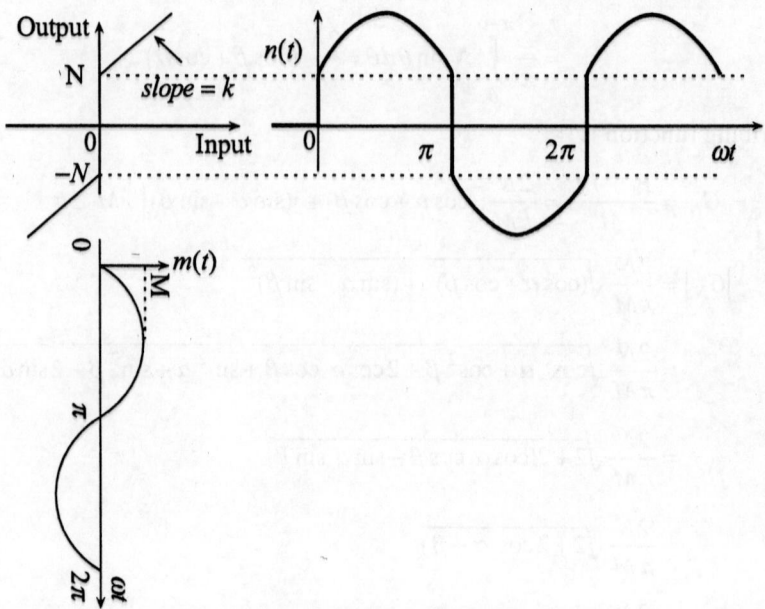

Figure 7.20 Input–output waveforms for combined Coulomb and viscous friction nonlinearity for sinusoidal input

The sinusoidal input to the nonlinear element is

$$m(t) \quad M \sin \quad t.$$

The output of the nonlinear element is periodic and for a period it can be represented as

$$n(t) = \begin{cases} N + kM \sin \omega t \; ; \; 0 < \omega t < \pi \\ -(N + kM \sin \omega t); \; \pi < \omega t < 2\pi \end{cases}.$$

By inspection, the average value is zero. $A_0 = 0$. The waveform has odd symmetry and hence the cosine terms are absent in the Fourier series expansion so that $A_1 = 0$.

$$B_1 = \frac{2}{\pi} \int_0^\pi n(t) \sin \omega t \; d(\omega t)$$

$$= \frac{2}{\pi} \int_0^\pi (N + kM \sin \omega t) \sin \omega t \; d(\omega t)$$

$$= \frac{2}{\pi} \int_0^\pi N \sin \theta \, d\theta + \frac{2}{\pi} \int_0^\pi kM \sin^2 \theta \, d\theta$$

$$= \frac{2N}{\pi} \{-\cos \theta\}_0^\pi + \frac{2kM}{\pi} \int_0^\pi \frac{(1 - \cos 2\theta)}{2} d\theta$$

$$= \frac{4N}{\pi} + \frac{kM}{\pi} \left\{ \theta - \frac{\sin 2\theta}{2} \right\}_0^{\pi}$$

$$= \frac{4N}{\pi} + \frac{kM}{\pi} \{\pi - 0\}$$

$$= \frac{4N}{\pi} + kM .$$

Describing function is

$$G_N = \frac{B_1 + jA_1}{M} = k + \frac{4N}{\pi M} .$$

7.6.9 Combined Coulomb and Viscous Friction with Dead Zone

The characteristics of a nonlinear element with combined coulomb and viscous friction for a sinusoidal input are shown in Figure 7.21. When the input is less than a, the output is zero. When the input is greater than a, positive, the output is $N + k(\text{input} - a)$.

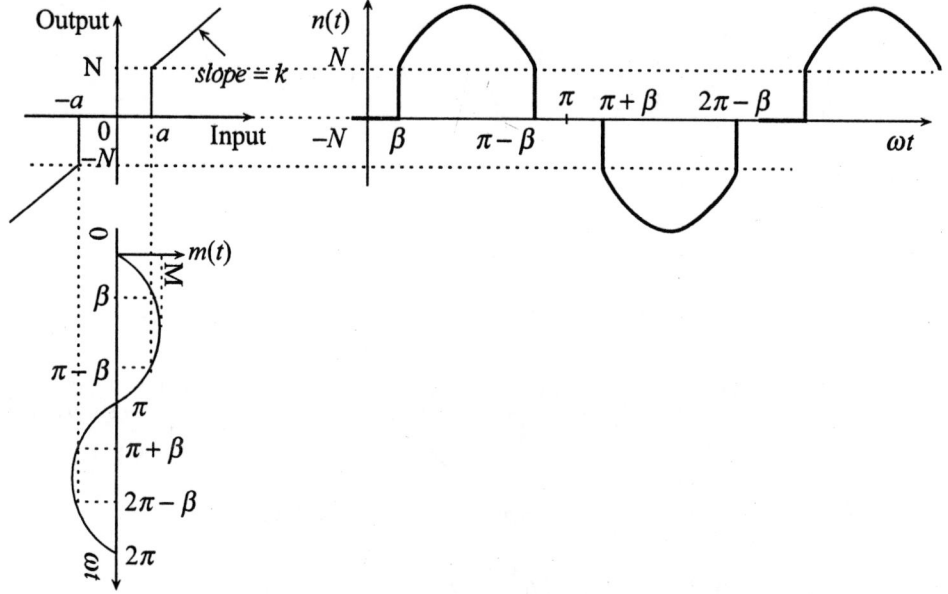

Figure 7.21 Input–output waveforms for combined Coulomb and viscous friction with dead zone for sinusoidal input

The sinusoidal input to the nonlinear element is

$$m(t) = M \sin \omega t .$$

The output of the nonlinear element is periodic and for a half period it can be represented as

$$n(t) = \begin{cases} 0; 0 < \omega t < \beta \\ N + k(M \sin \omega t - a) ; \beta < \omega t < \pi - \beta \\ 0; (\pi - \beta) < \omega t < (\pi + \beta) \end{cases}$$

$$a = M \sin \beta ; \ M \geq a .$$

By inspection, the average value is zero. $A_0 = 0$. The waveform has odd symmetry and hence the cosine terms are absent in the Fourier series expansion so that $A_1 = 0$.

$$B_1 = \frac{2}{\pi} \int_0^{\pi} n(t) \sin \omega t \ d(\omega t)$$

$$= \frac{2}{\pi} \int_{\beta}^{\pi-\beta} \{N + k(M \sin \omega t - a)\} \sin \omega t \ d(\omega t)$$

$$= \frac{2}{\pi} \int_{\beta}^{\pi-\beta} N \sin \theta d\theta + \frac{2k}{\pi} \int_{\beta}^{\pi-\beta} (M \sin^2 \theta - a \sin \theta) \ d\theta$$

$$= \frac{2}{\pi} \int_{\beta}^{\pi-\beta} N \sin \theta d\theta + \frac{2k}{\pi} \int_{\beta}^{\pi-\beta} \left[M \frac{(1 - \cos 2\theta)}{2} - M \sin \beta \sin \theta \right] d\theta$$

$$= \frac{4N}{\pi} \cos \beta + \frac{2kM}{\pi} \left[\frac{\theta}{2} - \frac{\sin 2\theta}{4} - \sin \beta (-\cos \theta) \right]_{\beta}^{\pi-\beta}$$

$$= \frac{4N}{\pi} \cos \beta + \frac{2kM}{\pi} \left[\frac{\theta}{2} - \frac{\sin 2\theta}{4} + \sin \beta \cos \theta) \right]_{\beta}^{\pi-\beta}$$

$$= \frac{4N}{\pi} \cos \beta + \frac{2kM}{\pi} \left[\frac{\pi - \beta}{2} - \frac{\sin 2(\pi - \beta)}{4} + \sin \beta \cos(\pi - \beta) - \frac{\beta}{2} + \frac{\sin 2\beta}{4} - \sin \beta \cos \beta \right]$$

$$= \frac{4N}{\pi} \cos \beta + \frac{2kM}{\pi} \left[\frac{\pi}{2} - \beta + \frac{\sin 2\beta}{4} + \frac{\sin 2\beta}{4} - 2 \sin \beta \cos \beta \right]$$

$$= \frac{4N}{\pi} \cos \beta + \frac{kM}{\pi} \left[\pi - 2\beta + \frac{\sin 2\beta}{2} + \frac{\sin 2\beta}{2} - 4 \sin \beta \cos \beta \right]$$

$$= \frac{4N}{\pi} \cos \beta + \frac{kM}{\pi} \left[\pi - 2\beta + \sin 2\beta - 2 \sin 2\beta \right]$$

$$= \frac{4N}{\pi} \cos \beta + \frac{kM}{\pi} \left[\pi - 2\beta - \sin 2\beta \right].$$

Describing function is

$$G_N = \frac{B_1 + jA_1}{M} == \frac{4N}{\pi M}\cos\beta + \frac{k}{\pi}[\pi - 2\beta - \sin 2\beta]\ ;\ M \geq a$$

$$G_N = 0\ ;\ M < a.$$

7.6.10 Backlash

In a mechanical gear system, there is a small angular gap between the teeth on the driving wheel and the driven wheel. Because of this, the driven gear wheel does not truthfully follow the driving gear wheel. When the driving gear rotates through a small angle less than the gap, the driven wheel does not rotate at all. This is called backlash.

The characteristics of a mechanical gear with backlash and its output response to a sinusoidal input are shown in Figure 7.22. The dead zone is $2a$, during which the driven gear wheel does not rotate while the driving gear wheel is still rotating. After the contact has been made between the two gear wheels, the driven gear wheel linearly follows the rotation of the driving gear wheel. When the driving gear starts rotating in the reverse direction through a distance $2a$, the driven gear wheel does not rotate again. Thereafter the driven wheel follows linearly with the rotation of the driving wheel. The slope of the linear portion is taken to be unity. The output is shown to be periodic and for half a period it can be represented as

$$n(t) = \begin{cases} (M\sin\omega t) - a; & 0 < \omega t < \pi/2 \\ (M - a); & \pi/2 < \omega t < \pi - \beta. \\ (M\sin\omega t) + a; & \pi - \beta < \omega t < \pi \end{cases}$$

It can be observed from the figure that

$$(M\sin\beta) + 2a = M \tag{1}$$

$$\beta = \sin^{-1}\left(1 - \frac{2a}{M}\right). \tag{2}$$

The Fourier series expansion for the output $n(t)$ has both sine and cosine components.

$$A_1 = \frac{2}{\pi}\int_0^\pi n(t)\cos\omega t\, d(\omega t)$$

$$= \frac{2}{\pi}\left[\int_0^{\pi/2}(M\sin\theta - a)\cos\theta\, d\theta + \int_{\pi/2}^{\pi-\beta}(M - a)\cos\theta\, d\theta + \int_{\pi-\beta}^{\pi}(M\sin\theta + a)\cos\theta\, d\theta\right]$$

$$= \frac{2}{\pi} [\int_0^{\pi/2} M \sin\theta \cos\theta d\theta - \int_0^{\pi/2} a \cos\theta d\theta + \int_{\pi/2}^{\pi-\beta} (M-a)\cos\theta d\theta$$

$$+ \int_{\pi-\beta}^{\pi} M \sin\theta \cos\theta d\theta + \int_{\pi-\beta}^{\pi} a \cos\theta d\theta]$$

$$= \frac{2}{\pi} [\int_0^{\pi/2} \frac{M}{2} \sin 2\theta \ d\theta - a \left\{ \sin\theta \right\}_0^{\pi/2} + (M-a) \left\{ \sin\theta \right\}_{\pi/2}^{\pi-\beta}$$

$$+ \int_{\pi-\beta}^{\pi} \frac{M}{2} \sin 2\theta \ d\theta + a \left\{ \sin\theta \right\}_{\pi-\beta}^{\pi}].$$

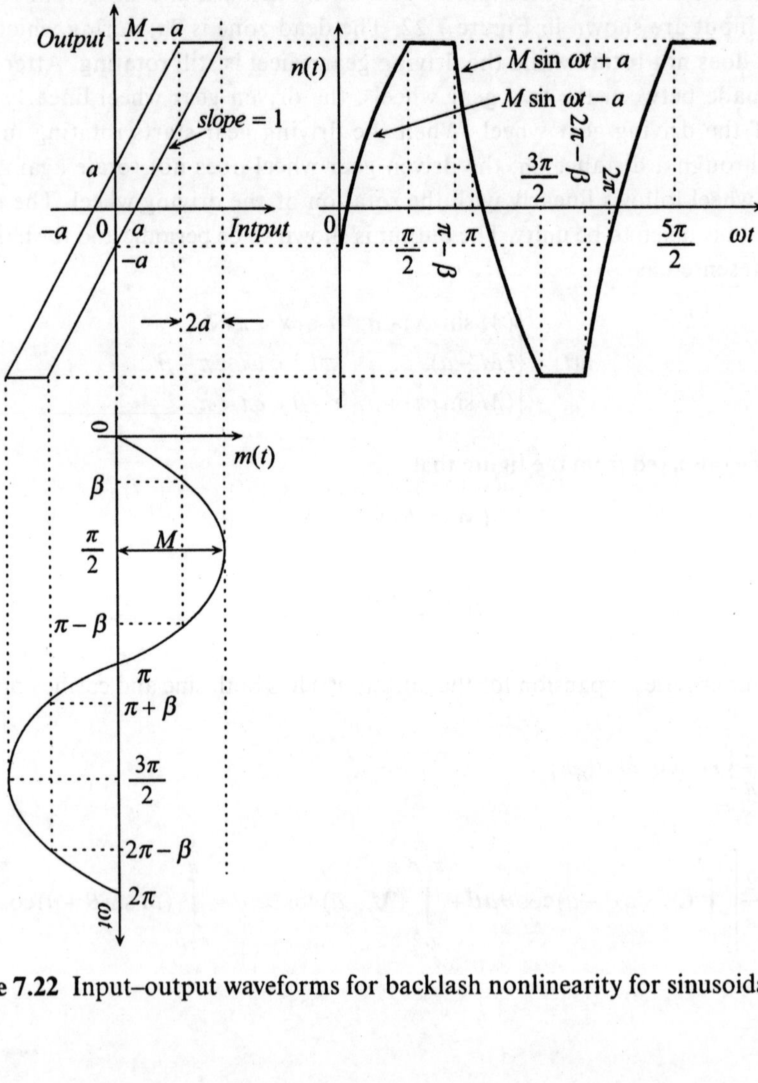

Figure 7.22 Input–output waveforms for backlash nonlinearity for sinusoidal input

$$= \frac{2}{\pi}\left[\frac{M}{4}\{-\cos 2\theta\}_0^{\pi/2} - a + (M-a)\{\sin\beta - 1\} + \frac{M}{4}\{-\cos 2\theta\}_{\pi-\beta}^{\pi} + a\{-\sin\beta\}\right]$$

$$= \frac{2}{\pi}\left[\frac{M}{4}\{2\} - a + M\sin\beta - M - a\sin\beta + a + \frac{M}{4}(\cos 2\beta - 1) - a\sin\beta\right]$$

$$= \frac{2}{\pi}\left[\frac{M}{4}(1+\cos 2\beta) + M\sin\beta - M - 2a\sin\beta\right].$$

From equation (1), $2a = M - M\sin\beta$.

Hence $A_1 = \frac{2}{\pi}\left[\frac{M}{4}(2\cos^2\beta) + M\sin\beta - M - (M - M\sin\beta)\sin\beta\right]$

$$= \frac{2}{\pi}\left[\frac{M}{2}\cos^2\beta + M\sin\beta - M - M\sin\beta + M\sin^2\beta\right]$$

$$= \frac{2}{\pi}\left[\frac{M}{2}\cos^2\beta - M(1-\sin^2\beta)\right]$$

$$= \frac{2}{\pi}\left[\frac{M}{2}\cos^2\beta - M\cos^2\beta\right]$$

$$= -\frac{M}{\pi}\cos^2\beta.$$

$$B_1 = \frac{2}{\pi}\int_0^{\pi} n(t)\sin\omega t\, d(\omega t)$$

$$= \frac{2}{\pi}\left[\int_0^{\pi/2}(M\sin\theta - a)\sin\theta d\theta + \int_{\pi/2}^{\pi-\beta}(M-a)\sin\theta d\theta + \int_{\pi-\beta}^{\pi}(M\sin\theta + a)\sin\theta d\theta\right]$$

$$= \frac{2}{\pi}[\int_0^{\pi/2} M\sin^2\theta\, d\theta - \int_0^{\pi/2} a\sin\theta d\theta + \int_{\pi/2}^{\pi-\beta}(M-a)\sin\theta d\theta$$

$$+ \int_{\pi-\beta}^{\pi} M\sin^2\theta\, d\theta + \int_{\pi-\beta}^{\pi} a\sin\theta d\theta]$$

$$= \frac{2}{\pi}[\int_0^{\pi/2}\frac{M}{2}(1-\cos 2\theta)\, d\theta - a\{-\cos\theta\}_0^{\pi/2} + (M-a)\{-\cos\theta\}_{\pi/2}^{\pi-\beta}$$

$$+ \int_{\pi-\beta}^{\pi}\frac{M}{2}(1-\cos 2\theta)\, d\theta + a\{-\cos\theta\}_{\pi-\beta}^{\pi}]$$

$$= \frac{2}{\pi}\left[\frac{M}{2}\left(\theta - \frac{\sin 2\theta}{2}\right)_0^{\pi/2} - a + (M-a)\cos\beta + \frac{M}{2}\left(\theta - \frac{\sin 2\theta}{2}\right)_{\pi-\beta}^{\pi} + a\{-\cos\beta + 1\}\right]$$

$$= \frac{2}{\pi}\left[\frac{M}{2}\left(\frac{\pi}{2}\right) - a + (M-a)\cos\beta + \frac{M}{2}\left(\pi - (\pi-\beta) - \frac{\sin 2\beta}{2}\right) - a\cos\beta + a\right]$$

$$= \frac{2}{\pi}\left[\frac{M}{2}\left(\frac{\pi}{2} + \beta\right) - \frac{M}{4}\sin 2\beta + M\cos\beta - 2a\cos\beta\right]$$

$$= \frac{2}{\pi}\left[\frac{M}{2}\left(\frac{\pi}{2} + \beta\right) - \frac{M}{4}\sin 2\beta + M\cos\beta - (M - M\sin\beta)\cos\beta\right]$$

$$= \frac{2}{\pi}\left[\frac{M}{2}\left(\frac{\pi}{2} + \beta\right) - \frac{M}{4}\sin 2\beta + M\cos\beta - M\cos\beta + M\sin\beta\cos\beta\right]$$

$$= \frac{2}{\pi}\left[\frac{M}{2}\left(\frac{\pi}{2} + \beta\right) - \frac{M}{4}\sin 2\beta + \frac{M}{2}\sin 2\beta\right]$$

$$= \frac{2}{\pi}\left[\frac{M}{2}\left(\frac{\pi}{2} + \beta\right) + \frac{M}{4}\sin 2\beta\right]$$

$$= \frac{2}{\pi}\left[\frac{M}{2}\left(\frac{\pi}{2} + \beta + \frac{1}{2}\sin 2\beta\right)\right].$$

Describing function is

$$G_N = \frac{B_1 + jA_1}{M} = \frac{\dfrac{M}{\pi}\left(\dfrac{\pi}{2} + \beta + \dfrac{1}{2}\sin 2\beta\right) - j\dfrac{M}{\pi}\cos^2\beta}{M}$$

$$= \frac{1}{\pi}\left(\frac{\pi}{2} + \beta + \frac{1}{2}\sin 2\beta - j\cos^2\beta\right)$$

From equation (2) $\beta = \sin^{-1}\left(1 - \dfrac{2a}{M}\right)$.

7.6.11 Square Nonlinearity

In this type of nonlinearity, the output varies as square of the input. This is shown in Figure 7.23.

Input → G_N → Output

$m(t) = M \sin \omega t$
$M\angle 0°$

$n(t) = km^2(t)$

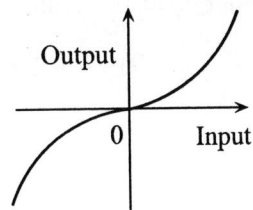

Output

0   Input

Figure 7.23 (a) Square nonlinearity   (b) Input–output characteristic

The sinusoidal input to the nonlinear element is

$$m(t) = M \sin \omega t \ .$$

The output of the nonlinear element can be represented as

$$n(t) = k\, m^2 (t)$$

$$= k\,(M \sin \omega t)^2 = \frac{kM^2}{2}(1 - \cos 2\omega t) \ .$$

Neglecting higher-order harmonics,

$$n(t) = \frac{kM^2}{2} \ .$$

Describing function is

$$G_N = \frac{(kM^2 / 2)}{M} = \frac{kM}{2} \ .$$

7.6.12 Cubic Nonlinearity

Cubic non-linearity is one in which the output varies as the cube of the input. This is shown in Figure 7.24.

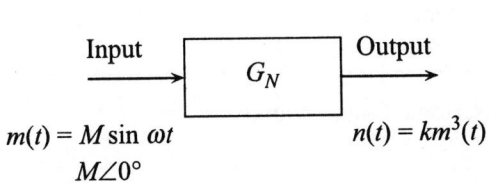

Input → G_N → Output

$m(t) = M \sin \omega t$
$M\angle 0°$

$n(t) = km^3(t)$

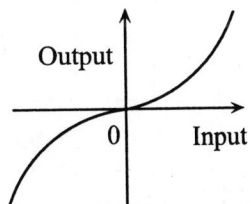

Output

0   Input

Figure 7.24 (a) Cubic nonlinearity   (b) Input–output characteristic

The sinusoidal input to the nonlinear element is

$$m(t) = M \sin \omega t \ .$$

The output of the nonlinear element can be represented as

$$n(t) = k\,m^3(t)$$

$$= k\,(M \sin \omega t)^3 = kM^3 \sin^3 \omega t = \frac{kM^3}{4}(3 \sin \omega t - \sin 3\omega t).$$

Neglecting higher-order harmonics,

$$n(t) = \frac{kM^3}{4}(3 \sin \omega t) = \frac{3kM^3}{4}\sin \omega t.$$

Describing function is

$$G_N = \frac{3kM^3}{4M} = \frac{3kM^2}{4}.$$

7.7 DESCRIBING FUNCTION ANALYSIS

The describing function for a nonlinear element is based on the assumption that the input to the nonlinear element is sinusoidal. If the input $r(t)$ is not zero, the input to the nonlinear element need not be sinusoidal. Hence the describing function analysis is applicable only when the input $r(t)$ is considered to be zero and the system is excited by some initial conditions.

A nonlinear system is shown in Figure 7.25, in which G_N is the describing function of the nonlinear element and $G_p(s)$ is the transfer function of the linear plant. A limit cycle in the output can be considered to be sinusoidal. Hence the input to the nonlinear element is sinusoidal. From the block diagram

$$C(s) = -C(s)G_N\,G_p(s).$$

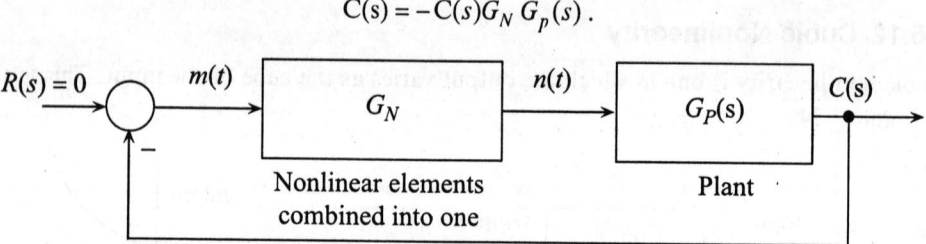

Figure 7.25 Describing function analysis

Hence $G_N\,G_p(s) = -1$.

Using this equation and the Nyquist criterion, the existence of a limit cycle can be predicted.

$$G_p(j\omega) = -1/G_N.$$

Draw the frequency response $G_P(j\omega)$ versus ω in the polar plane. This is nothing but the polar plot. Assume that the plant transfer function is a minimum phase function with all its poles and zeros lying on the left half of the s-plane. Hence the polar plot together with its mirror image constitutes the Nyquist plot. Also draw the locus of $(-1/G_N)$ versus M, in the same polar plane, considering that M is varied from zero to infinity. M is the peak value of the sinusoidal input to the nonlinear element. Such a diagram is shown in Figure 7.26.

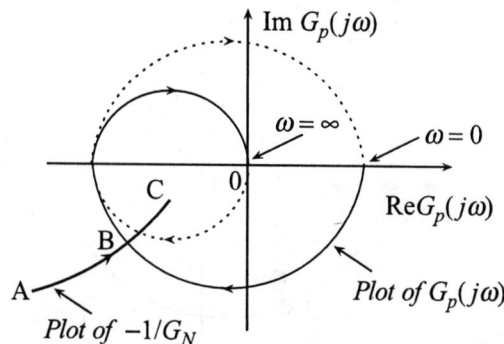

Figure 7.26 Describing function analysis using Nyquist criterion

In the stability analysis for a linear system using Nyquist criterion, the critical point is $(-1+j0)$ in the complex plane. For a nonlinear system, while using Nyquist criterion, the locus of $(-1/G_N)$ can be taken to be the locus of the many existing critical points. If the critical points of the $(-1/G_N)$ locus lie to the left of the $G_P(j\omega)$ locus, they are not enclosed and not encircled by the Nyquist plot and the system is stable. This is shown as segment AB in Figure 7.26 in which the polar plot is drawn for a minimum phase function. On the other hand, if the critical points of the $(-1/G_N)$ locus lie to the right of the $G_P(j\omega)$ locus, they are enclosed and encircled by the Nyquist plot so that the system is unstable. This is shown as segment BC in Figure 7.26. It can be observed that it is only necessary to draw the polar plot of $G_P(j\omega)$ rather than the Nyquist plot to examine the relative positions of the $(-1/G_N)$ plot, if $G_P(j\omega)$ is a minimum phase function. If this locus intersects with the polar plot, then the system exhibits a limit cycle. The point B thus corresponds to a limit cycle. The frequency of the oscillations is the value of ω on the $G_P(j\omega)$ locus at the intersecting point. The amplitude of limit cycle can be calculated by equating the magnitude of $-1/G_N$ and the magnitude of $G(j\omega)$ at the intersecting point.

The existence of limit cycles can also be checked, if the solution to the characteristic equation can be determined. The characteristic equation is

$$1 + G_N\, G_p(s) = 0 .$$

If the roots of this characteristic equation are with negative real parts, the system will not exhibit limit cycles.

7.8 STABILITY OF LIMIT CYCLE

A control system should not exhibit limit cycles. If a limit cycle exists, it is important to assess whether it is stable or unstable. Consider a $G_P(j\omega)$ locus and one $-1/G_N$ locus intersecting at two points P and Q as shown in Figure 7.27. These two intersecting points correspond to two possible limit cycles. The arrows on the $G_P(j\omega)$ locus indicate the direction in which the locus is traversed when the frequency is varied from $\omega = 0$ to $\omega = \infty$. The arrows on the $-1/G_N$ locus indicate the direction in which the locus is traversed when the amplitude of the input to the nonlinear element is increased from $M = 0$ to $= \infty$.

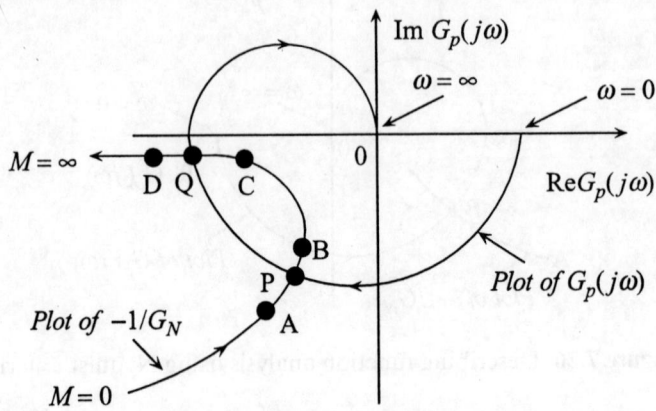

Figure 7.27 Stability of limit cycles

Consider that the system operates at P under the state of limit a cycle. Assume that due to a slight perturbation in the system, the input to the nonlinear element decreases so that the operating point is shifted to A. Since the point A is not enclosed by the $G_P(j\omega)$ locus, it is in the stable region. Hence the input to the nonlinear element gradually decreases and point A is further shifted to the left.

Next, suppose that due to a slight perturbation in the system, the operating point is shifted to point B on the $-1/G_N$ locus. Since point B is enclosed by the $G_P(j\omega)$ locus, it is in the unstable region. Hence the input to the nonlinear element increases steadily and the point moves towards Q. Thus the point P shows divergent characteristics and this corresponds to an unstable limit cycle.

Consider that the system now operates at the point Q under the state of limit cycle. Assume that due to a slight disturbance in the system, the input to the nonlinear element decreases so that the operating point is shifted to point C. Since point C is enclosed by the $G_P(j\omega)$ locus, it is in the unstable region. Hence the input to the nonlinear element steadily increases and point C is shifted back to Q.

Now suppose that due to a slight disturbance in the system, the operating point is shifted to point D on the $-1/G_N$ locus. Since the point D is not enclosed by the $G_P(j\omega)$ locus, it is in the stable region. Hence the input to the nonlinear element decreases steadily and the point moves back to Q. Thus the point Q shows convergent characteristics and this corresponds to a stable limit cycle.

The stability of a limit cycle can be assessed by inspection of the $-1/G_N$ locus and the $G_P(j\omega)$ locus. Assume that an observer traverses the $-1/G_N$ locus in the direction of increasing M. After crossing a limit cycle point, if he enters a region enclosed by the $G_P(j\omega)$ locus, then that limit cycle is unstable. On the other hand, if he moves out of the enclosing region, then the limit cycle is stable. Hence in Figure 7.27, the point P corresponds to an unstable limit cycle and the point Q represents a stable limit cycle.

Example 7.1 An ideal relay is introduced as a nonlinearity in a linear control system the transfer function of which is

$$G(s) = \frac{100}{s(s+6)}.$$

The relay has an output of 10 V. Check whether a limit cycle exists or not and assess the stability of the system.

Solution: The system representation is shown in Figure E7.1 (a).

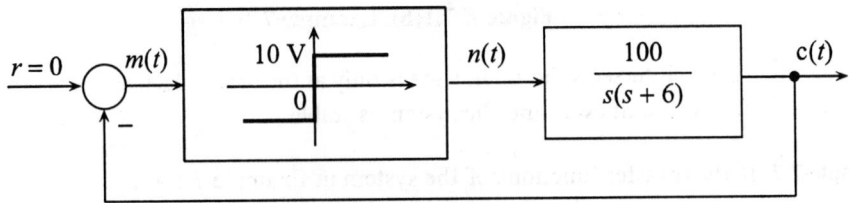

Figure E7.1 (a) Example 7.1

The input to the nonlinear element is

$$m(t) = M \sin \omega t.$$

The describing function of the ideal relay is

$$G_N = \frac{4N}{\pi M}; \text{where } N = 10$$

$$-\frac{1}{G_N} = -\frac{\pi M}{40}.$$

When $M = 0, -\dfrac{1}{G_N} = 0$ and when $M \to \infty, -\dfrac{1}{G_N} \to -\infty$.

The locus of $-1/G_N$ is the negative real axis and is shown in Figure E7.1(b).

$$G(s) = \frac{100}{s(s+6)}$$

$$G(j\omega) = \frac{100}{j\omega(j\omega+6)} = \frac{100}{\omega\sqrt{\omega^2+36}} \angle -90° - \tan^{-1}(\omega/6).$$

When $\omega \to 0$; $G(j\omega) \to \infty \angle -90°$; when $\omega \to \infty$; $G(j\omega) \to 0 \angle -180°$.

The locus of $G(j\omega)$ is also shown in Figure E7.1(b).

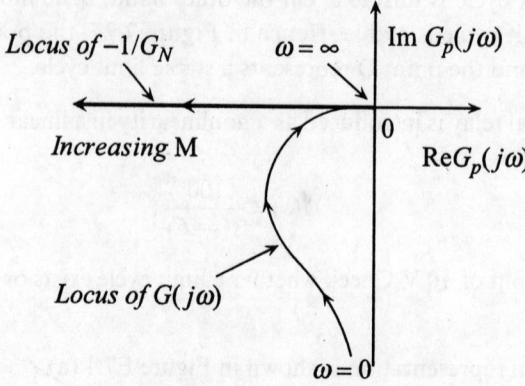

Figure E 7.1(b) Example 7.1

The locus of $-1/G_N$ intersects the $G(j\omega)$ locus only at the origin. Hence, there is no possibility of occurrence of a limit cycle and the system is stable.

Example 7.2 If the transfer functions of the system in Example 7.1 were

$$G(s) = \frac{500}{s(s+5)(s+20)},$$

check whether a limit cycle exists or not and assess its stability.

Solution:

$$G(s) = \frac{500}{s(s+5)(s+20)}$$

$$G(j\omega) = \frac{500}{j\omega(j\omega+5)(j\omega+20)}$$

$$= \frac{500}{\omega\sqrt{\omega^2 +25}\,\sqrt{\omega^2 +400}} \angle -90° - \tan^{-1}(\omega/5) - \tan^{-1}(\omega/20)$$

When $\omega \to 0$; $G(j\omega) \to \infty \angle -90°$;   when $\omega \to \infty$; $G(j\omega) \to 0 \angle -270°$.

The locus of $G(j\omega)$ is shown in Figure E7.2.
The describing function of the ideal relay is

$$G_N = \frac{4N}{\pi M}; \text{where } N = 10$$

$$-\frac{1}{G_N} = -\frac{\pi M}{40}.$$

When $M = 0, -\dfrac{1}{G_N} = 0$ and when $M \to \infty, -\dfrac{1}{G_N} \to -\infty$.

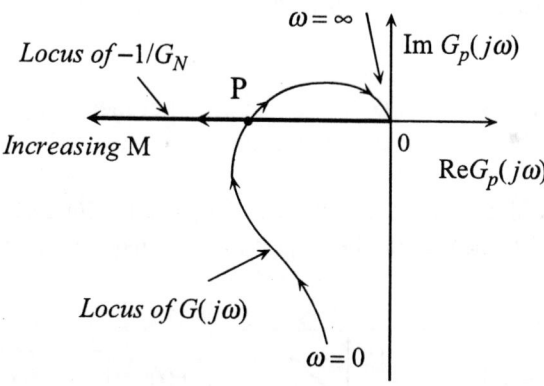

Figure E 7.2 Example 7.2

The locus of $-1/G_N$ is the negative real axis and is shown in Figure E7.2. It can be observed that $G(j\omega)$ locus intersects the locus of $-1/G_N$ at point P. Hence a limit cycle is predicted. The frequency ω of the limit cycle can be determined from the equation for the angle of $G(j\omega)$. Point P is in the negative real axis. Hence

$$-90^\circ - \tan^{-1}(\omega/5) - \tan^{-1}(\omega/20) = -180^\circ$$

$$\tan^{-1}(\omega/5) + \tan^{-1}(\omega/20) = 90^\circ$$

$$\frac{(\omega/5) + (\omega/20)}{1 - (\omega/5)(\omega/20)} = \tan 90^\circ = \infty$$

$$1 - (\omega/5)(\omega/20) = 0$$

$$1 - \frac{\omega^2}{100} = 0 \qquad \text{or } \omega = 10 \text{ rad/sec.}$$

The amplitude of the limit cycle can be calculated from the magnitude of $G(j\omega)$ at point P. At the intersecting point,

$$G(j\omega)\big|_{\omega=10} = \frac{500}{\omega\sqrt{\omega^2 + 25}\,\sqrt{\omega^2 + 400}}\bigg|_{\omega=10} \angle -180^\circ = -0.2 .$$

At the intersection point,

$$G_N = -\frac{1}{G(j\omega)} = \frac{1}{0.2}$$

$$G_N = \frac{4N}{\pi M} = \frac{40}{\pi M} = \frac{1}{0.2} \qquad \text{Or} \quad M = 2.55.$$

Hence the limit cycle has a magnitude of 2.55. The limit cycle is

$$c(t) = -m(t) = -2.55 \sin 10t.$$

As we traverse the locus of $(-1/G_N)$ in the direction of increasing M, we enter a region which is not enclosed by the $G(j\omega)$ locus. Hence the limit cycle is stable.

Example 7.3 A nonlinear system is shown in Figure E7.3(a). Check whether a limit cycle exists or not. If exists, determine the magnitude and frequency of the limit cycle. Also assess its stability.

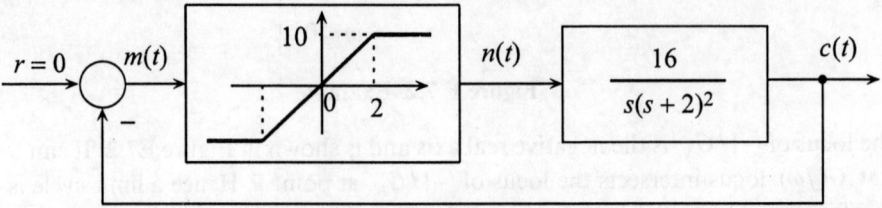

Figure E 7.3(a) Example 7.3

Solution: The nonlinearity is saturation. The input to the nonlinear element is

$$m(t) = M \sin \omega t.$$

The describing function for the nonlinear element is

$$G_N = \frac{k}{\pi} [2\beta + \sin 2\beta]; M \geq 2.$$

And $G_N = k; M < 2$

where k is the slope of the linear portion and $k = (10/2) = 5$

$$M \sin \beta = 2 \qquad or \ \sin \beta = \frac{2}{M}$$

$$-\frac{1}{G_N} = \frac{-\pi}{k[2\beta + \sin 2\beta]}.$$

When $M = 2$, $\sin \beta = 1$; $\beta = \frac{\pi}{2}$; so that $-\frac{1}{G_N} = -\frac{1}{k} = -\frac{1}{5} = -0.2$

When $M \to \infty, \beta \to 0$; hence $-\frac{1}{G_N} \to -\infty$.

Thus the locus of $-1/G_N$ is along the negative real axis, starting from 0.2 and approaching infinity. This is shown in Figure E 7.3(b).

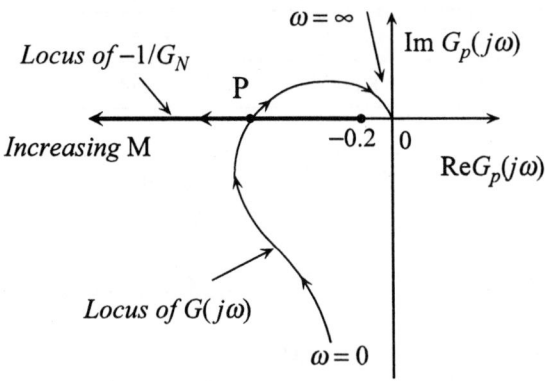

Figure E 7.3(b) Example 7.3

$$G(s) = \frac{16}{s(s+2)^2}$$

$$G(j\omega) = \frac{16}{j\omega(j\omega+2)^2} = \frac{16}{\omega(\omega^2+4)} \angle -90° - 2\tan^{-1}(\omega/2).$$

When $\omega \to 0, G(j\omega) \to \infty \angle -90°$ and when $\omega \to \infty, G(j\omega) \to 0 \angle -270°$.

The locus of $G(j\omega)$ is also shown in Figure 7.3(b). The intersecting point, if any, can be determined from the equation for the angle of $G(j\omega)$.

$$-90° - 2\tan^{-1}(\omega/2) = -180°$$

$$2\tan^{-1}(\omega/2) = 90° \qquad \text{Or } \tan^{-1}(\omega/2) = 45° \qquad \omega = 2 \text{ rad/s}.$$

$$G(j\omega)\big|_{\omega=2} = \frac{16}{\omega(\omega^2+4)}\bigg|_{\omega=2} \angle -180° = -1.$$

The two loci intersect at the point $(-1,0)$. Hence a limit cycle exists. The frequency of the limit cycle is 2 rad / s. The amplitude of the limit cycle can be calculated from the magnitude of $G(j\omega)$ at the point P.

$$G(j\omega)\big|_{\omega=2} = \frac{-1}{G_N} = \frac{-\pi}{k[2\beta + \sin 2\beta]} = -1.$$

Since k = 5, $2\beta + \sin 2\beta - 0.6283 = 0$.

This nonlinear equation in terms of β and can be solved using Newton–Raphson method:

$$f(\beta) = 2\beta + \sin 2\beta - 0.6283 \qquad f'(\beta) = 2 + 2\cos 2\beta = 2(1+\cos 2\beta) = 4\cos^2 \beta.$$

Iteration	$f(\beta)$	$f'(\beta)$	β
1	0.889	3.3934	0.4
2	−0.078	3.9243	0.138
3	−1.923 X 10⁻³	3.9010	0.1579
4	2.7402X10⁻⁵		0.1584

The value of β is 0.1584 radian or 9.08°. The value of M can be calculated from

$$\sin\beta = \frac{2}{M} \qquad or \ M = \frac{2}{\sin\beta} = \frac{2}{\sin 9.08°} = 12.67.$$

Hence the limit cycle is $c(t) = c(t) = -m(t) = -12.67\sin 2t$.

As we traverse the locus of $(-1/G_N)$ in the direction of increasing M, we enter a region which is not enclosed by the $G(j\omega)$ locus. The limit cycle is stable.

Example 7.4 A nonlinear system is shown in Figure E7.4(a). Find the range of k for which a limit cycle is predicted.

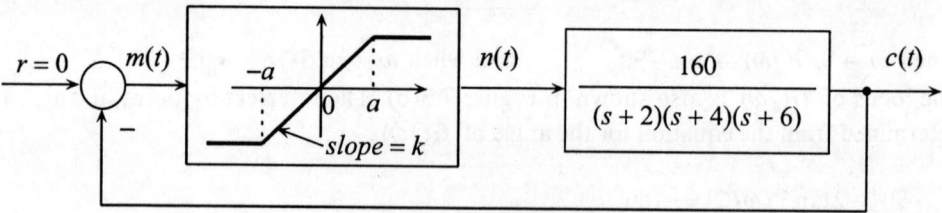

Figure E 7.4(a) Example 7.4

Solution: The nonlinearity is saturation. The input to the nonlinear element is assumed to be

$$m(t) = M\sin\omega t .$$

The describing function for the nonlinear element is

$$G_N = \frac{k}{\pi}[2\beta + \sin 2\beta]; M \geq a$$

and $G_N = k; M < a$,

where k is the slope of the linear portion and $a = M\sin\beta$ $\qquad$ or $\sin\beta = \dfrac{a}{M}$,

$$-\frac{1}{G_N} = \frac{-\pi}{k[2\beta + \sin 2\beta]}$$

When $M = a,$ $\quad \sin\beta = 1; \ \ \beta = \dfrac{\pi}{2};$ $\quad$ and $-\dfrac{1}{G_N} = -\dfrac{1}{k}$

When $M \to \infty, \beta \to 0$; hence $G_N \to 0$ and $-\dfrac{1}{G_N} \to -\infty$.

Thus the locus of $-1/G_N$ is along the negative real axis, starting from the point $Q = -1/k$ and approaching infinity. This is shown in Figure E 7.4(b).

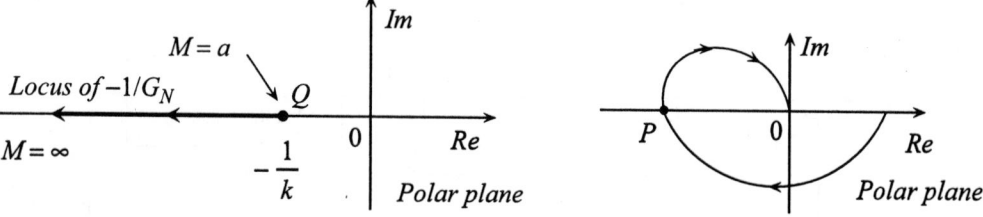

Figure E7.4 (b) Plot of $-1/G_N$ Figure E7.4 (c) Plot of $G(j\omega)$

$$G(s) = \frac{160}{(s+2)(s+4)(s+6)}$$

$$G(j\omega) = \frac{160}{(j\omega+2)(j\omega+4)(j\omega+6)}$$

$$= \frac{160}{\sqrt{\omega^2+4}\ \sqrt{\omega^2+16}\ \sqrt{\omega^2+36}} \angle -\tan^{-1}(\omega/2) - \tan^{-1}(\omega/4)\tan^{-1}(\omega/6).$$

When $\omega \to 0, G(j\omega) \to \dfrac{10}{3}\angle 0°$

When $\omega \to \infty, G(j\omega) \to 0\angle -270°$.

The locus of $G(j\omega)$ which is the polar plot is shown in Figure E7.4(c). The polar plot intersects the negative real axis at point P. This point can be found from the expression for the phase angle.

$$-\tan^{-1}(\omega/2) - \tan^{-1}(\omega/4) - \tan^{-1}(\omega/6) = -180^0$$

$$\tan^{-1}(\omega/4) + \tan^{-1}(\omega/6) = 180^0 - \tan^{-1}(\omega/2)$$

$$\frac{(\omega/4)+(\omega/6)}{1-(\omega^2/24)} = -(\omega/2).$$

Solving this equation, $\omega = \sqrt{44} = 6.633$ rad/s

$$G(j\omega)\big|_{\omega=\sqrt{44}} = \frac{160}{\sqrt{(48)(60)(80)}}\angle -180° = -\frac{1}{3}.$$

Point P is $\left(-\dfrac{1}{3}, 0\right)$. If the polar plot has to intersect the locus of $(-1/G_N)$, point P has to be to the left of Q. For this, the value of k must be greater than 3. Thus for a limit cycle to occur, $k > 3$.

Example 7.5 A nonlinear system is shown in Figure E7.5(a). Check for the possibility of a limit cycle.

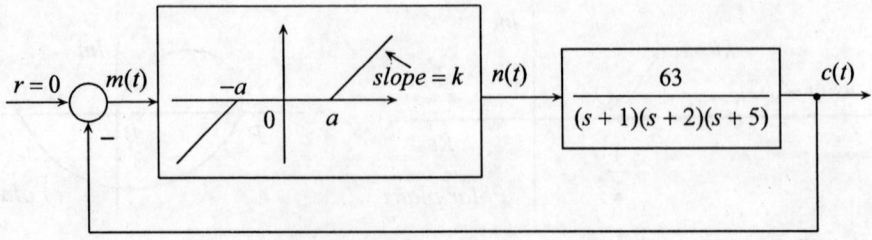

Figure E 7.5(a) Example 7.5

Solution: The nonlinearity is dead zone. The input to the nonlinear element is assumed to be

$$m(t) = M \sin \omega t.$$

The describing function for the nonlinear element is

$$G_N = \frac{k}{\pi}\left[\pi - 2\beta - \sin 2\beta\right]; M \geq a \qquad\qquad a = M \sin \beta$$

and $G_N = 0; M < a$,
where k is the slope of the linear portion of the input–output characteristics.

$$-\frac{1}{G_N} = \frac{-\pi}{k\left[\pi - 2\beta - \sin 2\beta\right]} \qquad\qquad \sin\beta = \frac{a}{M}$$

When $M = a$, $\sin \beta = 1$, $\beta = \dfrac{\pi}{2}$ and $-\dfrac{1}{G_N} = -\infty$

When $M \to \infty, \beta \to 0$; hence $\quad -\dfrac{1}{G_N} \to -\dfrac{1}{k}$.

Thus the locus of $-1/G_N$ is along the negative real axis, starting from $-\infty$ to $-1/k$. This is shown in Figure E 7.5(b).

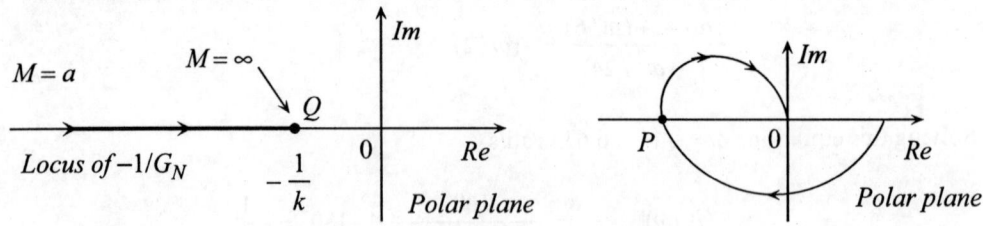

Figure E7.5 (b) Plot of $-1/G_N$ **Figure E7.5 (c)** Plot of $G(j\omega)$

$$G(s) = \frac{63}{(s+1)(s+2)(s+5)}$$

$$G(j\omega) = \frac{63}{(j\omega+1)(j\omega+2)(j\omega+5)}$$

$$= \frac{63}{\sqrt{\omega^2+1}\sqrt{\omega^2+4}\sqrt{\omega^2+25}} \angle -\tan^{-1}\omega - \tan^{-1}(\omega/2) - \tan^{-1}(\omega/5).$$

When $\omega \to 0, G(j\omega) \to 6.3\angle 0°$

When $\omega \to \infty, G(j\omega) \to 0 \angle -270°$.

The locus of $G(j\omega)$ which is the polar plot is shown in Figure E7.5(c). The point of intersection of the polar plot on the negative real axis can be found from the expression for the phase angle.

$$-\tan^{-1}\omega - \tan^{-1}(\omega/2) - \tan^{-1}(\omega/5) = -180°$$

$$\tan^{-1}(\omega/2) + \tan^{-1}(\omega/5) = 180° - \tan^{-1}\omega$$

$$\frac{(\omega/2)+(\omega/5)}{1-(\omega^2/10)} = -\omega.$$

Solving this equation, $\omega = \sqrt{17} = 4.123$ rad/s

$$G(j\omega)\big|_{\omega=\sqrt{17}} = \frac{63}{\sqrt{(18)(21)(42)}} \angle -180° = -\frac{1}{2}.$$

Point P on the polar plot is $(-\frac{1}{2}, 0)$. If the polar plot intersects the locus of $(-1/G_N)$, point P has to be to the left of Q. For this, the value of k must be greater than 2. Thus for a limit cycle to occur, $k > 2$.

Example 7.6 A nonlinear system consists of a dead zone as its nonlinearity shown in Figure E7.6(a). Check whether a limit cycle is predicted. When $k = 1$, if a limit cycle is predicted, find its amplitude and frequency. Also assess its stability.

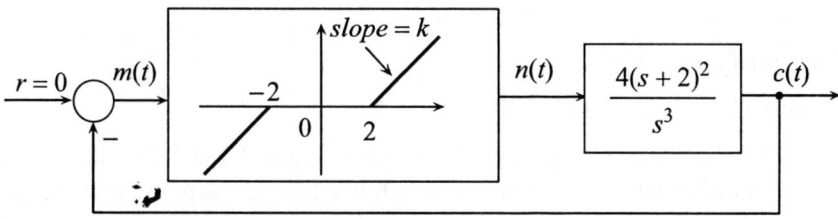

Figure E 7.6 (a) Example 7.6

Solution: The nonlinearity is dead zone. The input to the nonlinear element is assumed to be

$$m(t) = M \sin \omega t .$$

The describing function for the nonlinear element is

$$G_N = \frac{k}{\pi}[\pi - 2\beta - \sin 2\beta]; M \geq a \qquad\qquad a = M \sin \beta \qquad\qquad a = 2$$

and $G_N = 0; M < a$,
where k is the slope of the linear portion.

$$-\frac{1}{G_N} = \frac{-\pi}{k[\pi - 2\beta - \sin 2\beta]}.$$

When $M = a$, $\sin \beta = 1$, $\beta = \frac{\pi}{2}$ and $-\frac{1}{G_N} = -\infty$

When $M \to \infty, \beta \to 0$; hence $G_N \to 0$ and $-\frac{1}{G_N} \to -\frac{1}{k}$.

Thus the locus of $1/$ is along the negative real axis, starting from $-\infty$ to $-1/k$. This is shown in Figure E 7.6(b).

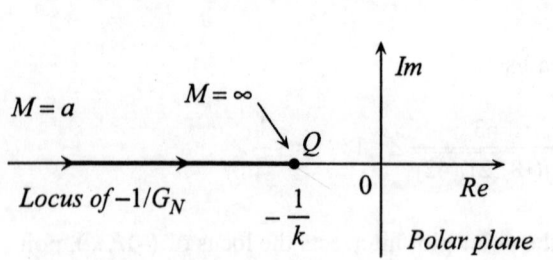

| Figure E7.6 (b) Plot of $-1/G_N$ | Figure E7.6 (c) Plot of $G(j\omega)$ |

$$G(s) = \frac{4(s+2)^2}{s^3}$$

$$G(j\omega) = \frac{4(j\omega+2)^2}{(j\omega)^3} = \frac{4(\omega^2+4)}{\omega^3} \angle 2\tan^{-1}(\omega/2) - 270°.$$

When $\omega \to 0, G(j\omega) \to \infty \angle -270°$

When $\omega \to \infty, G(j\omega) \to 0 \angle -90°$.

The locus of $G(j\omega)$ which is the polar plot is shown in Figure E7.6(c). The point of intersection of the polar plot on the negative real axis can be found from the expression for the phase angle.

$$2\tan^{-1}(\omega/2) - 270^{\circ} = -180^{\circ}$$

$$2\tan^{-1}(\omega/2) = 90^{\circ} \qquad \tan^{-1}(\omega/2) = 45^{\circ} \qquad \text{or } \omega = 2$$

$$G(j\omega)\big|_{\omega=2} = \frac{4(\omega^2 + 4)}{\omega^3}\bigg|_{\omega=2} \qquad \angle -180^{\circ} = -4 .$$

Point P on the polar plot is $(-4,0)$. If the polar plot intersects the locus of $(-1/G_N)$, point P has to be to the left of Q. For this, the value of $(1/k)$ must be less than 4 or $k > (1/4)$. Thus for a limit cycle to occur, $k > 1/4$.

When k = 1, $G_N = \dfrac{1}{\pi}[\pi - 2\beta - \sin 2\beta]$.

At the intersecting point,

$$-\frac{1}{G_N} = G(j\omega)\big|_{\omega=2}$$

$$\frac{-\pi}{\pi - 2\beta - \sin 2\beta} = -4$$

$$\pi - 2\beta - \sin 2\beta = \frac{\pi}{4} = 0.7854$$

Or $2\beta + \sin 2\beta - 2.3562 = 0$.

This equation can be solved using Newton–Raphson method.

Let $f(\beta) = 2\beta + \sin 2\beta - 2.3562$ $\qquad f'(\beta) = 2 + 2\cos 2\beta = 4\cos^2\beta$.

Iteration	$f(\beta)$	$f'(\beta)$	β
1	0.2434	1.9416	0.8
2	−0.0315	2.4396	0.6746
3	−0.3069x10^{-4}	2.3891	0.6875
4			0.6875

$$\beta = 0.6875 \text{ radian} = 39.39^{\circ}$$

$$M = \frac{a}{\sin\beta} = \frac{2}{\sin 39.39^{\circ}} = 3.15 .$$

Hence the limit cycle is $c(t) = -3.15\sin 2t$ and it is unstable.

Example 7.7 A nonlinear control system is shown in Figure E7.7(a). Using describing function analysis, investigate the possibility of a limit cycle in the system, and if it exists determine its amplitude and frequency when $k = 0.4$. Also assess its stability.

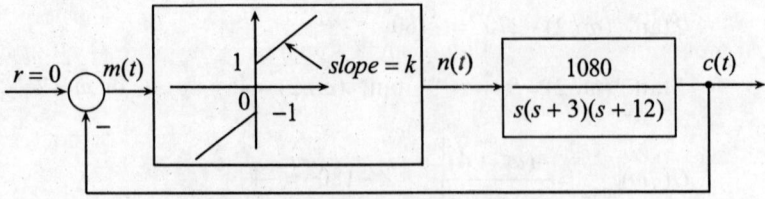

Figure E 7.7 (a) Example 7.7

Solution: The input to the nonlinear element is assumed to be

$$m(t) = M \sin \omega t \, .$$

The describing function for the nonlinear element is

$$G_N = k + \frac{4N}{\pi M}; \ N = 1 \text{ as shown in the figure.}$$

When $M \to 0$, $G_N \to \infty$ and $-\dfrac{1}{G_N} \to 0$

When $M \to \infty$, $G_N \to k$ and $-\dfrac{1}{G_N} \to -\dfrac{1}{k} \, .$

Thus the locus of $-1/G_N$ is along the negative real axis, starting from the origin to the point $-1/k$. This is shown in Figure E 7.7(b).

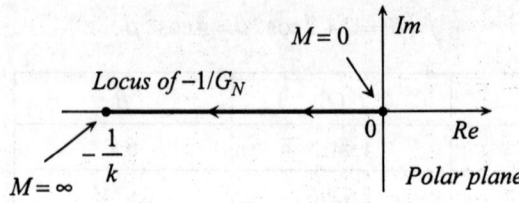

Figure E7.7 (b) Plot of $-1/G_N$ **Figure E7.7 (c) Plot of $G(j\omega)$**

$$G(s) = \frac{1080}{s(s+3)(s+12)}$$

$$G(j\ ) \quad \frac{1080}{j\omega(j\omega+3)(j\omega+12)}$$

$$= \frac{1080}{\omega\sqrt{\omega^2 + 9}\sqrt{\omega^2 + 144}} \angle -90^0 - \tan^{-1}(\omega/3) - \tan^{-1}(\omega/12) \, .$$

When $\omega \to 0, G(j\omega) \to \infty \angle -90^\circ$

When $\omega \to \infty, G(j\omega) \to 0 \angle -270^\circ \, .$

The locus of $G(j\omega)$ which is the polar plot is shown in Figure E7.7(c). The point of intersection of the polar plot on the negative real axis can be found from the expression for the phase angle.

$$-90° - \tan^{-1}(\omega/3) - \tan^{-1}(\omega/12) = -180°$$

$$\tan^{-1}(\omega/3) + \tan^{-1}(\omega/12) = 90°$$

$$\frac{(\omega/3) + (\omega+12)}{1 - \dfrac{\omega^2}{36}} = \infty$$

$$1 - \frac{\omega^2}{36} = 0 \qquad \text{or} \quad \omega = 6 \text{ rad/s}$$

$$G(j\omega)\big|_{\omega=6} = \frac{1080}{6\sqrt{45}\sqrt{180}} \angle -180° = -2.$$

Point P in Figure E 7.7(c) is (–2, 0). If the locus of $G(j\omega)$ intersects the locus of $-1/G_N$, then

$$(1/k) > 2 \qquad \text{or} \quad k < (1/2).$$

Hence for a limit cycle to exist, $k < \dfrac{1}{2}$, and for $k = 0.4$ a limit cycle is predicted.

At the intersection point of $-1/G_N$ locus with $G(j\omega)$ locus,

$$G_N = -\frac{1}{G(j\omega)}\bigg|_{\omega=6}$$

$$k + \frac{4N}{\pi M} = 2$$

When $k = 0.4$ and $N = 1$, $M = 0.796$.

Hence the limit cycle is $c(t) = -0.796\sin 6t$.

Superposing Figures E7.7 (a) and E 7.7(b), it can be observed that as the amplitude M is increased, the $-1/G_N$ locus moves out of the $G(j\omega)$ locus and is not enclosed. Thus the limit cycle is stable.

Example 7.8 Using the describing function analysis, find the range of k for which a limit is predicted for the system shown in Figure E7.8 (a).

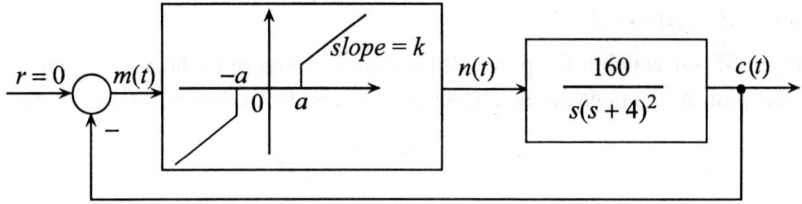

Figure E 7.8 (a) Example 7.8

Solution: The input to the nonlinear element is assumed to be

$$m(t) = M \sin \omega t .$$

The describing function for the nonlinear element is

$$G_N = \frac{4N}{\pi M}\cos \beta + \frac{k}{\pi}[\pi - 2\beta - \sin 2\beta]; \quad M \geq a \qquad a = M \sin \beta \qquad \sin \beta = \frac{a}{M} .$$

When $M \to a$, $\sin \beta \to 1$, $\beta \to \dfrac{\pi}{2}$ $G_N \to 0$ and $-\dfrac{1}{G_N} \to -\infty$

When $M \to \infty$, $\sin \beta \to 0$, $\beta \to 0$ $G_N \to k$ and $-\dfrac{1}{G_N} \to -\dfrac{1}{k}$.

Thus the locus of $-1/G_N$ is along the negative real axis, starting from $-\infty$ to $-1/k$. This is shown in Figure E 7.8(b).

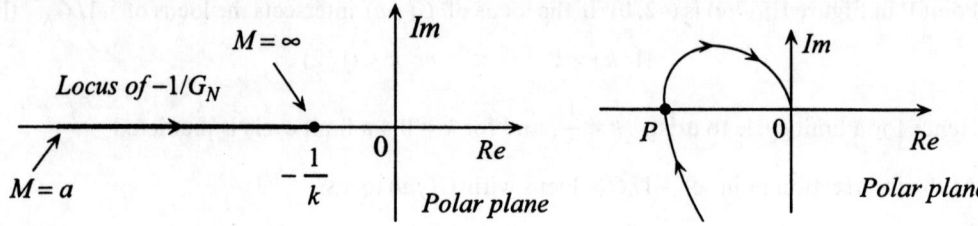

Figure E7.8 (b) Plot of $-1/G_N$ **Figure E7.8 (c)** Plot of $G(j\omega)$

$$G(s) = \frac{160}{s(s+4)^2}$$

$$G(j\omega) = \frac{160}{j\omega(j\omega+4)^2}$$

$$= \frac{160}{\omega(\omega^2 +16)} \angle -90° - 2\tan^{-1}(\omega/4) .$$

When $\omega \to 0, G(j\omega) \to \infty \angle -90°$

When $\omega \to \infty, G(j\omega) \to 0 \angle -270°$.

The locus of $G(j\omega)$ which is the polar plot is shown in Figure E7.8(c). The point of intersection of the polar plot on the negative real axis can be found from the expression for the phase angle.

$$-90° - 2\tan^{-1}(\omega/4) = -180°$$

$$2\tan^{-1}(\omega/4) = 90° \qquad \omega = 4 \text{ rad/s}$$

$$G(j\omega)\big|_{\omega=4} = \frac{160}{4(32)}\angle-180° = -1.25 .$$

Point P in Figure E 7.8(c) is $(-1.25,0)$. If the locus of $G(j\omega)$ intersects the locus of $-1/G_N$, then $(1/k)<1.25$ or $k>0.8$.

Hence for a limit cycle to exist, the range of k is $k>0.8$.

Example 7.9 Using the describing function analysis, check whether a limit is predicted for the system shown in Figure E7.9(a). If for $k = 2$, a limit cycle is predicted, determine its amplitude and frequency and assess its stability.

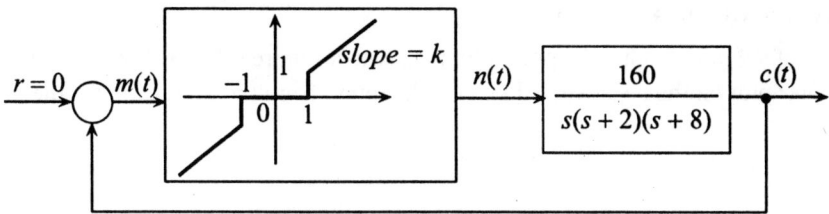

Figure E 7.9 (a) Example 7.9

Solution: The input to the nonlinear element is assumed to be sinusoidal:

$$m(t) = M\sin\omega t .$$

The describing function for the nonlinear element is

$$G_N = \frac{4N}{\pi M}\cos\beta + \frac{k}{\pi}\left[\pi - 2\beta - \sin 2\beta\right] ; \quad M \ge a \qquad a = M\sin\beta .$$

When $M \to a$, $\sin\beta \to 1$, $\beta \to \dfrac{\pi}{2}$ $G_N \to 0$ and $-\dfrac{1}{G_N} \to -\infty$

When $M \to \infty$, $\sin\beta \to 0$, $\beta \to 0$ $G_N \to k$ and $-\dfrac{1}{G_N} \to -\dfrac{1}{k}$.

Thus the locus of $-1/G_N$ is along the negative real axis, starting from $-\infty$ to $-1/k$. This is shown in Figure E 7.9(b).

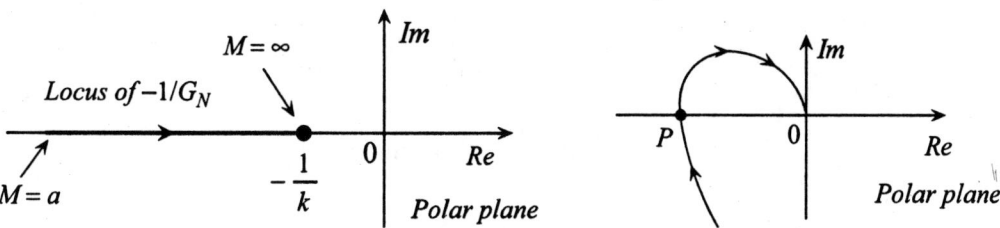

Figure E7.9 (b) Plot of -1/G_N **Figure E7.9 (c)** Plot of $G(j\omega)$

$$G(s) = \frac{160}{s(s+2)(s+8)}$$

$$G(j\omega) = \frac{160}{j\omega(j\omega+2)(j\omega+8)}$$

$$= \frac{160}{\omega\sqrt{\omega^2+4}\sqrt{\omega^2+64}} \angle -90^\circ - \tan^{-1}(\omega/2) - \tan^{-1}(\omega/8)$$

When $\omega \to 0, G(j\omega) \to \infty \angle -90^\circ$

When $\omega \to \infty, G(j\omega) \to 0 \angle -270^\circ$.

The locus of $G(j\omega)$ which is the polar plot is shown in Figure E7.9(c). The point of intersection of the polar plot on the negative real axis can be found from the expression for the phase angle.

$$-90^\circ - \tan^{-1}(\omega/2) - \tan^{-1}(\omega/8) = -180^\circ$$

$$\tan^{-1}(\omega/2) + \tan^{-1}(\omega/8) = 90^\circ$$

$$\frac{(\omega/2)+(\omega+8)}{1-\dfrac{\omega^2}{16}} = \infty$$

$$1 - \frac{\omega^2}{16} = 0 \qquad \text{or} \quad \omega = 4 \text{ rad/s}.$$

$$G(j\omega)\big|_{\omega=4} = \frac{160}{4\sqrt{20}\sqrt{80}} \angle -180^\circ = -1.$$

Point P in Figure E 7.9(c) is (–1,0). If the locus of $G(j\omega)$ intersects the locus of $(-1/G_N)$, then

$$(1/k) < 1 \qquad\qquad \text{or} \quad k > 1.$$

Hence for a limit cycle to exist, $k > 1$ and for the given value of $k = 2$, a limit cycle is predicted. At the intersection point of $(-1/G_N)$ locus with $G(j\omega)$ locus,

$$G_N = -\frac{1}{G(j\omega)}\bigg|_{\omega=4}$$

$$\frac{4N}{\pi M}\cos\beta + \frac{2}{\pi}[\pi - 2\beta - \sin 2\beta] = 1$$

Since $a = 1, N = 1$

$$a = M \sin \beta \quad Or \quad M = \frac{1}{\sin \beta}$$

$$Hence \quad \frac{4}{\pi} \sin \beta \cos \beta + \frac{2}{\pi} [\pi - 2\beta - \sin 2\beta] = 1$$

$$\frac{2}{\pi} \sin 2\beta + \frac{2}{\pi} [\pi - 2\beta - \sin 2\beta] = 1$$

$$\frac{2}{\pi} [\pi - 2\beta] = 1.$$

Solving this equation

$$\beta = \frac{\pi}{4}$$

$$M = \frac{1}{\sin \beta} = \frac{1}{\sin(\pi / 4)} = \sqrt{2}$$

Hence the limit cycle is

$c(t) = -\sqrt{2} \sin 4t$ and it is unstable.

Example 7.10 A system has an ON–OFF relay with dead zone as its nonlinearity. Obtain the describing function G_N of the nonlinearity and plot the locus of $-1/G_N$ in a polar plane.

Solution: The input–output characteristic of the ON–OFF relay with dead zone is shown in Figure E7.10(a). The sinusoidal input to the nonlinear element is

$$m(t) = M \sin \omega t .$$

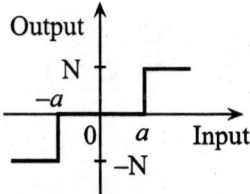

Figure E7.10 (a) Example 7.10

The describing function of the nonlinear element has already been derived as

$$G_N = \frac{4N}{\pi M} \cos \beta; \; M \geq a \quad a = M \sin \beta \qquad\qquad \sin \beta = \frac{a}{M}$$

$$G_N = \frac{4N}{\pi M} \cos \beta = \frac{4N}{\pi M} \sqrt{1 - \sin^2 \beta} = \frac{4N}{\pi M} \sqrt{1 - \left(\frac{a}{M}\right)^2} .$$

When $M = a$, $\sin\beta = 1$ $\cos\beta = 0$ $G_N = 0$ and $(-1/G_N) = -\infty$

When $M \to \infty$, $G_N \to 0$ and $(-1/G_N) \to -\infty$.

Hence when M is varied from a to ∞, G_N increases from zero, reaches a maximum finite non-zero value and then decreases back to zero. This maximum finite value can be determined from the expression for G_N.

Rearranging the expression for G_N,

$$G_N = \frac{4N}{\pi M}\sqrt{1-\left(\frac{a}{M}\right)^2}$$

$$= \frac{4N}{\pi M^2}\sqrt{M^2 - a^2}$$

$$= \frac{4N}{\pi}\frac{a}{M^2}\sqrt{\left(\frac{M}{a}\right)^2 - 1}$$

$$= \frac{4N}{\pi a}\frac{a^2}{M^2}\sqrt{\left(\frac{M}{a}\right)^2 - 1}$$

$$= \frac{4N}{\pi a}\frac{\sqrt{(M/a)^2 - 1}}{(M/a)^2}$$

$$= \frac{4N}{\pi a}\frac{\sqrt{x^2 - 1}}{x^2}, \text{ where } x = (M/a).$$

$$\frac{1}{G_N} = \frac{\pi a}{4N}\frac{x^2}{\sqrt{x^2 - 1}}.$$

Let $y = \dfrac{\pi a}{4N}\dfrac{x^2}{\sqrt{x^2 - 1}}$.

For maximum value of y,

$$\frac{dy}{dx} = \frac{\pi a}{4N}\left[\frac{\sqrt{x^2 - 1}(2x) - \frac{1}{2}x^2(x^2 - 1)^{-\frac{1}{2}}(2x)}{x^2 - 1}\right] = 0.$$

Solving this equation, $x = \sqrt{2}$ and $y_{max} = \dfrac{\pi a}{4N}\dfrac{x^2}{\sqrt{x^2 - 1}}\bigg|_{x=\sqrt{2}} = \dfrac{\pi a}{2N}$.

The locus of $(-1/G_N)$ is shown in Figure E7.10(b)

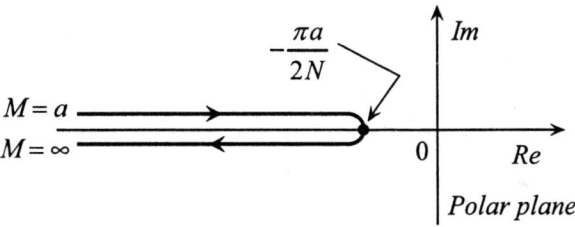

Figure E7.10 (b) Example 7.10

Example 7.11 A nonlinear system has an element with cubic nonlinearity as shown in Figure E7.11(a). Using describing function analysis, assess the stability of the system.

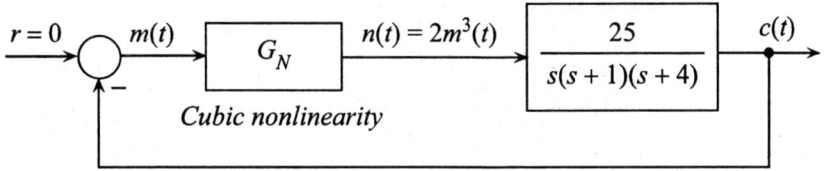

Figure E7.11 (a) Example 7.11

Solution: The input to the nonlinear element is assumed to be sinusoidal.

$$m(t) = M \sin \omega t .$$

The output of the nonlinear element is

$$n(t) = 2 m^3 (t) = 2M^3 \sin^3 \omega t = 2M^3 \frac{1}{4}(3\sin \omega t - \sin 3\omega t).$$

Considering only the fundamental component

$$n(t) = \frac{3}{2} M^3 \sin \omega t .$$

The describing function is

$$G_N = \frac{3}{2} \frac{M^3}{M} = \frac{3}{2} M^2 .$$

When $M \to 0, G_N \to 0$ and $-\dfrac{1}{G_N} \to -\infty$

When $M \to \infty, G_N \to \infty$ and $-\dfrac{1}{G_N} \to 0 .$

The locus of $-1/G_N$ is shown in Figure E7.11(b).

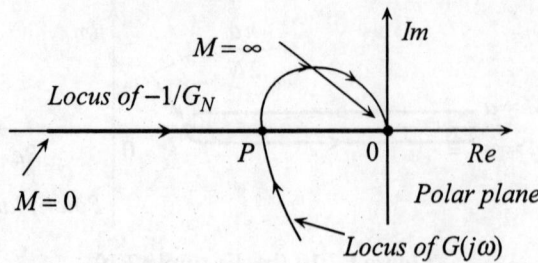

Figure E7.11 (b) Plots of $-1/G_N$ and $G(j\omega)$

$$G(s) = \frac{25}{s(s+1)(s+4)}$$

$$G(j\omega) = \frac{25}{j\omega(j\omega+1)(j\omega+4)}$$

$$= \frac{25}{\omega\sqrt{\omega^2+1}\sqrt{\omega^2+16}}\angle -90^0 -\tan^{-1}\omega -\tan^{-1}(\omega/4).$$

When $\omega \to 0, G(j\omega) \to \infty\angle -90^\circ$
When $\omega \to \infty, G(j\omega) \to 0\angle -270^\circ$.

The locus of $G(j\omega)$ which is the polar plot is also shown in Figure E7.11(b) superimposed over the locus of $-1/G_N$. It can be observed from Figure E 7.11(b) that the locus of $-1/G_N$ intersects the locus of $G(j\omega)$ and a limit cycle is predicted. The point of intersection of the $G(j\omega)$ locus with the negative real axis can be obtained from the expression for the phase angle of $G(j\omega)$.

$$-90^\circ -\tan^{-1}\omega -\tan^{-1}(\omega/4) = -180^\circ$$

$$\tan^{-1}\omega +\tan^{-1}(\omega/4) = 90^\circ$$

$$\frac{\omega+(\omega/4)}{1-\frac{\omega^2}{4}} = \infty .$$

$$1-\frac{\omega^2}{4} \quad \text{or} \quad \omega = 2 \text{ rad/s} .$$

The frequency of the limit cycle is $\omega = 2$ rad/s.

$$G(j\omega)\big|_{\omega=2} = \frac{25}{\omega\sqrt{\omega^2+1}\sqrt{\omega^2+16}}\Bigg|_{\omega=2} = \frac{25}{2\sqrt{5}\sqrt{20}}\angle-180° = -1.25 .$$

At the intersecting point,

$$G_N = -\frac{1}{G(j\omega)}\bigg|_{\omega=2}$$

$$\frac{3}{2}M^2 = \frac{1}{1.25} \qquad \text{or } M = 0.7303 .$$

Hence the limit cycle is $c(t) = -0.7303\sin 2t$ and it is unstable.

Example 7.12 The describing function of a hypothetical nonlinear element is

$$G_N = \frac{1}{M}\angle(-45° - \tan^{-1}\frac{1}{M})$$

The plant transfer function is

$$G(s) = \frac{1}{s} .$$

Check whether a limit cycle is predicted. If a limit cycle is predicted, determine its amplitude and frequency and assess its stability.

Solution: The describing function of the nonlinear element is

$$G_N = \frac{1}{M}\angle(-45° - \tan^{-1}\frac{1}{M}) \qquad\qquad -G_N = \frac{1}{M}\angle(135° - \tan^{-1}\frac{1}{M})$$

$$-\frac{1}{G_N} = M\angle\left(-135° + \tan^{-1}\frac{1}{M}\right) . \qquad\qquad (1)$$

When $M \to 0$, $-\dfrac{1}{G_N} \to 0\angle-45°$

When $M \to \infty$, $-\dfrac{1}{G_N} \to \infty\angle-135°$.

The locus of $(-1/G_N)$ is shown in Figure E7.12.

$$G(s) = \frac{1}{s} \qquad\qquad G(j\omega) = \frac{1}{j\omega} = \frac{1}{\omega}\angle-90° . \qquad\qquad (2)$$

The locus of $G(j\omega)$ is the negative imaginary axis.

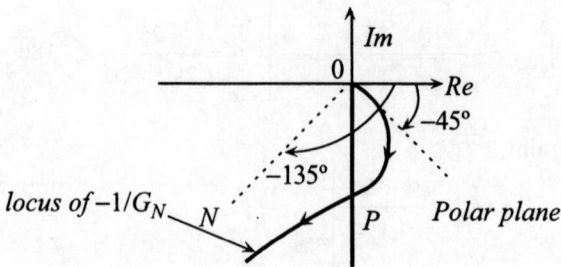

Figure E7.12 Example 7.12

It can be observed that there is an intersecting point and hence a limit cycle is predicted. Using equation (1), at the intersecting point,

$$-135° + \tan^{-1}\frac{1}{M} = -90°$$

$$\tan^{-1}\frac{1}{M} = 45° \qquad\qquad M = 1.$$

Using equation (2),

$$\frac{1}{\omega}\angle -90° = 1\angle -90° \qquad \text{or} \quad \omega = 1.$$

Hence the limit cycle is

$$c(t) = -1.0\ \sin t\ .$$

Example 7.13 For a system shown in Figure 7.13, check whether a limit cycle exists by inspecting the roots of the characteristic equation.

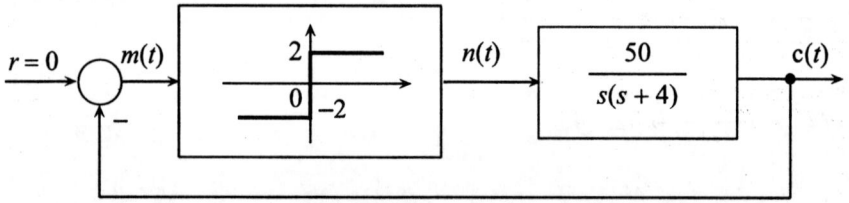

Figure E7.13 Example 7.13

Solution: The describing function of the nonlinear element preceding the plant is

$$G_N = \frac{4N}{\pi M} = \frac{8}{\pi M}\ .$$

The characteristic equation is

$$1 + G_N G(s) = 0$$

$$1 + \frac{8}{\pi M} \frac{50}{s(s+4)} = 0$$

$$\pi M s^2 + 4\pi M s + 400 = 0$$

$$s^2 + 4s + \frac{400}{\pi M} = 0 .$$

The roots of this equation are

$$s = -2 \pm \sqrt{4 - \frac{400}{\pi M}} .$$

Since there is always a negative real part, the system does not exhibit a limit cycle.

Example 7.14 For the system shown in Figure E7.14, check whether a limit cycle is predicted.

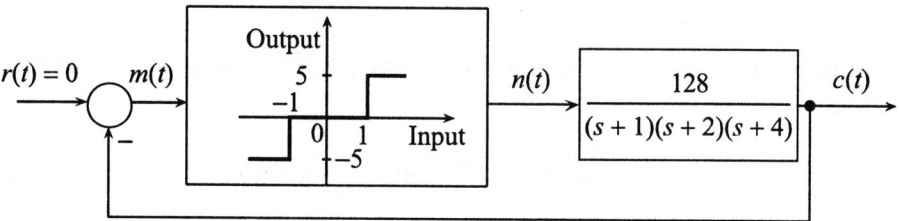

Figure E 7.14 Example 7.14

Solution: The describing function of the nonlinear element is

$$G_N = \frac{4N}{\pi M} \cos \beta; \; \beta = \sin^{-1} \frac{a}{M} .$$

In this case $a = 1$ and $N = 5$.

$$G_N = \frac{4N}{\pi M} \cos \beta; \; M \geq a \quad a = M \sin \beta \qquad\qquad \sin \beta = \frac{a}{M}$$

$$G_N = \frac{4N}{\pi M} \cos \beta = \frac{4N}{\pi M} \sqrt{1 - \sin^2 \beta} = \frac{4N}{\pi M} \sqrt{1 - \left(\frac{a}{M}\right)^2}$$

When $M = a$, $\sin \beta = 1$ $\cos \beta = 0$ $G_N = 0$ and $(-1/G_N) = -\infty$
When $M \to \infty$, $G_N \to 0$ and $(-1/G_N) \to -\infty$.
The critical point Q on the negative real axis is $(-\pi a / 2N) = -0.314$. This is shown in Figure E7.14(a).

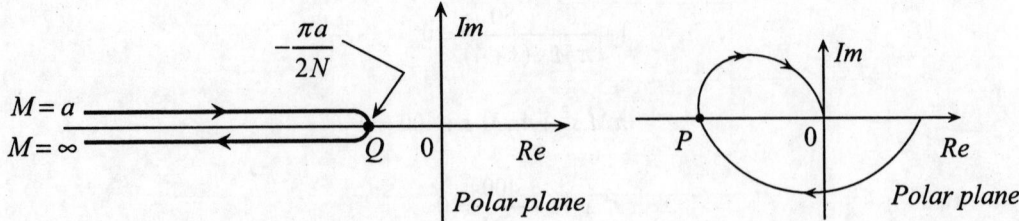

Figure E7.14 (a) Example 7.14   **Figure E7.14 (b)** Example 7.14

$$G(s) = \frac{128}{(s+1)(s+2)(s+4)}$$

$$G(j\omega) = \frac{128}{(j\omega+1)(j\omega+2)(j\omega+4)}$$

$$= \frac{128}{\sqrt{(\omega^2+1)(\omega^2+4)(\omega^2+16)}} \angle -\tan^{-1}\omega - \tan^{-1}\frac{\omega}{2} - \tan^{-1}\frac{\omega}{4}$$

To find the point of intersection P of the $G(j\omega)$ locus on the negative real axis, solve the equation

$$-\tan^{-1}\omega - \tan^{-1}\frac{\omega}{2} - \tan^{-1}\frac{\omega}{4} = -180^\circ.$$

$$\tan^{-1}\frac{\omega}{2} + \tan^{-1}\frac{\omega}{4} = 180^0 - \tan^{-1}\omega$$

$$\frac{\frac{\omega}{2}+\frac{\omega}{4}}{1-\frac{\omega^2}{8}} = -\omega.$$

Solving this equation, $\sqrt{14}$.

$$G(j\omega)\big|_{\omega=\sqrt{14}} = \frac{128}{\sqrt{(15)(18)(30)}} \angle -180^\circ = -1.422.$$

The intersection point is shown in Figure E7.14(b).
Since point Q lies to the right of P on the negative real axis, the $-1/G_N$ locus intersects the $G(j\omega)$ locus and hence limit cycle exists.

Example 7.15 A nonlinear system is shown in Figure E7.15. Check if a limit cycle is predicted. If a limit cycle exists, determine its magnitude and frequency.

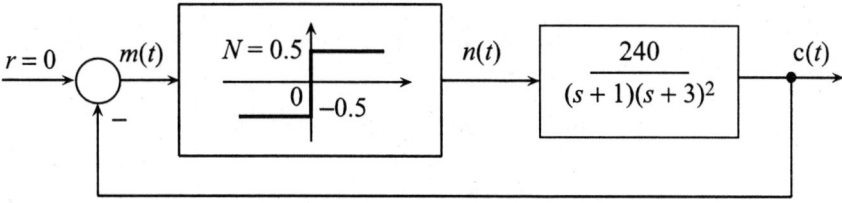

Figure E7.15 Example 7.15

Solution: The describing function of the nonlinear element is

$$G_N = \frac{4N}{\pi M} = \frac{2}{\pi M} \qquad\qquad -\frac{1}{G_N} = -\frac{\pi M}{2}.$$

When $M \to 0$, $-1/G_N \to 0$

When $M \to \infty$, $-1/G_N \to -\infty$.

The locus of $-1/G_N$ is shown in Figure E7.15(a).

$$G(s) = \frac{240}{(s+1)(s+3)^2}$$

$$(j\ ) \quad \frac{240}{(j\omega+1)(j\omega+3)}$$

$$= \frac{240}{\sqrt{(\omega^2 +1)\,(\omega^2 +9)}} \angle -\tan^{-1}\omega - 2\tan^{-1}\frac{\omega}{3}.$$

When $\omega \to 0$, $G(j\omega) \to 26.67\angle 0°$

When $\omega \to \infty$, $G(j\omega) \to 0\angle 270°$.

The locus of $G(j\omega)$ is shown in Figure E7.15(b).

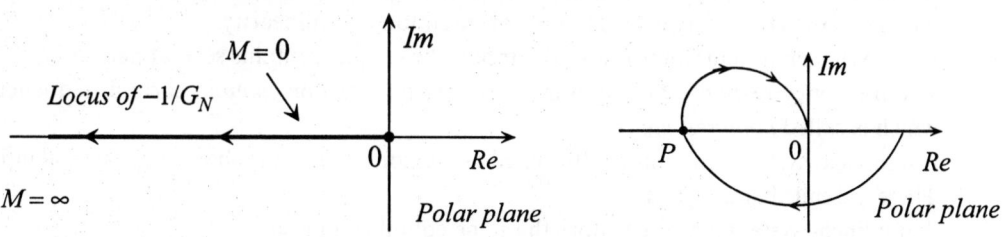

Figure E 7.15 (a) Example 7.15   **Figure E7.15** (b)

The intersection point P can be determined by solving the equation

$$-\tan^{-1}\omega - 2\tan^{-1}\frac{\omega}{3} = -180°.$$

$$\tan^{-1}\frac{\omega}{3} + \tan^{-1}\frac{\omega}{3} = 180° - \tan^{-1}\omega$$

$$\frac{\dfrac{\omega}{3} + \dfrac{\omega}{3}}{1 - \dfrac{\omega^2}{9}} = -\omega.$$

Solving this equation, $\omega = \sqrt{15}$.

$$G(j\omega)\Big|_{\sqrt{15}} = \frac{240}{\sqrt{(\omega^2 + 1)(\omega^2 + 9)}}\Bigg|_{\sqrt{15}} = -2.5.$$

The $G(j\omega)$ locus intersects the $-1/G_N$ locus at the point $(-2.5, 0)$. A limit cycle exists. At the intersecting point

$$-\frac{\pi M}{2} = -2.5 \qquad\qquad \text{Hence } M = 1.59.$$

The limit cycle is $c(t) = -1.59\sin\sqrt{15}\,t$.

SUMMARY

- For the analysis of a nonlinear system, the principle of superposition is not applicable.
- Theories and techniques applicable to linear systems for analysis and design are invalid for nonlinear systems.
- Nonlinearities in physical systems can be classified into *inherent* and *intentional*.
- Saturation, dead zone, hysteresis are certain properties inherent to nonlinear systems.
- Intentional nonlinearities are those which are deliberately introduced in a system for the convenience of simple operation and making the system cost effective.
- An ideal ON–OFF relay is an example of intentional nonlinearity.
- In a nonlinear system, for a sinusoidal input, the output contains several harmonics.
- The frequency response of a nonlinear system may exhibit *jump* at certain frequencies, which is called *jump resonance*.
- A nonlinear system can exhibit self-excited oscillations even in the absence of external input. These are called limit cycles.
- A nonlinear system may have more than one equilibrium point.
- The *describing function* of a nonlinear element is defined as the ratio of the fundamental component of the output in phasor form to the sinusoidal input in phasor form.
- The describing function of a nonlinear element is derived based on the assumption that only the magnitude of the fundamental component of the output is significant and the harmonics are negligibly small.

- This assumption is valid because most of the control systems behave as low pass filters and higher harmonics are attenuated.
- Using describing function analysis, existence of a limit cycle can be predicted, its magnitude and frequency can be determined and stability can be assessed.

PRACTICE PROBLEMS

PP 7.1 Use the describing function analysis to investigate the possibility of a limit cycle for the nonlinear system shown in Figure PP7.1. If a limit cycle is predicted, determine its amplitude and frequency and investigate its stability.

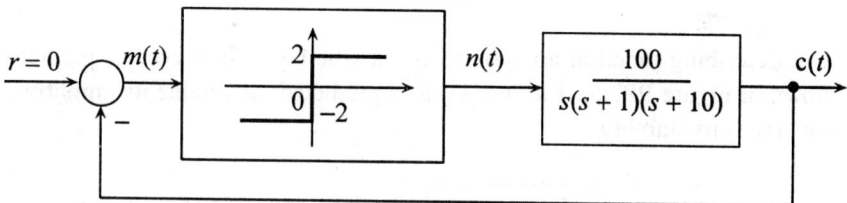

Figure PP7.1

PP 7.2 Use the describing function analysis to investigate the possibility of a limit cycle for the nonlinear system shown in Figure PP7.2. If a limit cycle is predicted, determine its amplitude and frequency and investigate its stability.

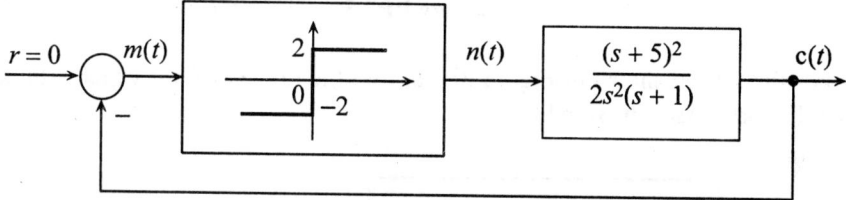

Figure PP7.2

PP 7.3 Use the describing function analysis to investigate the possibility of a limit cycle for the nonlinear system shown in Figure PP7.3. If a limit cycle is predicted, determine its amplitude and frequency and investigate its stability.

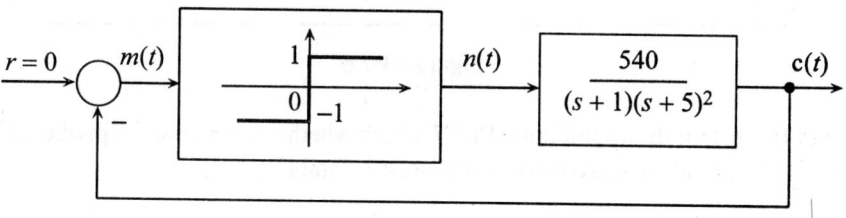

Figure PP7.3

PP 7.4 A nonlinear system is shown in Figure PP7.4. Find the range of k for which a limit cycle is predicted and assess the stability of the system.

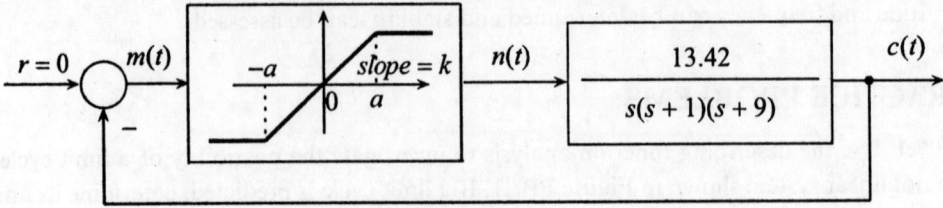

Figure PP7.4

PP 7.5 Use describing function analysis to check whether a limit cycle is predicted for the system shown in Figure PP7.5. If a limit cycle is predicted, determine its amplitude and frequency and assess its stability.

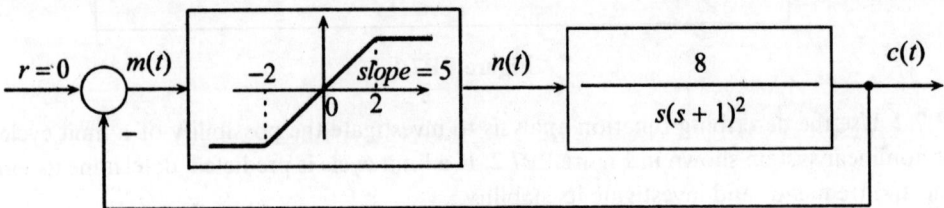

Figure PP7.5

PP 7.6 For the system shown in Figure PP7.6, find the range of k for which a limit cycle is predicted. If exists, assess its stability.

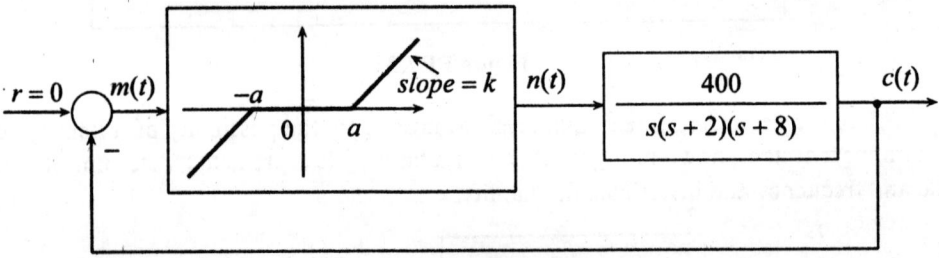

Figure PP7.6

PP 7.7 For the system shown in Figure PP7.7, check whether a limit cycle is predicted. If exists, determine its amplitude and frequency and assess its stability.

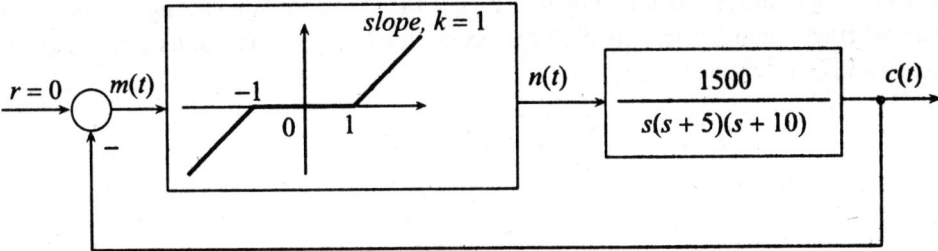

Figure PP7.7

PP 7.8 A nonlinear system is shown in Figure PP 7.8. Investigate the possibility of a limit cycle. If exists, dètermine its amplitude and frequency and assess its stability.

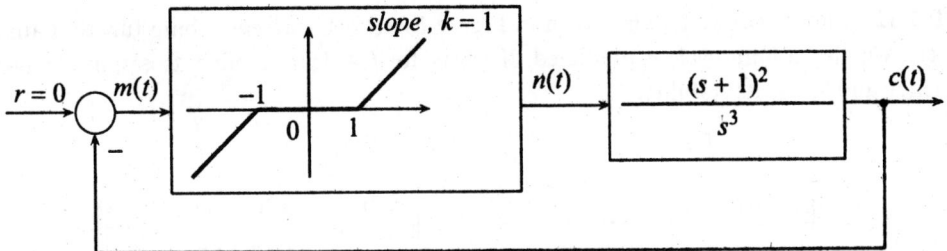

Figure PP7.8

PP 7.9 A system with a nonlinearity is shown in Figure PP7.9. Using describing function analysis, check whether a limit cycle is predicted. If exists, investigate its stability.

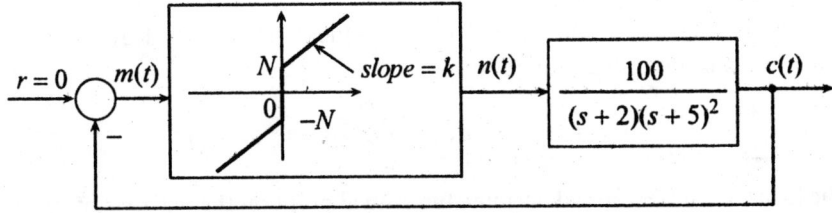

Figure PP7.9

PP 7.10 A nonlinear system is shown in Figure PP7.10. Using describing function analysis, check whether a limit cycle is predicted. If exists, determine its amplitude and frequency and assess its stability.

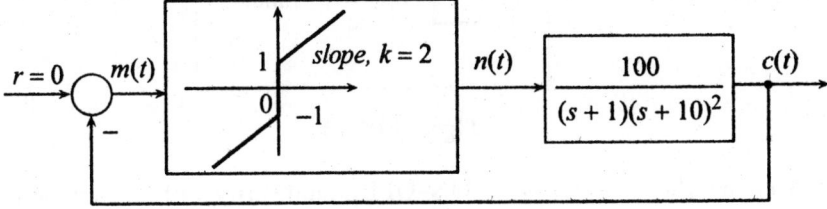

Figure PP7.10

PP 7.11 A nonlinear system is shown in Figure PP7.11. Using describing function analysis, check whether a limit cycle is predicted. If exists, for $k = 4$, determine its amplitude and frequency and assess its stability.

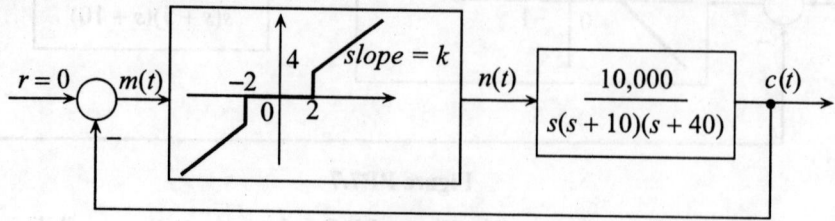

Figure PP7.11

PP 7.12 A nonlinear system is shown in Figure PP7.12. Using describing function analysis, check whether a limit cycle is predicted. If exists, for $k = 1$, determine its amplitude and frequency and assess its stability.

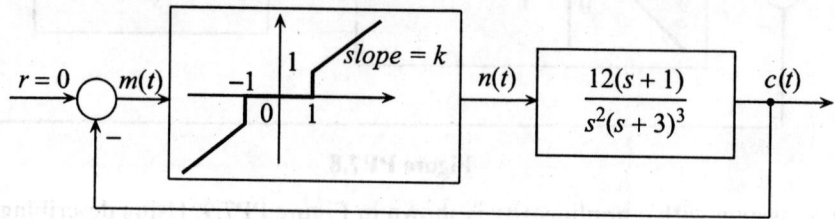

Figure PP7.12

PP 7.13 The input–output characteristic of an ideal relay with dead zone is shown in Figure PP7.13. The describing function is

$$G_N = \frac{4N}{\pi M} \cos \beta, \quad \text{where } \beta = \sin^{-1}(a/M).$$

Draw the locus of $-1/G_N$ and determine the critical point on the real axis of the polar plane.

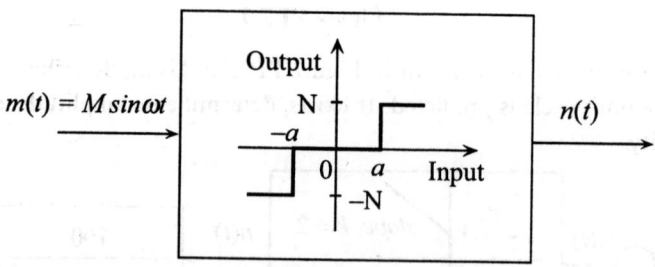

Figure PP7.13

PP 7.14 A system that consists of an ON–OFF nonlinearity with dead zone is shown in Figure PP7.14. Find the range of a, for which there is the possibility of a limit cycle.

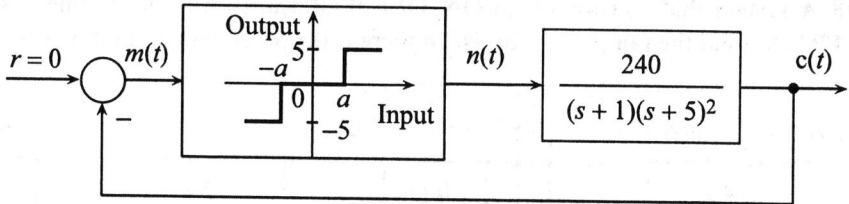

Figure PP7.14

PP 7.15 A system that consists of an ON–OFF nonlinearity with dead zone is shown in Figure PP7.15. Find the range of N for which there is the possibility of a limit cycle.

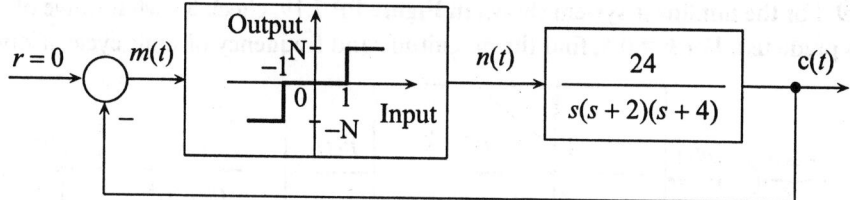

Figure PP7.15

PP 7.16 A hypothetical nonlinear element shown in Figure PP7.16 has the describing function

$$G_N = \frac{1}{M}\angle -45^0 .$$

Investigate the possibility of a limit cycle. If there exists one, determine its amplitude and frequency and assess its stability.

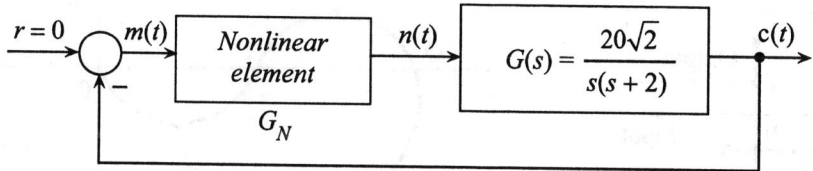

Figure PP7.16

PP 7.17 For the nonlinear system shown in Figure PP 7.17, check whether a limit cycle exists, by inspecting the roots of the characteristic equation.

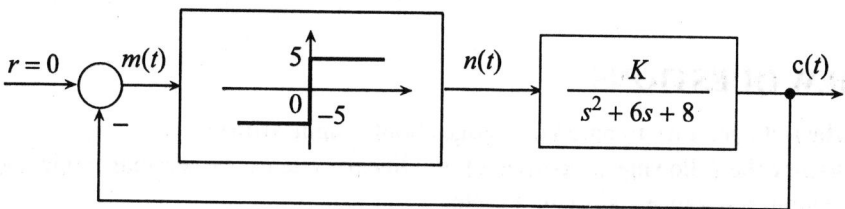

Figure PP 7.17

PP 7.18 A system that consists of an ON–OFF nonlinearity with dead zone is shown in Figure PP7.18. Find the range of N for which there is the possibility of a limit cycle.

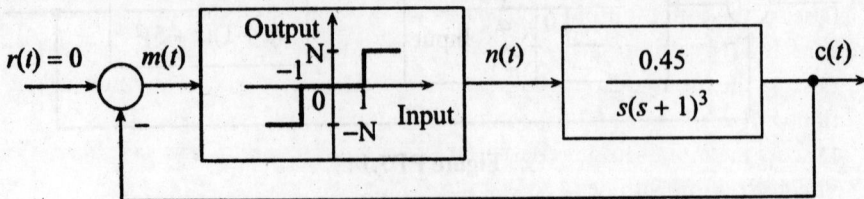

Figure PP7.18

PP 7.19 For the nonlinear system shown in Figure PP 7.19, check for what range of k, a limit cycle is predicted. For $k = 0.5$, find the magnitude and frequency of limit cycle, if one exists.

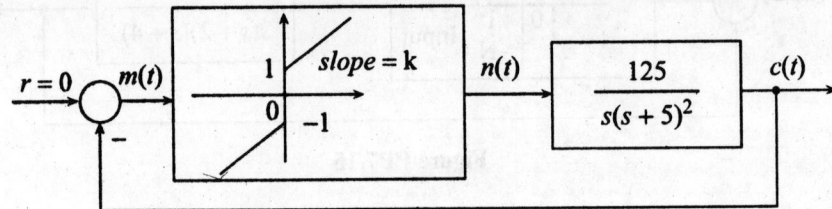

Figure PP7.19

PP 7.20 A nonlinear system consists of a nonlinear element shown in figure PP7.20 (a). The polar plot of the plant transfer function of the system is shown in Figure PP7.20 (b). Assess the stability of the limit cycles, if any.

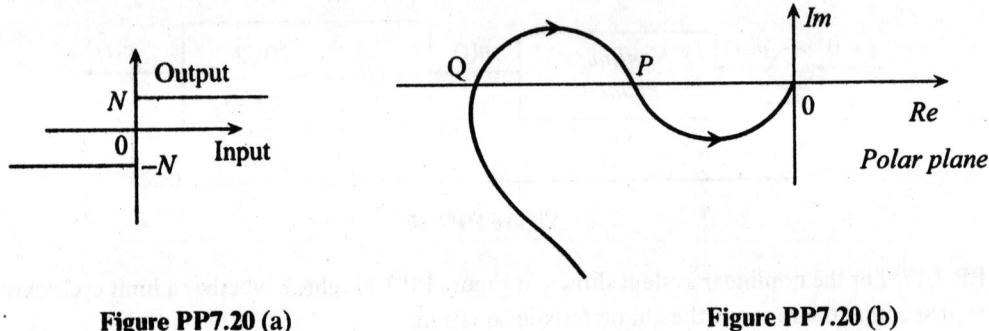

Figure PP7.20 (a)   Figure PP7.20 (b)

REVIEW QUESTIONS

7.1   Why is it necessary to have knowledge about nonlinearities?

7.2   Discuss the following properties of nonlinear systems: (a) sub-harmonic oscillations, (b) jump resonance and (c) limit cycle.

7.3   Which are the different classes of nonlinearities? Explain.

7.4 What is an ON–OFF nonlinearity? Explain.

7.5 Explain the input–output characteristics of an ideal relay with dead zone.

7.6 Discuss the input–output characteristics of an ON–OFF nonlinearity with hysteresis.

7.7 Discuss the input–output characteristics of following types of nonlinearities: (a) dead zone and (b) saturation.

7.8 Discuss the input–output characteristics of following types of nonlinearities: (a) saturation with dead zone and (b) combined Coulomb and viscous friction.

7.9 Discuss the input–output characteristics of a nonlinearity with combined Coulomb and viscous friction with dead zone.

7.10 Discuss the input–output characteristics of an ON–OFF nonlinearity with hysteresis and dead zone.

7.11 Define the describing function of a nonlinear element.

7.12 List the assumptions made in deriving the describing function of a nonlinear element.

7.13 Derive the describing function for an ideal relay.

7.14 Derive the describing function for an ideal relay with dead zone in its input–output characteristics.

7.15 Obtain the describing function for a nonlinear element with dead zone in its input–output characteristics.

7.16 Derive the describing function for a nonlinear element with saturation in its input–output characteristics.

7.17 Derive the describing function for a nonlinear element with saturation and dead zone in its input–output characteristics.

7.18 Obtain the describing function for an ideal relay with hysteresis in its input–output characteristics.

7.19 Obtain the describing function for a nonlinear element with hysteresis and dead zone in its input–output characteristics.

7.20 Derive the describing function for a nonlinear element with Coulomb and viscous friction.

7.21 Derive the describing function for a nonlinear element with Coulomb and viscous friction with dead zone in its input–output characteristics.

7.22 A nonlinear spring has input–output characteristics
$n(t) = k_1 m(t) + k_2 m^3(t)$, where $m(t) = M \sin \omega t$.
Derive the describing function for the spring.

7.23 In what way the occurrence of a limit cycle can be predicted, using describing function analysis?

7.24 Explain the method of assessing the stability of a limit cycle.

7.25 Discuss the method of determining the magnitude and frequency of a limit cycle, if it exists.

7.4 What is an ON-OFF nonlinearity? Explain.

7.5 Explain the input-output characteristics of an ideal relay with dead zone.

7.6 Discuss the input-output characteristics of an ON-OFF nonlinearity with hysteresis.

7.7 Discuss the input-output characteristics of following types of nonlinearities: (a) dead zone and (b) saturation.

7.8 Discuss the input-output characteristics of following types of nonlinearities: (a) saturation with dead zone and (b) combined Coulomb and viscous friction.

7.9 Discuss the input-output characteristics of a nonlinearity with combined Coulomb and viscous friction with dead zone.

7.10 Discuss the input-output characteristics of an ON-OFF nonlinearity with hysteresis and dead zone.

7.11 Define the describing function of a nonlinear element.

7.12 List the assumptions made in deriving the describing function of a nonlinear element.

7.13 Derive the describing function for an ideal relay.

7.14 Derive the describing function for an ideal relay with dead zone in its input-output characteristics.

7.15 Obtain the describing function for a nonlinear element with dead zone in its input-output characteristics.

7.16 Derive the describing function for a nonlinear element with saturation in its input-output characteristics.

7.17 Derive the describing function for a nonlinear element with saturation and dead zone in its input-output characteristics.

7.18 Obtain the describing function for an ideal relay with hysteresis in its input-output characteristics.

7.19 Obtain the describing function for a nonlinear element with hysteresis and dead zone in its input-output characteristics.

7.20 Derive the describing function for a nonlinear element with combined Coulomb and viscous friction.

7.21 Derive the describing function for a nonlinear element with combined Coulomb and viscous friction with dead zone in its input-output characteristics.

7.22 A nonlinear spring has its input-output characteristics. Derive the describing function for the spring.

7.23 In what way the occurrence of a limit cycle can be predicted using describing function analysis?

7.24 Explain the method of assessing the stability using describing function.

7.25 Discuss the method of determining the amplitude and frequency of a limit cycle if it exists.

8   Phase Plane Analysis

8.1 THE PHASE PLANE AND THE PHASE TRAJECTORY

The phase plane method of analysis is a graphical procedure for determining the nature of transient response of a system. It is applicable only to a first-order or to a second-order system. Consider a second-order system, the dynamics of which are described by the differential equation

$$\ddot{y} + 2\dot{y} + 4y = 0,$$

where y is the output of the system.

This equation represents the free response of the system as the input is zero.

Let $x_1 = y$ and $x_2 = \dot{x}_1 = \dot{y}$ be the two state variables. The given differential equation can be written down as

$$\dot{x}_2 + 2x_2 + 4x_1 = 0.$$

The state equations are

$$\dot{x}_1 = x_2 \quad \text{and} \quad \dot{x}_2 = -2x_2 - 4x_1.$$

The characteristic equation is

$$s^2 + 2s + 4 = 0.$$

This is equivalent to

$$s^2 + 2\delta\omega_n s + \omega_n^2 = 0.$$

The natural frequency of oscillation of the response is $\omega_n = 2$ rad/s and the damping ratio is $\delta = 0.5$. Since the damping ratio is less than unity, the system is under-damped and the system response is sinusoidal with exponential decay. Thus, for given initial conditions $x_1(0)$ and $x_2(0)$, $x_1(t)$ can be plotted as shown in Figure 8.1. Since $x_2(t)$ is the derivative of $x_1(t)$, it can also be plotted for a given initial condition $x_2(0)$. This is also shown in Figure 8.1.

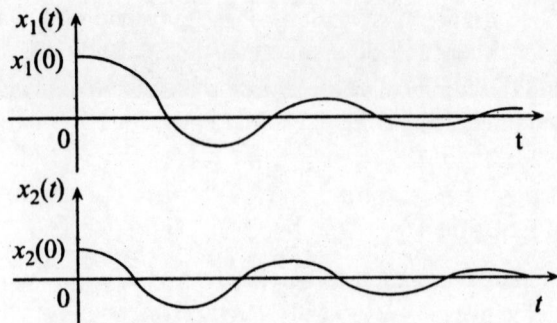

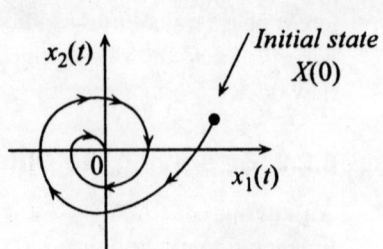

Figure 8.1 Plots of $x_1(t)$ and $x_2(t)$ versus t **Figure 8.2** Phase trajectory

The plots of $x_1(t)$ and $x_2(t)$ are oscillatory in nature and exponentially decay down to zero as t tends to infinity. From these plots, a sketch of $x_2(t)$ versus $x_1(t)$ can be drawn. This is shown in Figure 8.2 and is called the *phase trajectory*. This is also sometimes called the *state trajectory*. The $x_1 - x_2$ plane is called the *phase plane*. The variables $x_1(t)$ and $x_2(t)$ are state variables which are the rectangular co-ordinates in the phase plane. The initial state $(x_1(0), x_2(0))$ is shown as the starting point of the phase trajectory.

8.2 ANALYTICAL METHOD OF CONSTRUCTION OF PHASE TRAJECTORY

8.2.1 Using the Expression for the Slope of Phase Trajectory

Consider a differential equation that describes the dynamics of a second-order system

$$\ddot{y} + a\dot{y} + by = c . \tag{8.1}$$

Let $x_1 = y$ and $x_2 = \dot{x}_1 = \dot{y}$.
Hence the state equations are

$$\dot{x}_1 = x_2 \tag{8.2}$$

$$\dot{x}_2 = -ax_2 - bx_1 + c \tag{8.3}$$

$$\frac{d x_2}{d x_1} = \frac{(dx_2 / dt)}{(dx_1 / dt)} = \frac{\dot{x}_2}{\dot{x}_1} = \frac{-bx_1 - ax_2 + c}{x_2} = \frac{f(x_1, x_2)}{x_2} . \tag{8.4}$$

The solution of this first-order equation can be written as

$$x_2 = g(x_1) . \tag{8.5}$$

This can be plotted in the x_1-x_2 plane which gives the phase trajectory. It is to be observed that in the process of obtaining equation (8.5) from equations (8.2) and (8.3), the independent variable t has been eliminated and (dx_2 / dx_1) represents the slope of the tangent drawn at any point on the phase trajectory in the $x_1 - x_2$ plane. For different sets of initial conditions, different phase trajectories can be drawn and this family of phase trajectories is known as a *phase portrait*. Since the phase trajectories obtained using different initial conditions are unique, they do not cross one another.

8.2.2 By Solving the Differential Equation

An alternate method is to solve the differential equation (8.1) for $x_1(t) = y$ as a function of time t and then to obtain $x_2(t)$ as the derivative of $x_1(t)$. Then from the expressions of $x_1(t)$ and $x_2(t)$, the independent variable t can be eliminated. The process of elimination of t may not be always easy and sometimes may not be possible.

8.3 DIRECTION OF PHASE TRAJECTORY

The state variable x_2 is $x_2 = (dx_1 / dt)$. For positive values of x_2, (dx_1 / dt) is positive. This means that the value of x_1 increases with time t. Hence, for values of x_2 above the $x_1 - axis$, where x_2 is positive, the direction of the phase trajectory is in the increasing direction of x_1 and this is indicated by an arrow on the trajectory directed from left to right. Similarly for the portion of the trajectory below the $x_1 - axis$, where x_2 is negative, the direction of the arrow on the phase trajectory is from right to left. This is shown in Figure 8.3.

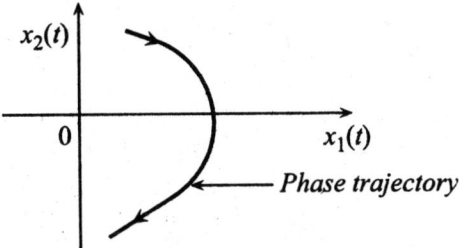

Figure 8.3 Direction of phase trajectory

8.4 PROPERTIES OF PHASE TRAJECTORY

1. From equation (8.4)

$$\frac{dx_2}{dx_1} = \frac{(dx_2 / dt)}{(dx_1 / dt)} = \frac{\dot{x}_2}{\dot{x}_1} = \frac{f(x_1, x_2)}{x_2},$$

where (dx_2 / dx_1) is the slope of the trajectory at any point on it.

On the x_1-axis, x_2 is zero. Thus at a point on the x_1 axis, $(dx_2 / dx_1) = \infty$, or $(dx_2 / dx_1) = \tan \theta;\ \theta = 90°$. This means that the phase trajectory crosses the x_1-axis at right angles. This is illustrated in Figure 8.4.

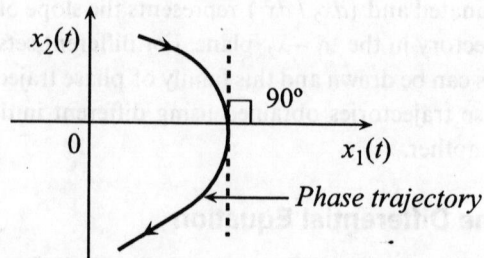

Figure 8.4 Phase trajectory crossing the x_1-axis

2. If $f(x_1, x_2)$ is a single-valued function, at any point on the phase trajectory, the slope has one and only one value, or the slope is unique. Hence two phase trajectories, drawn using two different initial conditions, are unique so that they do not intersect each other. Hence no two phase trajectories cross each other.

3. The slope of the trajectory at any point is

$$\frac{d x_2}{d x_1} = \frac{(dx_2 / dt)}{(dx_1 / dt)} = \frac{\dot{x}_2}{\dot{x}_1} = \frac{f(x_1, x_2)}{x_2}.$$

The slope of the trajectory at any point above the x_1-axis is

$$\frac{dx_2}{dx_1} = \frac{f(x_1, x_2)}{x_2}.$$

The slope of the trajectory at any point below the x_1-axis is

$$\frac{dx_2}{dx_1} = \frac{f(x_1, -x_2)}{-x_2}.$$

The phase trajectory will be symmetrical about the x_1-axis if these slopes are equal and opposite for all x_1. This is shown in Figure 8.5. Hence the phase trajectory will be symmetrical about the x_1-axis if

$$\frac{f(x_1, x_2)}{x_2} = -\frac{f(x_1, -x_2)}{-x_2}.$$

Or

$$f(x_1, x_2) = f(x_1, -x_2).$$

This shows that the phase trajectory is symmetrical about the x_1-axis if $f(x_1, x_2)$ is an even function of x_2.

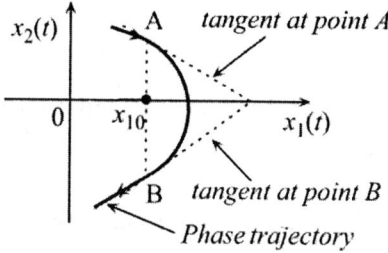

Figure 8.5 Symmetry about the
x_1 – axis

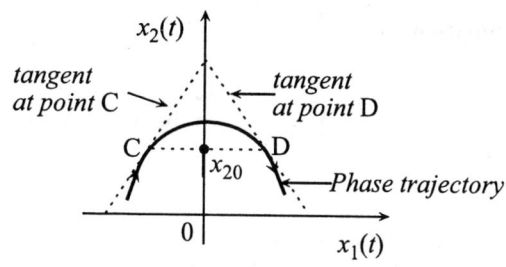

Figure 8.6 Symmetry about the
x_2 – axis

4. The slope of the trajectory at any point to the right of x_2 – axis is

$$\frac{dx_2}{dx_1} = \frac{f(x_1, x_2)}{x_2}.$$

The slope of the trajectory at any point to the left of x_2 – axis is

$$\frac{dx_2}{dx_1} = \frac{f(-x_1, x_2)}{x_2}.$$

The phase trajectory will be symmetrical about the x_2 – axis if these slopes are equal and opposite for all x_2. This is shown in Figure 8.6. Hence the phase trajectory will be symmetrical about the x_2 – axis if

$$\frac{f(x_1, x_2)}{x_2} = -\frac{f(-x_1, x_2)}{x_2}.$$

Or

$$f(x_1, x_2) = -f(-x_1, x_2).$$

This shows that the phase trajectory is symmetrical about the x_2 – axis if $f(x_1, x_2)$ is an odd function of x_1.

Following are some salient features:

1. A phase trajectory graphically shows the transient response of the system starting from an initial state to a final state as time approaches infinity.
2. The existence of a limit cycle can be investigated using the phase portrait.
3. The phase trajectory is generally drawn only for a first-order or second-order system with no input or with constant input.
4. Generally nonlinear systems cannot be easily solved using the analytical methods. In such cases, the phase plane method can be conveniently used with advantage.

Example 8.1 Draw the phase trajectory of a linear first-order system represented by the differential equation $\dot{y} + y = 0$.

Solution:

$$\dot{y} + y = 0$$

Let $x_1 = y$ and $x_2 = \dot{x}_1 = \dot{y}$ be the two state variables. Hence the given differential equation becomes

$$x_2 = -x_1 . \tag{1}$$

This represents a straight line with a slope of -1 and passing through the origin of the $x_1 - x_2$ plane and is shown in Figure E8.1.

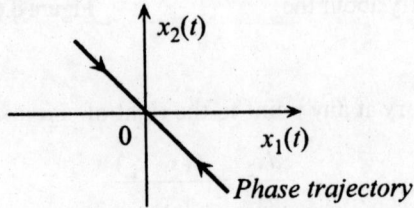

Figure E8.1 Example 8.1

Example 8.2 Draw the phase trajectory of a first-order nonlinear system represented by the differential equation $\dot{y} + 2y - 2y^3 = 0$.

Solution:

$$\dot{y} + 2y - 2y^3 = 0$$

Let $x_1 = y$ and $x_2 = \dot{x}_1 = \dot{y}$ be the two state variables. Hence the given differential equation becomes

$$x_2 + 2x_1 - 2x_1^3 = 0 .$$

Or $x_2 = -2x_1 + 2x_1^3 = -2x_1(1 - x_1^2) = -2x_1(1 - x_1)(1 + x_1) .$

The phase trajectory crosses the $x_1 - axis$ when $x_2 = 0$. In this case, the phase trajectory crosses the $x_1 - axis$ when $x_1 = 0, x_1 = 1$ and $x_1 = -1$ and the phase trajectory is shown in Figure E8.2.

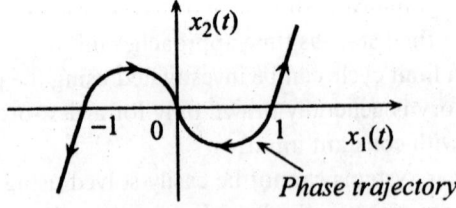

Figure E8.2 Example 8.2

Example 8.3 A linear second-order system is represented by the differential equation $\ddot{y} + k\,y = 0$. Draw the phase trajectory, using the expression for the slope of the trajectory at any point.

Solution:

$$\ddot{y} + ky = 0$$

Let $x_1 = y$ and $x_2 = \dot{x}_1 = \dot{y}$ be the two state variables. Hence the given differential equation becomes

$$\dot{x}_2 + kx_1 = 0 \qquad \text{or } \dot{x}_2 = -kx_1.$$

$$\frac{dx_2}{dx_1} = \frac{(dx_2 / dt)}{(dx_1 / dt)} = \frac{\dot{x}_2}{\dot{x}_1} = \frac{\dot{x}_2}{x_2} = \frac{-kx_1}{x_2}$$

This can be written as

$$x_2 \, dx_2 = -kx_1 \, dx_1 \, .$$

Integrating both sides

$$\frac{1}{2}x_2^2 = -\frac{k}{2}x_1^2 + \frac{1}{2}c^2 \qquad \text{or } x_2^2 + kx_1^2 = c^2 \, .$$

Or

$$\frac{x_1^2}{(c^2 / k)} + \frac{x_2^2}{c^2} = 1 \, .$$

This represents an ellipse with semi-major axis $a = c / \sqrt{k}$ and semi-minor axis $b = c$. This is shown in Figure E8.3. The constant c can be evaluated using the initial conditions $y(0)$ and $\dot{y}(0)$.

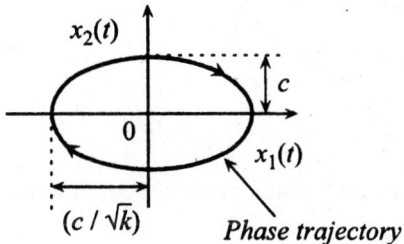

Figure E8.3 Example 8.3

Example 8.4 Repeat Example 8.3 by directly solving the given differential equation.

Solution:

$$\ddot{y} + ky = 0$$

Taking Laplace transforms,

$$s^2 Y(s) - s\, y(0) - \dot{y}(0) + kY(s) = 0$$

$$(s^2 + k)Y(s) = s\, y(0) + \overset{\bullet}{y}(0)$$

$$Y(s) = \frac{s\, y(0) + \overset{\bullet}{y}(0)}{(s^2 + k)}$$

$$Y(s) = \frac{s\, y(0)}{(s^2 + k)} + \frac{\overset{\bullet}{y}(0)}{(s^2 + k)} = y(0)\frac{s}{(s^2 + k)} + \frac{\overset{\bullet}{y}(0)}{\sqrt{k}}\frac{\sqrt{k}}{(s^2 + k)}.$$

Taking inverse Laplace transforms

$$y(t) = y(0)\cos\sqrt{k}\,t + \frac{\overset{\bullet}{y}(0)}{\sqrt{k}}\sin\sqrt{k}\,t$$

$$= M\sin(\sqrt{k}\,t + \theta) \quad \text{where } M\sin\theta = y(0) \text{ and } M\cos\theta = \frac{\overset{\bullet}{y}(0)}{\sqrt{k}}.$$

The constants M and θ can be evaluated if the initial conditions $y(0)$ and $\overset{\bullet}{y}(0)$ are known. Let $x_1 = y$ and $x_2 = \overset{\bullet}{x_1} = \overset{\bullet}{y}$ be the two state variables. Hence

$$x_1(t) = y(t) = M\sin(\sqrt{k}t + \theta) \tag{1}$$

$$x_2(t) = \overset{\bullet}{x_1}(t) = M\sqrt{k}\,\cos(\sqrt{k}t + \theta) \tag{2}$$

From equations (1) and (2),

$$\frac{x_1^2}{M^2} + \frac{x_2^2}{(M\sqrt{k})^2} = 1.$$

This represents an ellipse with semi-major axis $a = M$ and semi-minor axis $b = M/\sqrt{k}$. This is shown in Figure E8.4.

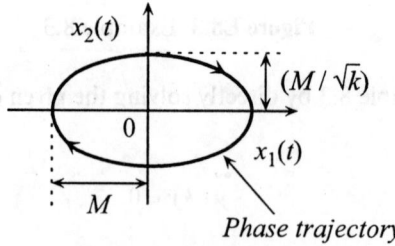

Figure E8.4 Example 8.4

Example 8.5 Draw the phase trajectory of a linear system represented by the differential equation $\ddot{y} + 4y = 2$.

Use the expression for the slope of the trajectory at any point.

Solution:

$$\ddot{y} + 4y = 2$$

Let $x_1 = y$ and $x_2 = \dot{x}_1 = \dot{y}$ be the two state variables. Hence the given differential equation becomes

$$\dot{x}_2 + 4x_1 = 2 \qquad or \ \dot{x}_2 = -4x_1 + 2.$$

$$\frac{dx_2}{dx_1} = \frac{(dx_2 / dt)}{(dx_1 / dt)} = \frac{\dot{x}_2}{\dot{x}_1} = \frac{\dot{x}_2}{x_2} = \frac{-4x_1 + 2}{x_2}$$

This can be written as

$$x_2 \, dx_2 = (-4x_1 + 2)dx_1 .$$

Integrating both sides,

$$\frac{1}{2}x_2^2 = -\frac{4}{2}x_1^2 + 2x_1 + \frac{1}{2}c .$$

Or

$$x_2^2 = -4x_1^2 + 4x_1 + c$$

$$x_2^2 = -4(x_1^2 - x_1 + \frac{1}{4}) + (c+1)$$

$$= -4(x_1 - \frac{1}{2})^2 + (c+1)$$

$$(x_1 - \frac{1}{2})^2 + (\frac{x_2}{2})^2 = \frac{(c+1)}{4} .$$

This represents a circle with co-ordinate axes x_1 and $(x_2 / 2)$ with radius equal to $\dfrac{\sqrt{(c+1)}}{2}$ and centre at $(\frac{1}{2}, 0)$. The constant c can be evaluated if the initial conditions $y(0)$ and $\dot{y}(0)$ are known. The phase trajectory is shown in Figure E8.5.

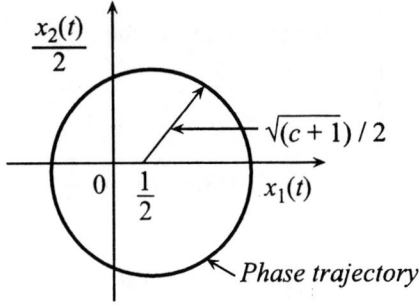

Figure E8.5 Example 8.5

Example 8.6 Repeat Example 8.5 by directly solving the given differential equation.

Solution:

$$\ddot{y} + 4y = 2$$

Taking Laplace transforms,

$$s^2 Y(s) - s\, y(0) - \dot{y}(0) + 4Y(s) = \frac{2}{s}$$

$$(s^2 + 4)Y(s) = s\, y(0) + \dot{y}(0) + \frac{2}{s}$$

$$Y(s) = \frac{s^2\, y(0) + s\, \dot{y}(0) + 2}{s(s^2 + 4)}.$$

Expanding in partial fractions,

$$Y(s) = \frac{(1/2)}{s} + \frac{\{y(0) - (1/2)\}s + \dot{y}(0)}{(s^2 + 4)}$$

$$= \frac{(1/2)}{s} + \{y(0) - (1/2)\}\frac{s}{(s^2 + 4)} + \frac{\dot{y}(0)}{2}\frac{2}{(s^2 + 4)}.$$

Taking inverse Laplace transforms,

$$y = \frac{1}{2} + \{y(0) - (1/2)\}\cos 2t + \frac{\dot{y}(0)}{2}\sin 2t$$

$$= \frac{1}{2} + M \sin(2t + \theta) \qquad \text{where } M \sin\theta = \{y(0) - (1/2)\} \text{ and } M\cos\theta = \frac{\dot{y}(0)}{2}.$$

The constants M and θ can be evaluated using the initial conditions.

Let $x_1 = y$ and $x_2 = \dot{x}_1 = \dot{y}$ be the two state variables.

Hence $x_1 = M \sin(2t + \theta) + \frac{1}{2}$.

$$x_1 = M \sin(2t + \theta) + \frac{1}{2} \qquad (1)$$

$$x_2 = \dot{x}_1 = 2M \cos(2t + \theta) \qquad (2)$$

Combining equations (1) and (2),

$$(x_1 - \frac{1}{2})^2 + \frac{x_2^2}{4} = M^2 .$$

This represents a circle with co-ordinate axes x_1 and $(x_2 / 2)$ with radius equal to M and centre at $(\frac{1}{2}, 0)$. The phase trajectory is shown in Figure E8.6.

The constant M can be evaluated if the initial conditions are known.

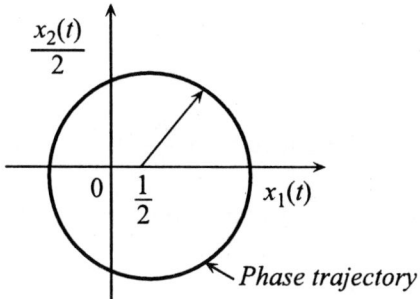

Figure E8.6 Example 8.6

Example 8.7 Draw the phase trajectory of a linear system represented by the differential equation $\ddot{y} + k = 0$.

Use the expression for the slope of the trajectory at any point. The initial conditions are $y(0) = y_0$ and $\dot{y}(0) = 0$.

Solution:

$$\ddot{y} + k = 0$$

Let $x_1 = y$ and $x_2 = \dot{x}_1 = \dot{y}$ be the two state variables. Hence the given differential equation becomes $\dot{x}_2 = -k$.

$$\frac{dx_2}{dx_1} = \frac{(dx_2 / dt)}{(dx_1 / dt)} = \frac{\dot{x}_2}{\dot{x}_1} = \frac{\dot{x}_2}{x_2} = \frac{-k}{x_2} \tag{1}$$

This can be written as

$$x_2 \, dx_2 = -k dx_1 .$$

Integrating both sides,

$$\frac{x_2^2}{2} = -kx_1 + \frac{C}{2}$$

$$x_2^2 = -2k x_1 + C . \tag{2}$$

The constant C can be evaluated, using the given initial conditions.

Let the initial conditions be $x_{10} = y(0) = y_0$ and $x_{20} = \dot{y}(0) = 0$.

Hence from equation (2),

$$0 = -2k\,x_{10} + C \qquad\qquad \text{or } C = 2k\,x_{10}\,.$$

Equation (2) now becomes

$$x_2^2 = -2kx_1 + 2k\,x_{10} = -2k(x_1 - x_{10})\,. \tag{3}$$

From equation (1),

$$\frac{dx_2}{dx_1} = \frac{-k}{x_2}\,.$$

For $k < 0$, let $k = -A$, where A is positive; then $\dfrac{dx_2}{dx_1} = \dfrac{A}{x_2}$.

The slope of the phase trajectory at any point is positive for the portion above the x_1-axis and negative for the portion below the x_1-axis. The phase portrait for different values of x_{10} are shown in Figure E8.7(a).

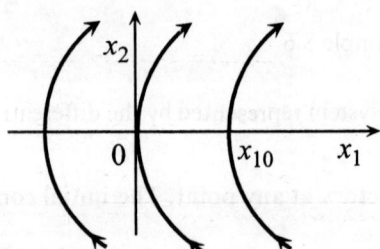
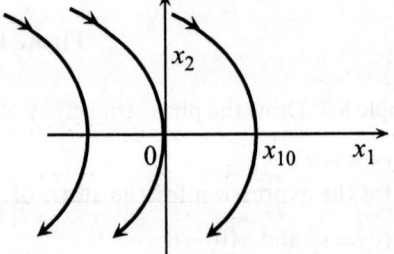

Figure E8.7 (a) Phase portrait for $k < 0$ **Figure E8.7** (b) Phase portrait for $k > 0$

For $k > 0$, $\dfrac{dx_2}{dx_1} = \dfrac{-k}{x_2}$.

The slope of the phase trajectory at any point is negative for the portion above the x_1-axis and positive for the portion below the x_1-axis. The phase portrait for different values of x_{10} are shown in Figure E8.7 (b).

Example 8.8 Repeat Example 8.7 by directly solving the given differential equation.

Solution:

$$\ddot{y} + k = 0$$

Integrating once,

$$\dot{y} = -kt + C_1\,. \tag{1}$$

Integrating once again,

$$y = -\frac{k}{2}t^2 + C_1 t + C_2\,. \tag{2}$$

The initial conditions are $x_{10} = y(0) = y_0$ and $x_{20} = \dot{y}(0) = 0$.

From equation (1), $\dot{y}(0) = C_1 = 0$.

From equation (2), $y(0) = C_2$ or $C_2 = x_{10}$.

Let $x_1 = y$ and $x_2 = \dot{x}_1 = \dot{y}$ be the two state variables. Hence from equation (2) and (1),

$$x_1 = -\frac{k}{2}t^2 + x_{10} \tag{3}$$

and
$$x_2 = -kt. \tag{4}$$

From equation (4)

$$t = -\frac{x_2}{k}. \tag{5}$$

Combining equations (3) and (5),

$$x_1 = -\frac{x_2^2}{2k} + x_{10} \qquad \text{Or} \qquad x_2^2 = -2k(x_1 - x_{10}).$$

This is the same expression that is obtained in Example 8.7 and the phase portrait is the same as shown in Figures E8.7(a) and E8.7(b).

Example 8.9 A linear system is represented by the differential equation

$$\ddot{y} + \dot{y} = 0 \text{ with the initial conditions } y(0) = 1 \text{ and } \dot{y}(0) = 2.$$

Find the equation for the phase trajectory.

Solution:

$$\ddot{y} + \dot{y} = 0 \text{ with the initial conditions } y(0) = 1 \text{ and } \dot{y}(0) = 2.$$

Let $x_1 = y$ and $x_2 = \dot{x}_1 = \dot{y}$ be the two state variables. Hence the given differential equation becomes

$$\dot{x}_2 + x_2 = 0 \qquad \text{or } \dot{x}_2 = -x_2.$$

$$\frac{dx_2}{dx_1} = \frac{(dx_2/dt)}{(dx_1/dt)} = \frac{\dot{x}_2}{\dot{x}_1} = \frac{\dot{x}_2}{x_2} = -1$$

$$dx_2 = -dx_1$$

Integrating both sides,

$$x_2 = -x_1 + C.$$ (1)

The initial conditions are $x_{10} = 1$ and $x_{20} = 2$.
From equation (1),

$$x_{20} = -x_{10} + C \qquad \text{or } C = x_{20} + x_{10} = 3.$$

Equation (1) now becomes

$$x_2 = -x_1 + 3.$$

The phase trajectory is a straight line as shown in Figure E8.9.

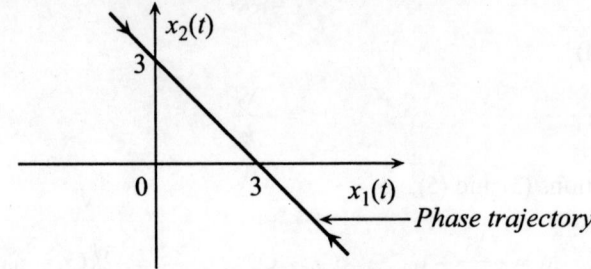

Figure E8.9 Example 8.9

Example 8.10 Repeat Example 8.9 by directly solving the given differential equation.

Solution:

$$\ddot{y} + \dot{y} = 0 \text{ with the initial conditions } y(0) = 1 \text{ and } \dot{y}(0) = 2.$$

Taking Laplace transforms,

$$s^2 Y(s) - sy(0) - \dot{y}(0) + sY(s) - y(0) = 0$$

$$s(s+1)Y(s) = sy(0) + \dot{y}(0) + y(0)$$

$$Y(s) = \frac{s+3}{s(s+1)} = \frac{3}{s} - \frac{2}{s+1}.$$

Taking inverse Laplace transforms,

$$y(t) = 3 - 2e^{-t}.$$

Hence

$$x_1 = y(t) = 3 - 2e^{-t}$$ (1)

$$x_2 = \dot{x_1} = 2e^{-t}.$$ (2)

Combining equations (1) and (2),

$$x_1 + x_2 = 3 \qquad \text{or } x_2 = -x_1 + 3.$$

This is the equation for the phase trajectory and is the same as that in Example 8.9. The phase trajectory is as shown in Figure E8.9.

Example 8.11 Obtain the equation for the phase portrait for the system described by the differential equation $\ddot{y} + a\dot{y} = K$.

Solution:

$$\ddot{y} + a\dot{y} = K$$

Let $x_1 = y$ and $x_2 = \dot{x}_1 = \dot{y}$ be the two state variables. Hence the given differential equation becomes

$$\dot{x}_2 + a x_2 = K \qquad \text{or } \dot{x}_2 = -a x_2 + K$$

$$\frac{dx_2}{dx_1} = \frac{(dx_2/dt)}{(dx_1/dt)} = \frac{\dot{x}_2}{\dot{x}_1} = \frac{\dot{x}_2}{x_2} = \frac{-a x_2 + K}{x_2}. \tag{1}$$

Let

$$z = K - a x_2 \tag{2}$$

$$x_2 = (K - z)/a. \tag{3}$$

Differentiating equation (2) with respect to x_2,

$$\frac{dz}{dx_2} = -a. \tag{4}$$

Combining equations (1), (2) and (3),

$$\frac{dx_2}{dx_1} = \frac{K - a x_2}{x_2} = \frac{z}{(K - z)/a} = \frac{az}{(K - z)} \tag{5}$$

$$\frac{dz}{dx_1} = \frac{dz}{dx_2}\frac{dx_2}{dx_1}.$$

Using equations (4) and (5),

$$\frac{dz}{dx_1} = -a\frac{az}{(K - z)}$$

$$= -a^2 \frac{z}{K - z}$$

$$\frac{dx_1}{dz} = \frac{1}{a^2}\frac{z - K}{z} = \frac{1}{a^2}\left(1 - \frac{K}{z}\right).$$

Integrating with respect to z,

$$x_1 = \frac{1}{a^2}[z - K \ln z] + C$$

$$= \frac{1}{a^2}[K - a x_2 - K \ln(K - a x_2)] + C.$$

If the initial state is (x_{10}, x_{20}),

$$C = x_{10} - \frac{1}{a^2}[K - a x_{20} - K \ln(K - a x_{20})].$$

The equation for the phase portrait is

$$x_1 = \frac{1}{a^2}[K - a x_2 - K \ln(K - a x_2)] + x_{10} - \frac{1}{a^2}[K - a x_{20} - K \ln(K - a x_{20})]$$

$$(x_1 - x_{10}) = \frac{1}{a^2}\left[-a(x_2 - x_{20}) - K\{\ln(K - a x_2) - \ln(K - a x_{20}\}\right]$$

$$= \frac{1}{a^2}\left[-a(x_2 - x_{20}) - K \ln\left(\frac{K - a x_2}{K - a x_{20}}\right)\right]$$

$$(x_2 - x_{20}) = -a(x_1 - x_{10}) - \frac{K}{a} \ln\left(\frac{K - a x_2}{K - a x_{20}}\right).$$

8.5 GRAPHICAL CONSTRUCTION OF PHASE TRAJECTORY

In general, a nonlinear equation that represents the free response of a second-order system can be written down as

$$\ddot{y} + f_1(y, \dot{y})\dot{y} + f_2(y, \dot{y})y = c. \tag{8.6}$$

Let $x_1 = y$ and $x_2 = \dot{x}_1 = \dot{y}$ be the two state variables. The given differential equation (8.6) becomes

$$\dot{x}_2 + f_1(x_1, x_2)x_2 + f_2(x_1, x_2)x_1 = c$$

$$\dot{x}_2 = -f_1(x_1, x_2)x_2 - f_2(x_1, x_2)x_1 + c.$$

$$\frac{d x_2}{d x_1} = \frac{(dx_2 / dt)}{(dx_1 / dt)} = \frac{\dot{x}_2}{x_2} = \frac{-f_1(x_1, x_2)x_2 - f_2(x_1, x_2)x_1 + c}{x_2}. \tag{8.7}$$

The solution of equation (8.7) can be graphically represented by a curve in the phase plane, for a given set of initial conditions $x_1(0)$ and $x_2(0)$, which is the phase trajectory.

8.5.1 Method of Isoclines

$$\frac{d x_2}{d x_1} = \frac{(dx_2 / dt)}{(dx_1 / dt)} = \frac{\dot{x}_2}{\dot{x}_2} = \frac{-f_1(x_1, x_2)x_2 - f_2(x_1, x_2)x_1 + c}{x_2}.$$

This represents the slope of the phase trajectory in the $x_1 - x_2$ plane at a given point (x_1, x_2). If this slope is m, then equation (8.7) can be written as

$$m = \frac{d x_2}{d x_1} = \frac{-f_1(x_1, x_2)x_2 - f_2(x_1, x_2)x_1 + c}{x_2}. \tag{8.8}$$

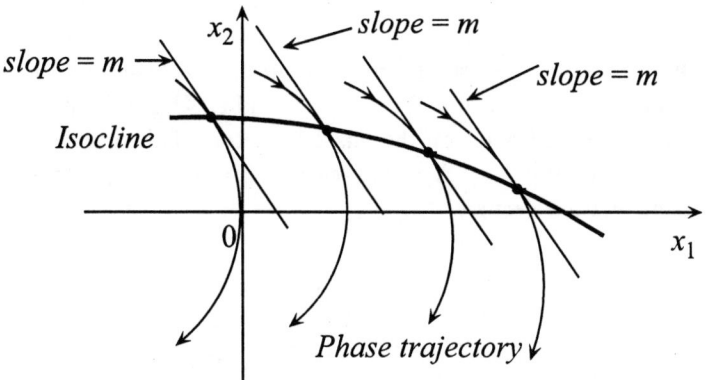

Figure 8.7 Phase trajectories and an isocline

Equation (8.8) represents a curve which crosses all trajectories at points where the slopes of the tangents drawn at these points are the same. In other words, this equation represents the locus of points on all the trajectories where the slopes are the same. Several such curves can be drawn and these are called *isoclines*. One such isocline is shown in Figure 8.7.

Phase trajectories can be drawn using isoclines. Using equation (8.8), an isocline is drawn for a certain value of m, say $m = m_1$. Several cross line are drawn on it having slope m_1 as shown in Figure 8.8. Another isocline is now drawn for a value of slope $m = m_2$ and many cross lines are drawn on it having slope m_2. Several such isoclines are thus drawn. The initial state (x_{10}, x_{20}) is located on an isocline. If this point cannot be located on the isocline drawn, an isocline can be drawn passing through this initial point. Starting from the initial point, a smooth curve is drawn so that it is always tangential to the cross lines. This is shown in Figure 8.8.

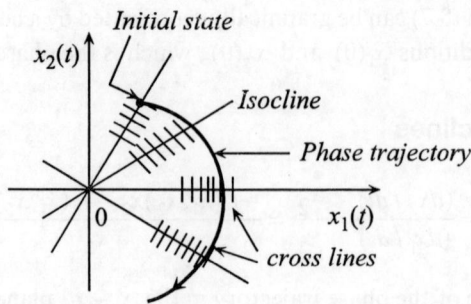

Figure 8.8 Construction of phase trajectory – Method of isoclines

In a better and more precise method, draw several isoclines first. Let I_1, I_2 and I_3 be three isoclines drawn for slopes m_1, m_2 and m_3 respectively and using equation (8.8). Let the initial point P, which represents the initial state (x_{10}, x_{20}), be on the isocline I_1. From this initial point, draw two line segments with slopes m_1 and m_2 as shown in Figure 8.9. Let these intersect the next isocline I_2 at points A and B. The midpoint of the segment AB is Q and is taken as the next point on the phase trajectory. From this point Q, draw two line segments with slopes m_2 and m_3 that meet the next isocline I_3 at points C and D. The midpoint of the segment CD is R and is taken as the next point on the phase trajectory. This process is continued to get more points on the phase trajectory. A smooth curve is drawn through these points P, Q, R etc. to get the complete phase trajectory.

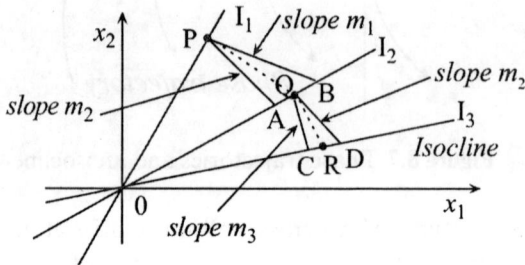

Figure 8.9 Construction of phase trajectory – Isocline method 2

Note that the line PB is tangential to the phase trajectory at the point P; the line QD is tangential to the phase trajectory at the point Q and so on.

Example 8.12 A system is represented by the differential equation $\ddot{y} + 0.5\,\dot{y} + y = 0$. Draw the phase trajectory starting from the initial state $(x_{10}, x_{20}) = (0, 2)$.

Solution:

Let $x_1 = y$ and $x_2 = \dot{x}_1 = \dot{y}$ be the two state variables. Hence the given differential equation becomes

$$\dot{x}_2 + 0.5x_2 + x_1 = 0 \qquad or \quad \dot{x}_2 = -0.5x_2 - x_1$$

$$m = \frac{dx_2}{dx_1} = \frac{\dot{x}_2}{\dot{x}_2} = \frac{-0.5x_2 - x_1}{x_2} = -0.5 - \frac{x_1}{x_2}.$$

$$\frac{x_1}{x_2} = -(m+0.5)$$

$$x_2 = -\frac{1}{(m+0.5)}x_1 \qquad or \quad x_2 = C\,x_1 \;\; where \; C = -\frac{1}{(m+0.5)}.$$

This represents the equation for a family of straight lines with slope $C = -1/(m+0.5)$ and passing through the origin. Hence in this case, the isoclines are straight lines. A table for the slopes of isoclines for different values of m and the slope of cross lines along each isocline can be prepared as shown in Table E8.12.

Let $C = \tan\phi$ and $m = \tan\theta$.

$$m = -\left(\frac{1}{c}\right) - 0.5$$

Isocline	Angle of isocline ϕ (degrees)	$C = \tan\phi$	$m = \tan\theta$	Angle of cross lines θ (degrees)
1	90	∞	-0.5	-26.56
2	70	2.747	-0.8639	-40.83
3	50	1.1917	-1.339	-53.25
4	30	0.5774	-2.2320	-65.87

Table E8.12 Example 8.12

At first isoclines, which are straight lines, are drawn in the $x_1 - x_2$ plane at angles 70^0, 50^0, 30^0, etc. and passing through the origin. The initial state is $P(x_{10}, x_{20}) = (0,2)$, which is located on the $x_2 - axis$. From this point, draw a line segment at an angle of $= -26.56^\circ$ and another at an angle of $\theta = -40.83°$ and let these intersect isocline-1 at points A and B as shown in Figure E8.12. The midpoint Q of segment AB is a point on the phase trajectory. From this point, draw a line at an angle of $\theta = -40.83^\circ$ and another at an angle of $\theta = -53.25^\circ$. They intersect the next isocline-2 at the points C and D. The midpoint R of the segment CD is another point on the phase trajectory. The process is thus continued and these points P, Q, R etc. are joined by a smooth curve. Take care to see that the line PB is tangential to the phase trajectory at point P; the line QD is tangential to the phase trajectory at point Q and so on. The initial portion of the phase trajectory is shown in Figure E8.12.

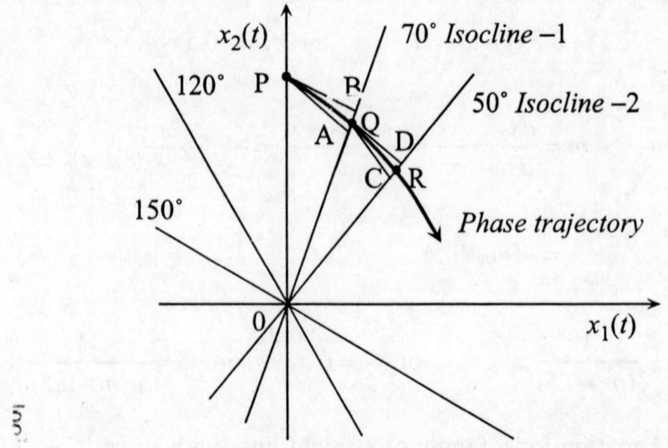

Figure E8.12 Example 8.12

Example 8.13 Draw the phase trajectory of a system represented by the differential equation $\ddot{y} + \dot{y} + y = 1$ given the initial state $(-4, 6.93)$.

Let $x_1 = y$ and $x_2 = \dot{x}_1 = \dot{y}$ be the two state variables. Hence the given differential equation becomes

$$\dot{x}_2 + x_2 + x_1 = 1 \quad or \quad \dot{x}_2 = -x_2 - x_1 + 1.$$

$$m = \frac{dx_2}{dx_1} = \frac{\dot{x}_2}{\dot{x}_2} = \frac{-x_2 - x_1 + 1}{x_2} = -1 - \frac{x_1}{x_2} + \frac{1}{x_2}$$

$$(m+1) = \frac{-x_1 + 1}{x_2} \qquad (m+1)x_2 = -x_1 + 1$$

$$x_2 = -\frac{1}{(m+1)} x_1 + \frac{1}{(m+1)} \quad or \quad x_2 = C\,x_1 + b$$

where $C = -\dfrac{1}{(m+1)}$ and $b = \dfrac{1}{(m+1)}$.

$$m = -\left(\frac{1}{c}\right) - 1$$

This represents the equation for a family of straight lines with slope $C = -1/(m+1)$. Hence in this case the isoclines are straight lines. When $x_2 = 0$, $x_1 = 1$. Hence all the isoclines are concurrent at a point on the $x_1 - axis$ at $x_1 = 1$.

The initial state is $(-4, 6.93)$. The slope of the isocline on which this point is located is

$$\theta = 180° - \tan^{-1}(6.93/4) = 120°.$$

Isocline	Angle of isocline ϕ (degrees)	$C = \tan\phi$	$m = \tan\theta$	Angle of cross lines on isocline θ (degrees)
1	120	−1.732	−0.4226	−22.9
2	90	∞	−1	−45
3	70	2.747	−1.364	−53.75
4	50	1.1917	−1.839	−61.47
5	30	0.5774	−2.7320	−69.89

Table E8.13 Example 8.13

A table for the slopes of isoclines for different values of m and the slope of cross lines along each isocline can be prepared as shown in Table E8.13.

Let $C = \tan\phi$ and $m = \tan\theta$.

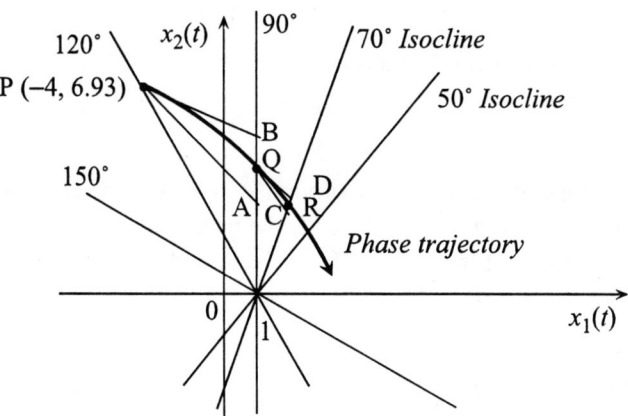

Figure E 8.13 Example 8.13

The initial point is located on the 120° isocline. From this point, draw two line segments, one at an angle of −22.9° and the other at an angle of −45°. Let these intersect at the 90° isocline at points A and B. The midpoint of AB is Q and it is a point on the phase trajectory. From point Q, draw two line segments, one at an angle of −45° and the other at an angle of −53.75°. Let these intersect at the 70° isocline at points C and D. The midpoint of CD is R and it is a point on the phase trajectory. Points on phase trajectory are thus obtained and they are joined by a smooth curve. The initial portion of the phase trajectory is shown in Figure E8.13.

8.5.2 Pell's Method

This method of constructing phase trajectory is applicable to second-order systems, the differential equation description of which can be represented as

$$\ddot{y} + g(\dot{y}) + f(y) = 0.$$

Let $x_1 = y$ and $x_2 = \dot{x}_1 = \dot{y}$ be the two state variables. Hence the given differential equation becomes

$$\dot{x}_2 + g(x_2) + f(x_1) = 0 \qquad \text{or } \dot{x}_2 = -g(x_2) - f(x_1).$$

$$\frac{dx_2}{dx_1} = \frac{\dot{x}_2}{x_2} = \frac{-g(x_2) - f(x_1)}{x_2}.$$

This is the slope of the trajectory at a given point. The constructional procedure is explained below.

Plot $g(x_2)$ versus x_2 and then $f(x_1)$ versus (x_1) as per the co-ordinate directions shown in Figure 8.10. Let $P(x_{10}, x_{20})$ be the initial state. The value of $f(x_{10})$ is AB. Make BC = AB along the x_1-axis. Corresponding to P, EF is $g(x_{20})$. Make CD = EF. Join DP. The slope of the line DP is

$$\frac{x_{20}}{g(x_{20}) + f(x_{10})}.$$

The slope of the line perpendicular to this is

$$\frac{-g(x_{20}) - f(x_{10})}{x_{20}}.$$

This is the required slope of the trajectory at point P. Now draw a very small line segment PQ perpendicular to DP. With Q as the next point, the process is repeated to draw the next line segment. A smooth curve is drawn starting from P and with all these line segments tangential to it. Thus the entire phase trajectory can be drawn.

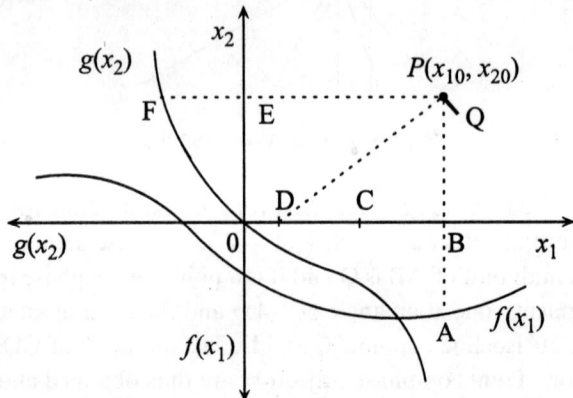

Figure 8.10 Pell's method of constructing phase trajectory

Example 8.14 Draw the phase trajectory for a second-order system represented by the differential equation $\ddot{y} + 2\dot{y} + y = 1$.
Using Pell's method and starting from the initial state (4, 2).

Solution:

$$\ddot{y} + 2\dot{y} + y = 1$$

Let $x_1 = y$ and $x_2 = \dot{x}_1 = \dot{y}$ be the two state variables. Hence the given differential equation becomes

$$\dot{x}_2 + 2x_2 + x_1 = 1 \quad \text{or} \quad \dot{x}_2 = -2x_2 - (x_1 - 1).$$

$$\frac{dx_2}{dx_1} = \frac{\dot{x}_2}{x_2} = \frac{-2x_2 - (x_1 - 1)}{x_2} = \frac{-g(x_2) - f(x_1)}{x_2}$$

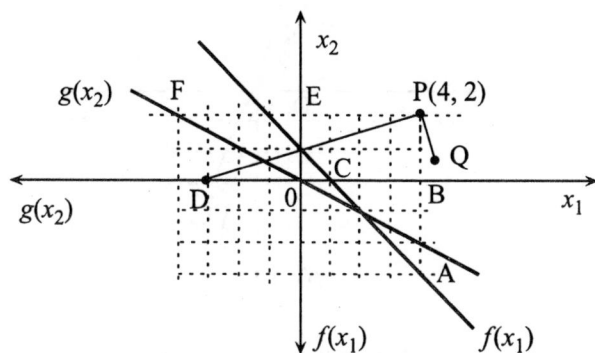

Figure E8.14 Example 8.14

Now $g(x_2) = 2x_2$ and $f(x_1) = (x_1 - 1)$. Plot $g(x_2)$ versus x_2 and $f(x_1)$ versus x_1 as shown in Figure E8.14. Locate the initial state P(4, 2). Make AB = BC and CD = EF. Draw the line DP and draw a small line segment PQ perpendicular to DP. With Q as the next point, the process is repeated to draw the next line segment. A smooth curve is drawn starting from P and with all these line segments tangential to it. Thus the entire phase trajectory can be drawn.

Example 8.15 Draw the phase trajectory for a second-order system represented by the differential equation $\ddot{y} + 0.25\dot{y} + y + 0.5y^3 = 0$, using Pell's method and starting from the initial state (0, 4).

Solution:

$$\ddot{y} + 0.25\dot{y} + y + 0.5y^3 = 0.$$

Let $x_1 = y$ and $x_2 = \dot{x}_1 = \dot{y}$ be the two state variables. Hence the given differential equation becomes

$$\dot{x}_2 = -0.25x_2 - (x_1 + 0.5x_1^3).$$

$$\frac{dx_2}{dx_1} = \frac{\dot{x}_2}{x_2} = \frac{-0.25x_2 - (x_1 + 0.5x_1^3)}{x_2} = \frac{-g(x_2) - f(x_1)}{x_2}$$

Now $g(x_2) = 0.25x_2$ and $f(x_1) = (x_1 + 0.5x_1^3)$. Plot $g(x_2)$ versus x_2 and $f(x_1)$ versus x_1 as shown in Figure E8.15. Locate the initial state P(0, 4). Mark the point D corresponding to $g(x_{20}) + f(x_{10})$. Draw the line DP and draw a small line segment PQ perpendicular to DP. With Q as the next point, the process is repeated to draw the next line segment. A smooth curve is drawn starting from P and with all these line segments tangential to it. By a similar procedure, the entire phase trajectory can be drawn.

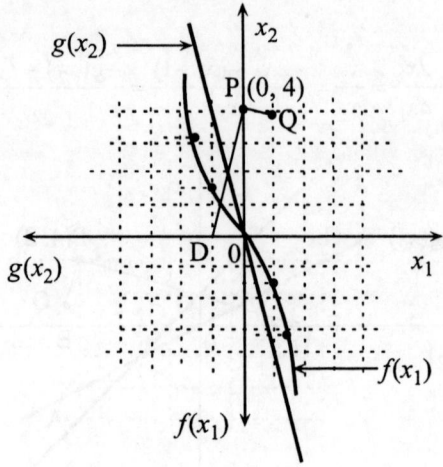

Figure E8.15 Example 8.15

8.5.3 Delta Method

In this method of graphical construction of phase trajectory, small segments of phase trajectory are drawn as small circular arcs. This method is applicable to a second-order system described by the differential equation

$$\ddot{y} + f(y, \dot{y}) + k\,y = 0.$$

Let $x_1 = y$ and $x_2 = \dot{y}$ be the two state variables. Hence the given differential equation becomes

$$\dot{x}_2 + f(x_1, x_2) + k\,x_1 = 0 \quad or \quad \dot{x}_2 = -f(x_1, x_2) - k\,x_1.$$

$$\frac{dx_2}{dx_1} = \frac{\dot{x}_2}{x_2} = \frac{-f(x_1, x_2) - k\,x_1}{x_2}$$

Assume that $f(x_1, x_2)$ is a constant in the vicinity of a point in the $x_1 - x_2$ plane so that it can be represented as

$$f(x_1, x_2) = k\delta \quad or \quad \delta = \frac{f(x_1, x_2)}{k}.$$

Hence $\dfrac{dx_2}{dx_1} = \dfrac{-k\,\delta - k x_1}{x_2} = \dfrac{-k(x_1 + \delta)}{x_2}$

$$\frac{x_2 \, dx_2}{k} = -(x_1 + \delta) dx_1 .$$

Integrating both sides,

$$\frac{1}{2k} x_2^2 = -\left(\frac{x_1^2}{2} + \delta x_1\right) + \frac{C}{2}$$

$$\frac{1}{k} x_2^2 = -\left(x_1^2 + 2\delta x_1 + \delta^2\right) + (C + \delta^2) .$$

This can be rearranged as

$$(\frac{x_2}{\sqrt{k}})^2 + (x_1 + \delta)^2 = (C + \delta^2) .$$

This is an equation for a circle in a rectangular co-ordinate system in the $x_1 - (x_2 / \sqrt{k})$ plane with centre at $(-\delta, 0)$ and radius equal to $\sqrt{C + \delta^2}$. Hence the construction of the phase trajectory can be made in circular arcs and is explained hereunder.

1. Given the initial point $P(x_{10}, x_{20})$ calculate the value of δ:

$$\delta = \frac{f(x_1, x_2)}{k} = \frac{f(x_{10}, x_{20})}{k} .$$

2. With $(-\delta, 0)$ as centre, draw a small arc PQ. Q is a point on the phase trajectory.
3. Taking Q as the next point, find the co-ordinates of Q and calculate δ.

The procedure is shown in Figure 8.11. The process is repeated to get several points on the trajectory and thus the phase trajectory can be drawn.

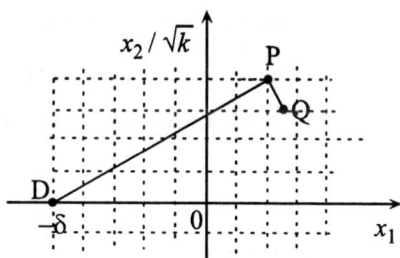

Figure 8.11 Delta method of drawing phase trajectory

Example 8.16 Draw the phase trajectory for a second-order system represented by the differential equation $\ddot{y} + y\dot{y} + 4y = 0$, using Delta method and starting from the initial state (4, 2).

Solution:

$$\ddot{y} + y\dot{y} + 4y = 0 .$$

Let $x_1 = y$ and $x_2 = \dot{x}_1 = \dot{y}$ be the two state variables. Hence the given differential equation becomes

$$\dot{x}_2 + x_1 x_2 + 4x_1 = 0 .$$

This equation can be written as

$$\dot{x}_2 + f(x_1, x_2) + k x_1 = 0 .$$

$$f(x_1, x_2) = x_1 x_2 \qquad k = 4$$

The initial point is P(4, 2).

$$\delta = \frac{f(x_1, x_2)}{k} = \frac{x_1 x_2}{4} = \frac{8}{4} = 2 .$$

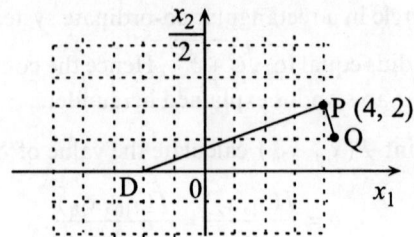

Figure E8.16 Example 8.16

Locate the point D(–δ, 0) on the x_1-axis. Mark the initial state P(4, 2). This is shown in Figure E8.16. With D as centre and DP as radius draw a small arc PQ. The allowable length of the arc is a compromise. With Q as the next point, repeat the process to get the next point on the trajectory. The phase trajectory can thus be completely drawn.

Example 8.17 Draw the phase trajectory for a second-order system represented by the differential equation $\ddot{y} + 0.25\,\dot{y} + y + 0.5y^3 = 0$, using Delta method and starting from the initial state (2, 4).

Solution:

$$\ddot{y} + 0.25\,\dot{y} + y + 0.5y^3 = 0$$

Let $x_1 = y$ and $x_2 = \dot{x}_1 = \dot{y}$ be the two state variables. Hence the given differential equation becomes

$$\dot{x}_2 + 0.25x_2 + x_1 + 0.5x_1^3 = 0 \qquad \text{or} \quad \dot{x}_2 + (0.5x_1^3 + 0.25x_2) + x_1 = 0 .$$

This equation can be written as

$$\dot{x}_2 + f(x_1, x_2) + k\,x_1 = 0 .$$

$$f(x_1, x_2) = 0.5x_1^3 + 0.25x_2 \qquad k = 1$$

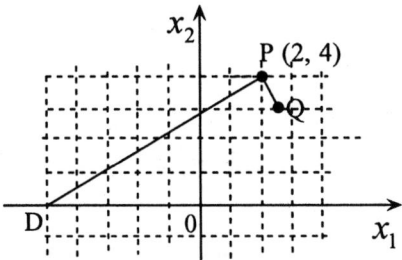

Figure E8.17 Example 8.17

The initial point is P(2, 4).

$$\delta = \frac{f(x_1, x_2)}{k} = 0.5x_1^3 + 0.25x_2 = 0.5(2)^3 + 0.25 \times 4 = 5.$$

Locate the point D(–5, 0) on the x_1-axis. Mark the initial state P(2, 4). This is shown in Figure E8.17. With D as centre and DP as radius, draw a small arc PQ. With Q as the next point, repeat the process to get the next point on the trajectory. The phase trajectory can thus be completely drawn.

Example 8.18 A nonlinear system is shown in Figure E8.18. Derive the equations for the phase trajectory. Also obtain the equation for the isoclines, in terms of e.

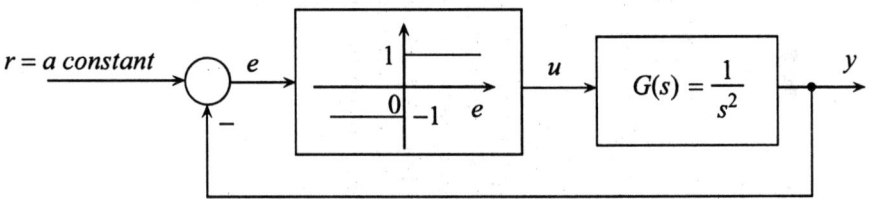

Figure E 8.18 Example 8.18

Solution: The input–output relation that describes the linear plant is

$$\frac{Y(s)}{U(s)} = \frac{1}{s^2} \qquad \text{or } s^2 Y(s) = U(s).$$

The differential equation that describe the plant is

$$\ddot{y} = u \qquad u = \begin{cases} 1; \ e > 0 \\ -1; \ e < 0 \end{cases}$$

$$e = r - y; \text{ where r is a constant.}$$

$$y = r - e \qquad \dot{y} = -\dot{e} \quad \text{and} \quad \ddot{y} = -\ddot{e}.$$

In terms of e, the differential equation becomes

$$\ddot{e} = -u.$$

Select

$$x_1 = e \text{ and } x_2 = \dot{x}_1 = \dot{e}.$$

Hence the equation becomes $\dot{x}_2 = -u$.

(A) For

$$e > 0 \text{ or } x_1 > 0, \, u = 1$$

The differential equation is

$$\dot{x}_2 = -1 \; ; \; x_1 > 0.$$

Equation for phase trajectory

$$\frac{dx_2}{dx_1} = \frac{\dot{x}_2}{x_2} = -\frac{1}{x_2} \qquad \text{or } x_2 dx_2 = -dx_1.$$

Integrating both sides,

$$\frac{x_2^2}{2} = -x_1 + \frac{C}{2} \qquad \text{or } x_2^2 = -2x_1 + C.$$

This is the equation for the trajectory. The constant C can be evaluated using initial states.

Equation for isoclines

$$m = \frac{\dot{x}_2}{x_2} = -\frac{1}{x_2} \qquad \text{or } x_2 = -\frac{1}{m}.$$

This is the equation for the isoclines and this represents a family of straight lines for different values of m. These lines are parallel to the $x_1 - axis$.

(B) For $e < 0$, $u = -1$

The differential equation is

$$\dot{x}_2 = 1 \; ; \; x_1 < 0.$$

Equation for phase trajectory

$$\frac{dx_2}{dx_1} = \frac{\dot{x}_2}{x_2} = \frac{1}{x_2} \qquad \text{or } x_2 dx_2 = dx_1.$$

Integrating both sides,

$$\frac{x_2^2}{2} = x_1 + \frac{C}{2} \qquad \text{or } x_2^2 = 2x_1 + C .$$

This is the equation for the trajectory. The constant C can be evaluated using initial states.

Equation for isoclines

$$m = \frac{\dot{x}_2}{x_2} = \frac{1}{x_2} \qquad \text{or } x_2 = \frac{1}{m} .$$

This is the equation for the isoclines and this represents a family of straight lines for different values of m. These lines are parallel to the $x_1 - axis$.

Example 8.19 A nonlinear system is shown in Figure E8.19. Obtain the equations for the isoclines, in terms of e, for the construction of phase trajectory.

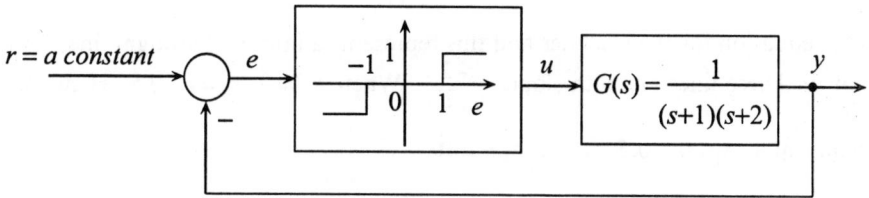

Figure E8.19 Example 8.19

Solution: The input–output relation that describes the linear plant is

$$\frac{Y(s)}{U(s)} = \frac{1}{(s+1)(s+2)} \qquad \text{or } (s^2 + 3s + 2)Y(s) = U(s) .$$

The differential equation that describe the plant is

$$\ddot{y} + 3\dot{y} + 2y = u \qquad u = \begin{cases} 1; \ e > 1 \\ -1; \ e < -1 \\ 0; -1 < e < 1 \end{cases}$$

$$e = r - y; \text{ where r is a constant.}$$

$$y = r - e \qquad \dot{y} = -\dot{e} \quad \text{and} \quad \ddot{y} = -\ddot{e} .$$

In terms of e, the differential equation becomes

$$-\ddot{e} - 3\dot{e} + 2(r - e) = u$$

$$\ddot{e} + 3\dot{e} + 2e = 2r - u \ .$$

Select
$$x_1 = e \text{ and } x_2 = \dot{x}_1 = \dot{e} \ .$$

Hence the equation becomes $\dot{x}_2 + 3x_2 + 2x_1 = 2r - u$.

(A) *For $e > 1$, $u = 1$*

The differential equation is

$$\dot{x}_2 + 3x_2 + 2x_1 = 2r - u = 2r - 1 \qquad ; \quad x_1 > 1 \ .$$

$$m = \frac{\dot{x}_2}{x_2} = \frac{-3x_2 - 2x_1 + 2r - 1}{x_2} = -3 - \frac{2x_1}{x_2} + \frac{2r - 1}{x_2}; \quad x_1 > 1$$

$$m + 3 = \frac{-2x_1}{x_2} + \frac{2r - 1}{x_2} \qquad \text{or } x_2 = \frac{-2}{m+3}x_1 + \frac{2r - 1}{m+3}$$

This is the equation for the isoclines and this represents a family of straight lines for different values of *m*. These lines are with slopes $\dfrac{-2}{m+3}$. When $x_2 = 0$, $x_1 = r - 0.5$. Hence these lines are concurrent at $x_1 = r - 0.5$ on the $x_1 - $axis

(B) *For $e < -1$, $u = -1$*

The differential equation is

$$\dot{x}_2 + 3x_2 + 2x_1 = 2r - u = 2r + 1 \qquad ; \quad x_1 < -1$$

$$m = \frac{\dot{x}_2}{x_2} = \frac{-3x_2 - 2x_1 + (2r + 1)}{x_2} = -3 - \frac{2x_1}{x_2} + \frac{2r + 1}{x_2}; \quad x_1 < -1$$

$$m + 3 = \frac{-2x_1 + (2r + 1)}{x_2} \qquad \text{or } x_2 = \frac{-2}{m+3}x_1 + \frac{2r + 1}{m+3} \ .$$

This is the equation for the isoclines and this represents a family of straight lines for different values of *m*. These lines are with slopes $\dfrac{-2}{m+3}$. When $x_2 = 0$, $x_1 = r + 0.5$. Hence these lines are concurrent at $x_1 = r + 0.5$ on the $x_1 - $axis .

(C) *For $-1 < e < 1$, $u = 0$*

The differential equation is

$$\dot{x}_2 + 3x_2 + 2x_1 = 2r; \quad -1 < x_1 < 1 \ .$$

$$m = \frac{\dot{x_2}}{x_2} = \frac{-3x_2 - 2x_1 + 2r}{x_2} = -3 - \frac{2x_1}{x_2} + \frac{2r}{x_2}; \quad -1 < x_1 < 1$$

$$m + 3 = \frac{-2x_1 + 2r}{x_2} \qquad or \ x_2 = \frac{-2}{m+3} x_1 + \frac{2r}{m+3}.$$

This is the equation for the isoclines and this represents a family of straight lines for different values of m. These lines are with slopes $\dfrac{-2}{m+3}$. When $x_2 = 0$, $x_1 = r$. Thus the isoclines are concurrent on the $x_1 -$ axis at the point $x_1 = r$ in the $x_1 - x_2$ plane .

Example 8.20 A nonlinear system is shown in Figure E8.20. Derive the equation for the isoclines.

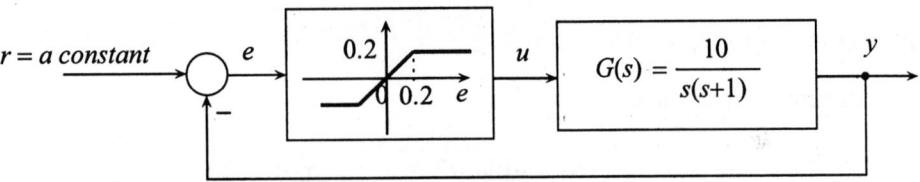

Figure E8.20 Example 8.20

Solution: The input–output relation that describes the linear plant is

$$\frac{Y(s)}{U(s)} = \frac{10}{s(s+1)} \quad or \ (s^2 + s)Y(s) = 10U(s).$$

The differential equation that describes the plant is

$$\ddot{y} + \dot{y} = 10u \qquad u = \begin{cases} 0.2; \ e > 1 \\ -0.2; \ e < -1 \\ e; -0.2 < e < 0.2 \end{cases}$$

$$e = r - y; \ \text{where r is a constant.}$$

$$\dot{y} = \dot{r} - \dot{e} \qquad \dot{y} = -\dot{e} \quad and \quad \ddot{y} = -\ddot{e} .$$

In terms of e, the differential equation becomes

$$\ddot{e} + \dot{e} = -10u .$$

Select $x_1 = e$ and $x_2 = \dot{x_1} = \dot{e}$.

Hence the equation becomes $\dot{x_2} + x_2 = -10u .$

(A) *For $e > 0.2$ or $x_1 > 0.2$, $u = 0.2$*

The differential equation is

$$\dot{x}_2 + x_2 = -10u = -2 \ .$$

$$m = \frac{\dot{x}_2}{x_2} = \frac{-x_2 - 2}{x_2} = -1 - \frac{2}{x_2}$$

$$m + 1 = \frac{-2}{x_2} \qquad or \ \ x_2 = \frac{-2}{m+1} \ .$$

This is the equation for the isoclines and this represents a family of straight lines for different values of m. These isoclines are parallel to the $x_1 - axis$.

(B) *For $e < -0.2$ or $x_1 < -0.2$, $u = -0.2$*

The differential equation is

$$\dot{x}_2 + x_2 = -10u = 2 \qquad ; \ \ x_1 < -0.2 \ .$$

$$m = \frac{\dot{x}_2}{x_2} = \frac{-x_2 + 2}{x_2} = -1 + \frac{2}{x_2}; \ \ x_1 < -0.2$$

$$m + 1 = \frac{2}{x_2} \qquad or \ \ x_2 = \frac{2}{m+1} \ .$$

This is the equation for the isoclines and this represents a family of straight lines for different values of m. These isoclines are parallel to the $x_1 - axis$.

(C) *For $-0.2 < e < 0.2$ or $-0.2 < x_1 < 0.2$, $u = e$*

The differential equation is

$$\dot{x}_2 + x_2 = -10e = -10x_1 \qquad .$$

$$m = \frac{\dot{x}_2}{x_2} = \frac{-x_2 - 10x_1}{x_2} = -1 - \frac{10x_1}{x_2}$$

$$m + 1 = \frac{-10x_1}{x_2} \qquad or \ \ x_2 = -\frac{10}{m+1}x_1 \ .$$

This is the equation for the isoclines and this represents a family of straight lines for different values of m and with the slope of $-10/(m+1)$. These isoclines are concurrent at the origin.

Example 8.21 Derive the equation for the isoclines for the construction of phase trajectory for the nonlinear system shown in Figure E8.21.

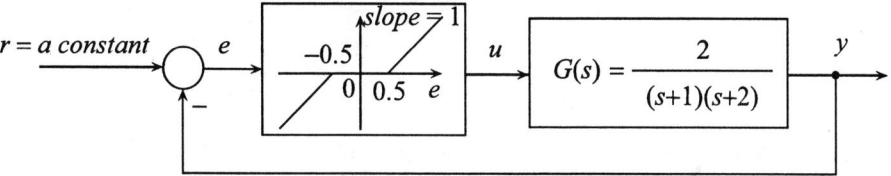

Figure E8.21 Example 8.21

Solution: The input–output relation that describes the linear plant is

$$\frac{Y(s)}{U(s)} = \frac{2}{(s+1)(s+2)} \qquad or \ (s^2 + 3s + 2)Y(s) = 2U(s).$$

The differential equation that describe the plant is

$$\ddot{y} + 3\dot{y} + 2y = 2u \qquad u = \begin{cases} (e-0.5); \ e > 0.5 \\ (e+0.5); \ e < -0.5 \\ 0; \ -0.5 < e < 0.5 \end{cases}$$

$$e = r - y; \text{ where r is a constant.}$$

$$y = r - e \qquad \dot{y} = -\dot{e} \qquad and \qquad \ddot{y} = -\ddot{e}.$$

In terms of e, the differential equation becomes

$$-\ddot{e} - 3\dot{e} + 2(r - e) = 2u$$

$$\ddot{e} + 3\dot{e} + 2e = 2r - 2u.$$

Select

$$x_1 = e \text{ and } x_2 = \dot{x}_1 = \dot{e}$$

Hence the equation becomes

$$\dot{x}_2 + 3x_2 + 2x_1 = 2r - 2u.$$

(A) *For e > 0.5, u = e − 0.5*

The differential equation is

$$\dot{x}_2 + 3x_2 + 2x_1 = 2r - 2(e - 0.5) = 2r - 2e + 1 = 2r - 2x_1 + 1; \quad x_1 > 0.5.$$

Or

$$\dot{x}_2 + 3x_2 + 4x_1 = 2r + 1; \quad x_1 > 0.5.$$

$$m = \frac{\dot{x}_2}{x_2} = \frac{-3x_2 - 4x_1 + 2r + 1}{x_2} = -3 - \frac{4x_1}{x_2} + \frac{2r+1}{x_2}; \quad x_1 > 0.5$$

$$m + 3 = \frac{-4x_1}{x_2} + \frac{2r+1}{x_2} \qquad or \quad x_2 = \frac{-4}{m+3}x_1 + \frac{2r+1}{m+3}.$$

This is the equation for the isoclines and this represents a family of straight lines for different values of m. These lines are with slopes $\dfrac{-4}{m+3}$. When $x_2 = 0$, $x_1 = 0.5r + 0.25$. Hence these lines are concurrent at the point $((0.5r + 0.25), 0)$.

(B) *For* $e < -0.5$, $u = e + 0.5$

The differential equation is

$$\dot{x}_2 + 3x_2 + 2x_1 = 2r - 2(e + 0.5) = 2r - 2e - 1 = 2r - 2x_1 - 1; \quad x_1 < -0.5.$$

Or
$$\dot{x}_2 + 3x_2 + 4x_1 = 2r - 1; \quad x_1 < -0.5.$$

$$m = \frac{\dot{x}_2}{x_2} = \frac{-3x_2 - 4x_1 + 2r - 1}{x_2} = -3 - \frac{4x_1}{x_2} + \frac{2r-1}{x_2}; \quad x_1 < -0.5$$

$$m + 3 = \frac{-4x_1}{x_2} + \frac{2r-1}{x_2} \qquad or \quad x_2 = \frac{-4}{m+3}x_1 + \frac{2r-1}{m+3}.$$

This is the equation for the isoclines and this represents a family of straight lines for different values of m. These lines are with slopes $\dfrac{-4}{m+3}$. When $x_2 = 0$, $x_1 = 0.5r - 0.25$. Hence these lines are concurrent at the point $((0.5r - 0.25), 0)$.

(C) *For* $-0.5 < e < 0.5$, $u = 0$

The differential equation is

$$\dot{x}_2 + 3x_2 + 2x_1 = 2r - 2u = 2r; \quad -0.5 < e < 0.5.$$

Or
$$\dot{x}_2 + 3x_2 + 2x_1 = 2r; \quad -0.5 < x_1 < 0.5.$$

$$m = \frac{\dot{x}_2}{x_2} = \frac{-3x_2 - 2x_1 + 2r}{x_2} = -3 - \frac{2x_1}{x_2} + \frac{2r}{x_2}; \quad -0.5 < x_1 < 0.5$$

$$m + 3 = \frac{-2x_1}{x_2} + \frac{2r}{x_2} \qquad or \quad x_2 = \frac{-2}{m+3}x_1 + \frac{2r}{m+3}.$$

This is the equation for the isoclines and this represents a family of straight lines for different values of m. These lines are with slopes $\dfrac{-2}{m+3}$. When $x_2 = 0$, $x_1 = r$. Hence these isoclines are concurrent at the point $(r,0)$.

Example 8.22 For the nonlinear system shown in Figure E8.22, obtain the equation for the isoclines for the construction of phase trajectory.

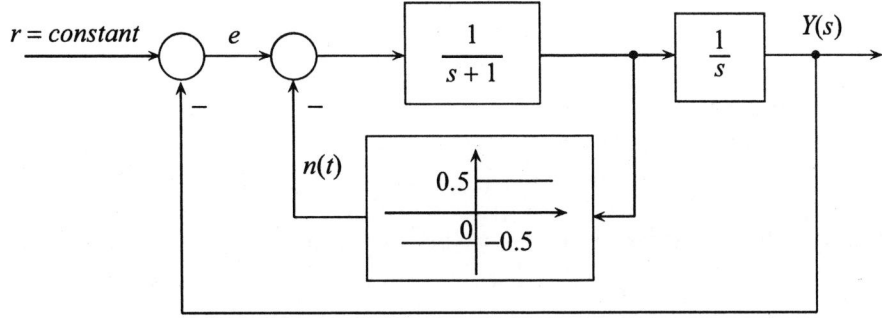

Figure E8.22 Example 8.22

Solution: The block diagram of the system is redrawn in Figure E8.22(a).

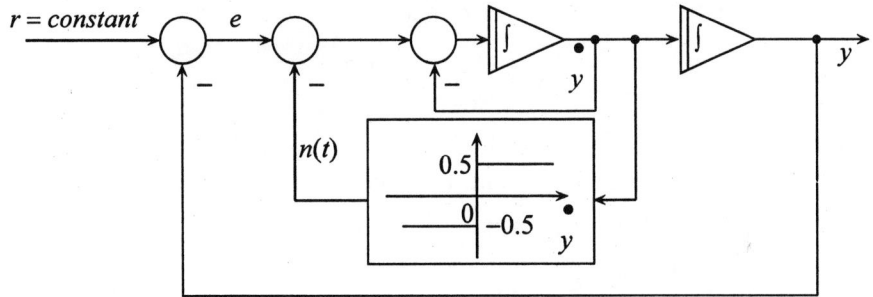

Figure E8.22 (a) Example E8.22

The input–output relation that describes the linear plant is

$$\ddot{y} = e - n(t) - \dot{y} \qquad\qquad \ddot{y} + \dot{y} = e - n(t) \tag{1}$$

$$n(t) = \begin{cases} 0.5; & \dot{y} > 0 \\ -0.5; & \dot{y} < 0 \end{cases} \tag{2}$$

The differential equation that describes the plant is

$$\ddot{y} + \dot{y} = e - n(t)$$

$$e = r - y; \text{ where r is a constant.}$$

$$\dot{y} = r - \dot{e} \qquad \dot{y} = -\dot{e} \quad \text{and} \quad \ddot{y} = -\ddot{e}.$$

In terms of e, the differential equation becomes

$$-\ddot{e} - \dot{e} = e - n(t)$$

$$\ddot{e} + \dot{e} = -e + n(t)$$

$$\ddot{e} + \dot{e} + e = n(t). \tag{3}$$

Select $x_1 = e$ and $x_2 = \dot{x}_1 = \dot{e}$.

Hence the equation (3) becomes

$$\dot{x}_2 + x_2 + x_1 = n(t). \tag{4}$$

(A) When $\dot{y} > 0$, $n(t) = 0.5$; or when $-\dot{e} > 0$ or $-x_2 > 0$ or $x_2 < 0$; $n(t) = 0.5$

Hence $\dot{x}_2 + x_2 + x_1 = 0.5; x_2 < 0$.

$$m = \frac{\dot{x}_2}{x_2} = -1 - \frac{x_1}{x_2} + \frac{0.5}{x_2}$$

$$m + 1 = -\frac{x_1}{x_2} + \frac{0.5}{x_2}$$

$$x_2 = \frac{-1}{m+1} x_1 + \frac{0.5}{m+1}.$$

This is the equation for a family of isoclines which are straight lines. The slope is $-1/(m+1)$.
When $x_2 = 0$, $x_1 = 0.5$. Hence the isoclines are concurrent at the point (0.5, 0).

(B) When $\dot{y} < 0$, $n(t) = -0.5$;   or when $-\dot{e} < 0$ or $-x_2 < 0$ or $x_2 > 0$; $n(t) = -0.5$

Hence $\dot{x}_2 + x_2 + x_1 = -0.5 ; x_2 > 0$.

$$m = \frac{\dot{x}_2}{x_2} = -1 - \frac{x_1}{x_2} - \frac{0.5}{x_2}$$

$$m + 1 = -\frac{x_1}{x_2} - \frac{0.5}{x_2}$$

$$x_2 = \frac{-1}{m+1} x_1 - \frac{0.5}{x_2}.$$

This is the equation for a family of isoclines which are straight lines. The slope is $-1/(m+1)$.
When $x_2 = 0$, $x_1 = -0.5$. Hence the isoclines are concurrent at the point (–0.5, 0).

Example 8.23 A second-order system is represented by the differential equation $\ddot{y} + |\dot{y}| = 1$.
Determine the equation for the phase trajectory. Also find the equation of the isoclines for the
construction of phase trajectory.

Solution: $\ddot{y} + |\dot{y}| = 1$

(A) For $\dot{y} > 0$, the differential equation is $\ddot{y} + \dot{y} = 1$.

Let $x_1 = y$ and $x_2 = \dot{x}_1 = \dot{y}$. The differential equation becomes

$$\dot{x}_2 + x_1 = 1 \qquad \text{or } \dot{x}_2 = -x_1 + 1.$$

Equation for phase trajectory

$$\frac{dx_2}{dx_1} = \frac{\dot{x}_2}{x_2} = \frac{-x_1 + 1}{x_2}$$

$$x_2 \, dx_2 = (-x_1 + 1) dx_1.$$

Integrating both sides,

$$\frac{x_2^2}{2} = -\frac{x_1^2}{2} + x_1 + \frac{C}{2}$$

$$x_2^2 = -x_1^2 + 2x_1 + C.$$

This is the equation for the phase trajectory. The constant C can be evaluated if the initial state is known.

Equation for isoclines

$$m = \frac{dx_2}{dx_1} = \frac{\dot{x}_2}{x_2} = \frac{-x_1 + 1}{x_2}$$

$$x_2 = \frac{-1}{m} x_1 + \frac{1}{m}.$$

Isoclines are straight lines with slope $(-1/m)$. When $x_2 = 0, x_1 = 1$. The isoclines are concurrent at the point (1, 0).

(B) For $\dot{y} < 0$ the differential equation is $\ddot{y} - \dot{y} = 1$.

Let $x_1 = y$ and $x_2 = \dot{x}_1 = \dot{y}$. The differential equation becomes

$$\dot{x}_2 - x_1 = 1 \qquad \text{or } \dot{x}_2 = x_1 + 1.$$

Equation for phase trajectory

$$\frac{dx_2}{dx_1} = \frac{\dot{x}_2}{x_2} = \frac{x_1 + 1}{x_2}$$

$$x_2 \, dx_2 = (x_1 + 1) dx_1.$$

Integrating both sides,

$$\frac{x_2^2}{2} = \frac{x_1^2}{2} + x_1 + \frac{C}{2}$$

$$x_2^2 = x_1^2 + 2x_1 + C.$$

This is the equation for phase trajectory. The constant C can be evaluated using the given initial state.

Equation for isoclines

$$m = \frac{dx_2}{dx_1} = \frac{\dot{x}_2}{x_2} = \frac{x_1 + 1}{x_2}$$

$$x_2 = \frac{1}{m}x_1 + \frac{1}{m}.$$

Isoclines are straight lines with slope $(1/m)$. When $x_2 = 0, x_1 = -1$. The isoclines are concurrent at the point $(-1, 0)$.

8.6 LIMIT CYCLES

A closed contour in a phase plane predicts the presence of a limit cycle. The closed contour indicates the periodic nature of the limit cycle. It is also to be isolated with the neighbouring trajectories either converging to it or diverging away from it. Hence a limit cycle in a phase plane is identified as an isolated closed contour. All closed contours do not represent limit cycles if they are not isolated. This is shown in Figure 8.12(a). A limit cycle is stable if all trajectories in the neighbourhood of it converge to it as time tends to infinity as shown in Figure 8.12(b). A limit cycle is unstable if all trajectories in the neighbourhood of it diverge away from it as time approaches infinity as shown in Figure 8.12(c).

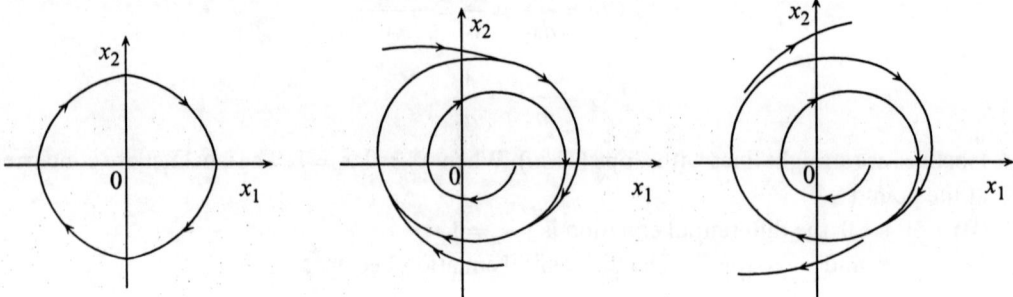

Figure 8.12 (a) Closed contour – not a limit cycle (b) stable limit cycle (c) unstable limit cycle

8.7 EQUILIBRIUM STATE

A system is called a *free system* if the input is zero. A free system, if it is time invariant, is called an *autonomous system*. An autonomous system is said to be in equilibrium state if all its states have settled down to constant values. These need not be zero values. Hence a system is in equilibrium state if none of its states vary with time.

Consider an autonomous system described by

$$\dot{X} = f(X).$$

This system is in equilibrium state if its states X_e are in equilibrium state and this happens only when

$$\dot{X}_e = f(X_e) = 0.$$

If the equilibrium state is not at origin, then a new set of state variables z_1 and $z_2 = \dot{z}_1$ can be defined such that in the $z_1 - z_2$ plane, the singular point is at the origin. The transformation from (x_1, x_2) to (z_1, z_2) can always be carried out. The method is illustrated in Example 8.27.

Example 8.24 Find the equilibrium state of the system represented by the state equation

$$\begin{bmatrix} \dot{x}_1 \\ \dot{x}_2 \end{bmatrix} = \begin{bmatrix} -1 & 2 \\ -3 & -5 \end{bmatrix} \begin{bmatrix} x_1 \\ x_2 \end{bmatrix}.$$

Solution: The state equations are

$$\dot{x}_1 = -x_1 + 2x_2$$

$$\dot{x}_2 = -3x_1 - 5x_2.$$

For the equilibrium state (x_{1e}, x_{2e}),

$$\dot{x}_{1e} = -x_{1e} + 2x_{2e} = 0$$

$$\dot{x}_{2e} = -3x_{1e} - 5x_{2e} = 0.$$

Solving these two equations,

$$x_{1e} = 0 \quad \text{and} \quad x_{2e} = 0.$$

The equilibrium state is at the origin of the $x_1 - x_2$ plane.

Example 8.25 For a nonlinear autonomous system, the state equations are

$$\dot{x}_1 = x_2 \quad \text{and} \quad \dot{x}_2 = -x_1 - x_1^2 - x_2.$$

Determine its equilibrium states.

Solution: Let (x_{1e}, x_{2e}) be the equilibrium state. At the equilibrium state,

$$\dot{x}_{1e} = 0 \text{ and } \dot{x}_{2e} = 0.$$

The given state equations can be written as

$$\dot{x}_{1e} = x_{2e} = 0 \tag{1}$$

$$\dot{x}_{2e} = -x_{1e} - x_{1e}^2 - x_{2e} = 0. \tag{2}$$

From equation (1), $x_{2e} = 0$.
From equation (2), $x_{1e}(1 + x_{1e}) = 0$ or $x_{1e} = 0$ and $x_{1e} = -1$.
There are two equilibrium states and they are

$$\begin{bmatrix} x_{1e} \\ x_{2e} \end{bmatrix} = \begin{bmatrix} 0 \\ 0 \end{bmatrix} \text{ and } \begin{bmatrix} x_{1e} \\ x_{2e} \end{bmatrix} = \begin{bmatrix} -1 \\ 0 \end{bmatrix}.$$

These equilibrium points, in the $x_1 - x_2$ plane, are shown in Figure E8.25.

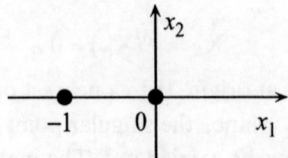

Figure E8.25 Equilibrium points – Example 8.25

Example 8.26 A system is represented by the state equations

$$\dot{x}_1 = -x_1 + x_1^2 x_2$$

$$\dot{x}_2 = -x_1 + x_2 .$$

Determine its equilibrium states.

Solution: Let (x_{1e}, x_{2e}) be the equilibrium state. At the equilibrium state,

$$\dot{x}_{1e} = 0 \text{ and } \dot{x}_{2e} = 0.$$

The given state equations can be written as

$$\dot{x}_{1e} = -x_{1e} + x_{1e}^2 x_{2e} = 0 \tag{1}$$

$$\dot{x}_{2e} = -x_{1e} + x_{2e} = 0 . \tag{2}$$

From equation (2), $x_{1e} = x_{2e}$. $\tag{3}$

From equation (1) $-x_{1e}(1 - x_{1e}^2) = 0$ $x_{1e} = 0$ or $x_{1e}^2 = 1$. $\tag{4}$

From equation (3), when $x_{1e} = 0$, $x_{2e} = 0$.

Hence one equilibrium state is $(x_{1e}, x_{2e}) = (0,0)$

From equation (4), $x_{1e}^2 = 1$ or $x_{1e} = \pm 1$.

When $x_{1e} = 1$, $x_{2e} = 1$ and when $x_{1e} = -1$, $x_{2e} = -1$.

There are two more equilibrium states which are $(x_{1e}, x_{2e}) = (1,1)$ and $(x_{1e}, x_{2e}) = (-1,-1)$

Example 8.27 Determine the equilibrium state of a system described by the state equations

$$\dot{x}_1 = -x_1 + 2x_2 + 1$$

$$\dot{x}_2 = -2x_1 - x_2 - 3 .$$

If the equilibrium state is not at origin, transform it to the origin of a new co-ordinate system.

Solution:

$$\dot{x}_1 = -x_1 + 2x_2 + 1$$

$$\dot{x}_2 = -2x_1 - x_2 - 3 .$$

To find the equilibrium state, substitute $\dot{x}_1 = 0$ and $\dot{x}_2 = 0$. Hence,

$$\dot{x}_{1e} = -x_{1e} + 2x_{2e} + 1 = 0 \tag{1}$$

$$\dot{x}_{2e} = -2x_{1e} - x_{2e} - 3 = 0 \qquad (2)$$

Solving equations (1) and (2), the equilibrium state is

$$(x_{1e}, x_{2e}) = (-1, -1).$$

It can be observed that the equilibrium state is not at origin.

Let $z_1 = x_1 - a$ and $z_2 = x_2 - b$ where a and b are constants.

Now $\dot{z}_1 = \dot{x}_1 = -(z_1 + a) + 2(z_2 + b) + 1 = 0$

$$\text{or} \quad -z_1 + 2z_2 = a - 2b - 1 \qquad (3)$$

$$\dot{z}_2 = \dot{x}_2 = -2(z_1 + a) - (z_2 + b) - 3 = 0$$

$$\text{or} \quad -2z_1 - z_2 = 2a + b + 3 \qquad (4)$$

In equation (3), let $a - 2b - 1 = 0$. $\qquad (5)$

In equation (4), let $2a + b + 3 = 0$. $\qquad (6)$

Solving equations (5) and (6), $a = -1$ and $b = -1$.

Hence $\qquad\qquad\qquad\qquad \dot{z}_1 = -z_1 + 2z_2 \qquad (7)$

and $\qquad\qquad\qquad\qquad \dot{z}_2 = -2z_1 - z_2 . \qquad (8)$

For equations (7) and (8), the equilibrium points are $(z_{1e}, z_{2e}) = (0, 0)$.

The transformation is $z_1 = x_1 - (-1) = x_1 + 1$ and $z_2 = x_2 - (-1) = x_2 + 1$.

In general, the transformation is $z_1 = x_1 - x_{1e}$ and $z_2 = x_2 - x_{2e}$.

Example 8.28 Determine the equilibrium state of a system described by the state equations

$$\dot{x}_1 = x_2$$

$$\dot{x}_2 = -x_1 - x_1^2(x_2 - 1).$$

If the equilibrium state is not at origin, transform it to the origin of a new co-ordinate system.

Solution:

$$\dot{x}_1 = x_2 \qquad (1)$$

$$\dot{x}_2 = -x_1 - x_1^2(x_2 - 1). \qquad (2)$$

To determine the equilibrium points,

$$\dot{x}_{1e} = \dot{x}_{2e} = 0 \qquad (3)$$

$$\dot{x}_{2e} = -x_{1e} - x_{1e}^2(x_{2e} - 1) = 0. \qquad (4)$$

Solving equations (3) and (4),

$$x_{2e} = 0 \quad \text{and} \quad x_{1e} = 0 \text{ or } 1.$$

Thus the equilibrium points are

$$(x_{1e}, x_{2e}) = (0,0) \text{ or } (1,0).$$

The equilibrium state at $(1, 0)$ can be shifted to the origin of a new co-ordinate system by using the transformation

$$z_1 = x_1 - 1 \text{ and } z_2 = x_2$$

$$\dot{z}_1 = \dot{x}_1 \text{ and } \dot{z}_2 = \dot{x}_2 .$$

Using equation (1),

$$\dot{z}_1 = \dot{x}_1 = x_2 = z_2 . \tag{5}$$

Using equation (2),

$$\dot{z}_2 = \dot{x}_2 = -x_1 - x_1^2 (x_2 - 1)$$

$$= -(z_1 + 1) - (z_1 + 1)^2 (z_2 - 1)$$

$$= -(z_1 + 1) + (z_1 + 1)^2 - z_2 (z_1 + 1)^2$$

$$= (z_1 + 1)\{-1 + z_1 + 1 - z_2 (z_1 + 1)\}$$

$$= (z_1 + 1)\{z_1 - z_2 (z_1 + 1)\}$$

$$= (z_1 + 1)(z_1 - z_1 z_2 - z_2) . \tag{6}$$

From equations (5) and (6), it can be observed that $(0, 0)$ is an equilibrium point in the $z_1 - z_2$ plane.

8.8 SINGULAR POINTS

At any point on the phase trajectory in the $x_1 - x_2$ plane, the slope is given by

$$m = \frac{dx_2}{dx_1} = \frac{\dot{x}_2}{\dot{x}_2} = \frac{\dot{x}_2}{\dot{x}_1} .$$

If at any point on the phase trajectory, $\dot{x}_1 = 0$ and $\dot{x}_2 = 0$, then the slope is indeterminable. Such points are called *singular points*. Hence the point on the phase trajectory where the slope is indeterminable is a singular point. It is stated earlier that equilibrium states are obtained by solving $\dot{x}_1 = 0$ and $\dot{x}_2 = 0$. Thus equilibrium states are singular points.

Some important observations can be made.

1. A nonlinear system can have more than one equilibrium state or singular point.

2. For a system, if one state equation is $\dot{x}_1 = x_2$, then at the singular point $\dot{x}_1 = x_2 = 0$ so that the singular point lies on the $x_1 - axis$.

The nature of phase trajectory in the vicinity of a singular point at origin is an indication of the stability of a system. A study of the nature of phase trajectory near the singular point together with the assessment of limit cycle may yield much information about the stability of nonlinear systems.

8.9 CLASSIFICATION OF SINGULAR POINTS

Consider a second-order autonomous system represented by the state equations

$$\dot{x}_1 = x_2 \tag{8.9}$$

and

$$\dot{x}_2 = f(x_1, x_2). \tag{8.10}$$

The singular state is assumed to be at origin. If equation (8.10) is nonlinear, it can be linearized about its singular point at origin, using Taylor series and can be written as

$$\dot{x}_2 = -ax_1 - bx_2. \tag{8.11}$$

Clearly, the singular state is at origin. Combining equations (8.9) and (8.11),

$$\ddot{x}_1 = -ax_1 - b\dot{x}_1$$

Or

$$\ddot{x}_1 + b\dot{x}_1 + ax_1 = 0.$$

In general a second-order system can be represented at the singular point at origin as

$$\ddot{y} + b\dot{y} + ay = 0.$$

Taking Laplace transforms,

$$s^2 - sy(0) - \dot{y}(0) + b\{sY(s) - y(0)\} + aY(s) = 0$$

$$(s^2 + bs + a)Y(s) = sy(0) + \dot{y}(0) + by(0)$$

$$Y(s) = \frac{sy(0) + by(0) + \dot{y}(0)}{(s^2 + bs + a)}.$$

The characteristic equation can be written as

$$(s^2 + bs + a) = (s - \lambda_1)(s - \lambda_2) = 0,$$

where λ_1 and λ_2 are the roots of the characteristic equation.

The nature of the response is decided by the location of the characteristic roots λ_1 and λ_2 in the s-plane.

Case 1: λ_1 and λ_2 are real and lie on the left half of s-plane

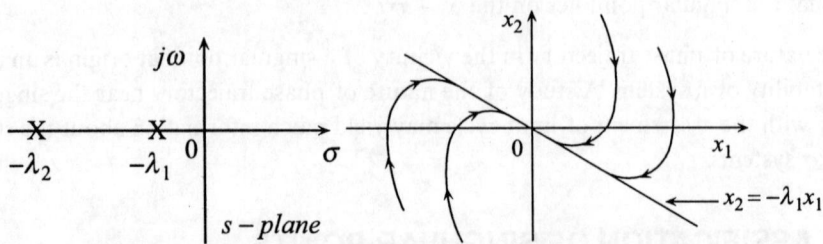

Figure 8.13 (a) Location of closed loop poles (b) stable node

The locations of the characteristic roots are shown in Figure 8.13(a). The response of the system is

$$x_1(t) = k_1\, e^{-\lambda_1 t} + k_2\, e^{-\lambda_2 t} \text{ and } x_2 = \dot{x}_1 = -k_1\, \lambda_1 e^{-\lambda_1 t} - k_2\, \lambda_2 e^{-\lambda_2 t}.$$

Both $x_1(t)$ and $x_2(t)$ approach zero as time t approaches infinity. All trajectories converge at the origin. The phase portrait in the neighbourhood of the singular point is shown in Figure 8.13(b). The term $e^{-\lambda_2 t}$ decays faster than the term $e^{-\lambda_1 t}$. Hence for larger values of t, $x_1(t) = k_1\, e^{-\lambda_1 t}$ and $x_2(t) = -\lambda_1 k_1\, e^{-\lambda_1 t} = -\lambda_1\, x_1(t)$. This is a straight line trajectory and all trajectories are tangential to this straight line trajectory at the origin. The singular point is called a *stable node*. The system is stable.

Case 2: λ_1 and λ_2 are real and lie on the right half of s-plane

The locations of the characteristic roots are shown in Figure 8.14(a). The response is

$x_1(t) = k_1\, e^{\lambda_1 t} + k_2\, e^{\lambda_2 t}$ and $x_2 = \dot{x}_1 = k_1\, \lambda_1 e^{\lambda_1 t} + k_2\, \lambda_2 e^{\lambda_2 t}$. For larger and larger values of time t, both $x_1(t)$ and $x_2(t)$ approach infinity. The phase portrait, in the vicinity of the singular point, is shown in Figure 8.14(b). All phase trajectories emerge out of the singular point and approach infinity. This type of singular point is called an *unstable node*. The system is unstable. The straight line trajectory is given by $x_2(t) = \lambda_1\, x_1(t)$.

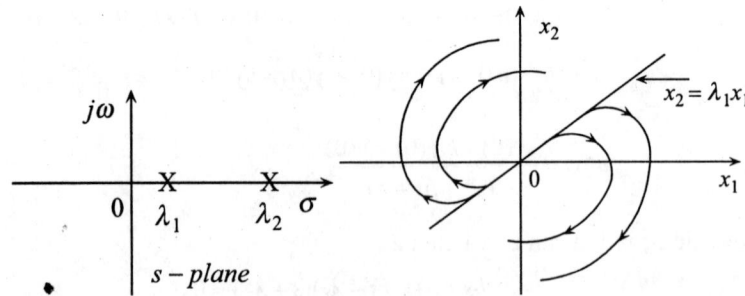

Figure 8.14 (a) Location of closed loop poles (b) unstable node

Case 3: λ_1 and λ_2 are complex and lie on the left half of s-plane

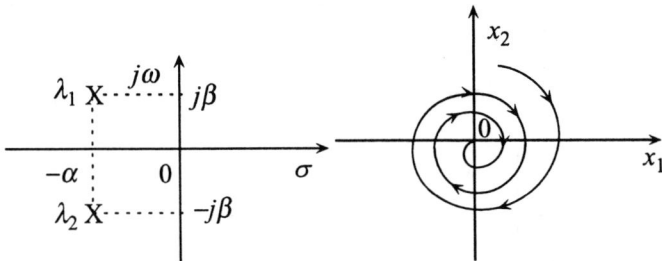

Figure 8.15 (a) Location of closed loop poles (b) stable focus

The characteristic roots are located on the left half of the s-plane as shown in Figure 8.15(a). Let the roots be

$$\lambda_1 = -\alpha + j\beta \text{ and } \lambda_2 = -\alpha - j\beta \,.$$

The system response is

$$x_1(t) = K_1 e^{-\alpha t} \sin(\beta t + \theta) \text{ and } x_2(t) = \dot{x}_1(t) = K_2 e^{-\alpha t} \sin(\beta t + \phi) \,.$$

Both the responses decay down to zero as time t approaches infinity. The phase trajectory is in the form of a spiral entering the singular point at origin as shown in figure 8.15(b). Such a singular point is called a *stable focus*. The system is stable.

Case 4: λ_1 and λ_2 are complex and lie on the right half of s-plane

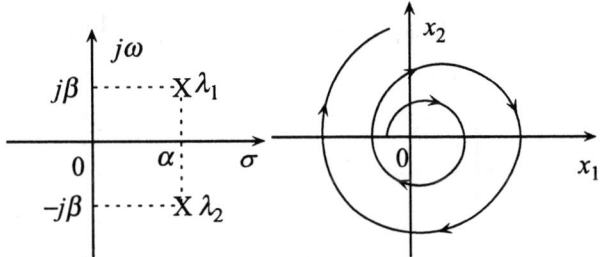

Figure 8.16 (a) Location of closed loop poles (b) unstable focus

The characteristic roots are located on the right half of the s-plane as shown in Figure 8.16(a). Let the roots be

$$\lambda_1 = \alpha + j\beta \text{ and } \lambda_2 = \alpha - j\beta \,.$$

The system response is

$$x_1(t) = k_1 e^{\alpha t} \sin(\beta t + \theta) \text{ and } x_2(t) = k_2 e^{\alpha t} \sin(\beta t + \phi) \,.$$

Both the responses increase without bound as time t approaches infinity. The phase trajectory is in the form of a spiral emerging from the singular point and approaching infinity as shown in Figure 8.16(b). Such a singular point is called an *unstable focus*. The system is unstable.

Case 5: λ_1 **and** λ_2 **are purely imaginary and lie on the** $j\omega$**-axis of s-plane**

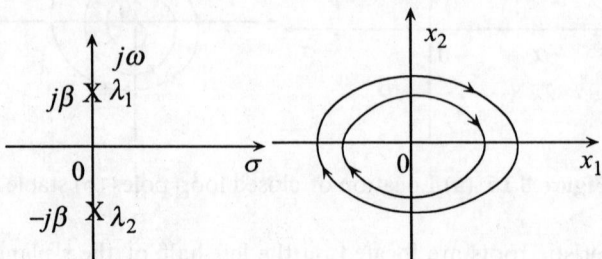

Figure 8.17 (a) Location of closed loop poles (b) centre or vortex

The two roots of the characteristic equation lie on the $j\omega$-axis as shown in Figure 8.17(a). The response is $x_1(t) = k_1 \sin(\beta t + \theta)$ and $x_2(t) = k_2 \cos(\beta t + \theta)$.

$$\left(\frac{x_1(t)}{k_1}\right)^2 + \left(\frac{x_2(t)}{k_2}\right)^2 = 1 .$$

This represents an ellipse and the phase trajectory is an ellipse as shown in Figure 8.17(b). The system generates unforced oscillations or limit cycles. The singular point is called a *centre* or *vortex*.

Case 6: λ_1 **and** λ_2 **are real with one positive and the other negative**

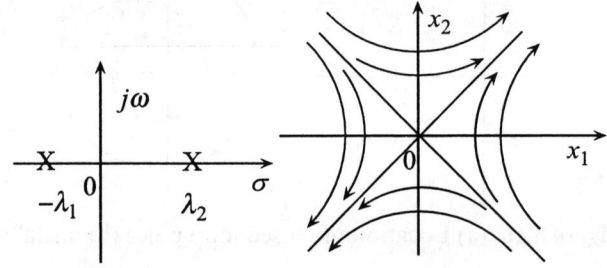

Figure 8.18 (a) Location of closed loop poles (b) saddle point

The locations of the characteristic roots are shown in Figure 8.18(a). The system response is

$$x_1(t) = k_1 e^{-\lambda_1 t} + k_2 e^{\lambda_2 t} \text{ and } x_2 = \dot{x}_1 = -k_1 \lambda_1 e^{-\lambda_1 t} + k_2 \lambda_2 e^{\lambda_2 t} .$$

In the expressions for both $x_1(t)$ and $x_2(t)$, the first term decays down to zero as time t approaches infinity whereas the second term increases without bounds. Hence both

$x_1(t)$ and $x_2(t)$ approach infinity for large values of time. The system is unstable. The phase trajectory, in the vicinity of the singular point, is shown in Figure 8.18(b). In this case, the singular point is called a *saddle point*.

Example 8.29 For the system given in Example 8.25, determine the type of the singular points.

Solution: The system is given as

$$\dot{x}_1 = x_2 \quad \text{and} \quad \dot{x}_2 = -x_1 - x_1^2 - x_2.$$

The singular points are $(0, 0)$ and $(-1, 0)$.

Let $f_1(x_1, x_2) = \dot{x}_1 = x_2$ and $f_2(x_1, x_2) = \dot{x}_2 = -x_1 - x_1^2 - x_2$.

$$\frac{\partial f_1}{\partial x_1} = 0 \qquad\qquad \frac{\partial f_1}{\partial x_2} = 1$$

$$\frac{\partial f_2}{\partial x_1} = -1 - 2x_1 \qquad\qquad \frac{\partial f_2}{\partial x_2} = -1$$

For the singular point $(0, 0)$,

$$J(0,0) = \begin{bmatrix} \dfrac{\partial f_1}{\partial x_1} = 0 & \dfrac{\partial f_1}{\partial x_2} = 1 \\[2mm] \dfrac{\partial f_2}{\partial x_1} = -1 - 2x_1 & \dfrac{\partial f_2}{\partial x_2} = -1 \end{bmatrix}_{(0,0)} = \begin{bmatrix} 0 & 1 \\ -1 & -1 \end{bmatrix}.$$

The Eigen values are obtained by solving

$$|\lambda I - J| = 0 \qquad \begin{vmatrix} \lambda & -1 \\ 1 & \lambda + 1 \end{vmatrix} = 0 \qquad \text{or} \quad \lambda^2 + \lambda + 1 = 0.$$

The Eigen values are $\lambda_1 = -0.5 + j0.866$ and $\lambda_2 = -0.5 - j0.866$. They are complex with negative real parts. The phase trajectory is spiral, converging at the origin. The singular point at the origin is a stable focus.

For the singular point $(-1, 0)$,

$$J(-1,0) = \begin{bmatrix} \dfrac{\partial f_1}{\partial x_1} = 0 & \dfrac{\partial f_1}{\partial x_2} = 1 \\[2mm] \dfrac{\partial f_2}{\partial x_1} = -1 - 2x_1 & \dfrac{\partial f_2}{\partial x_2} = -1 \end{bmatrix}_{(-1,0)} = \begin{bmatrix} 0 & 1 \\ 1 & -1 \end{bmatrix}.$$

The Eigen values are obtained by solving

$$|\lambda I - J| = 0 \qquad \begin{vmatrix} \lambda & -1 \\ -1 & \lambda + 1 \end{vmatrix} = 0 \qquad \text{or} \quad \lambda^2 + \lambda - 1 = 0.$$

The Eigen values are $\lambda_1 = 0.618$ and $\lambda_2 = -1.618$. They are real, one positive and the other negative. The singular point is a saddle point.

Example 8.30 For the system given in Example 8.27, determine the type of the singular points.

Solution: The system is given as

$$\dot{x_1} = -x_1 + 2x_2 + 1 \text{ and } \dot{x_2} = -2x_1 - x_2 - 3.$$

The singular point is $(-1, -1)$.

Let $f_1(x_1, x_2) = \dot{x_1} = -x_1 + 2x_2 + 1$ and $f_2(x_1, x_2) = \dot{x_2} = -2x_1 - x_2 - 3$.

$$\frac{\partial f_1}{\partial x_1} = -1 \qquad\qquad \frac{\partial f_1}{\partial x_2} = 2$$

$$\frac{\partial f_2}{\partial x_1} = -2 \qquad\qquad \frac{\partial f_2}{\partial x_2} = -1.$$

$$J(-1,-1) = \begin{bmatrix} \dfrac{\partial f_1}{\partial x_1} = -1 & \dfrac{\partial f_1}{\partial x_2} = 2 \\ \dfrac{\partial f_2}{\partial x_1} = -2 & \dfrac{\partial f_2}{\partial x_2} = -1 \end{bmatrix}_{(-1,-1)} = \begin{bmatrix} -1 & 2 \\ -2 & -1 \end{bmatrix}.$$

The Eigen values are obtained by solving

$$|\lambda I - J| = 0 \qquad \begin{vmatrix} \lambda + 1 & -2 \\ 2 & \lambda + 1 \end{vmatrix} = 0 \qquad \text{or } (\lambda + 1)^2 + 4 = 0.$$

The Eigen values are $\lambda_1 = -1 + j2$ and $\lambda_2 = -1 - j2$. They are complex with negative real parts. The phase trajectory is spiral, converging at the singular point. The singular point is a stable focus.

Example 8.31 For the system given in Example 8.28, determine the type of the singular points.

Solution: The system is given as

$$\dot{x_1} = x_2 \text{ and } \dot{x_2} = -x_1 - x_1^2(x_2 - 1).$$

The singular points are $(0, 0)$ and $(1, 0)$.

Let $f_1(x_1, x_2) = \dot{x_1} = x_2$ and $f_2(x_1, x_2) = \dot{x_2} = -x_1 - x_1^2(x_2 - 1)$.

$$\frac{\partial f_1}{\partial x_1} = 0 \qquad\qquad \frac{\partial f_1}{\partial x_2} = 1$$

$$\frac{\partial f_2}{\partial x_1} = -1 - 2x_1(x_2 - 1) \qquad\qquad \frac{\partial f_2}{\partial x_2} = -x_1^2.$$

For the singular point (0. 0),

$$J(0,0) = \begin{bmatrix} \dfrac{\partial f_1}{\partial x_1} = 0 & \dfrac{\partial f_1}{\partial x_2} = 1 \\[4mm] \dfrac{\partial f_2}{\partial x_1} = -1 - 2x_1(x_2 - 1) & \dfrac{\partial f_2}{\partial x_2} = -x_1^2 \end{bmatrix}_{(0,0)} = \begin{bmatrix} 0 & 1 \\ -1 & 0 \end{bmatrix}.$$

The Eigen values are obtained by solving

$$|\lambda I - J| = 0 \qquad \begin{vmatrix} \lambda & -1 \\ 1 & \lambda \end{vmatrix} = 0 \qquad \text{or } \lambda^2 + 1 = 0.$$

The Eigen values are $\lambda_1 = j_1$ and $\lambda_2 = -j_1$. They are purely imaginary and lie on the imaginary axis. The phase trajectory is elliptical. The singular point at the origin is a vortex.

For the singular point (1. 0),

$$J(1,0) = \begin{bmatrix} \dfrac{\partial f_1}{\partial x_1} = 0 & \dfrac{\partial f_1}{\partial x_2} = 1 \\[4mm] \dfrac{\partial f_2}{\partial x_1} = -1 - 2x_1(x_2 - 1) & \dfrac{\partial f_2}{\partial x_2} = -x_1^2 \end{bmatrix}_{(1,0)} = \begin{bmatrix} 0 & 1 \\ 1 & -1 \end{bmatrix}.$$

The Eigen values are obtained by solving

$$|\lambda I - J| = 0 \qquad \begin{vmatrix} \lambda & -1 \\ -1 & \lambda + 1 \end{vmatrix} = 0 \qquad \text{or } \lambda^2 + \lambda - 1 = 0.$$

The Eigen values are $\lambda_1 = 0.618$ and $\lambda_2 = -1.618$. They are real, one positive and the other negative. The singular point is a saddle point.

Example 8.32 Determine the type of singular points for the system of Example 8.26.

Solution: The system is given as

$$\dot{x}_1 = -x_1 + x_1^2 x_2 \quad \text{and} \quad \dot{x}_2 = -x_1 + x_2.$$

The singular points are (0, 0), (1, 1) and (−1, −1).

Let $f_1(x_1, x_2) = \dot{x}_1 = -x_1 + x_1^2 x_2$ $\qquad$ and $f_2(x_1, x_2) = \dot{x}_2 = -x_1 + x_2.$

$$\frac{\partial f_1}{\partial x_1} = -1 + 2x_1 x_2 \qquad\qquad \frac{\partial f_1}{\partial x_2} = x_1^2$$

$$\frac{\partial f_2}{\partial x_1} = -1 \qquad\qquad \frac{\partial f_2}{\partial x_2} = 1.$$

For the singular point $(0, 0)$,

$$J(0,0) = \begin{bmatrix} \dfrac{\partial f_1}{\partial x_1} = -1 + 2x_1 x_2 & \dfrac{\partial f_1}{\partial x_2} = x_1^2 \\[2ex] \dfrac{\partial f_2}{\partial x_1} = -1 & \dfrac{\partial f_2}{\partial x_2} = 1 \end{bmatrix}_{(0,0)} = \begin{bmatrix} -1 & 0 \\ -1 & 1 \end{bmatrix}.$$

The Eigen values are obtained by solving

$$|\lambda I - J| = 0 \qquad \begin{vmatrix} \lambda+1 & 0 \\ 1 & \lambda-1 \end{vmatrix} = 0 \qquad \text{or} \quad (\lambda+1)(\lambda-1) = 0.$$

The Eigen values are $\lambda_1 = -1$ and $\lambda_2 = 1$. They are real, one positive and the other negative. The singular point at the origin is a saddle point.

For the singular point $(1, 1)$,

$$J(1,1) = \begin{bmatrix} \dfrac{\partial f_1}{\partial x_1} = -1 + 2x_1 x_2 & \dfrac{\partial f_1}{\partial x_2} = x_1^2 \\[2ex] \dfrac{\partial f_2}{\partial x_1} = -1 & \dfrac{\partial f_2}{\partial x_2} = 1 \end{bmatrix}_{(1,1)} = \begin{bmatrix} 1 & 1 \\ -1 & 1 \end{bmatrix}.$$

The Eigen values are obtained by solving

$$|\lambda I - J| = 0 \qquad \begin{vmatrix} \lambda-1 & -1 \\ 1 & \lambda-1 \end{vmatrix} = 0 \qquad \text{or} \quad (\lambda-1)^2 + 1 = 0.$$

The Eigen values are $\lambda_1 = 1 + j_1$ and $\lambda_2 = 1 - j_1$. They are real complex with positive real parts. The phase trajectory is a spiral, emerging out of the singular point. The singular point is an unstable focus.

For the singular point $(-1, -1)$,

$$J(-1,-1) = \begin{bmatrix} \dfrac{\partial f_1}{\partial x_1} = -1 + 2x_1 x_2 & \dfrac{\partial f_1}{\partial x_2} = x_1^2 \\[2ex] \dfrac{\partial f_2}{\partial x_1} = -1 & \dfrac{\partial f_2}{\partial x_2} = 1 \end{bmatrix}_{(-1,-1)} = \begin{bmatrix} 1 & 1 \\ -1 & 1 \end{bmatrix}.$$

The Eigen values are obtained by solving

$$|\lambda I - J| = 0 \qquad \begin{vmatrix} \lambda-1 & -1 \\ 1 & \lambda-1 \end{vmatrix} = 0 \qquad \text{or} \quad (\lambda-1)^2 + 1 = 0.$$

The Eigen values are $\lambda_1 = 1 + j_1$ and $\lambda_2 = 1 - j_1$. They are real complex with positive real parts. The phase trajectory is a spiral, emerging out of the singular point. The singular point is an unstable focus.

Example   8.33 A   nonlinear   system   is   represented   by   the   state   equations

$\dot{x}_1 = x_2$ and $\dot{x}_2 = -x_1 - k x_2$. Discuss the types of singularities for different ranges of values of k.

Solution: To find the singular point, let $\dot{x}_1 = 0$ and $\dot{x}_2 = 0$. This will yield $x_{1e} = 0$ and $x_{2e} = 0$.

The singular point is (0, 0).

$$J(0,0) = \begin{bmatrix} \dfrac{\partial f_1}{\partial x_1} = 0 & \dfrac{\partial f_1}{\partial x_2} = 1 \\[2mm] \dfrac{\partial f_2}{\partial x_1} = -1 & \dfrac{\partial f_2}{\partial x_2} = -k \end{bmatrix}_{(0,0)} = \begin{bmatrix} 0 & 1 \\ -1 & -k \end{bmatrix},$$

The Eigen values are obtained by solving

$$|\lambda I - J| = 0 \qquad \begin{vmatrix} \lambda & -1 \\ 1 & \lambda + k \end{vmatrix} = 0 \qquad \text{or} \quad \lambda^2 + k\lambda + 1 = 0$$

$$\lambda = -\frac{k}{2} \pm \sqrt{\frac{k^2}{4} - 1} = -\frac{k}{2} \pm \frac{1}{2}\sqrt{k^2 - 4}.$$

The Eigen values depend on the value of k.

When $k > 2$, the Eigen values are real and negative. They lie on the negative real axis. The singular point at origin is a stable node.

For $0 < k < 2$, the Eigen values are complex conjugates with negative part. The singular point at origin is a stable focus.

When $k = 0$, the Eigen values are purely imaginary and conjugates and they lie on the imaginary axis. The singular point at origin is a vortex.

For $(-2 < k < 0)$, the Eigen values are complex with positive real part. The singular point at origin is an unstable focus.

When $k < -2$, the Eigen values are real and positive. They lie on the positive real axis. The singular point at origin is an unstable node.

SUMMARY

*   Phase plane analysis is a graphical procedure for determining the nature of transient response of a closed loop system.

- It is applicable only to a first-order or second-order system, the dynamics of which are described by a differential equation.
- The phase plane is a two dimensional $x_1 - x_2$ plane with the state variable x_1 along the horizontal axis and $x_2 = \dot{x}_1$ along the vertical axis.
- Phase trajectory is a plot of x_2 versus x_1 in the phase plane.
- For the portion of the phase trajectory above the x_1 axis, it is directed from left to right and for the portion of the phase trajectory below the x_1 axis, it is directed from right to left.
- All phase trajectories cross the x_1 axis at right angles. In other words, the slope of the tangent drawn at the x_1 axis crossing point of a phase trajectory is infinity.
- A family of phase trajectories is known as phase portrait.
- No two phase trajectories cross each other.
- The phase trajectory is symmetrical about the $x_1 - axis$ if $f(x_1, x_2)$ is an even function of x_2; that is if $f(x_1, x_2) = f(x_1, -x_2)$.
- The phase trajectory is symmetrical about the $x_2 - axis$ if $f(x_1, x_2)$ is an odd function of x_1; that is if $f(x_1, x_2) = -f(-x_1, x_2)$.
- A nonlinear differential equation representing a nonlinear system cannot be easily solved to study the nature of the transient response of the system. In such a case, phase plane analysis can be conveniently used with advantage.
- Phase trajectories can be drawn using analytical methods. But quite often these methods prove very difficult.
- Phase trajectories can be constructed graphically. Method of isoclines, Pell's method and Delta method are some graphical methods of construction of phase trajectories.
- An isocline is the locus of the points on phase trajectories where the slopes are the same.
- If a free system, that is a system in which the input is zero, is time invariant, then it is called an autonomous system.
- An autonomous system is said to be in equilibrium if all its states have settled down to constant or zero values. An autonomous system is in equilibrium state if its states do not vary with time.
- A second-order autonomous system is said to be in equilibrium state if for the two state variables x_1 and x_2

$$\dot{x}_1 = 0 \text{ and } \dot{x}_2 = 0.$$

- For a nonlinear system, there can be more than one equilibrium state.
- At any point on the phase trajectory if $\dot{x}_1 = 0$ and $\dot{x}_2 = 0$ then the slope at that point is indeterminable. Such a point is known as singular point. Equilibrium states are singular points.
- The nature of a phase portrait in the vicinity of a singular point depends on the locations of the closed loop poles in the s-plane.
- If both the closed loop poles are on the negative real axis, then the singular point is called a stable node.
- If both the closed loop poles lie on the positive real axis, then the singular point is called an unstable node.

- If the closed loops are complex with negative real parts, then the singular point is known as a stable focus.
- If the closed loops are complex with positive real parts, then the singular point is known as an unstable focus.
- When both the poles are on the imaginary axis, the singular point is called a vortex or centre.
- When one closed loop pole lies on the negative real axis and the other on the positive real axis, then the singular point is called a saddle point.

PRACTICE PROBLEMS

PP 8.1 Draw the phase trajectory for a linear system described by the differential equation

$$\ddot{y} + 4y = 0 \qquad y(0) = 1 \text{ and } \dot{y}(0) = 4\sqrt{6} \ .$$

Use the expression for the slope of the trajectory at any point.

PP 8.2 Draw the phase trajectory for a linear system, by directly solving the differential equation representation

$$\ddot{y} + 4y = 0 \qquad y(0) = 1 \text{ and } \dot{y}(0) = 4\sqrt{6} \ .$$

PP 8.3 Draw the phase trajectory for a linear system, by using the expression for the slope of the phase trajectory.

$$\ddot{y} + 4y = 1 \qquad y(0) = 0 \text{ and } \dot{y}(0) = 4 \ .$$

PP 8.4 Draw the phase trajectory for a linear system, by directly solving the differential equation representation

$$\ddot{y} + 4y = 1 \qquad y(0) = 0 \text{ and } \dot{y}(0) = 4 \ .$$

PP 8.5 Draw the phase trajectory for a linear system, by using the expression for the slope of the phase trajectory.

$$\ddot{y} + 2 = 0 \qquad y(0) = 4 \text{ and } \dot{y}(0) = 0 \ .$$

PP 8.6 Draw the phase trajectory for a linear system, by directly solving the differential equation representation

$$\ddot{y} + 2 = 0 \qquad y(0) = 4 \text{ and } \dot{y}(0) = 0 \ .$$

PP 8.7 Draw the phase trajectory for a linear system, by using the expression for the slope of the phase trajectory.

$$\ddot{y} + 4\dot{y} = 0 \qquad y(0) = 1 \text{ and } \dot{y}(0) = 3 \ .$$

PP 8.8 Draw the phase trajectory for a linear system, by directly solving the differential equation representation

$$\ddot{y} + 4\dot{y} = 0 \qquad y(0) = 1 \text{ and } \dot{y}(0) = 3 \ .$$

PP 8.9 Obtain the equation for the phase trajectory for a system described by the differential equation $\ddot{y} + 2\dot{y} = 4$ with initial state $y(0) = 4$ and $\dot{y}(0) = 0$.

PP 8.10 For the system of PP8.9, show that the isoclines are parallel to the $x_1 - axis$ in the phase plane.

PP 8.11 Draw an initial portion of the phase trajectory starting from the state $(3, 0)$ using the method of isoclines for a system represented by the differential equation $\ddot{y} + 0.25\,\dot{y} + y = 0$.

PP 8.12 Draw an initial portion of the phase trajectory starting from the state $(0, 3)$ using the method of isoclines for a system represented by the differential equation $\ddot{y} + 0.25\,\dot{y} + y = 2$.

PP 8.13 Show graphically, using Pell's method, a portion of the phase trajectory starting from the point $(0, 4)$ for the system represented by the differential equation $\ddot{y} + 2\,\dot{y} + y = 0$.

PP 8.14 Using Pell's method, show a portion of the phase trajectory starting from the initial point $(0, 3)$ for a system represented by the differential equation $\ddot{y} + 0.5\,\dot{y} + 0.5y = 1$.

PP 8.15 Using Delta method, show a portion of the phase trajectory starting from the initial point $(2, 4)$ for a system represented by the differential equation $\ddot{y} + 0.25\,\dot{y} + y + 0.5y^3 = 1$.

PP 8.16 Show graphically, using Delta method, a portion of the phase trajectory starting from the point $(1, 4)$ for the system represented by the differential equation $\ddot{y} + 0.5\,\dot{y} + 0.5\,y\,\dot{y} + 0.25y = 1$.

PP 8.17 A second-order system is represented by the differential equation $\ddot{y} + \dot{y} + |y| = 1$. Find the equation for the isoclines for constructing a phase trajectory.

PP 8.18 A second-order system is represented by the differential equation $\ddot{y} + |\dot{y}| = 1$.

Determine the equation for the isoclines for the construction of a phase trajectory.

PP 8.19 A nonlinear system is shown in Figure PP8.19. Derive the equation for the isoclines, in the $e - \dot{e}$ plane.

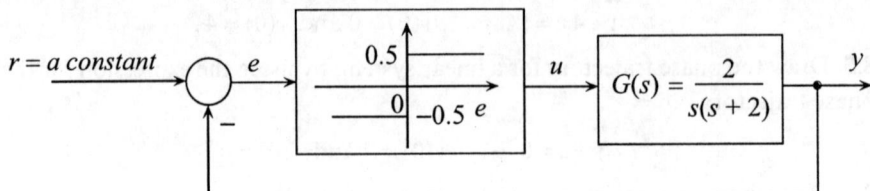

Figure PP 8.19

PP 8.20 A nonlinear system is shown in Figure PP8.20. Obtain the equations for the isoclines, in the $e - \dot{e}$ plane.

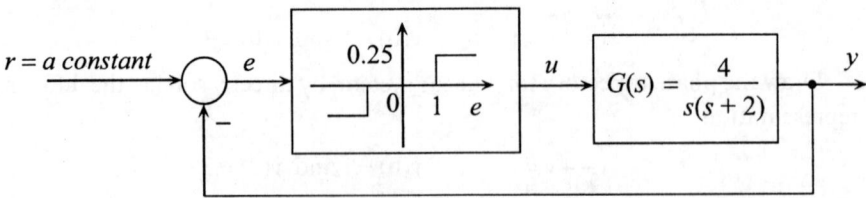

Figure PP 8.20

PP 8.21 For the nonlinear system shown in Figure PP8.21, derive the equation for the isoclines, in the $e - \dot{e}$ plane.

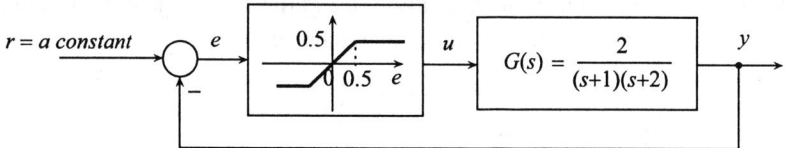

Figure PP 8.21

PP 8.22 For the nonlinear system shown in Figure PP8.22, derive the equation for the isoclines, in the $e - \dot{e}$ plane.

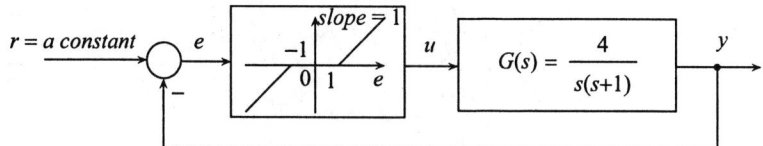

Figure PP 8.22

PP 8.23 For the nonlinear system shown in Figure PP8.23, obtain the equation for the isoclines for the construction of phase trajectory.

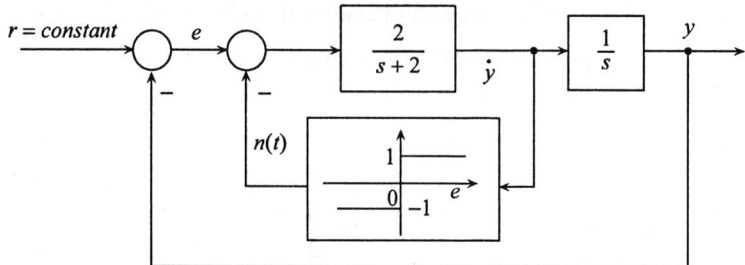

Figure PP 8.23

PP 8.24 For the nonlinear system shown in Figure PP8.24, obtain the equation for the isoclines for the construction of phase trajectory.

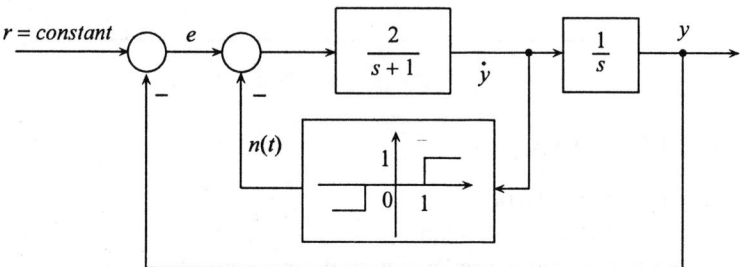

Figure P.P 8.24

PP 8.25 Find the equilibrium states and their nature in the vicinity of equilibrium state, for a system represented by the differential equation $\ddot{y} + y^2(\dot{y}-1) + y = 0$.

PP 8.26 Determine the equilibrium states and their nature in the vicinity of equilibrium state, for a nonlinear system represented by the state equations

$$\dot{x}_1 = -x_1 + x_2$$

$$\dot{x}_2 = -(x_1 + x_2)\sin x_1 - x_2.$$

PP 8.27 Determine the singular points and their nature in the vicinity of equilibrium state, for a nonlinear system represented by the differential equation $\ddot{\theta} + 2\dot{\theta} + 2\cos\theta = 0$.

PP 8.28 Determine the equilibrium state of a system represented by the state equations

$$\frac{dx_1}{dt} = -x_1 + 2x_2 - 5$$

$$\frac{dx_2}{dt} = -2x_1 + x_2 - 7$$

If the equilibrium state is not at origin, transform it to the origin adopting a new co-ordinate system.

PP 8.29 Determine the singular points and their nature in the vicinity of equilibrium state, for a nonlinear system represented by the state equations $\dot{x}_1 = -x_1 + x_1^2 x_2$ and $\dot{x}_2 = -(x_1 + x_2)$.

PP 8.30 Determine the singular points and their nature in the vicinity of equilibrium state, for a nonlinear system represented by the differential equation $\ddot{y} - \dot{y} + 2(\dot{y})^3 + y + y^2 = 0$.

PP 8.31 For the system shown in Figure PP8.31, determine the nature of singular points, in the $e - \dot{e}$ plane.

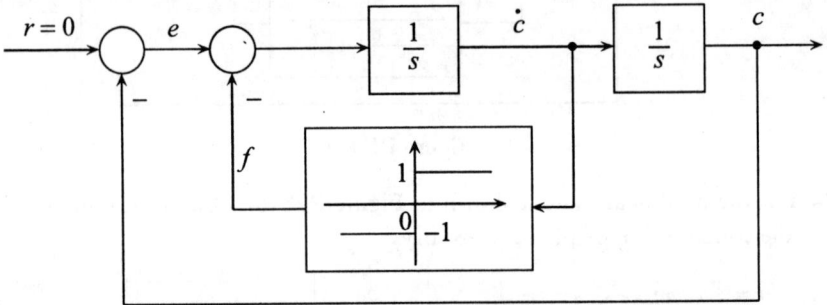

Figure PP 8.31

REVIEW QUESTIONS

8.1   What is a phase plane? Explain.
8.2   What is a phase trajectory? Explain.
8.3   Explain a method of obtaining an equation for a phase trajectory.

8.4 Establish the directions of phase trajectory above and below the $x_1 - axis$.

8.5 Write down the properties of phase trajectories.

8.6 Show that a phase trajectory crosses the $x_1 - axis$ at 90^0.

8.7 Two phase trajectories do not cross each other. Explain the reason.

8.8 Establish the condition for a phase trajectory to be symmetrical about the $x_1 - axis$.

8.9 Establish the condition for a phase trajectory to be symmetrical about the $x_2 - axis$.

8.10 What is an isocline? Explain.

8.11 Discuss the method of constructing the phase trajectory by the use of isoclines.

8.12 Explain the Pell's method of constructing a phase trajectory.

8.13 Explain the Delta method of constructing a phase trajectory.

8.14 Define the equilibrium state of a system.

8.15 What are singular points? Explain.

8.16 Discuss the classification of singular points.

8.17 Discuss the type of singular point if the two closed loop poles are real and negative.

8.18 What is the type of singular point if the two closed loop poles are real and positive? Explain.

8.19 What is the type of singular point if the two closed loop poles are complex with negative real parts? Show that the phase trajectory is spiral in nature.

8.20 If the two closed loop poles are complex with positive real parts, what is the type of singular point?

8.21 The two closed loop poles of a second order system are purely imaginary. Show that the phase trajectory is in the form of an ellipse.

8.22 Discuss the nature of singular point of a second order system, if one closed loop pole is real and negative and the other real and positive.

9 Stability

9.1 ASYMPTOTIC STABILITY AND BIBO STABILITY

A linear system is said to be stable if its impulse response decays down to zero as time approaches infinity. A linear system is asymptotically stable if all the Eigen values of the system lie on the left half of the s-plane. It is BIBO (bounded input bounded output) stable if all the poles of the system lie on the left half of the s-plane. The poles of the transfer function of a system are a subset of Eigen values. Hence, if a system is asymptotically stable, it is also BIBO stable.

Consider a linear system represented by the state model with input r and output y:

$$\dot{X} = AX + Br \qquad\qquad y = CX .$$

The Eigen values are obtained by solving the equation

$$|\lambda I - A| = 0 .$$

The transfer function of a system is given by

$$T(s) = C[sI - A]^{-1} B .$$

If a system is not completely controllable or completely observable, there will be cancellation of factors in $[sI - A]^{-1} B$ or $C[sI - A]^{-1}$. Hence a system which is BIBO stable may not be asymptotically stable but an asymptotically stable system is also BIBO stable.

Example 9.1 The system matrix of a linear time-invariant system is

$$A = \begin{bmatrix} 0 & 1 \\ -6 & -5 \end{bmatrix}.$$

Check whether the system is stable.

Solution: The Eigen values are obtained by solving the equation

$$|\lambda I - A| = 0 \quad \text{or} \quad \begin{vmatrix} \lambda & -1 \\ 6 & \lambda+5 \end{vmatrix} = 0$$

$$\lambda^2 + 5\lambda + 6 = 0 \quad \text{or} \quad (\lambda+2)(\lambda+3) = 0.$$

The Eigen values are $\lambda = -2$ and $\lambda = -3$. These lie on the negative real axis and so the system is asymptotically stable.

Example 9.2 The system matrix of a linear time-invariant system is

$$A = \begin{bmatrix} 0 & 1 \\ 6 & -1 \end{bmatrix}.$$

Check whether the system is stable.

Solution: The Eigen values are obtained by solving the equation

$$|\lambda I - A| = 0 \quad \text{or} \quad \begin{vmatrix} \lambda & -1 \\ -6 & \lambda+1 \end{vmatrix} = 0$$

$$\lambda^2 + \lambda - 6 = 0 \quad \text{or} \quad (\lambda-2)(\lambda+3) = 0.$$

The Eigen values are $\lambda = 2$ and $\lambda = -3$. One Eigen value is real and positive and so the system is unstable.

Example 9.3 The system matrix of a linear time-invariant system is

$$A = \begin{bmatrix} -2 & 1 & 0 \\ 0 & 0 & 2 \\ 0 & -1 & -2 \end{bmatrix}.$$

Check whether the system is asymptotically stable.

Solution: The Eigen values are obtained by solving the equation

$$|\lambda I - A| = 0 \quad \text{or} \quad \begin{vmatrix} \lambda+2 & -1 & 0 \\ 0 & \lambda & -2 \\ 0 & 1 & \lambda+2 \end{vmatrix} = 0$$

$$(\lambda+2)(\lambda^2 + 2\lambda + 2) = 0 \quad .$$

The Eigen values are $\lambda_1 = -2$, $\lambda_2 = -1 + j1$ and $\lambda_3 = -1 - j1$. One Eigen value is real and negative and the other two are complex conjugates with negative real parts. Hence the system is asymptotically stable.

Example 9.4 The system matrix of a linear time-invariant system is

$$A = \begin{bmatrix} -2 & 1 & 1 \\ 0 & 0 & 5 \\ 0 & -1 & 2 \end{bmatrix}.$$

Check whether the system is asymptotically stable.

Solution: The Eigen values are obtained by solving the equation

$$|\lambda I - A| = 0 \quad \text{or} \quad \begin{vmatrix} \lambda+2 & -1 & -1 \\ 0 & \lambda & -5 \\ 0 & 1 & \lambda-2 \end{vmatrix} = 0$$

$$(\lambda+2)(\lambda^2 - 2\lambda + 5) = 0 \ .$$

The Eigen values are $\lambda_1 = -2$, $\lambda_2 = 1 + j2$ and $\lambda_3 = 1 - j2$. One Eigen value is real and negative. The other two are complex conjugates with positive real parts. Hence the system is unstable.

9.2 SIGN DEFINITENESS

(A) Positive definiteness: A scalar function $F(X) = F(x_1, x_2, x_3, ..., x_n)$ is *positive definite* if $F(X)$ is continuous and is positive at all points in the state space except at the origin, where the function is zero.

For example $F(X) = x_1^2 + x_2^2$ is positive definite because it is always positive at all points in the $x_1 - x_2$ plane and is zero only when $x_1 = 0$ and $x_2 = 0$. Similarly $F(X) = x_1^2 + x_2^2 + x_3^2$ is positive definite because it is always positive at all points in the three-dimensional plane and is zero only when $x_1 = 0$, $x_2 = 0$ and $x_3 = 0$.

(B) Positive semi-definiteness: A scalar function $F(X)$ is *positive semi-definite* if $F(X)$ is continuous and is positive at all points in the state space except at a few points in addition to the origin, where the function is zero.

For example $F(X) = (x_1 + 2x_2)^2 + x_3^2$ is positive semi-definite because it is positive in the three-dimensional plane but can be zero when $x_1 = -2x_2$ *and* $x_3 = 0$. A semi-definite function can be zero at several points in addition to the origin.

(C) Negative definiteness: A scalar function $F(X)$ is *negative definite* if $F(X)$ is continuous and is negative at all points in the state space except at the origin where the function is zero. If $F(X)$ is positive definite, then $-F(X)$ is negative definite.

The scalar function $F(X) = -x_1^2 - x_2^2$ is negative definite in the two-dimensional $x_1 - x_2$ plane.

(D) Negative semi-definiteness: A scalar function $F(X)$ is *negative semi-definite* if $F(X)$ is continuous and is negative at all points in the state space except at a few points in addition to the origin, where the function is zero.

$F(X) = -(x_1 + x_2)^2 - x_3^2$ is negative semi-definite, because it is negative at all points but it is zero at the origin and also when $x_1 = -x_2$ *and* $x_3 = 0$.

(E) Sign indefiniteness: A scalar function $F(X)$ is said to be *indefinite* if $F(X)$ can be positive, negative or zero anywhere in the state space.

$F(X) = x_1^2 x_2 - x_1 x_2$ is indefinite.

9.3 QUADRATIC FORMS

A scalar function $F(X)$ is said to be in *quadratic form* if it can be expressed as

$$F(X) = X^T P X,$$

where X is a vector given by

$$X = \begin{bmatrix} x_1 & x_2 & x_3 & \cdots & x_n \end{bmatrix}^T$$

and P is a real symmetric matrix.

Consider a vector $X = \begin{bmatrix} x_1 & x_2 & x_3 \end{bmatrix}^T$.

A scalar function

$$F(X) = p_{11}x_1^2 + p_{22}x_2^2 + p_{33}x_3^2 + 2p_{12}\,x_1x_2 + 2p_{23}\,x_2\,x_3 + 2p_{31}\,x_3x_1$$

can be expressed as

$$F(X) = \begin{bmatrix} x_1 & x_2 & x_3 \end{bmatrix} \begin{bmatrix} p_{11} & p_{12} & p_{13} \\ p_{21} & p_{22} & p_{23} \\ p_{31} & p_{32} & p_{33} \end{bmatrix} \begin{bmatrix} x_1 \\ x_2 \\ x_3 \end{bmatrix} = X^T P X,$$

where P is a real symmetric matrix with $p_{ij} = p_{ji}; i \neq j$.
Then $F(X)$ is said to be in *quadratic form*.

9.4 SYLVESTER'S CRITERION FOR SIGN DEFINITENESS

The *sign definiteness* of a scalar function $F(X)$ in quadratic form can be determined using the *Sylvester's criterion*.

1. A necessary and sufficient condition for a function in quadratic form $F(X) = X^T P X$, where P is an $(n \times n)$ real symmetric matrix, to be *positive definite* is that $|P| > 0$ and the *successive principal* minors of P are positive.

    Consider a real symmetric matrix

    $$P = \begin{bmatrix} p_{11} & p_{12} & p_{13} \\ p_{21} & p_{22} & p_{23} \\ p_{31} & p_{32} & p_{33} \end{bmatrix}; \ p_{ij} = p_{ji}; \ i \neq j.$$

    By Sylvester's criterion, $F(X)$ is **positive definite** if

    $$p_{11} > 0 \qquad \begin{vmatrix} p_{11} & p_{12} \\ p_{12} & p_{22} \end{vmatrix} > 0 \quad \text{and} \quad |P| > 0.$$

2. A necessary and sufficient condition for a function in quadratic form $F(X) = X^T P X$ to be *positive semi-definite* is that $|P| = 0$ and **all the *principal minors*** of P are nonnegative.

    $$P = \begin{bmatrix} p_{11} & p_{12} & p_{13} \\ p_{21} & p_{22} & p_{23} \\ p_{31} & p_{32} & p_{33} \end{bmatrix}; \ p_{ij} = p_{ji}; \ i \neq j$$

Sylvester's criterion for $F(X)$ to be *positive semi-definite* is that

$$p_{11} \geq 0 \qquad\qquad p_{22} \geq 0 \qquad\qquad p_{33} \geq 0$$

$$\begin{vmatrix} p_{11} & p_{12} \\ p_{21} & p_{22} \end{vmatrix} \geq 0 \qquad \begin{vmatrix} p_{22} & p_{23} \\ p_{32} & p_{33} \end{vmatrix} \geq 0 \qquad \begin{vmatrix} p_{11} & p_{13} \\ p_{31} & p_{33} \end{vmatrix} \geq 0 \quad \text{and} \quad |P| = 0.$$

It is to be noted that all the principal minors are to be checked and not just the successive minors only.

3. A necessary and sufficient condition for a function in quadratic form $F(X) = X^T P X$ to be *negative definite* is that $|P|$ is positive if n is even and negative if n is odd and the *successive principal minors* of even order are positive and the *successive principal minors* of odd order are negative.

$$P = \begin{bmatrix} p_{11} & p_{12} & p_{13} \\ p_{21} & p_{22} & p_{23} \\ p_{31} & p_{32} & p_{33} \end{bmatrix}; \; p_{ij} = p_{ji}; \; i \neq j$$

Sylvester's criterion for $F(X)$ to be *negative definite* is that

$$p_{11} < 0 \qquad \begin{vmatrix} p_{11} & p_{12} \\ p_{21} & p_{22} \end{vmatrix} > 0 \qquad |P| < 0.$$

It is to be noted that $|P| > 0$ when n is even. and $|P| < 0$ when n is odd.

4. A necessary and sufficient condition for a function in quadratic form $F(X) = X^T P X$ to be *negative semi-definite* is that $|P| = 0$ and all the *principal minors* of even order are nonnegative and all the *principal minors* of odd order are nonpositive.

$$P = \begin{bmatrix} p_{11} & p_{12} & p_{13} \\ p_{21} & p_{22} & p_{23} \\ p_{31} & p_{32} & p_{33} \end{bmatrix}; \; p_{ij} = p_{ji}; \; i \neq j$$

Sylvester's criterion for $F(X)$ to be *positive semi-definite* is that

$$p_{11} \leq 0 \qquad\qquad p_{22} \leq 0 \qquad\qquad p_{33} \leq 0$$

$$\begin{vmatrix} p_{11} & p_{12} \\ p_{21} & p_{22} \end{vmatrix} \geq 0 \qquad \begin{vmatrix} p_{22} & p_{23} \\ p_{32} & p_{33} \end{vmatrix} \geq 0 \qquad \begin{vmatrix} p_{11} & p_{13} \\ p_{31} & p_{33} \end{vmatrix} \geq 0 \quad \text{and} \quad |\;| = 0.$$

Note:
(a) Changing the sign of each and every term of a positive definite function $F(X)$ yields a negative definite function. Thus $F(X)$ is **negative definite** if $-F(X)$ is positive definite.
(b) $F(X)$ is **negative definite** if successive principal minors alternate in sign with the first principal minor being negative.
   For an even order P, $p_{11} < 0$ and the successive principal minors alternate in sign, with $|P| > 0$.

For an odd order P, $p_{11} < 0$ and the successive principal minors alternate in sign with $|P| < 0$.

(c) $F(X)$ is said to be indefinite, if at points in the state space, it can have positive, negative or zero values.

Example 9.5 Find the sign definiteness of the scalar function $F(X) = x_1^2 + x_2^2 + 2x_1x_2$.

Solution:

$$F(X) = \begin{bmatrix} x_1 & x_2 \end{bmatrix} \begin{bmatrix} 1 & 1 \\ 1 & 1 \end{bmatrix} \begin{bmatrix} x_1 \\ x_2 \end{bmatrix}$$

$$P = \begin{bmatrix} 1 & 1 \\ 1 & 1 \end{bmatrix} \qquad p_{11} = 1 > 0 \qquad |P| = \begin{vmatrix} 1 & 1 \\ 1 & 1 \end{vmatrix} = 0.$$

It is a positive semi-definite function.

Alternatively, $F(X) = x_1^2 + x_2^2 + 2x_1x_2 = (x_1 + x_2)^2$. It is always positive except when $x_1 = -x_2$ and when $x_1 = x_2 = 0$. When $x_1 = -x_2$, $F(X) = 0$. Hence it is a positive semi-definite function.

Example 9.6 Determine the sign definiteness of the scalar function

$$F(X) = 3x_1^2 + 4x_2^2 + 2x_3^2 + 2x_1x_2 - 4x_2x_3 - 2x_1x_3.$$

Solution:

$$F(X) = \begin{bmatrix} x_1 & x_2 & x_3 \end{bmatrix} \begin{bmatrix} 3 & 1 & -1 \\ 1 & 4 & -2 \\ -1 & -2 & 2 \end{bmatrix} \begin{bmatrix} x_1 \\ x_2 \\ x_3 \end{bmatrix}$$

$$P = \begin{bmatrix} 3 & 1 & -1 \\ 1 & 4 & -2 \\ -1 & -2 & 2 \end{bmatrix} \quad p_{11} = 3 > 0 \quad \begin{vmatrix} 3 & 1 \\ 1 & 4 \end{vmatrix} = 11 > 0 \quad |P| = \begin{vmatrix} 3 & 1 & -1 \\ 1 & 4 & -2 \\ -1 & -2 & 2 \end{vmatrix} = 10 > 0.$$

All principal minors of P are positive and $|P|$ is positive. Hence the scalar function is positive definite.

Example 9.7 Find the sign definiteness of the scalar function

$$F(X) = -5x_1^2 - 4x_2^2 - 2x_3^2 - 2x_1x_2 + 2x_2x_3 + 4x_1x_3.$$

$$F(X) = \begin{bmatrix} x_1 & x_2 & x_3 \end{bmatrix} \begin{bmatrix} -5 & -1 & 2 \\ -1 & -4 & 1 \\ 2 & 1 & -2 \end{bmatrix} \begin{bmatrix} x_1 \\ x_2 \\ x_3 \end{bmatrix}$$

$$P = \begin{bmatrix} -5 & -1 & 2 \\ -1 & -4 & 1 \\ 2 & 1 & -2 \end{bmatrix} \quad p_{11} = -5 < 0 \quad \begin{vmatrix} -5 & -1 \\ -1 & -4 \end{vmatrix} = 19 > 0 \quad |P| = \begin{vmatrix} -5 & -1 & 2 \\ -1 & -4 & 1 \\ 2 & 1 & -2 \end{vmatrix} = -21 < 0.$$

Principal minors alternate in sign. Hence $F(X)$ is negative definite.

Example 9.8 Find the sign definiteness of the scalar function

$$F(X) = -x_1^2 - 3x_2^2 - 5x_3^2 + 2x_1x_2 - 4x_2x_3 - 2x_1x_3.$$

Solution:

$$P = \begin{bmatrix} -1 & 1 & -1 \\ 1 & -3 & -2 \\ -1 & -2 & -5 \end{bmatrix} \quad p_{11} = -1 < 0 \qquad \begin{vmatrix} -1 & 1 \\ 1 & -3 \end{vmatrix} = 2 > 0 \quad |P| = \begin{vmatrix} -1 & 1 & -1 \\ 1 & -3 & -2 \\ -1 & -2 & -5 \end{vmatrix} = 1 > 0.$$

The first principal minor is negative, the second principal minor is positive and $|P| > 0$. Hence $F(X)$ is sign indefinite.

Example 9.9 Find the sign definiteness of the scalar function

$$F(X) = x_1^2 + 4x_2^2 + 4x_1x_2 + 4x_2x_3 + 2x_1x_3.$$

Solution:

$$P = \begin{bmatrix} 1 & 2 & 1 \\ 2 & 4 & 2 \\ 1 & 2 & 0 \end{bmatrix} \quad p_{11} = 1 > 0 \qquad \begin{vmatrix} 1 & 2 \\ 2 & 4 \end{vmatrix} = 0 \qquad |P| = \begin{vmatrix} 1 & 2 & 1 \\ 2 & 4 & 2 \\ 1 & 2 & 0 \end{vmatrix} = 0 \quad .$$

It may look as though it is a positive semi-definite function. But we have to check other principal minors also.

$$p_{22} = 4 > 0 \qquad\qquad p_{33} = 0 \qquad\qquad \begin{vmatrix} 4 & 2 \\ 2 & 0 \end{vmatrix} = -4 < 0 \qquad \begin{vmatrix} 1 & 1 \\ 1 & 0 \end{vmatrix} = -1 < 0$$

Two principal minors of even order are negative so that $F(X)$ possesses sign indefiniteness.

Example 9.10 Determine the sign definiteness of the scalar function

$$F(X) = x_1^2 + 2x_2^2 + 4x_3^2 + 2x_1x_2 - 4x_2x_3 - 4x_1x_3.$$

Solution:

$$P = \begin{bmatrix} 1 & 1 & -2 \\ 1 & 2 & -2 \\ -2 & -2 & 4 \end{bmatrix} \quad p_{11} = 1 > 0 \qquad \begin{vmatrix} 1 & 1 \\ 1 & 2 \end{vmatrix} = 1 > 0 \qquad |P| = 0.$$

It may look as though it is positive semi-definite. Before we conclude, we have to check other principal minors also.

$$p_{22} = 2 > 0 \qquad p_{33} = 4 > 0 \qquad \begin{vmatrix} 2 & -2 \\ -2 & 4 \end{vmatrix} = 4 > 0 \qquad \begin{vmatrix} 1 & -2 \\ -2 & 4 \end{vmatrix} = 0.$$

All principal minors are nonnegative. Hence $F(X)$ is a positive semi-definite function.

Example 9.11 Find the sign definiteness of the scalar function

$$F(X) = -x_1^2 - 2x_2^2 - 4x_3^2 - 2x_1x_2 + 4x_2x_3 + 4x_1x_3 .$$

Solution:

$$P = \begin{bmatrix} -1 & -1 & 2 \\ -1 & -2 & 2 \\ 2 & 2 & -4 \end{bmatrix}$$

$$P_{11} = -1 < 0 \qquad \begin{vmatrix} -1 & -1 \\ -1 & -2 \end{vmatrix} = 1 > 0 \qquad |P| = \begin{vmatrix} -1 & -1 & 2 \\ -1 & -2 & 2 \\ 2 & 2 & -4 \end{vmatrix} = 0$$

$$P_{22} = -2 < 0 \qquad P_{33} = -4 < 0 \qquad \begin{vmatrix} -2 & 2 \\ 2 & -4 \end{vmatrix} = 4 > 0 \qquad \begin{vmatrix} -1 & 2 \\ 2 & -4 \end{vmatrix} = 0 .$$

Principal minors of odd order are nonpositive. Principal minors of even order are nonnegative and $|P| = 0$. Hence $F(X)$ is negative semi-definite.

Example 9.12 Find the condition for which the given function $F(X)$ is positive definite. $F(X) = ax_1^2 - x_1^3 + x_2^2 + 3x_3^2$, where a is a positive constant.

Solution:

$$F(X) = ax_1^2 - x_1^3 + x_2^2 + 3x_3^2$$

The given function can be written as

$$F(X) = x_1^2(a - x_1) + x_2^2 + 3x_3^2 .$$

Obviously $F(X)$ is positive definite if $x_1 < a$.

Example 9.13 Find the condition for which the given function $F(X)$ is positive definite. $F(X) = ax_1^2 + x_2^2 + 2x_3^2 - 2ax_1x_2 + 2x_2x_3$, where a is a positive constant.

Solution: The given function can be written as

$$F(X) = \begin{bmatrix} x_1 & x_2 & x_3 \end{bmatrix} \begin{bmatrix} a & -a & 0 \\ -a & 1 & 1 \\ 0 & 1 & 2 \end{bmatrix} \begin{bmatrix} x_1 \\ x_2 \\ x_3 \end{bmatrix} \qquad P = \begin{bmatrix} a & -a & 0 \\ -a & 1 & 1 \\ 0 & 1 & 2 \end{bmatrix} .$$

For $F(X)$ to be positive definite,

$$a > 0; \qquad \begin{vmatrix} a & -a \\ -a & 1 \end{vmatrix} = a - a^2 > 0 \quad \text{and} \quad \begin{vmatrix} a & -a & 0 \\ -a & 1 & 1 \\ 0 & 1 & 2 \end{vmatrix} = a - 2a^2 > 0$$

Or $a > 0; \quad a(1 - a) > 0$ and $a(1 - 2a) > 0$.

This yields $a > 0$ *and* $a < 0.5$ This yields $a > 0$ and $a < 0.5$.
Hence $F(X)$ is positive definite if $0 < a < 0.5$.

9.5 LYAPUNOV STABILITY

In the case of nonlinear systems, it is more meaningful to discuss about the stability of equilibrium points rather than the system stability. It is important to consider different regions around an equilibrium point in the state space while considering the stability of an equilibrium point. Stability in a region in the immediate neighbourhood of equilibrium point is called *stability in the small* or *local stability*. Stability in a larger region around the equilibrium point is called *stability in the large*. Stability of equilibrium point in the entire state space is called *global stability*. While discussing about the stability of equilibrium point, the path of phase trajectory starting from an initial point is analysed, when the system is subjected to perturbations. Lyapunov stability requires that the system phase trajectory, starting from an initial state, remains in a bounded region around the equilibrium state.

Assume that the equilibrium state is at the origin of the state space. Consider a spherical region $S(R)$ of radius R surrounding the origin of the state space. In an n-dimensional state space, it can be represented as

$$x_1^2 + x_2^2 + x_3^2 + \ldots + x_n^2 = R^2.$$

This can also be represented as

$$\|X\| = R,$$

where $\|X\|$ is called the *Euclidean Norm*.

The following are some important definitions with respect to Lyapunov stability.

1. A system is said to be stable in the sense of Lyapunov, if for $R > 0$ there exists a δ such that $0 < \delta < R$ and the phase trajectory starting from any initial state X_0 within the region $S(\delta)$ around the equilibrium state does not leave the region $S(R)$, as time t approaches infinity. This is illustrated in Figure 9.1.

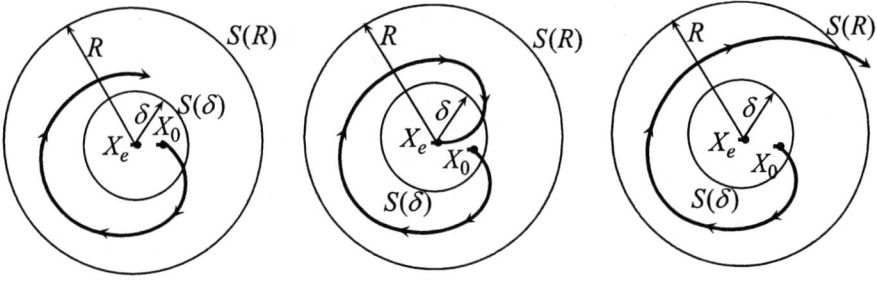

Figure 9.1 Lyapunov stability   **Figure 9.2** Asymptotic stability   **Figure 9.3** Instability

2. A system is said to be *asymptotically stable* if for $R > 0$ there exists a δ such that $0 < \delta < R$ and the phase trajectory starting from an initial state X_0 within the region $S(\delta)$ around the equilibrium state does not leave the region $S(R)$ at any time and finally returns to the equilibrium state as time t increases indefinitely. This is shown in Figure 9.2.

3. If the state of a system starting from a region within $S(\delta)$ remains within a small region $S(R)$, $0 < \delta < R$, when subjected to a small disturbance, it is said to have *local stability* or *stability in the small* and if the state returns to the equilibrium state as time approaches infinity, it is said to be *locally asymptotically stable*.

4. If the state of a system starting from a region within $S(\delta)$ remains within a large region $S(R)$, $0 < \delta < R$, when subjected to a large disturbance, it is said to have *global stability* or *stability in the large* and if the state returns to the equilibrium state as time approaches infinity, it is said to be *globally asymptotically stable*.

5. The equilibrium state of a system is unstable in the sense of Lyapunov if for a specified region $S(R)$, the phase trajectory starting from an initial state within a region $S(\delta)$ leaves the region $S(R)$. This is shown in figure 9.3.

9.6 FIRST METHOD OF LYAPUNOV

The first method proposed by Lyapunov, in determining the stability of nonlinear system, is based on the linearization of nonlinear equations about the equilibrium state. If there is more than one equilibrium state, each is investigated separately.

Consider a the nonlinear system described by

$$\dot{X} = f(X).$$

Let X_e be an equilibrium state and let $Z = (X - X_e)$. Using Taylor's series expansion of $f(X)$ about the equilibrium state X_e and neglecting higher-derivative terms,

$$\dot{Z} = \frac{\partial f}{\partial X}\Big|_{X=X_e} Z$$

$$= J(X_e)Z,$$

$$\text{where } J(X_e) = \frac{\partial f}{\partial X}\Big|_{X=X_e} = \begin{bmatrix} \dfrac{\partial f_1}{\partial x_1} & \dfrac{\partial f_1}{\partial x_2} & \dfrac{\partial f_1}{\partial x_3} & \cdots & \dfrac{\partial f_1}{\partial x_n} \\[2mm] \dfrac{\partial f_2}{\partial x_1} & \dfrac{\partial f_2}{\partial x_2} & \dfrac{\partial f_2}{\partial x_3} & \cdots & \dfrac{\partial f_2}{\partial x_n} \\[2mm] & \cdots & & & \\ & \cdots & & & \\ & \cdots & & & \\ \dfrac{\partial f_n}{\partial x_1} & \dfrac{\partial f_n}{\partial x_2} & \dfrac{\partial f_n}{\partial x_3} & \cdots & \dfrac{\partial f_n}{\partial x_n} \end{bmatrix}_{X=X_e}$$

1. If all the Eigen values of $J(X_e)$ are with negative real parts, then the equilibrium state in question is locally asymptotically stable.
2. If at least one Eigen value of $J(X_e)$ has positive real part, then the equilibrium state under consideration is unstable.
3. If at least one Eigen value of $J(X_e)$ is with zero real part, then it is a critical case. In this case, local stability of the equilibrium state is to be assessed taking into consideration higher-order derivatives also in the Taylor's series expansion of $f(X)$.

Since the linearization of the system is about the equilibrium state, the investigation of stability is only about local stability or stability in the small.

Example 9.14 A nonlinear system is represented by the state equation

$$\dot{x}_1 = -x_1 + 0.5x_2$$

$$\dot{x}_2 = x_1 + x_1 x_2 - x_2^2 .$$

Check whether the equilibrium state of the system is stable. Use the first method of Lyapunov.

Solution:

$$\dot{x}_1 = -x_1 + 0.5x_2$$

$$\dot{x}_2 = x_1 + x_1 x_2 - x_2^2 .$$

The equilibrium state is obtained by solving the equations

$$\dot{x}_1 = -x_{1e} + 0.5x_{2e} = 0 \tag{1}$$

$$\dot{x}_2 = x_{1e} + x_{1e}x_{2e} - x_{2e}^2 = 0 . \tag{2}$$

From equation (1),

$$x_{1e} = 0.5x_{2e} . \tag{3}$$

Substituting equation (3) in equation (2),

$$0.5x_{2e} + 0.5x_{2e}^2 - x_{2e}^2 = 0 \quad \text{or} \qquad 0.5x_{2e}(1 - x_{2e}) = 0 . \tag{4}$$

Hence $x_{2e} = 0$ \qquad or \quad $x_{2e} = 1$

When $x_{2e} = 0, x_{1e} = 0$, and when $x_{2e} = 1, x_{1e} = 0.5$.

The equilibrium states are $(0, 0)$ and $(0.5, 1)$.

Let $f_1(x_1, x_2) = \dot{x}_1 = -x_1 + 0.5x_2$.

$$f_2(x_1, x_2) = \dot{x}_2 = x_1 + x_1 x_2 - x_2^2$$

$$J(x_1, x_2) = \begin{bmatrix} \dfrac{\partial f_1}{\partial x_1} & \dfrac{\partial f_1}{\partial x_2} \\ \dfrac{\partial f_2}{\partial x_1} & \dfrac{\partial f_2}{\partial x_2} \end{bmatrix} = \begin{bmatrix} -1 & 0.5 \\ 1 + x_2 & x_1 - 2x_2 \end{bmatrix}$$

$$J(0,0) = \begin{bmatrix} \dfrac{\partial f_1}{\partial x_1} & \dfrac{\partial f_1}{\partial x_2} \\ \dfrac{\partial f_2}{\partial x_1} & \dfrac{\partial f_2}{\partial x_2} \end{bmatrix}_{(0,0)} = \begin{bmatrix} -1 & 0.5 \\ 1 & 0 \end{bmatrix}.$$

The characteristic equation is

$$\begin{vmatrix} \lambda+1 & -0.5 \\ -1 & \lambda \end{vmatrix} = 0 \text{ or } \lambda^2 + \lambda - 0.5 = 0$$

The Eigen values are $\lambda = 0.366$ and $\lambda = -1.366$.
One Eigen value is positive and so the equilibrium state (0, 0) is unstable.

$$J(0.5,1) = \begin{bmatrix} \dfrac{\partial f_1}{\partial x_1} & \dfrac{\partial f_1}{\partial x_2} \\ \dfrac{\partial f_2}{\partial x_1} & \dfrac{\partial f_2}{\partial x_2} \end{bmatrix}_{(0.5,1)} = \begin{bmatrix} -1 & 0.5 \\ 2 & -1.5 \end{bmatrix}.$$

The characteristic equation is

$$\begin{vmatrix} \lambda+1 & -0.5 \\ -2 & \lambda+1.5 \end{vmatrix} = 0 \qquad \text{or} \quad \lambda^2 + 2.5\lambda + 0.5 = 0$$

The Eigen values are $\lambda = -0.2192$ and $\lambda = -2.2808$.
Both the Eigen values are negative and so the equilibrium state (0.5, 1) is locally asymptotically stable in the sense of Lyapunov.

9.7 SECOND METHOD OF LYAPUNOV

The second method of Lyapunov, for investigating the stability, does not require any linearization or solution of the linear or nonlinear equations. This method is based on the fact that if an autonomous system that has asymptotically stable equilibrium state is subjected to a perturbation, the stored energy in the system decreases as time increases until it finally reaches a minimum value at the equilibrium state. On the other hand, if the internal energy is continuously increasing without bound, the system response becomes unbounded and the system tends towards instability.

The concept of Lyapunov's stability depends on determining whether an energy function for a system is continuously decreasing or increasing.

Consider a mechanical system with a mass, a damper and a nonlinear spring shown in Figure 9.4. The displacement of mass is x and its velocity is $\dot{x}$.

The reactive force due to acceleration of mass is

$$f_M = M\ddot{x}.$$

The damping force is

$$f_B = B \dot{x}.$$

Let the nonlinear spring force be

$$f_K = K_1 x + K_2 x^3.$$

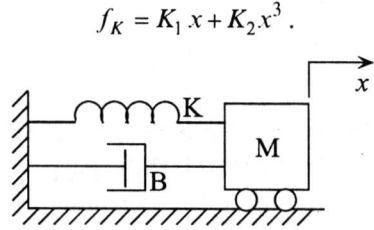

Figure 9.4 Nonlinear mechanical system

The differential equation that describes the dynamics of motion is given by

$$M \ddot{x} + B \dot{x} + K_1 x + K_2 x^3 = 0.$$

The damper absorbs energy and does not store it. The energy stored by the mass due to acceleration is

$$E_B = \frac{1}{2} M (\text{velocity})^2 = \frac{1}{2} M (\dot{x})^2.$$

The energy stored by the spring is

$$E_K = \int_0^x (K_1 y + K_2 y^3) dy = \frac{1}{2} K_1 x^2 + \frac{1}{4} K_2 x^4.$$

The total energy of the system is given by

$$E = \frac{1}{2} M (\dot{x})^2 + \frac{1}{2} K_1 x^2 + \frac{1}{4} K_2 x^4.$$

Let $x_1 = x$ and $x_2 = \dot{x}$.

Hence $E = \frac{1}{2} M x_2^2 + \frac{1}{2} K_1 x_1^2 + \frac{1}{4} K_2 x_1^4$.

This is an energy function and is positive definite. The simplest expression for a fictitious energy function is

$$V(X) = \sum_{i=1}^{n} x_i^2 = x_1^2 + x_2^2 + x_3^2 + \dots + x_n^2.$$

Obviously, since the energy is positive, $V(X) > 0$ for all X except at the origin where it is zero. This is called the Lyapunov function. If the energy is continuously decreasing, the rate of change of energy is negative and the system must eventually settle down to an equilibrium point. In other words, if the energy is continuously decreasing $\dot{V}(X)$ is negative definite.

$$V(X) = x_1^2 + x_2^2 + x_3^2 + \ldots + x_n^2 \text{ is a positive definite function.}$$

$$\dot{V}(X) = \frac{\partial V}{\partial x_1}\frac{dx_1}{dt} + \frac{\partial V}{\partial x_2}\frac{dx_2}{dt} + \frac{\partial V}{\partial x_3}\frac{dx_3}{dt} + \ldots + \frac{\partial V}{\partial x_n}\frac{dx_n}{dt}$$

$$= \frac{\partial V}{\partial x_1}\dot{x}_1 + \frac{\partial V}{\partial x_2}\dot{x}_2 + \frac{\partial V}{\partial x_3}\dot{x}_3 + \ldots + \frac{\partial V}{\partial x_n}\dot{x}_n$$

If $\dot{V}(X)$ is negative definite, the system will settle down to an equilibrium point and the system is stable in the sense of Lyapunov. This is the concept of second method of Lyapunov. This method is also called the direct method of Lyapunov.

9.8 LYAPUNOV STABILITY THEOREM

For an autonomous system represented by the state equation

$$\dot{X} = f(X); \quad f(0) = 0 ,$$

if there exists a scalar function $V(X)$ such that

(i) $V(X)$ is positive definite and
(ii) $V(X)$ has continuous partial derivative at the origin

then the equilibrium state at origin is

(a) stable if $\dot{V}(X)$ is negative semi-definite
(b) asymptotically stable if $\dot{V}(X)$ is negative definite
(c) globally asymptotically stable if $\dot{V}(X)$ is negative definite and in addition

$$V(X) \to \infty \text{ as } \|X\| \to \infty .$$

The scalar function $V(X)$ is called a *Lyapunov function*. If $\dot{V}(X)$ is negative semi-definite, then it should be possible to show that for asymptotic stability, the derivative function identically vanishes only at the origin.

9.9 LYAPUNOV INSTABILITY THEOREM

For an autonomous system represented by the state equation

$$\dot{X} = f(X); \quad f(0) = 0 ,$$

if there exists a scalar function $V(X)$ such that

(i) $V(X)$ is positive definite and
(ii) $V(X)$ has continuous partial derivative at the origin,

then the equilibrium state at origin is unstable if $\dot{V}(X)$ is positive definite.

Note:

1. The direct method of Lyapunov is applicable to all dynamic systems whether linear or nonlinear and continuous- or discrete-time systems.

2. The reason for the two theorems is that if the origin is unstable, it will be impossible to find a scalar function $V(X)$ which satisfies the stability theorem. But a scalar function that satisfies the instability theorem will exist and, if it can be found, will prove the instability of the equilibrium point at origin.

3. The direct method of Lyapunov stability analysis suffers from the common difficulty of finding a proper Lyapunov function for a given system. There is no general method to select a proper Lyapunov function for a given nonlinear system.

4. The Lyapunov function for a system is not unique. In general, different choices of $V(X)$ will indicate different stability regions.

5. Once a Lyapunov function is chosen and the stability of a nonlinear system is established within a certain region, there is no assurance that outside this region the system is unstable.

6. After choosing a Lyapunov function and applying the stability theorem, if it is found that the stability cannot be established, it does not mean that the equilibrium state is unstable. It only means that the selected Lyapunov function is not a proper Lyapunov function for the system under consideration for investigating its stability.

Example 9.15 A nonlinear system is represented by the state equations

$$\dot{x}_1 = x_2$$

$$\dot{x}_2 = -x_1 - x_2 + x_1 x_2 .$$

Check the stability of the system using Lyapunov stability criterion.

Solution: The given system is

$$\dot{x}_1 = x_2$$

$$\dot{x}_2 = -x_1 - x_2 + x_1 x_2 .$$

The equilibrium states are obtained by solving the equations

$$\dot{x}_1 = 0 \text{ and } \dot{x}_2 = 0 .$$

This gives $x_{2e} = 0$ and $x_{1e} = 0$.

Hence the equilibrium state is the origin of the $x_1 - x_2$ plane.

Select a Lyapunov function $V(X) = x_1^2 + x_2^2$ which is positive definite.

$$\dot{V}(X) = \frac{\partial V}{\partial x_1} \frac{dx_1}{dt} + \frac{\partial V}{\partial x_2} \frac{dx_2}{dt}$$

$$= \frac{\partial V}{\partial x_1} \dot{x}_1 + \frac{\partial V}{\partial x_2} \dot{x}_2$$

$$= 2x_1 \dot{x_1} + 2x_2 \dot{x_2}$$

$$= 2x_1 x_2 + 2x_2(-x_1 - x_2 + x_1 x_2)$$

$$= -2x_2^2 + 2x_1 x_2^2 = -2x_2^2(1 - x_1).$$

$\dot{V}(X)$ is negative definite if $x_1 < 1$. Hence the equilibrium state at origin is stable in the sense of Lyapunov if $x_1 < 1$.

Example 9.16 A nonlinear system is represented by the differential equation

$$\ddot{y} + \dot{y}(y^2 + (\dot{y})^2) + y = 0.$$

Check the stability of the system using Lyapunov stability criterion.

Solution: The given system is

$$\ddot{y} + \dot{y}(y^2 + (\dot{y})^2) + y = 0.$$

Select $x_1 = y$ and $x_2 = \dot{x_1} = \dot{y}$.

The given differential equation can be written as

$$\dot{x_2} + x_2(x_1^2 + x_2^2) + x_1 = 0.$$

Hence the state equations are

$$\dot{x_1} = x_2 \quad \text{and} \quad \dot{x_2} = -x_2(x_1^2 + x_2^2) - x_1.$$

The equilibrium states are obtained by solving the equations

$$\dot{x_1} = 0 \text{ and } \dot{x_2} = 0.$$

This gives the equilibrium state $x_{2e} = 0$ and $x_{1e} = 0$. Hence the equilibrium state is the origin of the $x_1 - x_2$ plane.

Select a Lyapunov function $V(X) = x_1^2 + x_2^2$ which is positive definite.

$$\dot{V}(X) = \frac{\partial V}{\partial x_1} \frac{dx_1}{dt} + \frac{\partial V}{\partial x_2} \frac{dx_2}{dt}$$

$$= \frac{\partial V}{\partial x_1} \dot{x_1} + \frac{\partial V}{\partial x_2} \dot{x_2}$$

$$= 2x_1 \dot{x_1} + 2x_2 \dot{x_2}$$

$$= 2x_1 x_2 + 2x_2[-x_2(x_1^2 + x_2^2) - x_1]$$

$$= -2x_2^2(x_1^2 + x_2^2).$$

$\dot{V}(X)$ is negative definite and hence the equilibrium state at origin is asymptotically stable. Since $\|V(X)\| \to \infty$ as $\|X\| \to \infty$, the system is *asymptotically stable in the large* in the sense of Lyapunov.

Example 9.17 Using Lyapunov stability theorem, check the stability of equilibrium state of a nonlinear system represented by the state equations

$$\dot{x}_1 = -x_1^3 - x_2$$

$$\dot{x}_2 = x_1 - x_2 \,.$$

Solution: The given system is

$$\dot{x}_1 = -x_1^3 - x_2$$

$$\dot{x}_2 = x_1 - x_2 \,.$$

The equilibrium state is obtained by solving the equations

$$\dot{x}_1 = 0 \text{ or } -x_{1e}^3 - x_{2e} = 0$$

and $$\dot{x}_2 = 0 \text{ or } x_{1e} - x_{2e} = 0 \,.$$

Hence $x_{1e} = 0$ and $x_{2e} = 0$. The equilibrium state is at origin.

Choose the Lyapunov function as

$V(X) = x_1^2 + x_2^2$ which is positive definite.

$$\dot{V}(X) = \frac{\partial V}{\partial x_1}\dot{x}_1 + \frac{\partial V}{\partial x_2}\dot{x}_2$$

$$= 2x_1 \dot{x}_1 + 2x_2 \dot{x}_2$$

$$= 2x_1(-x_1^3 - x_2) + 2x_2(x_1 - x_2)$$

$$= -2x_1^4 - 2x_2^2 \,.$$

$\dot{V}(X)$ is negative definite and so the equilibrium state at origin is asymptotically stable. Since $\|V(X)\| \to \infty$ as $\|X\| \to \infty$, the system is *asymptotically stable in the large* in the sense of Lyapunov.

Example 9.18 Using Lyapunov stability theorem, assess the stability of equilibrium state of a nonlinear system represented by the differential equation

$$\ddot{y} - a(1 - y^2)\dot{y} + y = 0 \,.$$

Solution: The given system is

$$\ddot{y} - a(1 - y^2)\dot{y} + y = 0 \,.$$

Choose $x_1 = y$ and $x_2 = \dot{x}_1 = \dot{y}$.

$$\dot{x}_1 = x_2 \,.$$

From the given differential equation

$$\dot{x}_2 = a(1 - x_1^2)x_2 - x_1 .$$

The equilibrium state is obtained by solving the equations

$$\dot{x}_1 = 0 \quad \text{or} \quad x_{2e} = 0$$

and

$$\dot{x}_2 = 0 \quad \text{or} \quad a(1 - x_{1e}^2)x_{2e} - x_{1e} = 0 .$$

Hence $x_{1e} = 0$ and $x_{2e} = 0$. The equilibrium state is at origin.

Choose the Lyapunov function as

$V(X) = x_1^2 + x_2^2$ which is positive definite.

$$\dot{V}(X) = \frac{\partial V}{\partial x_1}\dot{x}_1 + \frac{\partial V}{\partial x_2}\dot{x}_2$$

$$= 2x_1 \dot{x}_1 + 2x_2 \dot{x}_2$$

$$= 2x_1 x_2 + 2x_2(a x_2 - a x_1^2 x_2 - x_1)$$

$$= 2ax_2^2(1 - x_1^2) .$$

$\dot{V}(X)$ is negative definite if (ii) $a < 0$ and $|x_1| < 1$ or

(ii) if $a > 0$ and $|x_1| > 1$.

Under these conditions, the equilibrium state at origin is asymptotically stable.

Example 9.19 Assess the stability of equilibrium state of a nonlinear system represented by the state equations

$$\dot{x}_1 = -x_1 + x_2 + x_1(x_1^2 + x_2^2)$$

$$\dot{x}_2 = -x_1 - x_2 + x_2(x_1^2 + x_2^2) .$$

Solution: The system representation is

$$\dot{x}_1 = -x_1 + x_2 + x_1(x_1^2 + x_2^2)$$

$$\dot{x}_2 = -x_1 - x_2 + x_2(x_1^2 + x_2^2) .$$

The equilibrium state can be obtained by solving the equations

$$\dot{x}_1 = 0 \quad \text{or} \quad -x_{1e} + x_{2e} + x_{1e}(x_{1e}^2 + x_{2e}^2) = 0 \tag{1}$$

$$\dot{x}_2 = 0 \quad \text{or} \quad -x_{1e} - x_{2e} + x_{2e}(x_{1e}^2 + x_{2e}^2) = 0 . \tag{2}$$

ultiplying equation (1) by x_{2e},

$$-x_{1e}x_{2e} + x_{2e}^2 + x_{1e}x_{2e}(x_{1e}^2 + x_{2e}^2) = 0 . \tag{3}$$

Multiplying equation (2) by x_{1e},

$$-x_{1e}^2 - x_{1e} x_{2e} + x_{1e} x_{2e} (x_{1e}^2 + x_{2e}^2) = 0. \tag{4}$$

Subtracting equation (4) from equation (3),

$$(x_{1e}^2 + x_{2e}^2) = 0. \tag{5}$$

Substituting equation (5) in equation (1),

$$x_{1e} = x_{2e}. \tag{6}$$

Substituting equation (5) in equation (2),

$$x_{1e} = -x_{2e}. \tag{7}$$

From equations (6) and (7),

$$x_{1e} = x_{2e} = 0.$$

Hence the equilibrium state is at origin of the x_1 x_2 plane.

Choose the Lyapunov function as

$V(X) = x_1^2 + x_2^2$ which is positive definite.

$$\dot{V}(X) = \frac{\partial V}{\partial x_1} \dot{x}_1 + \frac{\partial V}{\partial x_2} \dot{x}_2$$

$$= 2x_1 \dot{x}_1 + 2x_2 \dot{x}_2$$

$$= 2x_1[-x_1 + x_2 + x_1(x_1^2 + x_2^2)] + 2x_2[-x_1 - x_2 + x_2(x_1^2 + x_2^2)]$$

$$= 2[-x_1^2 + x_1 x_2 + x_1^2(x_1^2 + x_2^2) - x_1 x_2 - x_2^2 + x_2^2(x_1^2 + x_2^2)]$$

$$= 2[-(x_1^2 + x_2^2) + (x_1^2 + x_2^2)(x_1^2 + x_2^2)]$$

$$= -2(x_1^2 + x_2^2)[1 - (x_1^2 + x_2^2)].$$

$\dot{V}(X)$ is negative definite if $(x_1^2 + x_2^2) < 1$ and the equilibrium state at origin is locally asymptotically stable if this condition is satisfied. The region of stability in the $x_1 - x_2$ plane is as shown in Figure E9.19.

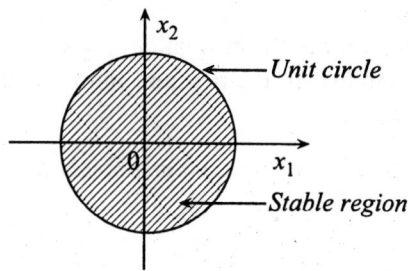

Figure E9.19 Example 9.19 Region of stability

Example 9.20 Assess the stability of equilibrium state of a nonlinear system represented by the state equations

$$\dot{x}_1 = -x_1 + x_1^2 x_2$$

$$\dot{x}_2 = -x_2.$$

Solution: The given system is represented by

$$\dot{x}_1 = -x_1 + x_1^2 x_2$$

$$\dot{x}_2 = -x_2.$$

It is clearly known that $x_{1e} = 0$ and $x_{2e} = 0$. The equilibrium state is at origin. Select a Lyapunov function $V(X) = x_1^2 + x_2^2$ which is positive definite.

$$\dot{V}(X) = \frac{\partial V}{\partial x_1}\dot{x}_1 + \frac{\partial V}{\partial x_2}\dot{x}_2$$

$$= 2x_1\,\dot{x}_1 + 2x_2\,\dot{x}_2$$

$$= 2x_1(-x_1 + x_1^2 x_2) + 2x_2(-x_2)$$

$$= -2x_1^2(1 - x_1 x_2) - 2x_2^2.$$

$\dot{V}(X)$ is negative definite if $x_1 x_2 < 1$. Hence the equilibrium state at origin is stable in the sense of Lyapunov, if $x_1 x_2 < 1$. The region of stability in the $x_1 - x_2$ plane is shown in Figure E9.18.

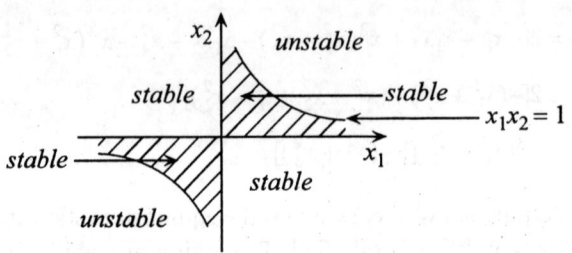

Figure E9.20 Example 9.20 Region of stability

In the second and fourth quadrants, the inequality $x_1 x_2 < 1$ is true for all values of x_1 and x_2 and these are regions of stability. In the first and third quadrants, the region of stability is restricted as shown in the figure.

Example 9.21 Using Lyapunov stability theorem, assess the stability of a nonlinear system represented by

$$\dot{x}_1 = x_2$$

$$\dot{x}_2 = -2x_1 - 2x_1^3 - 3x_2.$$

Solution: The system is represented by

$$\dot{x}_1 = x_2$$

$$\dot{x}_2 = -2x_1 - 2x_1^3 - 3x_2 = -2x_1(1 + x_1^2) - 3x_2 .$$

Clearly the equilibrium state is at the origin. Select a Lyapunov function $V(X) = x_1^2 + x_2^2$ which is positive definite.

$$\dot{V}(X) = \frac{\partial V}{\partial x_1} \dot{x}_1 + \frac{\partial V}{\partial x_2} \dot{x}_2$$

$$= 2x_1 \dot{x}_1 + 2x_2 \dot{x}_2$$

$$= 2x_1(x_2) + 2x_2(-2x_1 - 3x_2 - 2x_1^3)$$

$$= -2x_1 x_2 - 6x_2^2 - 4x_1^3 x_2 .$$

$\dot{V}(X)$ is sign indefinite. Hence the chosen function $V(X)$ is not a proper Lyapunov function and the stability of the system cannot be ascertained using the chosen function.

Choose another function

$$V(X) = a x_1^4 + b x_1^2 + c x_2^2$$

$$\dot{V}(X) = \frac{\partial V}{\partial x_1} \dot{x}_1 + \frac{\partial V}{\partial x_2} \dot{x}_2$$

$$= 4a x_1^3 \dot{x}_1 + 2b x_1 \dot{x}_1 + 2c x_2 \dot{x}_2$$

$$= 4a x_1^3 x_2 + 2b x_1 x_2 + 2c x_2(-2x_1 - 3x_2 - 2x_1^3)$$

$$= 4a x_1^3 x_2 + 2b x_1 x_2 - 4c x_1 x_2 - 6c x_2^2 - 4c x_1^3 x_2 .$$

If $a = c$, the first and the last terms vanish. If $b = 2$ and $c = 1$, the second and the third terms vanish. Hence when $a = 1$, $b = 2$ and $c = 1$, $\dot{V}(X) = -6x_2^2$, which is negative semi-definite. Hence the equilibrium state at origin is stable in the sense of Lyapunov. The Lyapunov function is $V(X) = x_1^4 + 2x_1^2 + x_2^2$.

It can be observed that the selection of a proper Lyapunov function, to prove the stability, is somewhat difficult.

Example 9.22 Using Lyapunov stability theorem, assess the stability of a nonlinear system represented by

$$\dot{x}_1 = 2x_1 + x_2$$

$$\dot{x}_2 = -x_1 + x_2^3 .$$

Solution: The given system representation is

$$\dot{x}_1 = 2x_1 + x_2$$

$$\dot{x}_2 = -x_1 + x_2^3.$$

To determine the equilibrium state

$$\dot{x}_1 = 0 \ \text{ or } \ 2x_{1e} + x_{2e} = 0 \ \text{ or } \ x_{2e} = -2x_{1e}$$

$$\dot{x}_2 = 0 \quad \text{or} \quad -x_{1e} + x_{2e}^3 = 0 \qquad \text{or} \quad -x_{1e} - 8x_{1e}^3 = 0 \qquad \text{or} \quad -x_{1e}(1 + 8x_{1e}^2) = 0.$$

The solution is $x_{1e} = 0$ and $x_{2e} = 0$.

The equilibrium state is at origin.

Select a Lyapunov function $V(X) = x_1^2 + x_2^2$ which is positive definite.

$$\dot{V}(X) = \frac{\partial V}{\partial x_1}\dot{x}_1 + \frac{\partial V}{\partial x_2}\dot{x}_2$$

$$= 2x_1\dot{x}_1 + 2x_2\dot{x}_2$$

$$= 2x_1(2x_1 + x_2) + 2x_2(-x_1 + x_2^3)$$

$$= 4x_1^2 + 2x_1 x_2 - 2x_1 x_2 + 2x_2^4$$

$$= 4x_1^2 + 2x_2^4.$$

$\dot{V}(X)$ is positive definite. Hence the equilibrium state at origin is not stable in the sense of Lyapunov.

Example 9.23 Using Lyapunov stability theorem, assess the stability of a nonlinear system represented by

$$\dot{x}_1 = x_2$$

$$\dot{x}_2 = -x_2 - \sin x_1.$$

Solution: The given system representation is

$$\dot{x}_1 = x_2$$

$$\dot{x}_2 = -x_2 - \sin x_1.$$

To determine the equilibrium state

$$\dot{x}_1 = 0 \ \text{ or } \ x_{2e} = 0$$

$$\dot{x}_2 = 0 \qquad \text{or} \ \sin x_{1e} = 0 \qquad \text{or} \ x_{1e} = 0 \qquad \text{or} \ x_{1e} = \pi.$$

The solution is $x_{1e} = 0$ and $x_{2e} = 0$. The equilibrium state is at origin and at $(\pi, 0)$.

Select a Lyapunov function $V(X) = \dfrac{1}{2}x_2^2 + \displaystyle\int_0^{x_1} \sin y\, dy = \dfrac{1}{2}x_2^2 + 1 - \cos x_1$, which is positive definite.

$$\dot{V}(X) = \frac{\partial V}{\partial x_1}\dot{x}_1 + \frac{\partial V}{\partial x_2}\dot{x}_2$$

$$= (\sin x_1)\dot{x}_1 + x_2\,\dot{x}_2$$

$$= x_2 \sin x_1 + x_2(-x_2 - \sin x_1)$$

$$= -x_2^2 .$$

$\dot{V}(X)$ is negative semi-definite. Hence the equilibrium state at origin is stable in the sense of Lyapunov.

9.10 KRASOVSKII'S THEOREM

Consider an autonomous system with the equilibrium state at origin

$$\dot{X} = f(X);\ f(0) = 0 .$$

Let

$$J(X) = \frac{\partial f}{\partial X} = \begin{bmatrix} \dfrac{\partial f_1}{\partial x_1} & \dfrac{\partial f_1}{\partial x_2} & \dfrac{\partial f_1}{\partial x_3} & \cdots & \dfrac{\partial f_1}{\partial x_n} \\[2mm] \dfrac{\partial f_2}{\partial x_1} & \dfrac{\partial f_2}{\partial x_2} & \dfrac{\partial f_2}{\partial x_3} & \cdots & \dfrac{\partial f_2}{\partial x_n} \\[2mm] \cdots & & & & \\ \cdots & & & & \\ \cdots & & & & \\ \dfrac{\partial f_n}{\partial x_1} & \dfrac{\partial f_n}{\partial x_2} & \dfrac{\partial f_n}{\partial x_3} & \cdots & \dfrac{\partial f_n}{\partial x_n} \end{bmatrix}$$

and let $\hat{J}(X) = J^*(X) + J(X)$,

where $J^*(X)$ is the conjugate transpose of $J(X)$.

Krasovskii's theorem states that if $\hat{J}(X)$ is negative definite, then the equilibrium state at origin is asymptotically stable. A proper Lyapunov function is $V(X) = f^*(X)f(X)$.

Proof: $\dot{X} = f(X);\ f(0) = 0$

Choose a Lyapunov function $V(X) = f^*(X)f(X)$. $\qquad\qquad$ (9.1)

$$\dot{f}(X) = \frac{\partial f}{\partial X}\dot{X} = J(X)f(X) \qquad\qquad (9.2)$$

Differentiating equation (9.1),

$$\dot{V}(X) = \dot{f}^*(X)f(X) + f^*(X)\dot{f}(X).$$

Using equation (9.2)

$$\dot{V}(X) = [J(X)f(X)]^* f(X) + f^*(X)J(X)f(X)$$
$$= f^*(X)J^*(X)f(X) + f^*(X)J(X)f(X)$$
$$= f^*(X)[J^*(X) + J(X)]f(X). \tag{9.3}$$

From equation (9.3), it can be concluded that if $[J^*(X) + J(X)]$ is negative definite then $\dot{V}(X)$ is negative definite and the equilibrium state at origin is asymptotically stable.

The applicability of this method is limited in practice since the matrix $[J^*(X) + J(X)]$, for many systems, do not satisfy the negative definiteness.

Example 9.24 Using Krasovskii's method, assess the stability of a system represented by the state equations

$$\dot{x}_1 = x_2 \quad \text{and} \quad \dot{x}_2 = -x_2 - x_1^3.$$

Solution:

$$\dot{x}_1 = x_2 \quad \text{and} \quad \dot{x}_2 = -x_2 - x_1^3$$

Clearly the equilibrium state is at origin.

Let

$$f_1 = \dot{x}_1 = x_2 \quad \text{and} \quad f_2 = \dot{x}_2 = -x_2 - x_1^3.$$

$$J(X) = \frac{\partial f}{\partial X} = \begin{bmatrix} \dfrac{\partial f_1}{\partial x_1} & \dfrac{\partial f_1}{\partial x_2} \\ \dfrac{\partial f_2}{\partial x_1} & \dfrac{\partial f_2}{\partial x_2} \end{bmatrix} = \begin{bmatrix} 0 & 1 \\ -3x_1^2 & -1 \end{bmatrix}.$$

Since the elements of $J(X)$ are real,

$$J^T(X) = \begin{bmatrix} 0 & -3x_1^2 \\ 1 & -1 \end{bmatrix}$$

$$\hat{J}(X) = J^T(X) + J(X) = \begin{bmatrix} 0 & -3x_1^2 \\ 1 & -1 \end{bmatrix} + \begin{bmatrix} 0 & 1 \\ -3x_1^2 & -1 \end{bmatrix} = \begin{bmatrix} 0 & 1 - 3x_1^2 \\ 1 - 3x_1^2 & -2 \end{bmatrix}.$$

Since $\hat{J}(X)$ is negative semi-definite, the equilibrium state at origin is stable in the sense of Lyapunov. The Lyapunov function is

$$V(X) = f^*(X)f(X) = \begin{bmatrix} x_2 & -x_2 - x_1^3 \end{bmatrix} \begin{bmatrix} x_2 \\ -x_2 - x_1^3 \end{bmatrix} = x_2^2 + (x_2 + x_1^3)^2,$$

which is positive definite.

Example 9.25 Using Krasovskii's method, assess the stability of a system represented by the state equations

$$\dot{x_1} = -x_1 \quad \text{and} \quad \dot{x_2} = x_1 - x_2 - x_2^3.$$

Solution:

$$\dot{x_1} = -x_1 \quad \text{and} \quad \dot{x_2} = x_1 - x_2 - x_2^3 = x_1 - x_2(1 + x_2^2)$$

Clearly the equilibrium state is at origin.

Let

$$f_1 = \dot{x_1} = -x_1 \quad \text{and} \quad f_2 = \dot{x_2} = x_1 - x_2 - x_2^3.$$

$$J(X) = \frac{\partial f}{\partial X} = \begin{bmatrix} \dfrac{\partial f_1}{\partial x_1} & \dfrac{\partial f_1}{\partial x_2} \\[2mm] \dfrac{\partial f_2}{\partial x_1} & \dfrac{\partial f_2}{\partial x_2} \end{bmatrix} = \begin{bmatrix} -1 & 0 \\ 1 & -1 - 3x_2^2 \end{bmatrix}.$$

Since the elements of $J(X)$ are real,

$$J^T(X) = \begin{bmatrix} -1 & 1 \\ 0 & -1 - 3x_2^2 \end{bmatrix}$$

$$\hat{J}(X) = J^T(X) + J(X) = \begin{bmatrix} -1 & 1 \\ 0 & -1 - 3x_2^2 \end{bmatrix} + \begin{bmatrix} -1 & 0 \\ 1 & -1 - 3x_2^2 \end{bmatrix} = \begin{bmatrix} -2 & 1 \\ 1 & -2 - 6x_2^2 \end{bmatrix}.$$

$\hat{J}(X)$ is negative definite. Hence the equilibrium state at origin is stable in the sense of Lyapunov.

The Lyapunov function is

$$V(X) = f^*(X)f(X) = \begin{bmatrix} -x_1 & x_1 - x_2 - x_2^3 \end{bmatrix} \begin{bmatrix} -x_1 \\ x_1 - x_2 - x_2^3 \end{bmatrix} = x_1^2 + (x_1 - x_2 + x_2^3)^2,$$

which is positive definite.

Example 9.26 Using Krasovskii's method, assess the stability of a system represented by the state equations

$$\dot{x_1} = -5x_1 + 4x_2 \quad \text{and} \quad \dot{x_2} = x_1 - 2x_2 - x_2^5.$$

Solution:

$$\dot{x_1} = -5x_1 + 4x_2 \quad \text{and} \quad \dot{x_2} = x_1 - 2x_2 - x_2^5$$

To find the equilibrium state

$$\dot{x_1} = -5x_{1e} + 4x_{2e} = 0 \tag{1}$$

$$\dot{x_2} = x_{1e} - 2x_{2e} - x_{2e}^5 = 0. \tag{2}$$

From equation (2), $\qquad\qquad 5x_{1e} - 10x_{2e} - 5x_{2e}^5 = 0$. $\qquad\qquad$ (3)

Combining equations (1) and (3), $-6x_{2e} - 5x_{2e}^5 = 0 \quad$ or $\quad x_{2e}(6 + 5x_{2e}^4) = 0$.

$x_{2e} = 0$ and $x_{1e} = 0$.

The equilibrium state is at origin.

Let $\qquad\qquad \dot{f_1} = x_1 = -5x_1 + 4x_2 \quad$ and $\quad f_2 = \dot{x_2} = x_1 - 2x_2 - x_2^5$.

$$J(X) = \frac{\partial}{\partial} = \begin{bmatrix} \dfrac{\partial f_1}{\partial x_1} & \dfrac{\partial f_1}{\partial x_2} \\[2mm] \dfrac{\partial f_2}{\partial x_1} & \dfrac{\partial f_2}{\partial x_2} \end{bmatrix} = \begin{bmatrix} 5 & 4 \\ 1 & -2-5 \end{bmatrix}.$$

Since the elements of $J(X)$ are real,

$$J^T(X) = \begin{bmatrix} -5 & 1 \\ 4 & -2-5x_2^4 \end{bmatrix}$$

$$\hat{J}(X) = J^T(X) + J(X) = \begin{bmatrix} -5 & 1 \\ 4 & -2-5x_2^4 \end{bmatrix} + \begin{bmatrix} -5 & 4 \\ 1 & -2-5x_2^4 \end{bmatrix} = \begin{bmatrix} -10 & 5 \\ 5 & -4-10x_2^4 \end{bmatrix}.$$

$\hat{J}(X)$ is negative semi-definite. Hence the system is stable in the sense of Lyapunov.

The Lyapunov function is

$$V(X) = f^*(X)f(X) = \begin{bmatrix} -5x_1 + 4x_2 & x_1 - 2x_2 - x_2^5 \end{bmatrix} \begin{bmatrix} -5x_1 + 4x_2 \\ x_1 - 2x_2 - x_2^5 \end{bmatrix}$$

$$= (-5x_1 + 4x_2)^2 + (x_1 - 2x_2 - x_2^5)^2$$

which is positive definite.

9.11 VARIABLE GRADIENT METHOD

This is a method for finding a proper Lyapunov function for a given nonlinear system. The method is based on selecting $\dot{V}(X)$ first in terms of some free parameters and then choosing these free parameters so that $V(X)$ is positive definite. Let the nonlinear system be represented as

$$\dot{X} = f(X).$$

If $V(X)$ is a proper Lyapunov function

$$\dot{V}(X) = \frac{\partial V}{\partial x_1}\dot{x_1} + \frac{\partial V}{\partial x_2}\dot{x_2} + \dots + \frac{\partial V}{\partial x_n}\dot{x_n}.$$

This can be written as

$$\dot{V}(X) = \begin{bmatrix} \dfrac{\partial V}{\partial x_1} & \dfrac{\partial V}{\partial x_2} & - - - & \dfrac{\partial V}{\partial x_n} \end{bmatrix} \begin{bmatrix} \dot{x}_1 \\ \dot{x}_2 \\ -- \\ \dot{x}_n \end{bmatrix} = G^T \dot{X},$$

where $G(X)$ is a column vector given by

$$G(X) = \begin{bmatrix} \dfrac{\partial V}{\partial x_1} \\ \dfrac{\partial V}{\partial x_2} \\ ... \\ \dfrac{\partial V}{\partial x_n} \end{bmatrix} = \begin{bmatrix} g_1(X) \\ g_2(X) \\ ... \\ g_n(X) \end{bmatrix}.$$

$G(X)$ is first chosen as a general function of the form

$$G(X) = \begin{bmatrix} \dfrac{\partial V}{\partial x_1} \\ \dfrac{\partial V}{\partial x_2} \\ ... \\ \dfrac{\partial V}{\partial x_n} \end{bmatrix} = \begin{bmatrix} g_1(X) \\ g_2(X) \\ ... \\ g_n(X) \end{bmatrix} = \begin{bmatrix} a_{11}x_1 + a_{12}x_2 + ... + a_{1n}x_n \\ a_{21}x_1 + a_{22}x_2 + ... + a_{2n}x_n \\ ... \\ a_{n1}x_1 + a_{n2}x_2 + ... + a_{nn}x_n \end{bmatrix}.$$

The parameters of G(X) are chosen so that $\dot{V}(X)$ is negative definite. The coefficients a_{ij} are undetermined quantities. These quantities can be constants, functions of time or functions of state variables. Many a_{ij} can be taken to be zero for making $\dot{V}(X)$ negative definite. The quantity a_{ij} is taken to be a constant for the sake of convenience. The remaining unknown parameters are determined using the Curl condition.

$$\frac{\partial G_i}{\partial X_j} = \frac{\partial}{\partial X_j}\left(\frac{\partial V}{\partial X_i}\right) = \frac{\partial^2 V}{\partial X_j \, \partial X_i}$$

and

$$\frac{\partial G_j}{\partial X_i} = \frac{\partial}{\partial X_i}\left(\frac{\partial V}{\partial X_j}\right) = \frac{\partial^2 V}{\partial X_i \, \partial X_j}.$$

Since $\dfrac{\partial^2 V}{\partial X_j \, \partial X_i} = \dfrac{\partial^2 V}{\partial X_i \, \partial X_j}$, $\dfrac{\partial G}{\partial X}$ is a symmetric matrix. This is the Curl condition.

Having determined $\dot{V}(X)$, which is negative definite, find out the Lyapunov function $V(X)$ as a line integral

$$V(X) = \int_0^X G^T(X)\,dX$$

$$= \int_0^{x_1} g_1(y_1, 0, 0, \ldots, 0, 0)\,dy_1 + \int_0^{x_2} g_2(x_1, y_2, 0, \ldots, 0, 0)\,dy_2$$

$$+ \int_0^{x_3} g_3(x_1, x_2, y_3, 0, \ldots, 0, 0)\,dy_3 + \ldots + \int_0^{x_n} g_n(x_1, x_2, x_3, \ldots, x_{n-1}, y_n)\,dy_n$$

Procedure:

Step 1: Write down the given state equations

$$\dot{X} = f(X).$$

Step 2: Write down $G(X)$ as

$$G(X) = \begin{bmatrix} g_1(X) \\ g_2(X) \\ \ldots \\ \ldots \\ g_n(X) \end{bmatrix} = \begin{bmatrix} a_{11}x_1 + a_{12}x_2 + \ldots + a_{1n}x_n \\ a_{21}x_1 + a_{22}x_2 + \ldots + a_{2n}x_n \\ \ldots \\ \ldots \\ a_{n1}x_1 + a_{n2}x_2 + ----- + a_{nn}x_n \end{bmatrix}.$$

Step 3: Find out $\dot{V}(X)$.

$$\dot{V}(X) = \begin{bmatrix} g_1(X) & g_2(X) \ldots g_n(X) \end{bmatrix} \begin{bmatrix} \dot{x}_1 \\ \dot{x}_2 \\ - \\ \dot{x}_n \end{bmatrix}.$$

Step 4: Adjust the parameters of $\dot{V}(X)$ to make it negative definite.

Step 5: Apply the Curl condition to the $G(X)$ matrix to determine the remaining unknown parameters.

$$\frac{\partial G_i}{\partial X_j} = \frac{\partial G_j}{\partial X_i}.$$

Step 6: Obtain $V(X)$ by evaluating the line integral

$$V(X) = \int_0^{x_1} g_1(y_1, 0, 0, \ldots, 0, 0)\,dy_1 + \int_0^{x_2} g_2(x_1, y_2, 0, \ldots, 0, 0)\,dy_2$$

$$+ \int_0^{x_3} g_3(x_1, x_2, y_3, 0, \ldots, 0, 0)\,dy_3 + \ldots + \int_0^{x_n} g_n(x_1, x_2, x_3, \ldots, x_{n-1}, y_n)\,dy_n$$

Example 9.27 Find out a proper Lyapunov function using variable gradient method for assessing the stability of equilibrium state of a nonlinear system described by

$$\dot{x}_1 = -x_1 \qquad \dot{x}_2 = -x_2 + x_1.x_2^2 .$$

Solution:

Step 1: Write down the given state equations

$$\dot{x}_1 = -x_1 \qquad \dot{x}_2 = -x_2 + x_1.x_2^2 .$$

Clearly the equilibrium state is at origin.

Step 2: Write down $G(X)$ as

$$G(X) = \begin{bmatrix} g_1(X) \\ g_2(X) \end{bmatrix} = \begin{bmatrix} a_{11}x_1 + a_{12}x_2 \\ a_{21}x_1 + a_{22}x_2 \end{bmatrix} .$$

Step 3: Find out $\dot{V}(X)$

$$\dot{V}(X) = \begin{bmatrix} g_1(X) & g_2(X) \end{bmatrix} \begin{bmatrix} \dot{x}_1 \\ \dot{x}_2 \end{bmatrix}$$

$$= \begin{bmatrix} a_{11}x_1 + a_{12}x_2 & a_{21}x_1 + a_{22}x_2 \end{bmatrix} \begin{bmatrix} \dot{x}_1 \\ \dot{x}_2 \end{bmatrix}$$

$$= (a_{11}x_1 + a_{12}x_2)(-x_1) + (a_{21}x_1 + a_{22}x_2)(-x_2 + x_1 x_2^2)$$

$$= -a_{11}x_1^2 - a_{12}x_1 x_2 - a_{21}x_1 x_2 - a_{22}x_2^2 + a_{21}x_1^2 x_2^2 + a_{22}x_1 x_2^3 .$$

Step 4: Remove the terms involving terms of sign indefiniteness and make $\dot{V}(X)$ negative definite.

$$a_{12} = a_{21} = 0 .$$

Hence $\dot{V}(X) = -a_{11}x_1^2 - a_{22}x_2^2 + a_{22}x_1 x_2^3 = -a_{11}x_1^2 - a_{22}x_2^2(1 - x_1 x_2) .$

If $x_1 x_2 < 1$, $\dot{V}(X)$ is negative definite and the equilibrium state at origin is locally asymptotically stable.

Step 5: Apply the Curl condition to the $G(X)$ matrix to determine the remaining unknown parameters.

$$G(X) = \begin{bmatrix} g_1(X) \\ g_2(X) \end{bmatrix} = \begin{bmatrix} a_{11}x_1 \\ a_{22}x_2 \end{bmatrix}$$

$$\frac{\partial g_1}{\partial x_2} = x_1 \frac{\partial a_{11}}{\partial x_2}$$

$$\frac{\partial g_2}{\partial x_1} = x_2 \frac{\partial a_{22}}{\partial x_1}.$$

The Curl condition is

$$x_1 \frac{\partial a_{11}}{\partial x_2} = x_2 \frac{\partial a_{22}}{\partial x_1}.$$

To satisfy the Curl condition, a_{11} and a_{22} have to be constants.

Step 6: Obtain $V(X)$ by evaluating the line integral

$$V(X) = \int_0^{x1} g_1(y_1, 0) dy_1 + \int_0^{x2} g_2(x_1, y_2) dy_2$$

$$= \int_0^{x_1} a_{11} y_1 \, dy_1 + \int_0^{x_2} a_{22} y_2 \, dy_2$$

$$= \frac{1}{2} a_{11} x_1^2 + \frac{1}{2} a_{22} x_2^2, \text{ where } a_{11} \text{ and } a_{22} \text{ are positive constants.}$$

Example 9.28 Find out a proper Lyapunov function using variable gradient method for assessing the stability of equilibrium state of a nonlinear system described by

$$\dot{x}_1 = x_2 \qquad\qquad \dot{x}_2 = -x_2 - x_1^3.$$

Solution:

Step 1: Write down the given state equations

$$\dot{x}_1 = x_2 \qquad\qquad \dot{x}_2 = -x_2 - x_1^3.$$

Clearly the equilibrium state is at origin.

Step 2: Write down $G(X)$ as

$$G(X) = \begin{bmatrix} g_1(X) \\ g_2(X) \end{bmatrix} = \begin{bmatrix} a_{11}x_1 + a_{12}x_2 \\ a_{21}x_1 + a_{22}x_2 \end{bmatrix}.$$

Step 3: Find out $\dot{V}(X)$

$$\dot{V}(X) = \begin{bmatrix} g_1(X) & g_2(X) \end{bmatrix} \begin{bmatrix} \dot{x}_1 \\ \dot{x}_2 \end{bmatrix}$$

$$= \begin{bmatrix} a_{11}x_1 + a_{12}x_2 & a_{21}x_1 + a_{22}x_2 \end{bmatrix} \begin{bmatrix} \dot{x}_1 \\ \dot{x}_2 \end{bmatrix}$$

$$= (a_{11}x_1 + a_{12}x_2)(x_2) + (a_{21}x_1 + a_{22}x_2)(-x_2 - x_1^3)$$

$$= a_{11}x_1x_2 + a_{12}x_2^2 - a_{21}x_1x_2 - a_{22}x_2^2 - a_{21}x_1^4 - a_{22}x_1^3x_2$$

$$= x_1x_2(a_{11} - a_{21} - a_{22}x_1^2) + x_2^2(a_{12} - a_{22}) - a_{21}x_1^4.$$

Step 4: Remove the terms involving terms of sign indefiniteness and make $\dot{V}(X)$ negative definite.

Hence $a_{11} - a_{21} - a_{22}x_1^2 = 0$ $\qquad$ $a_{12} < a_{22}$

$$a_{11} = a_{21} + a_{22}x_1^2 .$$

Choose $a_{12} = 1$ and $a_{22} = 2$.

Hence $\dot{V}(X) = -x_2^2 - a_{21}x_1^4$, which is negative definite and the equilibrium state at origin is asymptotically stable.

Step 5: Apply the Curl condition to the $G(X)$ matrix to determine the remaining unknown parameters.

$$G(X) = \begin{bmatrix} g_1(X) \\ g_2(X) \end{bmatrix} = \begin{bmatrix} a_{11}x_1 + a_{12}x_2 \\ a_{21}x_1 + a_{22}x_2 \end{bmatrix} = \begin{bmatrix} (a_{21} + a_{22}x_1^2)x_1 + a_{12}x_2 \\ a_{21}x_1 + a_{22}x_2 \end{bmatrix} = \begin{bmatrix} a_{21}x_1 + 2x_1^3 + x_2 \\ a_{21}x_1 + 2x_2 \end{bmatrix} .$$

The Curl condition is $\dfrac{\partial g_1}{\partial x_2} = \dfrac{\partial g_2}{\partial x_1}$.

$$x_1 \frac{\partial a_{21}}{\partial x_2} + 1 = a_{21} + x_1 \frac{\partial a_{21}}{\partial x_1} .$$

If $a_{21} = 1$, this condition is satisfied.

Step 6: $\begin{bmatrix} g_1(X) \\ g_2(X) \end{bmatrix} = \begin{bmatrix} x_1 + 2x_1^3 + x_2 \\ x_1 + 2x_2 \end{bmatrix}$

Obtain $V(X)$ by evaluating the line integral

$$V(X) = \int_0^{x_1} g_1(y_1, 0)\, dy_1 + \int_0^{x_2} g_2(x_1, y_2)\, dy_2$$

$$= \int_0^{x_1} (y_1 + 2y_1^3)\, dy_1 + \int_0^{x_2} (x_1 + 2y_2)\, dy_2$$

$$= (\frac{1}{2}x_1^2 + \frac{1}{2}x_1^4) + (x_1 x_2 + x_2^2) = \frac{1}{2}x_1^4 + \frac{1}{4}x_1^2 + (\frac{1}{2}x_1 + x_2)^2 .$$

This is a positive definite function.

Example 9.29 Determine the stability of all the equilibrium states of a system represented by

$$\dot{x}_1 = x_2 \text{ and } \dot{x}_2 = x_1 - x_2 - x_1^2 .$$

Solution:

$$\dot{x}_1 = x_2 \text{ and } \dot{x}_2 = x_1 - x_2 - x_1^2$$

The equilibrium states are obtained by solving $\dot{x}_1 = 0$ and $\dot{x}_2 = 0$. This gives

$$x_{2e} = 0 \text{ and } x_{1e}(1 - x_{1e}) = 0.$$

The equilibrium states are (0, 0) and (1, 0).

Equilibrium state at origin

$$f(X) = \begin{bmatrix} \dot{x}_1 \\ \dot{x}_2 \end{bmatrix} = \begin{bmatrix} x_2 \\ x_1 - x_2 - x_1^2 \end{bmatrix}$$

$$F(X) = \begin{bmatrix} 0 & 1 \\ 1 - 2x_1 & -1 \end{bmatrix} \qquad F^T(X) = \begin{bmatrix} 0 & 1 - 2x_1 \\ 1 & -1 \end{bmatrix}$$

$$\hat{F}(X) = \begin{bmatrix} 0 & 2 - 2x_1 \\ 2 - 2x_1 & -2 \end{bmatrix}.$$

$\hat{F}(X)$ is negative semi-definite so that the equilibrium state at origin is stable in the sense of Lyapunov.

Equilibrium state at (1, 0)

Shifting the coordinate axes, the new coordinates are

$$x_{1n} = x_1 - 1 \text{ and } x_{2n} = x_2$$

$$\dot{x}_{1n} = \dot{x}_1 = x_2 = x_{2n}$$

$$\dot{x}_{2n} = \dot{x}_2 = -x_2 + x_1 - x_1^2 = -x_{2n} + (x_{1n} + 1) - (x_{1n} + 1)^2 = -x_{2n} - x_{1n} - x_{1n}^2$$

$$f(X_n) = \begin{bmatrix} \dot{x}_{1n} \\ \dot{x}_{2n} \end{bmatrix} = \begin{bmatrix} x_{2n} \\ -x_{1n} - x_{2n} - x_{1n}^2 \end{bmatrix}$$

$$J(X_n) = \begin{bmatrix} 0 & 1 \\ -1 - 2x_{1n} & -1 \end{bmatrix} \qquad J^T(X) = \begin{bmatrix} 0 & -1 - 2x_{1n} \\ 1 & -1 \end{bmatrix}$$

$$\hat{J}(X_n) = \begin{bmatrix} 0 & -2x_{1n} \\ -2x_{1n} & -2 \end{bmatrix}.$$

$\hat{J}(X_n)$ is negative semi-definite so that the equilibrium state at origin is stable in the sense of Lyapunov.

9.12 LYAPUNOV STABILITY ANALYSES FOR LINEAR CONTINU-OUS TIME SYSTEMS

It was stated earlier that the Lyapunov function for a nonlinear system is not unique and it is difficult to find a proper Lyapunov function to establish the stability of the system. But in the case of a linear time-invariant continuous-time system, it is possible to find a proper Lyapunov function to assess the stability of the system.

Consider an autonomous system represented by

$$\dot{X} = AX .$$

Let $V(X) = X^T P X$ be a proper Lyapunov function where P is positive definite.

$$\dot{V}(X) = (\dot{X})^T PX + X^T P \dot{X}$$

$$= (AX)^T PX + X^T PAX$$

$$= X^T A^T PX + X^T PAX$$

$$= X^T (A^T P + PA)X$$

$$= -X^T QX .$$

The linear system is asymptotically stable if $\dot{V}(X)$ is negative definite. If $(A^T P + PA) = -Q$ for some positive definite matrix Q, then the system is asymptotically stable. If a positive definite matrix Q is selected, then the matrix P can be determined from the equation

$$A^T P + PA = -Q .$$

The system is asymptotically stable if the matrix P is positive definite. The simplest positive definite matrix Q is $Q = I$ where I is the identity matrix.

Example 9.30 A linear time-invariant system is represented by the state equation

$$\dot{X} = \begin{bmatrix} 0 & 1 \\ -1 & -2 \end{bmatrix} X + \begin{bmatrix} 0 \\ 1 \end{bmatrix} r .$$

Using Lyapunov stability criterion, assess the stability of the system.

Solution:

$$\dot{X} = \begin{bmatrix} 0 & 1 \\ -1 & -2 \end{bmatrix} X + \begin{bmatrix} 0 \\ 1 \end{bmatrix} r$$

$$A = \begin{bmatrix} 0 & 1 \\ -1 & -2 \end{bmatrix}$$

$$A^T P + PA = -I$$

$$\begin{bmatrix} 0 & -1 \\ 1 & -2 \end{bmatrix} \begin{bmatrix} p_1 & p_2 \\ p_2 & p_3 \end{bmatrix} + \begin{bmatrix} p_1 & p_2 \\ p_2 & p_3 \end{bmatrix} \begin{bmatrix} 0 & 1 \\ -1 & -2 \end{bmatrix} = \begin{bmatrix} -1 & 0 \\ 0 & -1 \end{bmatrix}$$

$$\begin{bmatrix} -p_2 & -p_3 \\ p_1 - 2p_2 & p_2 - 2p_3 \end{bmatrix} + \begin{bmatrix} -p_2 & p_1 - 2p_2 \\ -p_3 & p_2 - 2p_3 \end{bmatrix} = \begin{bmatrix} -1 & 0 \\ 0 & -1 \end{bmatrix}$$

$$\begin{bmatrix} -2p_2 & p_1 - 2p_2 - p_3 \\ p_1 - 2p_2 - p_3 & 2p_2 - 4p_3 \end{bmatrix} = \begin{bmatrix} -1 & 0 \\ 0 & -1 \end{bmatrix}.$$

This equation yields three simultaneous equations

$$-2p_2 = -1 \tag{1}$$

$$p_1 - 2p_2 - p_3 = 0 \tag{2}$$

$$2p_2 - 4p_3 = -1. \tag{3}$$

Solving these equations,

$$p_1 = \frac{3}{2} \qquad p_2 = \frac{1}{2} \quad \text{and} \quad p_3 = \frac{1}{2}$$

$$P = \begin{bmatrix} \dfrac{3}{2} & \dfrac{1}{2} \\ \dfrac{1}{2} & \dfrac{1}{2} \end{bmatrix}.$$

Since P is positive definite, the system is asymptotically stable. The Lyapunov function is

$$V(X) = \frac{3}{2}x_1^2 + \frac{1}{2}x_2^2 + x_1 x_2.$$

Example 9.31 A linear time-invariant system is represented by the state equation

$$\dot{X} = \begin{bmatrix} -1 & 1 \\ -1 & -1 \end{bmatrix} X + \begin{bmatrix} 1 \\ 1 \end{bmatrix} r.$$

Using Lyapunov stability criterion, assess the stability of the system.

Solution:

$$\dot{X} = \begin{bmatrix} -1 & 1 \\ -1 & -1 \end{bmatrix} X + \begin{bmatrix} 1 \\ 1 \end{bmatrix} r$$

$$A = \begin{bmatrix} -1 & 1 \\ -1 & -1 \end{bmatrix}$$

$$A^T P + PA = -I$$

$$\begin{bmatrix} -1 & -1 \\ 1 & -1 \end{bmatrix} \begin{bmatrix} p_1 & p_2 \\ p_2 & p_3 \end{bmatrix} + \begin{bmatrix} p_1 & p_2 \\ p_2 & p_3 \end{bmatrix} \begin{bmatrix} -1 & 1 \\ -1 & -1 \end{bmatrix} = \begin{bmatrix} -1 & 0 \\ 0 & -1 \end{bmatrix}$$

$$\begin{bmatrix} -p_1 - p_2 & -p_2 - p_3 \\ p_1 - p_2 & p_2 - p_3 \end{bmatrix} + \begin{bmatrix} -p_1 - p_2 & p_1 - p_2 \\ -p_2 - p_3 & p_2 - p_3 \end{bmatrix} = \begin{bmatrix} -1 & 0 \\ 0 & -1 \end{bmatrix}$$

$$\begin{bmatrix} -2p_1 - 2p_2 & p_1 - 2p_2 - p_3 \\ p_1 - 2p_2 - p_3 & 2p_2 - 2p_3 \end{bmatrix} = \begin{bmatrix} -1 & 0 \\ 0 & -1 \end{bmatrix}.$$

This equation yields three simultaneous equations

$$-2p_1 - 2p_2 = -1 \tag{1}$$

$$p_1 - 2p_2 - p_3 = 0 \tag{2}$$

$$2p_2 - 2p_3 = -1 . \tag{3}$$

Solving these equations,

$$p_1 = \frac{1}{2} \qquad p_2 = 0 \quad \text{and} \quad p_3 = \frac{1}{2}$$

$$P = \begin{bmatrix} \frac{1}{2} & 0 \\ 0 & \frac{1}{2} \end{bmatrix}.$$

Since P is positive definite, the system is asymptotically stable. The Lyapunov function is

$$V(X) = \frac{1}{2}x_1^2 + \frac{1}{2}x_2^2 .$$

Example 9.32 A linear time-invariant system is represented by the state equation

$$\dot{X} = \begin{bmatrix} -1 & 0 \\ 1 & 2 \end{bmatrix} X + \begin{bmatrix} 1 \\ -1 \end{bmatrix} r .$$

Using Lyapunov stability criterion, assess the stability of the system.

Solution:

$$\dot{X} = \begin{bmatrix} -1 & 0 \\ 1 & 2 \end{bmatrix} X + \begin{bmatrix} 1 \\ -1 \end{bmatrix} r .$$

$$A = \begin{bmatrix} -1 & 0 \\ 1 & 2 \end{bmatrix}$$

$$A^T P + PA = -I$$

$$\begin{bmatrix} -1 & 1 \\ 0 & 2 \end{bmatrix} \begin{bmatrix} p_1 & p_2 \\ p_2 & p_3 \end{bmatrix} + \begin{bmatrix} p_1 & p_2 \\ p_2 & p_3 \end{bmatrix} \begin{bmatrix} -1 & 0 \\ 1 & 2 \end{bmatrix} = \begin{bmatrix} -1 & 0 \\ 0 & -1 \end{bmatrix}$$

$$\begin{bmatrix} -p_1 + p_2 & -p_2 + p_3 \\ 2p_2 & 2p_3 \end{bmatrix} + \begin{bmatrix} -p_1 + p_2 & 2p_2 \\ -p_2 + p_3 & 2p_3 \end{bmatrix} = \begin{bmatrix} -1 & 0 \\ 0 & -1 \end{bmatrix}$$

$$\begin{bmatrix} -2p_1 + 2p_2 & p_2 + p_3 \\ p_2 + p_3 & 4p_3 \end{bmatrix} = \begin{bmatrix} -1 & 0 \\ 0 & -1 \end{bmatrix}.$$

This equation yields three simultaneous equations

$$-2p_1 + 2p_2 = -1 \tag{1}$$

$$p_2 + p_3 = 0 \tag{2}$$

$$4p_3 = -1. \tag{3}$$

Solving these equations,

$$p_1 = \frac{3}{4} \qquad p_2 = \frac{1}{4} \quad \text{and} \quad p_3 = -\frac{1}{4}$$

$$P = \begin{bmatrix} \dfrac{3}{4} & \dfrac{1}{4} \\ \dfrac{1}{4} & -\dfrac{1}{4} \end{bmatrix}.$$

Since P is not positive definite. The system is not stable in the sense of Lyapunov.

SUMMARY

- A system is said to be stable if its impulse response decays down to zero as time approaches infinity.
- A system is asymptotically stable if all the Eigen values of the system lie on the left half of the s-plane.
- A system is bounded input bounded output (BIBO) stable if all the poles of the system lie on the left half of the s-plane.
- The poles of the transfer function of a system are a subset of Eigen values. Hence if a system is asymptotically stable, it is also BIBO stable. But a system which is BIBO stable may not be asymptotically stable.
- A scalar function $F(X)$ expressed as $F(X) = X^T P X$, where P is a real symmetric matrix is said to be in quadratic form.
- A function is positive definite if it is positive at all points in the state space except at the origin, where it is zero.
- A negative definite function is one which is negative at all points in the state space except at the origin, where it is zero.
- The sign definiteness of a scalar function $F(X)$ in quadratic form can be determined using the Sylvester's criterion.
- Lyapunov stability requires that the system phase trajectory, starting from an initial state, remains in a bounded region around the equilibrium state.
- In an asymptotically stable system, the phase trajectory starting from an initial state returns to the equilibrium state.
- Lyapunov stability theorem is based on the principle that in an autonomous system if the total energy of the system continuously decreases, then the phase trajectory starting from an initial state finally approaches an equilibrium point.

- Lyapunov function is a positive definite scalar function representing total energy of the system.
- In the first method of Lyapunov for determining the stability of a nonlinear system, the nonlinear equation is first linearized about the equilibrium point. If the Eigen values are with negative real parts, then the system is locally stable in the sense of Lyapunov.
- In the second method of Lyapunov, first a proper Lyapunov function is chosen. If the derivative of this function, evaluated on the phase trajectory, is negative definite, then the system is asymptotically stable.
- If the derivative of the Lyapunov function evaluated on the phase trajectory is negative semi-definite, then the system is stable in the sense of Lyapunov.
- A major problem in applying Lyapunov's second method to nonlinear systems is to find a proper Lyapunov function.
- Lyapunov stability theorem is applicable to linear time-invariant systems as well.
- Using Krasovskii's theorem, a nonlinear system can be checked for its Lyapunov stability.
- Variable gradient method is used to generate a proper Lyapunov function for a given system.

PRACTICE PROBLEMS

PP 9.1 Find the sign definiteness of the scalar function

$$F(X) = x_2^2 + 2x_1 x_2 .$$

PP 9.2 Find the sign definiteness of the scalar function

$$F(X) = -x_1^2 - 3x_2^2 - 6x_3^2 + 2x_1 x_2 - 4x_2 x_3 - 2x_1 x_3 .$$

PP 9.3 Find the sign definiteness of the scalar function

$$F(X) = x_1^2 + 3x_2^2 + x_3^2 + 4x_1 x_2 + 2x_1 x_3 .$$

PP 9.4 Find the sign definiteness of the scalar function

$$F(X) = x_1^2 + 2x_2^2 + x_3^2 + 2x_1 x_2 + 6x_2 x_3 + 4x_1 x_3 .$$

PP 9.5 Find the sign definiteness of the scalar function

$$F(X) = -2x_1^2 - 4x_2^2 - 8x_3^2 - 4x_1 x_2 + 8x_2 x_3 + 8x_1 x_3 .$$

PP 9.6 Find the sign definiteness of the scalar function

$$F(X) = x_1^2 + x_2^2 + x_3^2 - 2x_1 x_2 + 2x_2 x_3 - 2x_1 x_3 .$$

PP 9.7 Using Lyapunov function, check the stability of equilibrium state at origin of a non-linear system represented by the state equations

$$\dot{x}_1 = -x_1^3 - x_2$$

$$\dot{x}_2 = x_1 - x_2 .$$

PP 9.8 Using Lyapunov stability theorem, assess the stability of equilibrium state of a non-linear system represented by the differential equation

$$\ddot{y} + a(1 + y^2)\dot{y} + y = 0 .$$

PP 9.9 Using Lyapunov stability theorem, assess the stability of equilibrium state of a non-linear system represented by the differential equation

$$\ddot{y} + (\dot{y})^3 + y = 0 .$$

PP 9.10 A nonlinear system is represented by the state equations

$$\dot{x}_1 = x_2 \text{ and } \dot{x}_2 = -a\, x_2 - b\, x_2^3 - c\, x_1 .$$

Choose a Lyapunov function $V(X) = k_1\, x_1^2 + k_2\, x_2^2$. Under what condition the equilibrium state is stable in the sense of Lyapunov?

PP 9.11 Check the Lyapunov stability of a system represented by

$$\dot{x}_1 = -2x_2 - 3x_1^3 \text{ and } \dot{x}_2 = 2x_1 - 3x_2^3 .$$

PP 9.12 Check the Lyapunov stability of a system represented by

$$\ddot{y} + a\,\dot{y} + b\,(\dot{y})^3 + cy = 0 .$$

Use the Lyapunov function $V(X) = x_1^2 + x_2^2$.

PP 9.13 A system is represented by

$$\dot{x}_1 = -x_1 + 2x_1^2 x_2 \text{ and } \dot{x}_2 = -x_2 - 2x_1^3 .$$

Check the stability of equilibrium state at origin.

PP 9.14 Check the stability of equilibrium state of a system represented by

$$\dot{x}_1 = x_2 \text{ and } \dot{x}_2 = -x_2 - x_1^3 .$$

Use the Lyapunov function $V(X) = x_1^4 + (x_1 + x_2)^2 + x_2^2$.

PP 9.15 Check the stability of equilibrium state of a system represented by

$$\dot{x}_1 = x_2 \text{ and } \dot{x}_2 = -x_1 - x_2 - x_1^5 .$$

Choose a proper Lyapunov function.

PP 9.16 Using Krasovskii's method, determine the stability of equilibrium state at origin of a system described by

$$\dot{x}_1 = -x_1 + x_2 \text{ and } \dot{x}_2 = x_1 - x_2 - x_2^3 .$$

PP 9.17 Using Krasovskii's method, determine the stability of equilibrium state at origin of a system described by

$$\dot{x}_1 = -6x_1 + 2x_2 \text{ and } \dot{x}_2 = 2x_1 - 6x_2 - 2x_2^3 .$$

PP 9.18 Using Krasovskii's method, determine the stability of equilibrium state at origin of a system described by

$$\dot{x}_1 = x_2 \text{ and } \dot{x}_2 = -x_1 - x_2 - (x_1 + x_2)^2.$$

PP 9.19 Using the variable gradient method, determine a Lyapunov function for the nonlinear system described by the state equations

$$\dot{x}_1 = -x_1 + 2x_1^2 x_2 \text{ and } \dot{x}_2 = -x_2.$$

PP 9.20 Using the variable gradient method, determine a Lyapunov function for the nonlinear system described by the state equations

$$\dot{x}_1 = -2x_1 \text{ and } \dot{x}_2 = -2x_2 + 2x_1 x_2^2.$$

PP 9.21 A system is represented by the state equation

$$\dot{X} = \begin{bmatrix} -2 & 1 \\ 0 & -2 \end{bmatrix} X + \begin{bmatrix} 1 \\ 0 \end{bmatrix} r.$$

Using Lyapunov stability criterion, assess the stability of the system.

PP 9.22 A system is represented by the state equation

$$\dot{X} = \begin{bmatrix} -1 & 1 \\ 2 & -3 \end{bmatrix} X + \begin{bmatrix} 1 \\ 1 \end{bmatrix} r.$$

Using Lyapunov stability criterion, assess the stability of the system.

PP 9.23 A system is represented by the state equation

$$\dot{X} = \begin{bmatrix} 0 & 1 \\ -1 & -1 \end{bmatrix} X + \begin{bmatrix} 1 \\ 0 \end{bmatrix} r.$$

Using Lyapunov stability criterion, assess the stability of the system.

PP 9.24 A system is represented by the state equation

$$\dot{X} = \begin{bmatrix} -2 & 1 \\ -1 & 1 \end{bmatrix} X + \begin{bmatrix} 0 \\ 1 \end{bmatrix} r.$$

Using Lyapunov stability criterion, assess the stability of the system.

PP 9.25 An autonomous nonlinear system is represented by the state equations

$$\dot{x}_1 = x_2 \text{ and } \dot{x}_2 = -x_1^3 - x_2^3.$$

Assess the stability of the system by using the direct method of Lyapunov. Verify the result using Krasovskii's method.

PP 9.26 Using Krasovskii's method, assess the stability of the equilibrium state of the system given by the state equations

$$\dot{x}_1 = x_1 + 3x_2 \quad \dot{x}_2 = -3x_1 - 2x_2 - 3x_3 \text{ and } \dot{x}_3 = x_1.$$

PP 9.27 Determine the stability of all the equilibrium states of a system represented by

$$\dot{x}_1 = x_2 \text{ and } \dot{x}_2 = -x_1 - x_2 + x_1^2 .$$

REVIEW QUESTIONS

9.1 Define the stability of a linear system.

9.2 Distinguish between BIBO stability and asymptotic stability.

9.3 A system is BIBO stable. Is it asymptotic stable too? Explain.

9.4 What is meant by sign definiteness of a scalar function?

9.5 What is meant by quadratic form of a scalar function?

9.6 Discuss the method to test the sign definiteness of a scalar function.

9.7 Discuss the different types of stability with respect to regions around the equilibrium point.

9.8 Define stability, asymptotic stability and instability in the sense of Lyapunov.

9.9 Explain the first method of Lyapunov to determine the stability of an equilibrium point.

9.10 Explain the second method of Lyapunov to determine the stability of an equilibrium point.

9.11 State and explain the Lyapunov stability theorem.

9.12 State and explain the Lyapunov instability theorem.

9.13 State and prove the Krasovskii's theorem.

9.14 Discuss the various steps involved in finding a Lyapunov function for a system represented by $\dot{X} = F(X)$, by variable gradient method.

9.15 Explain how the Lyapunov stability theorem is applied to a linear time-invariant system.

Optimal Control Theory

10.1 INTRODUCTION

It is quite natural that a system is to be designed based on certain performance criteria like minimum energy consumption, minimum fuel consumption, minimum error, maximum efficiency, maximum precision and quality. It is necessary that a system is designed under certain restrictions or constraints to minimize or maximize a performance measure or index. This is referred to as optimal design. Minimization or maximization of a performance index subject to certain constraints is optimization.

An *optimal control system* is one in which the design is based on finding an *optimal control signal* which optimizes a cost function, composed of performance measures, subject to certain dynamic constraints and boundary conditions. If the optimal control signal is found out based on initial state and other system parameters, the control is said to be open loop. On the other hand, if the optimal control signal is found based on the current state, then it is said to be closed loop or feedback control law. In the closed loop method, the current states of the process or plant are fed back to the controller which processes it and produces the necessary optimum control signal.

In optimal control, the performance measure is mostly composed of quadratic functions and such a control is called a *quadratic optimal control*. The choice of quadratic functions introduces mathematical convenience of analyzing and tracking the system.

10.2 CONCEPT OF OPTIMAL CONTROL

Consider a linear plant as shown in Figure 10.1. Let us consider that the error in the system is the performance measure. For the system, the error ratio is

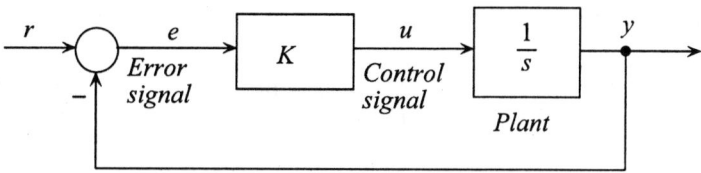

Figure 10.1 Concept of optimization

$$\frac{E(s)}{R(s)} = \frac{1}{1+\dfrac{K}{s}} = \frac{s}{s+K}$$

$$E(s) = \frac{s}{s+K} R(s).$$

For a unit step input, $r(t) = 1$ so that $R(s) = 1/s$. Hence

$$E(s) = \frac{1}{s+K} \quad \text{and} \quad e(t) = e^{-Kt}.$$

If the quadratic performance measure is taken as the integral square error,

$$J = \int_0^\infty e^2(t)\,dt = \int_0^\infty e^{-2Kt}\,dt = \frac{1}{2K}.$$

This shows that for J to be a minimum, K must take a maximum value and J tends to zero as K tends to infinity. Making the value of K very large has physical constraints and is not practically feasible. To overcome this difficulty, a performance measure $u^2(t)$ which is a measure of the energy consumption is included in the performance index which is modified as

$$J = \int_0^\infty [e^2(t) + u^2(t)]\,dt.$$

Usually a weighting index R, which is a scalar variable, is associated with the energy consumption component $u^2(t)$ that introduces weightage or penalty for different control signals. The performance measure J is thus modified as

$$J = \int_0^\infty [e^2(t) + Ru^2(t)]\,dt.$$

From Figure 10.1, $u(t) = Ke^{-Kt}$ so that the performance measure is

$$J = \int_0^\infty [e^{-2Kt} + RK^2 e^{-2Kt}]\,dt = (1+RK^2)\int_0^\infty e^{-2Kt}\,dt = \frac{1+RK^2}{2K} = \frac{1}{2K} + \frac{RK}{2}.$$

For minimum J,

$$\frac{\partial J}{\partial K} = -\frac{1}{2K^2} + \frac{R}{2} = 0.$$

Solving this, $K = 1/\sqrt{R}$.

$$\frac{\partial^2 J}{\partial K^2} = \frac{1}{K^3} > 0$$

Hence $K = 1/\sqrt{R}$ gives a minimum value for J.

$$J_{min} = \sqrt{R} \ .$$

This is the optimum value for J.

If a large value of R is chosen it means that more weightage is given to the magnitude of control signal rather than the error. A suitable value for J is chosen to strike a trade-off between the error and the control signals.

10.3 PERFORMANCE INDICES

A performance criterion J is a measure of the quality of the system behaviour. Usually the performance measure is minimized or maximized by using a proper control input signal. This control input signal is generally comprised of a feedback of state variables.

10.3.1. Minimum Time Problem

The objective is to minimize the time taken in transferring the system state from the initial state $X(t_0)$ to a specified state $X(t_f)$.

$$J = \int_{t_0}^{t_f} dt = (t_f - t_0) \ .$$

10.3.2. Minimum Energy Problem

Energy expended is proportional to the control effort. The problem is to minimize the energy expended in transferring the system from the initial state $X(t_0)$ to a specified state $X(t_f)$.

$$J = \int_{t_0}^{t_f} [u_1^2(t) + u_2^2(t) + ... + u_n^2(t)] dt = \int_{t_0}^{t_f} U^T(t)U(t) dt \ .$$

For giving different weightages or penalties to different control efforts, a weighting matrix R can be introduced. In such a case, the performance index J can be modified as

$$J = \int_{t_0}^{t_f} [R_1 u_1^2(t) + R_2 u_2^2(t) + ... + R_n u_n^2(t)] dt = \int_{t_0}^{t_f} U^T(t) RU(t) dt \ ,$$

where R is a positive definite matrix or a diagonal matrix

$$R = \begin{bmatrix} R_1 & 0 & 0 & ... & 0 \\ 0 & R_2 & 0 & ... & 0 \\ & & & ... & \\ 0 & 0 & 0 & ... & R_n \end{bmatrix}.$$

10.3.3. Minimum Terminal Error Problem

The objective is to minimize the square of the norm of the error between the final state $X(t_f)$ and the desired state X_d.

$$J = [x_1(t_f) - x_{d1}]^2 + [x_2(t_f) - x_{d2}]^2 + \ldots + [x_n(t_f) - x_{dn}]^2$$
$$= [X(t_f) - X_d]^T [X(t_f) - X_d].$$

A modified performance index is

$$J = [X(t_f) - X_d]^T H [X(t_f) - X_d],$$

where H is a positive definite or positive semi-definite matrix.

If the desired state is the origin, then the performance measure is

$$J = [X(t_f)]^T H [X(t_f)].$$

10.3.4. State Regulator Problem

The problem is to transfer a system from the initial state $X(t_0)$ to a specified state $X(t_f)$ with minimum integral square error.

$$J = \int_{t_0}^{t_f} [X(t) - X_d]^T Q [X(t) - X_d] \, dt.$$

If the desired state is the origin, then the performance measure is

$$J = \int_{t_0}^{t_f} X^T(t) Q X(t) \, dt.$$

In addition to this, if it is desired to include the minimum terminal error problem and the minimum energy problem, then the performance measure is

$$J = X^T(t_f) H X(t_f) + \int_{t_0}^{t_f} [X^T(t) Q X(t) + U^T(t) R U(t)] dt.$$

10.3.5. Output Regulator Problem

In this case, the objective is to make the output as close to the desired output or reference input as possible in the time interval $[t_0, t_f]$. The performance measure is

$$J = Y^T(t_f) H Y(t_f) + \int_{t_0}^{t_f} [Y^T(t) Q Y(t) + U^T(t) R U(t)] dt.$$

10.3.6. Tracking Problem

In this case, the objective is to make the state $X(t)$ of the system as close as possible to the desired state X_d in the interval $[t_0, t_f]$. The performance measure is

$$J = E^T(t_f) H E(t_f) + \int_{t_0}^{t_f} [E^T(t) Q E(t) + U^T(t) R U(t)] dt,$$

where $E(t) = X(t) - X_d$.

10.4 RELATION BETWEEN QUADRATIC PERFORMANCE INDEX AND LYAPUNOV FUNCTION

Consider an autonomous system described by the state equation

$$\dot{X} = f(X) = AX.$$

It is desired to transfer an initial state $X(t_0)$ to the origin so that $X(\infty) = 0$. While driving the initial state to the origin, let the quadratic performance index that is to be optimized be

$$J = \int_{t_0}^{\infty} X^T(t) Q X(t) dt, \tag{10.1}$$

where Q is a positive definite or positive semi-definite real symmetric matrix. Choose a Lyapunov function and its derivative function as

$$V(X) = X^T P X ; \quad V(0) = 0 \tag{10.2}$$

$$\dot{V}(X) = -X^T Q X. \tag{10.3}$$

Using equations (10.1) and (10.3), the given performance index can be written as

$$J = \int_{t_0}^{\infty} X^T(t) Q X(t) dt = -\int_{t_0}^{\infty} \dot{V}(X) dt$$

$$= -V(X(\infty)) + V(X(t_0))$$

$$= V(X(t_0)). \tag{10.4}$$

Comparing equation (10.4) with equation (10.2),

$$J = V(X(t_0)) = X^T(t_0) P X(t_0). \tag{10.5}$$

Equation (10.5) indicates that the given performance index is equal to the value of Lyapunov function $V(X)$ at $t = t_0$.

10.5 STATE REGULATOR DESIGN USING LYAPUNOV EQUATION

The objective of optimal control is to maintain the state of a system at a desired state, quite often the equilibrium state, and at the same time to optimize a given performance measure. An optimal control problem is to design a feedback control law $U(t) = -K\,X(t)$ that achieves this result. Consider a linear time-invariant system represented by the state equation

$$\dot{X} = AX + BU \cdot \tag{10.6}$$

Let the performance measure that is to be optimized be

$$J = \int_0^\infty [X^T Q\, X + U^T R\, U]dt\,, \tag{10.7}$$

where Q is a positive definite or positive semi-definite real symmetric matrix and R is a positive definite real symmetric matrix or a diagonal matrix.

Let the control law be

$$U(t) = -K\,X(t) \cdot \tag{10.8}$$

where $K = \begin{bmatrix} k_{11} & k_{12} & ... & k_{1n} \\ k_{21} & k_{22} & ... & k_{2n} \\ & ... & \\ k_{p1} & k_{p2} & ... & k_{pn} \end{bmatrix}$ and $X(t) = \begin{bmatrix} x_1 & x_2 & x_3 & ... & x_n \end{bmatrix}^T$.

K is a $p \times n$ matrix corresponding to p inputs and n state variables.

From equations (10.6) and (10.8),

$$\dot{X} = AX + B[-KX] = [A - BK]X \cdot \tag{10.9}$$

Assume that the matrix K exists such that the matrix $[A - BK]$ is stable. Substituting equation (10.8) in equation (10.7), the performance index now can be written as

$$J = \int_0^\infty [X^T Q\, X + (-KX)^T R\,(-KX)]dt$$

$$= \int_0^\infty [X^T Q\, X + X^T K^T R\, K\, X]dt$$

$$= \int_0^\infty X^T (Q + K^T R\, K)X\, dt \cdot \tag{10.10}$$

Choose a Lyapunov function

$$V(X) = \int_t^{\infty} X^T (Q + K^T RK) X \, dt \ .$$

Then $\dot{V}(X) = X^T (Q + K^T RK) X \big|_t^{\infty}$

$$= X^T (\infty)(Q + K^T RK) X(\infty) - X^T (t)(Q + K^T RK) X(t) \ .$$

Since $X(\infty) \to 0$,

$$\dot{V}(X) = -X^T (Q + K^T RK) X \ . \tag{10.11}$$

In general, a Lyapunov function for a linear time-invariant system can be expressed in quadratic form as

$$V(X) = X^T PX \ . \tag{10.12}$$

Derivative of equation (10.12) is

$$\dot{V}(X) = (\dot{X})^T PX + X^T P \dot{X} \ . \tag{10.13}$$

Substituting equation (10.9) in equation (10.13),

$$\dot{V}(X) = [(A - BK) X]^T PX + X^T P(A - BK) X$$

$$= X^T (A - BK)^T PX + X^T P(A - BK) X$$

$$= X^T \left[(A - BK)^T P + P(A - BK) \right] X \ . \tag{10.14}$$

Comparing equations (10.11) and (10.14),

$$(A - BK)^T P + P(A - BK) = -(Q + K^T RK)$$

$$\text{Or } (A - BK)^T P + P(A - BK) + K^T RK = -Q \ . \tag{10.15}$$

If the matrix $(A - BK)$ is stable, that is if its Eigen values are with negative real parts, then the matrix P can be determined by solving the equation (10.15). Sufficient conditions for optimal controllability are that the pair of matrices (A, B) provides complete controllability and the pair (A, C) provides complete observability. Equation (10.15) is often referred to as Lyapunov equation for optimal regulator design.

Design steps

1. Using the given matrices A and B, determine the matrix P from the equation

$$(A - BK)^T P + P(A - BK) + K^T RK = -Q$$

2. Substitute the matrix P in the equation

$$J = X^T(0) P X(0).$$

This equation for performance measure is a function of the elements of the feedback gain matrix K.

3. To determine the minimum value of J,

$$\frac{\partial J}{\partial k_{ij}} = 0 \text{ and } \frac{\partial^2 J}{\partial k_{ij}^2} > 0.$$

Solve for optimum values of k_{ij} that decides the feedback optimal control law that minimizes the performance index defined in equation (10.7).

4. Evaluate the minimum value of J.

10.6 RICCATI EQUATION

The reduced matrix Riccati equation is a modification of Lyapunov equation (10.15). Riccati equation can be solved in a more convenient way. Since R is a positive definite matrix, it can be expressed as

$$R = T^T T.$$
(10.16)

Substituting equation (10.16) in equation (10.15),

$$A^T P - K^T B^T P + PA - PBK + K^T T^T TK = -Q.$$

This equation can be written as

$$A^T P + PA + \left[\left(TK - (T^T)^{-1} B^T P\right)^T \left(TK - (T^T)^{-1} B^T P\right) - PBK\right] = -Q.$$
(10.17)

Minimization of J with respect to K requires the minimization of

$$X^T \left[\left(TK - (T^T)^{-1} B^T P\right)^T \left(TK - (T^T)^{-1} B^T P\right) - PBK\right] X.$$

This will be minimum when

$$TK - (T^T)^{-1} B^T P = 0.$$

Or

$$K = T^{-1}(T^T)^{-1} B^T P$$

$$= (T^T T)^{-1} B^T P$$

$$= R^{-1} B^T P.$$
(10.18)

Equation (10.18) decides the feedback control law that optimises the performance index given in equation (10.7). Equation (10.17) now becomes

$$A^T P + PA - PBR^{-1} B^T P = -Q.$$
(10.19)

This is called the matrix Riccati equation for determining the optimal control law. The performance measure is the value of the Lyapunov function at $t = 0$. Hence

$$J_{min} = X(0)^T PX(0) . \tag{10.20}$$

Design Procedure
1. From the given A, B, Q and R matrices, check whether the matrix $[A - BK]$ is stable; that is the Eigen values of $[A - BK]$ are negative or with negative real parts.
2. Solve for the matrix P using the relation

$$A^T P + PA - PBR^{-1}B^T P = -Q .$$

3. Obtain the feedback gain matrix

$$K = R^{-1}B^T P .$$

4. Substitute P in the performance index

$$J_{min} = X(0)^T PX(0) .$$

Note: Let $Q = S^T S$. To check whether the matrix $[A - BK]$ is stable, it is sufficient to see that the matrix $M = \begin{bmatrix} S^T & A^T S^T & (A^T)^2 S^T ... (A^T)^{n-1} S^T \end{bmatrix}$ has full rank.

Example 10.1 A system is represented by the state equation

$$\dot{X} = \begin{bmatrix} 0 & 1 \\ 0 & 0 \end{bmatrix} X + \begin{bmatrix} 0 \\ 1 \end{bmatrix} u .$$

Determine an optimal control law $u = -KX$ that minimizes the performance index

$$J = \frac{1}{2}\int_0^\infty (X^T X)\,dt .$$

Given that the initial state $X(0) = \begin{bmatrix} 1 & 1 \end{bmatrix}^T$.
  The undamped natural frequency of the system is to be 1 rad/se

Solution:

$$\dot{X} = \begin{bmatrix} 0 & 1 \\ 0 & 0 \end{bmatrix} X + \begin{bmatrix} 0 \\ 1 \end{bmatrix} u$$

$$A = \begin{bmatrix} 0 & 1 \\ 0 & 0 \end{bmatrix} \qquad B = \begin{bmatrix} 0 \\ 1 \end{bmatrix}$$

The performance index is

$$J = \frac{1}{2}\int_0^\infty (X^T X)\,\mathrm{dt} = \frac{1}{2}\int_0^\infty (X^T Q X)\,\mathrm{dt};\ Q = \begin{bmatrix} 1 & 0 \\ 0 & 1 \end{bmatrix} \quad R = 0.$$

The Lyapunov equation is

$$[A - BK]^T P + P[A - BK] + K^T RK = -Q$$

$$[A - BK] = \begin{bmatrix} 0 & 1 \\ 0 & 0 \end{bmatrix} - \begin{bmatrix} 0 \\ 1 \end{bmatrix}[k_1 \quad k_2] = \begin{bmatrix} 0 & 1 \\ -k_1 & -k_2 \end{bmatrix}.$$

With state feedback, the characteristic equation is

$$\begin{vmatrix} s & -1 \\ k_1 & s + k_2 \end{vmatrix} = 0 \quad Or \quad s^2 + k_2 s + k_1 = 0 \equiv s^2 + 2\delta\omega_n s + \omega_n^2.$$

It is given that $\omega_n = 1$. Hence $k_1 = \omega_n^2 = 1$.

Let $P = \begin{bmatrix} a & b \\ b & c \end{bmatrix}.$

Since R = 0, the Lyapunov equation is

$$[A - BK]^T P + P[A - BK] = -Q$$

$$\begin{bmatrix} 0 & -1 \\ 1 & -k_2 \end{bmatrix}\begin{bmatrix} a & b \\ b & c \end{bmatrix} + \begin{bmatrix} a & b \\ b & c \end{bmatrix}\begin{bmatrix} 0 & 1 \\ -1 & -k_2 \end{bmatrix} = \begin{bmatrix} -1 & 0 \\ 0 & -1 \end{bmatrix}$$

$$\begin{bmatrix} -b & -c \\ a - k_2 b & b - k_2 c \end{bmatrix} + \begin{bmatrix} -b & a - k_2 b \\ -c & b - k_2 c \end{bmatrix} = \begin{bmatrix} -1 & 0 \\ 0 & -1 \end{bmatrix}$$

$$\begin{bmatrix} -2b & a - k_2 b - c \\ a - k_2 b - c & 2b - 2k_2 c \end{bmatrix} = \begin{bmatrix} -1 & 0 \\ 0 & -1 \end{bmatrix}.$$

The three simultaneous equations are

$$2b = 1 \tag{1}$$

$$a - k_2 b - c = 0 \tag{2}$$

$$2b - 2k_2 c = -1 \tag{3}$$

Solving these equations,

$$a = \frac{k_2}{2} + \frac{1}{k_2} \qquad\qquad b = \frac{1}{2} \qquad\qquad c = \frac{1}{k_2}$$

Hence $P = \begin{bmatrix} \dfrac{k_2}{2} + \dfrac{1}{k_2} & \dfrac{1}{2} \\[2mm] \dfrac{1}{2} & \dfrac{1}{k_2} \end{bmatrix}$.

The performance index is

$$J = \frac{1}{2}\Big[X^T(0)\, P\, X(0) \Big]$$

$$= \frac{1}{2}[1 \quad 1]\begin{bmatrix} \dfrac{k_2}{2} + \dfrac{1}{k_2} & \dfrac{1}{2} \\[2mm] \dfrac{1}{2} & \dfrac{1}{k_2} \end{bmatrix}\begin{bmatrix} 1 \\ 1 \end{bmatrix} = \frac{k_2}{4} + \frac{1}{k_2} + \frac{1}{2}.$$

For the optimal value of k_2,

$$\frac{\partial J}{\partial k_2} = \frac{1}{4} - \frac{1}{k_2^2} = 0.$$

Hence the optimal value of k_2 is

$$k_2 = 2.$$

The optimal control law is

$$u = -[k_1 \quad k_2]\begin{bmatrix} x_1 \\ x_2 \end{bmatrix} = -[1 \quad 2]\begin{bmatrix} x_1 \\ x_2 \end{bmatrix} = -x_1 - 2x_2.$$

The minimum value of J is $J_{\min} = \dfrac{k_2}{4} + \dfrac{1}{k_2} + \dfrac{1}{2}\Big|_{k_2=2} = 1.5$.

Example 10.2 A linear system is represented by the state equation

$$\dot{X} = \begin{bmatrix} 0 & 1 \\ 0 & 0 \end{bmatrix}X + \begin{bmatrix} 0 \\ 1 \end{bmatrix}u \qquad X(0) = \begin{bmatrix} 1 & 0 \end{bmatrix}.$$

Design an optimal control law $u = -KX$ that minimizes the performance index

$$J = \frac{1}{2}\int_0^\infty (X^T X)\,\mathrm{dt}.$$

The damping ratio of the unit step response of the system is to be 0.5.

Solution:

$$\dot{X} = \begin{bmatrix} 0 & 1 \\ 0 & 0 \end{bmatrix}X + \begin{bmatrix} 0 \\ 1 \end{bmatrix}u$$

$$A = \begin{bmatrix} 0 & 1 \\ 0 & 0 \end{bmatrix} \qquad B = \begin{bmatrix} 0 \\ 1 \end{bmatrix}$$

The performance index is

$$J = \frac{1}{2}\int_0^\infty (X^T X)\,dt = \frac{1}{2}\int_0^\infty (X^T Q X)\,dt;\; Q = \begin{bmatrix} 1 & 0 \\ 0 & 1 \end{bmatrix} \qquad R = 0 .$$

The modified Lyapunov equation is

$$[A - BK]^T P + P[A - BK] + K^T RK = -Q$$

$$[A - BK] = \begin{bmatrix} 0 & 1 \\ 0 & 0 \end{bmatrix} - \begin{bmatrix} 0 \\ 1 \end{bmatrix}[k_1 \quad k_2] = \begin{bmatrix} 0 & 1 \\ -k_1 & -k_2 \end{bmatrix}.$$

With the state feedback, the characteristic equation is

$$\begin{vmatrix} s & -1 \\ k_1 & s + k_2 \end{vmatrix} = 0 \quad \text{Or} \quad s^2 + k_2 s + k_1 = 0 \equiv s^2 + 2\delta\omega_n s + \omega_n^2 .$$

It is given that $\delta = 0.5$. Hence $k_2 = \omega_n$ and $k_1 = \omega_n^2$.

Let $k_2 = k$ so that $k_1 = k^2$.

$$\text{Now } [A - BK] = \begin{bmatrix} 0 & 1 \\ -k^2 & -k \end{bmatrix} \qquad \text{Let } P = \begin{bmatrix} a & b \\ b & c \end{bmatrix}$$

Since R = 0, the modified Lyapunov equation is

$$[A - BK]^T P + P[A - BK] + K^T RK = -Q$$

$$\begin{bmatrix} 0 & -k^2 \\ 1 & -k \end{bmatrix}\begin{bmatrix} a & b \\ b & c \end{bmatrix} + \begin{bmatrix} a & b \\ b & c \end{bmatrix}\begin{bmatrix} 0 & 1 \\ -k^2 & -k \end{bmatrix} = \begin{bmatrix} -1 & 0 \\ 0 & -1 \end{bmatrix}$$

$$\begin{bmatrix} -k^2 b & -k^2 c \\ a - kb & b - kc \end{bmatrix} + \begin{bmatrix} -k^2 b & a - kb \\ -k^2 c & b - kc \end{bmatrix} = \begin{bmatrix} -1 & 0 \\ 0 & -1 \end{bmatrix}$$

$$\begin{bmatrix} -2k^2 b & a - kb - k^2 c \\ a - kb - k^2 c & 2b - 2kc \end{bmatrix} = \begin{bmatrix} -1 & 0 \\ 0 & -1 \end{bmatrix}.$$

The three simultaneous equations are

$$2k^2 b = 1 \tag{1}$$

$$a - kb - k^2 c = 0 \tag{2}$$

$$2b - 2kc = -1 . \tag{3}$$

Solving these equations,

$$a = \frac{1}{k} + \frac{k}{2} \qquad\qquad b = \frac{1}{2k^2} \qquad\qquad c = \frac{1+k^2}{2k^3}.$$

Hence $P = \begin{bmatrix} \dfrac{1}{k} + \dfrac{k}{2} & \dfrac{1}{2k^2} \\ \dfrac{1}{2k^2} & \dfrac{1+k^2}{2k^3} \end{bmatrix}.$

The performance index is

$$J = \frac{1}{2}\left[X^T(0) P\, X(0) \right]$$

$$= \frac{1}{2}[1 \quad 0] \begin{bmatrix} \dfrac{1}{k} + \dfrac{k}{2} & \dfrac{1}{2k^2} \\ \dfrac{1}{2k^2} & \dfrac{1+k^2}{2k^3} \end{bmatrix} \begin{bmatrix} 1 \\ 0 \end{bmatrix} = \frac{1}{2k} + \frac{k}{4}.$$

For the optimal value of k,

$$\frac{\partial J}{\partial k} = -\frac{1}{2k_2^2} + \frac{1}{4} = 0 \qquad\qquad k = \sqrt{2}.$$

Hence the optimal value of k is

$$k = \sqrt{2}.$$

The optimal control law is

$$u = -[2 \quad \sqrt{2}] \begin{bmatrix} x_1 \\ x_2 \end{bmatrix} = -2x_1 - \sqrt{2}x_2.$$

The minimum value of J is $J_{min} = \dfrac{1}{2k} + \dfrac{k}{4}\bigg|_{k=\sqrt{2}} = 0.707$.

Example 10.3 A state equation for a first order system is given as $\dot{x} = -x + u$. The initial state is $x(0) = 1$. Design an optimal feedback control signal $u = -k\,x$ which minimizes the performance measure

$$J = \int_0^\infty (x^2 + u^2)\,dt.$$

Solution: $\dot{x} = -x + u$

The Lyapunov equation is

$$[A - BK]^T P + P[A - BK] + K^T R K = -Q .$$

In this case $A = -1$; $B = 1$; $Q = 1$ and $R = 1$.

Hence $(-1 - K)P + P(-1 - K) + K^2 = -1$

Or $K^2 - 2(1 + K)P + 1 = 0$.

Solving this equation

$$P = \frac{K^2 + 1}{2(1 + K)} .$$

The performance measure

$$J = X^T(0) P X(0) = P = \frac{K^2 + 1}{2(1 + K)} .$$

For minimum J

$$\frac{\partial J}{\partial K} = \frac{1}{2} \left[\frac{(1 + K)2K - (K^2 + 1)}{(1 + K)^2} \right] = 0 .$$

Solving this equation, $K = \sqrt{2} - 1$. The optimal control law is

$$u = -(\sqrt{2} - 1)x$$

$$J_{min} = \frac{K^2 + 1}{2(1 + K)} \bigg|_{K = \sqrt{2} - 1} = \sqrt{2} - 1 = 0.414 .$$

Example 10.4 A system is represented by the state equation

$$\dot{X} = \begin{bmatrix} 0 & 1 \\ 0 & 0 \end{bmatrix} X + \begin{bmatrix} 0 \\ 1 \end{bmatrix} u \qquad X(0) = [1 \quad 0]^T .$$

Using Lyapunov equation, design an optimal control law that minimizes the performance index

$$J = \int_0^\infty (x_1^2 + x_2^2 + u^2) dt .$$

Solution:

$$\dot{X} = \begin{bmatrix} 0 & 1 \\ 0 & 0 \end{bmatrix} X + \begin{bmatrix} 0 \\ 1 \end{bmatrix} u \qquad X(0) = [1 \quad 0]^T$$

$$J = \int_0^\infty (x_1^2 + x_2^2 + u^2)\,dt = \int_0^\infty (X^T Q X + u^T R u)\,dt\,,$$

where $Q = \begin{bmatrix} 1 & 0 \\ 0 & 1 \end{bmatrix}$ and $R = 1$.

The Lyapunov equation for the design of optimal control law is

$$[A - BK]^T P + P[A - BK] + K^T R K = -Q$$

$$[A - BK] = \begin{bmatrix} 0 & 1 \\ 0 & 0 \end{bmatrix} - \begin{bmatrix} 0 \\ 1 \end{bmatrix} \begin{bmatrix} k_1 & k_2 \end{bmatrix} = \begin{bmatrix} 0 & 1 \\ -k_1 & -k_2 \end{bmatrix}.$$

Hence the Lyapunov equation is

$$\begin{bmatrix} 0 & -k_1 \\ 1 & -k_2 \end{bmatrix} \begin{bmatrix} a & b \\ b & c \end{bmatrix} + \begin{bmatrix} a & b \\ b & c \end{bmatrix} \begin{bmatrix} 0 & 1 \\ -k_1 & -k_2 \end{bmatrix} + \begin{bmatrix} k_1 \\ k_2 \end{bmatrix} (1) \begin{bmatrix} k_1 & k_2 \end{bmatrix} = \begin{bmatrix} -1 & 0 \\ 0 & -1 \end{bmatrix}$$

$$\begin{bmatrix} -k_1 b & -k_1 c \\ a - k_2 b & b - k_2 c \end{bmatrix} + \begin{bmatrix} -k_1 b & a - k_2 b \\ -k_1 c & b - k_2 c \end{bmatrix} + \begin{bmatrix} k_1^2 & k_1 k_2 \\ k_1 k_2 & k_2^2 \end{bmatrix} = \begin{bmatrix} -1 & 0 \\ 0 & -1 \end{bmatrix}$$

$$\begin{bmatrix} -2k_1 b + k_1^2 & a - k_2 b - k_1 c + k_1 k_2 \\ a - k_2 b - k_1 c + k_1 k_2 & 2b - 2k_2 c + k_2^2 \end{bmatrix} = \begin{bmatrix} -1 & 0 \\ 0 & -1 \end{bmatrix}.$$

The three equations are

$$-2k_1 b + k_1^2 = -1 \tag{1}$$

$$a - k_2 b - k_1 c + k_1 k_2 = 0 \tag{2}$$

$$2b - 2k_2 c + k_2^2 = -1. \tag{3}$$

Solving these equations

$$a = \frac{k_2}{2k_1} + \frac{(k_1^2 + k_1 + 1)}{2k_2} \qquad b = \frac{k_1^2 + 1}{2k_1} \quad \text{and} \quad c = \frac{k_2}{2} + \frac{(k_1^2 + k_1 + 1)}{2k_1 k_2}.$$

$$P = \begin{bmatrix} a & b \\ b & c \end{bmatrix} = \begin{bmatrix} \dfrac{k_2}{2k_1} + \dfrac{(k_1^2 + k_1 + 1)}{2k_2} & \dfrac{k_1^2 + 1}{2k_1} \\[3mm] \dfrac{k_1^2 + 1}{2k_1} & \dfrac{k_2}{2} + \dfrac{(k_1^2 + k_1 + 1)}{2k_1 k_2} \end{bmatrix}.$$

The performance measure is

$$J = X^T(0) P X(0) = \begin{bmatrix} 1 & 0 \end{bmatrix} \begin{bmatrix} a & b \\ b & c \end{bmatrix} \begin{bmatrix} 1 \\ 0 \end{bmatrix} = a = \frac{k_2}{2k_1} + \frac{(k_1^2 + k_1 + 1)}{2k_2}$$

For minimization of J

$$\frac{\partial J}{\partial k_1} = \frac{\partial}{\partial k_1}\left[\frac{k_2}{2k_1} + \frac{(k_1^2 + k_1 + 1)}{2k_2}\right] = \frac{-k_2}{2k_1^2} + \frac{(2k_1 + 1)}{2k_2} = 0.$$

This gives $\dfrac{k_2}{2k_1^2} = \dfrac{(2k_1 + 1)}{2k_2}$

Or

$$k_2^2 = k_1^2(2k_1 + 1) \qquad (4)$$

$$\frac{\partial J}{\partial k_2} = \frac{\partial}{\partial k_2}\left[\frac{k_2}{2k_1} + \frac{(k_1^2 + k_1 + 1)}{2k_2}\right] = \frac{1}{2k_1} - \frac{(k_1^2 + k_1 + 1)}{2k_2^2} = 0.$$

This gives $\dfrac{1}{2k_1} = \dfrac{(k_1^2 + k_1 + 1)}{2k_2^2}$.

Or

$$k_2^2 = k_1(k_1^2 + k_1 + 1). \qquad (5)$$

Solving equation (4) and (5),

$$k_1 = 1 \text{ and } k_2 = \sqrt{3}.$$

The optimum feedback control law is

$$u = -[1 \qquad \sqrt{3}]X .$$

Minimum value of J is

$$_{\min} = \frac{k_2}{2k_1} + \frac{(k_1 + k_1 + 1)}{2k_2}\Bigg|_{k_1=1,\ k_2=\sqrt{3}} = \frac{\sqrt{3}}{2} + \frac{3}{2\sqrt{3}} = \sqrt{} .$$

Example 10.5 Repeat Example 10.4 using Riccati equation.

Solution:

$$\dot{X} = \begin{bmatrix} 0 & 1 \\ 0 & 0 \end{bmatrix} X + \begin{bmatrix} 0 \\ 1 \end{bmatrix} u \qquad X(0) = [1 \qquad 0]^T$$

$$J = \int_0^\infty (x_1^2 + x_2^2 + u^2)\,dt$$

$$x_1^2 + x_2^2 = [x_1 \quad x_2]\begin{bmatrix} 1 & 0 \\ 0 & 1 \end{bmatrix}\begin{bmatrix} x_1 \\ x_2 \end{bmatrix} \qquad Q = \begin{bmatrix} 1 & 0 \\ 0 & 1 \end{bmatrix} \text{ and } R = 1 .$$

The Riccati equation is

$$A^T P + PA - PBR^{-1}B^T P = -Q .$$

$$A^T P + PA = \begin{bmatrix} 0 & 0 \\ 1 & 0 \end{bmatrix} \begin{bmatrix} a & b \\ b & c \end{bmatrix} + \begin{bmatrix} a & b \\ b & c \end{bmatrix} \begin{bmatrix} 0 & 1 \\ 0 & 0 \end{bmatrix} = \begin{bmatrix} 0 & 0 \\ a & b \end{bmatrix} + \begin{bmatrix} 0 & a \\ 0 & b \end{bmatrix} = \begin{bmatrix} 0 & a \\ a & 2b \end{bmatrix}$$

$$PBR^{-1}B^T P = \begin{bmatrix} a & b \\ b & c \end{bmatrix} \begin{bmatrix} 0 \\ 1 \end{bmatrix} (1) [0 \quad 1] \begin{bmatrix} a & b \\ b & c \end{bmatrix} = \begin{bmatrix} b \\ c \end{bmatrix} [b \quad c] = \begin{bmatrix} b^2 & bc \\ bc & c^2 \end{bmatrix}.$$

The Riccati equation is

$$\begin{bmatrix} 0 & a \\ a & 2b \end{bmatrix} - \begin{bmatrix} b^2 & bc \\ bc & c^2 \end{bmatrix} = \begin{bmatrix} -1 & 0 \\ 0 & -1 \end{bmatrix}$$

$$\begin{bmatrix} -b^2 & a-bc \\ a-bc & 2b-c^2 \end{bmatrix} = \begin{bmatrix} -1 & 0 \\ 0 & -1 \end{bmatrix}.$$

The equations are

$$b^2 = 1 \tag{1}$$

$$a - bc = 0 \tag{2}$$

$$2b - c^2 = -1. \tag{3}$$

Solving these equations,

$$a = \sqrt{3} \qquad b = 1 \qquad c = \sqrt{3}.$$

Hence $P = \begin{bmatrix} \sqrt{3} & 1 \\ 1 & \sqrt{3} \end{bmatrix}.$

$$K = R^{-1}B^T P = (1)[0 \quad 1] \begin{bmatrix} \sqrt{3} & 1 \\ 1 & \sqrt{3} \end{bmatrix} = [1 \quad \sqrt{3}].$$

The optimum feedback control law is

$$u = -[1 \quad \sqrt{3}]X$$

$$J_{min} = X^T(0) \, P X(0) = [1 \quad 0] \begin{bmatrix} \sqrt{3} & 1 \\ 1 & \sqrt{3} \end{bmatrix} \begin{bmatrix} 1 \\ 0 \end{bmatrix} = \sqrt{3}.$$

Example 10.6 A linear system is represented by the state equation

$$\dot{X} = \begin{bmatrix} 0 & 1 \\ 0 & -2 \end{bmatrix} X + \begin{bmatrix} 0 \\ 1 \end{bmatrix} u \quad X(0) = [1 \quad 0]^T.$$

Using Riccati equation, design an optimal control law that minimizes the performance index

$$J = \int_0^\infty (2x_1^2 + 2u^2)\, dt .$$

Solution:

$$\dot{X} = \begin{bmatrix} 0 & 1 \\ 0 & -2 \end{bmatrix} X + \begin{bmatrix} 0 \\ 1 \end{bmatrix} u \quad X(0) = [1 \quad 0]^T$$

$$J = \int_0^\infty (2x_1^2 + 2u^2)\, dt$$

$$2x_1^2 = [x_1 \quad x_2] \begin{bmatrix} 2 & 0 \\ 0 & 0 \end{bmatrix} \begin{bmatrix} x_1 \\ x_2 \end{bmatrix} \qquad\qquad Q = \begin{bmatrix} 2 & 0 \\ 0 & 0 \end{bmatrix} \text{ and } R = 2$$

The Riccati equation is

$$A^T P + PA - PBR^{-1}B^T P = -Q$$

$$A^T P + PA = \begin{bmatrix} 0 & 0 \\ 1 & -2 \end{bmatrix} \begin{bmatrix} a & b \\ b & c \end{bmatrix} + \begin{bmatrix} a & b \\ b & c \end{bmatrix} \begin{bmatrix} 0 & 1 \\ 0 & -2 \end{bmatrix}$$

$$= \begin{bmatrix} 0 & 0 \\ a-2b & b-2c \end{bmatrix} + \begin{bmatrix} 0 & a-2b \\ 0 & b-2c \end{bmatrix} = \begin{bmatrix} 0 & a-2b \\ a-2b & 2b-4c \end{bmatrix}$$

$$PBR^{-1}B^T P = \begin{bmatrix} a & b \\ b & c \end{bmatrix} \begin{bmatrix} 0 \\ 1 \end{bmatrix} \left(\frac{1}{2}\right) [0 \quad 1] \begin{bmatrix} a & b \\ b & c \end{bmatrix} = \frac{1}{2} \begin{bmatrix} b \\ c \end{bmatrix} [b \quad c] = \frac{1}{2} \begin{bmatrix} b^2 & bc \\ bc & c^2 \end{bmatrix}$$

The Riccati equation is

$$\begin{bmatrix} 0 & a-2b \\ a-2b & 2b-4c \end{bmatrix} - \frac{1}{2} \begin{bmatrix} b^2 & bc \\ bc & c^2 \end{bmatrix} = \begin{bmatrix} -2 & 0 \\ 0 & 0 \end{bmatrix}$$

The equations are

$$\frac{1}{2}b^2 = 2 \tag{1}$$

$$a - 2b - \frac{1}{2}bc = 0 \tag{2}$$

$$2b - 4c - \frac{1}{2}c^2 = 0 . \tag{3}$$

Solving these equations

$$a = 4.9 \quad b = 2 \qquad\qquad c = 0.9 .$$

Hence $P = \begin{bmatrix} 4.9 & 2 \\ 2 & 0.9 \end{bmatrix}$.

$$K = R^{-1}B^T P = \left(\frac{1}{2}\right)[0 \quad 1]\begin{bmatrix} 4.9 & 2 \\ 2 & 0.9 \end{bmatrix} = [1 \quad 0.45].$$

The optimum feedback control law is

$$u = -[1 \quad 0.45]X$$

$$J_{min} = X^T(0)\,PX(0) = [1 \quad 0]\begin{bmatrix} 4.9 & 2 \\ 2 & 0.9 \end{bmatrix}\begin{bmatrix} 1 \\ 0 \end{bmatrix} = 4.9.$$

Example 10.7 A state model for a linear system is

$$\dot{X} = \begin{bmatrix} 0 & 1 \\ 0 & -1 \end{bmatrix}X + \begin{bmatrix} 1 \\ 1 \end{bmatrix}u \quad X(0) = [1 \quad 0]^T$$

Using Riccati equation, design an optimal control law that minimizes the performance index

$$J = \int_0^\infty (x_1^2 + x_2^2 + u^2)\,dt.$$

Solution:

$$\dot{X} = \begin{bmatrix} 0 & 1 \\ 0 & -1 \end{bmatrix}X + \begin{bmatrix} 1 \\ 1 \end{bmatrix}u \quad X(0) = [1 \quad 0]^T$$

$$J = \int_0^\infty (x_1^2 + x_2^2 + u^2)\,dt \qquad Q = \begin{bmatrix} 1 & 0 \\ 0 & 1 \end{bmatrix} \qquad R = 1$$

The Riccati equation is

$$A^T P + PA - PBR^{-1}B^T P = -Q$$

$$A^T P + PA = \begin{bmatrix} 0 & 0 \\ 1 & -1 \end{bmatrix}\begin{bmatrix} a & b \\ b & c \end{bmatrix} + \begin{bmatrix} a & b \\ b & c \end{bmatrix}\begin{bmatrix} 0 & 1 \\ 0 & -1 \end{bmatrix}$$

$$= \begin{bmatrix} 0 & 0 \\ a-b & b-c \end{bmatrix} + \begin{bmatrix} 0 & a-b \\ 0 & b-c \end{bmatrix} = \begin{bmatrix} 0 & a-b \\ a-b & 2b-2c \end{bmatrix}.$$

$$PBR^{-1}B^T P = \begin{bmatrix} a & b \\ b & c \end{bmatrix}\begin{bmatrix} 1 \\ 1 \end{bmatrix}(1)[1 \quad 1]\begin{bmatrix} a & b \\ b & c \end{bmatrix}$$

$$= \begin{bmatrix} a+b \\ b+c \end{bmatrix}[a+b \quad b+c] = \begin{bmatrix} (a+b)^2 & (a+b)(b+c) \\ (a+b)(b+c) & (b+c)^2 \end{bmatrix}.$$

The Riccati equation is

$$\begin{bmatrix} -(a+b)^2 & a-b-(a+b)(b+c) \\ a-b-(a+b)(b+c) & 2b-2c-(b+c)^2 \end{bmatrix} = \begin{bmatrix} -1 & 0 \\ 0 & -1 \end{bmatrix}.$$

The three equations are

$$(a+b)^2 = 1 \tag{1}$$

$$a-b-(a+b)(b+c) = 0 \tag{2}$$

$$2b-2c-(b+c)^2 = -1. \tag{3}$$

From equation (1)

$$(a+b)=1 \qquad a=1-b. \tag{4}$$

From equation (2)

$$a-b-(b+c) = 0 \text{ Or } c = a - 2b$$

Using equation (4), $c = (1-b) - 2b = 1 - 3b$ \hfill (5)

$$b+c = 1-2b. \tag{6}$$

Substituting equations (5) and (6) in equation (3),

$$2b - 2(1-3b) - (1-2b)^2 = -1$$

$$8b - 2 - (1 - 4b + 4b^2) = -1.$$

Or $$b^2 - 3b + 0.5 = 0. \tag{7}$$

Solving this equation, $b = 2.823$ or $b = 0.177$.

From equation (4), $a = 1 - b = -1.823$ or $a = 0.823$.

Hence $a = 0.823$ and $b = 0.177$.

From equation (5), $c = 1 - 3b = 1 - 3(0.177) = 0.469$.

$$P = \begin{bmatrix} a & b \\ b & c \end{bmatrix} = \begin{bmatrix} 0.823 & 0.177 \\ 0.177 & 0.469 \end{bmatrix}$$

$$K = R^{-1} B^T P = (1)[1 \quad 1] \begin{bmatrix} 0.823 & 0.177 \\ 0.177 & 0.469 \end{bmatrix} = [1 \quad 0.646]$$

$$J_{min} = X^T(0) P X(0) = [1 \quad 0] \begin{bmatrix} 0:823 & 0.177 \\ 0.177 & 0.469 \end{bmatrix} \begin{bmatrix} 1 \\ 0 \end{bmatrix} = 0.823$$

Example 10.8 A system is represented by a state model

$$\dot{X} = \begin{bmatrix} 0 & 1 \\ -1 & -2 \end{bmatrix} X + \begin{bmatrix} 0 \\ 1 \end{bmatrix} u \quad X(0) = [1 \quad -1]^T.$$

Design an optimal feedback control law that minimizes the performance measure

$$J = \frac{1}{2} \int_0^\infty (2x_1^2 + 2x_2^2 + u^2) \, dt.$$

Solution:

$$\dot{X} = \begin{bmatrix} 0 & 1 \\ -1 & -2 \end{bmatrix} X + \begin{bmatrix} 0 \\ 1 \end{bmatrix} u \quad X(0) = [1 \quad -1]^T$$

$$J = \frac{1}{2} \int_0^\infty (2x_1^2 + 2x_2^2 + u^2) \, dt \qquad Q = \begin{bmatrix} 2 & 0 \\ 0 & 2 \end{bmatrix} \qquad R = 1.$$

The Riccati equation is

$$A^T P + PA - PBR^{-1}B^T P = -Q$$

$$A^T P + PA = \begin{bmatrix} 0 & -1 \\ 1 & -2 \end{bmatrix} \begin{bmatrix} a & b \\ b & c \end{bmatrix} + \begin{bmatrix} a & b \\ b & c \end{bmatrix} \begin{bmatrix} 0 & 1 \\ -1 & -2 \end{bmatrix}$$

$$= \begin{bmatrix} -b & -c \\ a-2b & b-2c \end{bmatrix} + \begin{bmatrix} -b & a-2b \\ -c & b-2c \end{bmatrix} = \begin{bmatrix} -2b & a-2b-c \\ a-2b-c & 2b-4c \end{bmatrix}$$

$$PBR^{-1}B^T P = \begin{bmatrix} a & b \\ b & c \end{bmatrix} \begin{bmatrix} 0 \\ 1 \end{bmatrix} (1)[0 \quad 1] \begin{bmatrix} a & b \\ b & c \end{bmatrix} = \begin{bmatrix} b \\ c \end{bmatrix} [b \quad c] = \begin{bmatrix} b^2 & bc \\ bc & c^2 \end{bmatrix}.$$

Hence the Riccati equation is

$$\begin{bmatrix} -2b & a-2b-c \\ a-2b-c & 2b-4c \end{bmatrix} - \begin{bmatrix} b^2 & bc \\ bc & c^2 \end{bmatrix} = \begin{bmatrix} -2 & 0 \\ 0 & -2 \end{bmatrix}.$$

The three equations are

$$-2b - b^2 = -2 \tag{1}$$

$$a - 2b - c - bc = 0 \tag{2}$$

$$2b - 4c - c^2 = -2. \tag{3}$$

Solving these equations, equation (1) can be written as

$$b^2 + 2b - 2 = 0 \qquad\qquad b = 0.732.$$

Substituting the value of b in equation (3),

$$c^2 + 4c - 3.464 = 0 \qquad c = 0.732.$$

Substituting the values of b and c in equation (2),

$$a = 2b + c + bc = 2.732$$

$$P = \begin{bmatrix} a & b \\ b & c \end{bmatrix} = \begin{bmatrix} 2.732 & 0.732 \\ 0.732 & 0.732 \end{bmatrix}$$

$$K = R^{-1} B^T P = (1)[0 \quad 1] \begin{bmatrix} 2.732 & 0.732 \\ 0.732 & 0.732 \end{bmatrix} = [0.732 \quad 0.732].$$

The optimal control law is

$$u = -0.732 x_1 - 0.732 x_2$$

$$J_{min} = \frac{1}{2} X^T(0) P X(0) = \frac{1}{2} [1 \quad -1] \begin{bmatrix} 2.732 & 0.732 \\ 0.732 & 0.732 \end{bmatrix} \begin{bmatrix} 1 \\ -1 \end{bmatrix} = 1.$$

Example 10.9 A system is represented by a state model

$$\dot{X} = \begin{bmatrix} 0 & 1 \\ -1 & -2 \end{bmatrix} X + \begin{bmatrix} 0 \\ 1 \end{bmatrix} u \quad X(0) = [1 \quad -1]^T \qquad y = [2 \quad 0] X.$$

Design an optimal feedback control law that minimizes the performance measure

$$J = \frac{1}{2} \int_0^\infty (Y^T Y + U^T U) dt$$

Solution:

$$\dot{X} = \begin{bmatrix} 0 & 1 \\ -1 & -2 \end{bmatrix} X + \begin{bmatrix} 0 \\ 1 \end{bmatrix} u \qquad X(0) = [1 \quad -1]^T \qquad y = [2 \quad 0] X$$

$$J = \frac{1}{2} \int_0^\infty (Y^T Y + U^T U) dt$$

$$CA = [2 \quad 0] \begin{bmatrix} 0 & 1 \\ -1 & -2 \end{bmatrix} = [0 \quad 2]$$

The observability matrix $\begin{bmatrix} C \\ CA \end{bmatrix} = \begin{bmatrix} 2 & 0 \\ 0 & 2 \end{bmatrix}$ has rank 2. Hence the system is completely observable.

$$y = CX = [2 \quad 0] \begin{bmatrix} x_1 \\ x_2 \end{bmatrix}$$

$$Y^T Y = (CX)^T (CX) = X^T C^T CX = X^T QX .$$

Hence the performance measure is

$$J = \frac{1}{2} \int_0^\infty (X^T QX + U^T RU) dt ,$$

where $Q = C^T C = \begin{bmatrix} 2 \\ 0 \end{bmatrix} [2 \quad 0] = \begin{bmatrix} 4 & 0 \\ 0 & 0 \end{bmatrix}$ $R = 1$

The Riccati equation is

$$A^T P + PA - PBR^{-1}B^T P = -Q$$

$$A^T P + PA = \begin{bmatrix} 0 & -1 \\ 1 & -2 \end{bmatrix} \begin{bmatrix} a & b \\ b & c \end{bmatrix} + \begin{bmatrix} a & b \\ b & c \end{bmatrix} \begin{bmatrix} 0 & 1 \\ -1 & -2 \end{bmatrix}$$

$$= \begin{bmatrix} -b & -c \\ a-2b & b-2c \end{bmatrix} + \begin{bmatrix} -b & a-2b \\ -c & b-2c \end{bmatrix} = \begin{bmatrix} -2b & a-2b-c \\ a-2b-c & 2b-4c \end{bmatrix}$$

$$PBR^{-1}B^T P = \begin{bmatrix} a & b \\ b & c \end{bmatrix} \begin{bmatrix} 0 \\ 1 \end{bmatrix} (1)[0 \quad 1] \begin{bmatrix} a & b \\ b & c \end{bmatrix} = \begin{bmatrix} b \\ c \end{bmatrix} [b \quad c] = \begin{bmatrix} b^2 & bc \\ bc & c^2 \end{bmatrix}.$$

Hence the Riccati equation is

$$\begin{bmatrix} -2b & a-2b-c \\ a-2b-c & 2b-4c \end{bmatrix} - \begin{bmatrix} b^2 & bc \\ bc & c^2 \end{bmatrix} = \begin{bmatrix} -4 & 0 \\ 0 & 0 \end{bmatrix}.$$

The three equations are

$$-2b - b^2 = -4 \tag{1}$$

$$a - 2b - c - bc = 0 \tag{2}$$

$$2b - 4c - c^2 = 0 . \tag{3}$$

From equation (1),

$$b^2 + 2b - 4 = 0 .$$

Solving this equation, $b = \sqrt{5} - 1 = 1.236$

From equation (3),

$$c + 4c - 2.472 = 0 .$$

Solving this equation, $c = -2 + \sqrt{6.472} = 0.544 .$

Substituting the values of b and c in equation (2),

$$a = 2b + c + bc = 3.688 .$$

Hence $P = \begin{bmatrix} 3.688 & 1.236 \\ 1.236 & 0.544 \end{bmatrix}$.

$$K = R^{-1}B^T P = (1)[0 \quad 1] \begin{bmatrix} 3.688 & 1.236 \\ 1.236 & 0.544 \end{bmatrix} = [1.236 \quad 0.544]$$

$$u = -1.236x_1 - 0.544x_2$$

$$J_{min} = \frac{1}{2}X^T(0) P X(0) = \frac{1}{2}[1 \quad -1] \begin{bmatrix} 3.688 & 1.236 \\ 1.236 & 0.544 \end{bmatrix} \begin{bmatrix} 1 \\ -1 \end{bmatrix} = 0.88 .$$

Example 10.10 A state model for a linear system is

$$\dot{X} = \begin{bmatrix} 0 & 1 \\ 0 & -1 \end{bmatrix}X + \begin{bmatrix} 0 \\ 1 \end{bmatrix}u \qquad X(0) = [1 \quad -1]^T \qquad Y = \begin{bmatrix} 1 & 1 \\ 1 & -1 \end{bmatrix}X$$

Design an optimal feedback control law that minimizes the performance measure

$$J = \frac{1}{2}\int_0^\infty (Y^T Y + U^T U)\,dt .$$

Solution:

$$\dot{X} = \begin{bmatrix} 0 & 1 \\ 0 & -1 \end{bmatrix}X + \begin{bmatrix} 0 \\ 1 \end{bmatrix}u \qquad X(0) = [1 \quad -1]^T \qquad Y = \begin{bmatrix} 1 & 1 \\ 1 & -1 \end{bmatrix}X$$

$$J = \frac{1}{2}\int_0^\infty (Y^T Y + U^T U)\,dt \quad R = 1$$

$$CA = \begin{bmatrix} 1 & 1 \\ 1 & -1 \end{bmatrix}\begin{bmatrix} 0 & 1 \\ 0 & -1 \end{bmatrix} = \begin{bmatrix} 0 & 0 \\ 0 & 2 \end{bmatrix}.$$

The observability matrix is

$$\begin{bmatrix} C \\ CA \end{bmatrix} = \begin{bmatrix} 1 & 1 \\ 1 & -1 \\ 0 & 0 \\ 0 & 2 \end{bmatrix}.$$

It has rank 2. Hence the system is completely observable.

$$Y^T Y = (CX)^T (CX) = X^T C^T CX = X^T Q X .$$

Hence $Q = C^T C = \begin{bmatrix} 1 & 1 \\ 1 & -1 \end{bmatrix} \begin{bmatrix} 1 & 1 \\ 1 & -1 \end{bmatrix} = \begin{bmatrix} 2 & 0 \\ 0 & 2 \end{bmatrix}$.

The Riccati equation is

$$A^T P + PA - PBR^{-1}B^T P = -Q$$

$$A^T P + PA = \begin{bmatrix} 0 & 0 \\ 1 & -1 \end{bmatrix} \begin{bmatrix} a & b \\ b & c \end{bmatrix} + \begin{bmatrix} a & b \\ b & c \end{bmatrix} \begin{bmatrix} 0 & 1 \\ 0 & -1 \end{bmatrix}$$

$$= \begin{bmatrix} 0 & 0 \\ a-b & b-c \end{bmatrix} + \begin{bmatrix} 0 & a-b \\ 0 & b-c \end{bmatrix} = \begin{bmatrix} 0 & a-b \\ a-b & 2b-2c \end{bmatrix}.$$

$$PBR^{-1}B^T P = \begin{bmatrix} a & b \\ b & c \end{bmatrix} \begin{bmatrix} 0 \\ 1 \end{bmatrix} (1)[0 \quad 1] \begin{bmatrix} a & b \\ b & c \end{bmatrix} = \begin{bmatrix} b \\ c \end{bmatrix} [b \quad c] = \begin{bmatrix} b^2 & bc \\ bc & c^2 \end{bmatrix}.$$

Hence $\begin{bmatrix} 0 & a-b \\ a-b & 2b-2c \end{bmatrix} - \begin{bmatrix} b^2 & bc \\ bc & c^2 \end{bmatrix} = \begin{bmatrix} -2 & 0 \\ 0 & -2 \end{bmatrix}$.

The three equations are

$$b^2 = 2 \tag{1}$$

$$a - b - bc = 0 \tag{2}$$

$$2b - 2c - c^2 = -2. \tag{3}$$

Solving these equations

$$a = 3.414 \qquad b = 1.414 \qquad c = 1.414.$$

Hence $P = \begin{bmatrix} 3.414 & 1.414 \\ 1.414 & 1.414 \end{bmatrix}$.

$$K = R^{-1}B^T P = (1)[0 \quad 1] \begin{bmatrix} 3.414 & 1.414 \\ 1.414 & 1.414 \end{bmatrix} = [1.414 \quad 1.414].$$

The optimal control law is

$$u = -1.414x_1 - 1.414x_2$$

$$J_{min} = \frac{1}{2} X^T(0) P X(0) = \frac{1}{2}[1 \quad -1] \begin{bmatrix} 3.414 & 1.414 \\ 1.414 & 1.414 \end{bmatrix} \begin{bmatrix} 1 \\ -1 \end{bmatrix} = 1.$$

SUMMARY

- Optimal design of a control system is based on minimization or maximization of a performance index, subject to certain constraints.
- An optimal control system is one in which the design is carried out by determining an optimal control signal that optimizes a performance index subject to certain constraints.
- In closed loop optimal control, the current values of the state variables in the process are fed back to the controller which produces the necessary optimal control signal.
- In most optimal control systems, the performance index is a quadratic function of state variables and inputs or outputs and such a control is called quadratic optimal control.
- There can be different performance measures like minimum time, minimum energy, minimum integral square error in state variables and outputs etc.
- An optimal state regulator can be designed using the second method of Lyapunov.
- A more convenient and easier method is to use Riccati equation for the design of optimal control law.

PRACTICE PROBLEMS

PP 10.1 A linear system is represented by the state equation

$$\dot{X} = \begin{bmatrix} 0 & 1 \\ 0 & 0 \end{bmatrix} X + \begin{bmatrix} 0 \\ 1 \end{bmatrix} u .$$

Given that the initial state $X(0) = \begin{bmatrix} 1 & 1 \end{bmatrix}^T$, design a control law $u = -KX$ that minimizes the performance index

$$J = \frac{1}{2} \int_0^\infty (X^T X) dt .$$

The undamped natural frequency of the system is to be $\sqrt{2}$ rad/sec.

PP 10.2 A linear time-invariant system is represented by the state equation

$$\dot{X} = \begin{bmatrix} 0 & 1 \\ 0 & 0 \end{bmatrix} X + \begin{bmatrix} 0 \\ 1 \end{bmatrix} u .$$

Given that the initial state $X(0) = \begin{bmatrix} 1 & 0 \end{bmatrix}^T$, design a control law $u = -KX$ that minimizes the performance index

$$J = \frac{1}{2} \int_0^\infty (2x_1^2 + 2x_2^2)) dt .$$

The damping ratio of the response of the system is to be 0.5.

PP 10.3 A first-order system is represented by the differential equation

$$\dot{y} = -2y + u \qquad y(0) = 1 .$$

Determine an optimal control law that minimizes the performance index

$$J = \int_0^\infty (y^2 + u^2) \, dt$$

PP 10.4 A state equation for a linear system is

$$\dot{X} = \begin{bmatrix} 0 & 0 \\ 0 & -1 \end{bmatrix} X + \begin{bmatrix} 1 \\ 1 \end{bmatrix} u \qquad X(0) = [1 \qquad 0]^T .$$

Design a control law $u = -KX$ that minimizes the performance index

$$J = \frac{1}{2} \int_0^\infty (x_1^2 + x_2^2 + u^2) \, dt .$$

PP 10.5 For a system represented by

$$\dot{X} = \begin{bmatrix} 0 & 0 \\ 0 & 1 \end{bmatrix} X + \begin{bmatrix} 1 \\ 1 \end{bmatrix} u \qquad\qquad X(0) = [1 \qquad 0]^T .$$

Design an optimal control law $u = -KX$ that minimizes

$$J = \frac{1}{2} \int_0^\infty (x_1^2 + u^2) \, dt .$$

PP 10.6 A system is represented by the state equation

$$\dot{X} = \begin{bmatrix} 0 & 0 \\ 0 & 1 \end{bmatrix} X + \begin{bmatrix} 1 \\ 1 \end{bmatrix} u .$$

Design an optimal control law that minimizes the performance index

$$J = \frac{1}{2} \int_0^\infty (2x_1^2 + x_2^2 + 2u^2) \, dt .$$

PP 10.7 A linear system is represented by a state equation

$$\dot{X} = \begin{bmatrix} 0 & 1 \\ -1 & -2 \end{bmatrix} X + \begin{bmatrix} 0 \\ 1 \end{bmatrix} u \qquad X(0) = [1 \qquad 0]^T .$$

Design an optimal control law that minimizes the performance measure

$$J = \int_0^\infty (0.5x_1^2 + x_2^2 + u^2) \, dt .$$

PP 10.8 A state equation that represents a linear system is

$$\dot{X} = \begin{bmatrix} 0 & 1 \\ -1 & -2 \end{bmatrix} X + \begin{bmatrix} 0 \\ 1 \end{bmatrix} u \qquad X(0) = [1 \quad -1]^T .$$

Design an optimal control law that minimizes the performance index

$$J = \frac{1}{2} \int_0^\infty (2x_1^2 + 2x_2^2 + 2u^2) dt .$$

PP 10.9 For a plant represented by a state model

$$\dot{X} = \begin{bmatrix} 0 & 1 \\ -1 & -2 \end{bmatrix} X + \begin{bmatrix} 0 \\ 1 \end{bmatrix} u \qquad y = [1 \quad 0]X \qquad X(0) = [1 \quad 0]^T ,$$

design an optimal control law that minimizes the performance measure

$$J = \frac{1}{2} \int_0^\infty (Y^T Y + U^T U) dt .$$

PP 10.10 For a system represented by a state model

$$\dot{X} = \begin{bmatrix} 0 & 1 \\ 0 & 0 \end{bmatrix} X + \begin{bmatrix} 0 \\ 1 \end{bmatrix} u \qquad X(0) = [1 \quad -1]^T ,$$

design, using Lyapunov equation, an optimal control law that minimizes the performance measure

$$J = \int_0^\infty (4x_1^2 + u^2) dt .$$

PP 10.11 Repeat PP10.10 using Riccati equation.

PP 10. 12 For a linear system represented by a state equation

$$\dot{X} = \begin{bmatrix} -1 & 0 \\ 1 & 0 \end{bmatrix} X + \begin{bmatrix} 1 \\ 0 \end{bmatrix} u ,$$

design an optimal control law that minimizes the performance index

$$J = \int_0^\infty (x_1^2 + x_2^2 + u^2) dt .$$

PP 10.13 A system is represented by a state model

$$\dot{X} = \begin{bmatrix} 0 & 1 \\ 0 & 0 \end{bmatrix} X + \begin{bmatrix} 1 \\ 1 \end{bmatrix} u \qquad y = \begin{bmatrix} 1 & 0 \\ 0 & 2 \end{bmatrix} X \qquad X(0) = [1 \quad 0]^T .$$

Design an optimal control law that minimizes the performance measure

$$J = \int_0^\infty (y_1^2 + y_2^2 + u^2)\, dt .$$

PP 10.14 A system is represented by a state model

$$\dot{X} = \begin{bmatrix} 0 & 1 \\ 0 & -1 \end{bmatrix} X + \begin{bmatrix} 0 \\ 1 \end{bmatrix} u \qquad y = \begin{bmatrix} 1 & 1 \\ 1 & -1 \end{bmatrix} X \qquad X(0) = [1 \qquad 0]^T$$

Design an optimal control law that minimizes the performance measure

$$J = \int_0^\infty (y_1^2 + 2y_2^2 + u^2)\, dt$$

REVIEW QUESTIONS

10.1 Explain what is meant by optimal control system.

10.2 What is meant by quadratic optimal control? Explain.

10.3 Explain the concept of optimal control, with a specific example.

10.4 Write down some performance measures which are considered in optimal controller design.

10.5 A system is represented by a state model

$$\dot{X} = AX + BU \qquad\qquad y = CX .$$

What are the conditions that are to be satisfied for the design of an optimal control law?

10.6 Establish the relation between a quadratic performance measure and a Lyapunov function.

10.7 A system is represented by a state model $\dot{X} = AX + BU$. Obtain the design equation, using the second method of Lyapunov, for determining a state feedback control law that minimizes the performance measure

$$J = \int_0^\infty (X^T QX + U^T RU)\, dt .$$

10.8 Explain the steps in designing a feedback optimal control law, using the second method of Lyapunov.

10.9 A system is represented by a state model $\dot{X} = AX + BU$. Obtain the Riccati equation and explain the design steps in determining a state feedback control law that minimizes the performance measure

$$J = \int_0^\infty (X^T QX + U^T RU)\, dt .$$

10.10 Explain the steps in designing a feedback optimal control law using the Riccati equation.

Answers

Chapter 1

1.1
$$\begin{bmatrix} \dfrac{dv_c}{dt} \\[2mm] \dfrac{di_L}{dt} \end{bmatrix} = \begin{bmatrix} -\dfrac{1}{R_1 C} & -\dfrac{1}{L} \\[2mm] \dfrac{1}{L} & -\dfrac{R_2}{L} \end{bmatrix} \begin{bmatrix} v_c \\ i_L \end{bmatrix} + \begin{bmatrix} \dfrac{1}{R_1 C} \\ 0 \end{bmatrix} v_S \qquad v_0 = \begin{bmatrix} 1 & 0 \end{bmatrix} \begin{bmatrix} v_c \\ i_L \end{bmatrix}$$

1.2 $x_1 = i_L; x_2 = \dot{x}_1 = \dfrac{di_L}{dt}$ $\quad A = \begin{bmatrix} 0 & 1 \\[2mm] -\dfrac{(R_1 + R_2)}{R_1 LC} & -\dfrac{(L + R_1 R_2 C)}{R_1 LC} \end{bmatrix}$ $\quad B = \begin{bmatrix} 0 \\[2mm] \dfrac{1}{R_1 LC} \end{bmatrix}$ $\quad C = [R_2 \quad L]$

1.3 $x_1 = i_L; x_2 = v_{C1}; x_3 = v_{C2}$ $\quad A = \begin{bmatrix} -2 & -1 & 0 \\ 2 & -0.5 & -0.5 \\ 0 & 0.5 & -0.5 \end{bmatrix}$ $\quad B = \begin{bmatrix} 1 \\ 0 \\ 0 \end{bmatrix}$ $\quad C = [0 \quad 0 \quad 1]$

1.4 $x_1 = v_{C2}; x_2 = \dot{x}_1 = \dot{v}_{C2}; x_3 = \dot{x}_2 = \ddot{v}_{C2}$ $\quad A = \begin{bmatrix} 0 & 1 & 0 \\ 0 & 0 & 1 \\ -1 & -5 & -3 \end{bmatrix}$ $\quad B = \begin{bmatrix} 0 \\ 0 \\ 1 \end{bmatrix}$ $\quad C = [1 \quad 0 \quad 0]$

1.5 $x_1 = i_L; x_2 = v_C$ $\quad A = \begin{bmatrix} -1 & -1 \\ 3 & -1.5 \end{bmatrix}$ $\quad B = \begin{bmatrix} 1 \\ 0 \end{bmatrix}$ $\quad C = [0 \quad 1]$

1.6 $x_1 = v_C; x_2 = \dot{x}_1 = \dot{v}_C$ $\quad A = \begin{bmatrix} 0 & 1 \\ -4.5 & -2.5 \end{bmatrix}$ $\quad B = \begin{bmatrix} 0 \\ 3 \end{bmatrix}$ $\quad C = [1 \quad 0]$

1.7 $x_1 = v_C; x_2 = i_L$ $A = \begin{bmatrix} -1 & -4 \\ 2 & -8 \end{bmatrix}$ $B = \begin{bmatrix} 4 \\ 0 \end{bmatrix}$ $C = \begin{bmatrix} 0 & 4 \end{bmatrix}$

1.8 $x_1 = i_L; x_2 = \dot{x}_1 = \dfrac{di_L}{dt}$ $A = \begin{bmatrix} 0 & 1 \\ -16 & -9 \end{bmatrix}$ $B = \begin{bmatrix} 0 \\ 9 \end{bmatrix}$ $C = \begin{bmatrix} 4 & 0 \end{bmatrix}$

1.9 $x_1 = \omega; x_2 = \dot{\omega}; \; A = \begin{bmatrix} 0 & 1 \\ -\dfrac{R_f B}{L_f J} & -(\dfrac{R_f}{L_f} + \dfrac{B}{J}) \end{bmatrix}$ $B = \begin{bmatrix} 0 \\ \dfrac{k_t}{L_f J} \end{bmatrix}$ $C = \begin{bmatrix} 0 & 1 \end{bmatrix}$

1.10 $\dfrac{d}{dt}\begin{bmatrix} \omega \\ \dot{\omega} \end{bmatrix} = \begin{bmatrix} 0 & 1 \\ -\dfrac{R_a B}{L_a J} & -(\dfrac{R_a}{L_a} + \dfrac{B}{J}) \end{bmatrix}\begin{bmatrix} \omega \\ \dot{\omega} \end{bmatrix} + \begin{bmatrix} \dfrac{k_t}{L_a J} & 0 \\ 0 & -\dfrac{R_a}{L_a J} \end{bmatrix}\begin{bmatrix} v_a \\ T_L \end{bmatrix}$ $y = \begin{bmatrix} 1 & 0 \end{bmatrix}$

1.11 $x_1 = y_1 \; ; \; x_2 = \dot{y}_1 = \dot{x}_1 \; ; \; x_3 = y_2 \; ; \; x_4 = \dot{y}_2 = \dot{x}_3$

$A = \begin{bmatrix} 0 & 1 & 0 & 0 \\ -(\dfrac{k_1 + k_2}{M_1}) & -\dfrac{B_1}{M_1} & \dfrac{k_2}{M_1} & 0 \\ 0 & 0 & 0 & 1 \\ \dfrac{k_2}{M_2} & 0 & -\dfrac{k_2}{M_2} & 0 \end{bmatrix}$ $B = \begin{bmatrix} 0 \\ \dfrac{1}{M_1} \\ 0 \\ 0 \end{bmatrix}$ $C = \begin{bmatrix} 0 & 0 & 1 & 0 \end{bmatrix}$

1.12 $x_1 = y_2 \; ; \; x_2 = \dot{y}_2 = \dot{x}_1 \; ; \; x_3 = \ddot{y}_2 = \dot{x}_2 \; ; \; x_4 = \dddot{y}_2 = \dot{x}_3$

$A = \begin{bmatrix} 0 & 1 & 0 & 0 \\ 0 & 0 & 1 & 0 \\ 0 & 0 & 0 & 1 \\ \dfrac{-k_1 k_2}{M_1 M_2} & \dfrac{-B_1 k_2}{M_1 M_2} & \dfrac{-(M_1 k_2 + M_2 k_1 + M_2 k_2)}{M_1 M_2} & -\dfrac{B_1}{M_1} \end{bmatrix}$

$$B = \begin{bmatrix} 0 \\ 0 \\ 0 \\ \dfrac{k_2}{M_1 M_2} \end{bmatrix} \qquad C = [1 \quad 0 \quad 0 \quad 0]$$

1.13 $x_1 = \dot{\theta}_1$; $x_2 = k(\theta_1 - \theta_2)$; $x_3 = \dot{\theta}_2$

$$A = \begin{bmatrix} \dfrac{-B_1}{J_1} & \dfrac{-1}{J_1} & 0 \\ k & 0 & -k \\ 0 & \dfrac{1}{J_2} & \dfrac{-(B_2 + B_3)}{J_2} \end{bmatrix} \qquad B = \begin{bmatrix} \dfrac{1}{J_1} \\ 0 \\ 0 \end{bmatrix} \qquad C = [0 \quad 0 \quad 1]$$

1.14 $x_1 = \dot{\theta}_2$; $x_2 = \ddot{\theta}_2 = \dot{x}_1$; $x_3 = \dddot{\theta}_2 = \dot{x}_2$

$$A = \begin{bmatrix} 0 & 1 & 0 \\ 0 & 0 & 1 \\ \dfrac{-k}{J_1 J_2}(B_1 + B_2 + B_3) & \dfrac{-1}{J_1 J_2}\left(kJ_1 + kJ_2 + B_1(B_2 + B_3)\right) & -\left(\dfrac{B_1}{J_1} + \dfrac{(B_2 + B_3)}{J_2}\right) \end{bmatrix}$$

$$B = \begin{bmatrix} 0 \\ 0 \\ \dfrac{k}{J_1 J_2} \end{bmatrix} \qquad C = [0 \quad 0 \quad 1]$$

1.15 $A = \begin{bmatrix} 0 & 1 & 0 \\ 0 & -1 & 1 \\ 0 & 0 & -5 \end{bmatrix} \qquad B = \begin{bmatrix} 0 \\ 1 \\ 2 \end{bmatrix} \qquad C = [5 \quad 0 \quad 0]$

1.16 $A = \begin{bmatrix} 0 & 1 \\ -6 & -3 \end{bmatrix}$ $\qquad B = \begin{bmatrix} 0 \\ 2 \end{bmatrix}$ $\qquad C = \begin{bmatrix} 1 & 0 \end{bmatrix}$

1.17 $A = \begin{bmatrix} 0 & 1 \\ -6 & -3 \end{bmatrix}$ $\qquad B = \begin{bmatrix} 1 \\ -1 \end{bmatrix}$ $\qquad C = \begin{bmatrix} 1 & 0 \end{bmatrix}$

1.18 $A = \begin{bmatrix} 0 & 1 & 0 \\ 0 & 0 & 1 \\ -2 & -5 & -4 \end{bmatrix}$ $\quad B = \begin{bmatrix} 0 \\ 0 \\ 3 \end{bmatrix}$ $\quad C = \begin{bmatrix} 1 & 0 & 0 \end{bmatrix}$

1.19 $A = \begin{bmatrix} 0 & 1 & 0 \\ 0 & 0 & 1 \\ -3 & -7 & -5 \end{bmatrix}$ $\quad B = \begin{bmatrix} 0 \\ 2 \\ -7 \end{bmatrix}$ $\quad C = \begin{bmatrix} 1 & 0 & 0 \end{bmatrix}$

1.20 $A = \begin{bmatrix} 0 & 1 & 0 \\ 0 & 0 & 1 \\ -3 & -7 & -5 \end{bmatrix}$ $\quad B = \begin{bmatrix} 1 \\ -3 \\ 11 \end{bmatrix}$ $\quad C = \begin{bmatrix} 1 & 0 & 0 \end{bmatrix}$

1.21 $\begin{bmatrix} \dot{x}_1 - 1 \\ \dot{x}_2 + 4 \end{bmatrix} = \begin{bmatrix} 0 & 1 \\ -5 & -4 \end{bmatrix} \begin{bmatrix} x_1 - 1 \\ x_2 - 1 \end{bmatrix}$

1.22 $\begin{bmatrix} \dot{x}_1 \\ \dot{x}_2 \end{bmatrix} = \begin{bmatrix} 0 & 1 \\ -1 & 0 \end{bmatrix} \begin{bmatrix} x_1 \\ x_2 \end{bmatrix}$

1.23 $\begin{bmatrix} \dot{x}_1 \\ \dot{x}_2 \end{bmatrix} = \begin{bmatrix} 0 & 1 \\ -1 & 0 \end{bmatrix} \begin{bmatrix} x_1 \\ x_2 \end{bmatrix}$

Chapter 2

2.1 $\begin{bmatrix} \dfrac{2}{s+1} & \dfrac{5}{s+1} \\ \dfrac{s+3}{(s+1)(s+2)} & \dfrac{2s+7}{(s+1)(s+2)} \end{bmatrix}$

2.2 $(-1, -2)$; $\begin{bmatrix} e^{-t} & 0 \\ e^{-t} - e^{-2t} & e^{-2t} \end{bmatrix}$; $\dfrac{s+3}{(s+1)(s+2)}$

2.3 $(-1, -2)$; $\begin{bmatrix} e^{-t} & 0 \\ e^{-t} - e^{-2t} & e^{-2t} \end{bmatrix}$; $\dfrac{s+3}{(s+1)(s+2)}$

2.4 $\begin{bmatrix} e^{-t} & 0 \\ 0 & e^{-2t} \end{bmatrix}$

2.5 $\begin{bmatrix} e^{-t} & te^{-t} \\ 0 & e^{-t} \end{bmatrix}$

2.6 $(-1, -2, -3)$; $\dfrac{1}{(s+1)(s+2)(s+3)} \begin{bmatrix} s^2 + 6s + 10 & 2s + 7 & s + 6 \\ -2 & s^2 + 4s + 1 & 2s \\ -(s+2) & -(s+2) & s(s+2) \end{bmatrix}$

2.7 $\begin{bmatrix} 2.5e^{-t} - 2e^{-2t} + 0.5e^{-3t} & 2.5e^{-t} - 3e^{-2t} + 0.5e^{-3t} & 2.5e^{-t} - 4e^{-2t} + 1.5e^{-3t} \\ -e^{-t} + 2e^{-2t} - e^{-3t} & -e^{-t} + 3e^{-2t} - e^{-3t} & -e^{-t} + 4e^{-2t} - 3e^{-3t} \\ -0.5e^{-t} + 0.5e^{-3t} & -0.5e^{-t} + 0.5e^{-3t} & -0.5e^{-t} + 1.5e^{-3t} \end{bmatrix}$

2.8 $\dfrac{2(s+6)}{(s+1)(s+2)(s+3)}$

2.9 $\dot{X} = \begin{bmatrix} 0 & 1 \\ -3 & -4 \end{bmatrix} X + \begin{bmatrix} 0 \\ 1 \end{bmatrix} r$ $\qquad y = \begin{bmatrix} 3 & 2 \end{bmatrix} X$

2.10 $\dot{X} = \begin{bmatrix} -4 & 1 \\ -3 & 0 \end{bmatrix} X + \begin{bmatrix} 2 \\ 3 \end{bmatrix} r$ $\qquad y = \begin{bmatrix} 1 & 0 \end{bmatrix} X$

2.11 $\dot{X} = \begin{bmatrix} -1 & 0 \\ 0 & -3 \end{bmatrix} X + \begin{bmatrix} 1 \\ 1 \end{bmatrix} r$ $\qquad y = \begin{bmatrix} 0.5 & 1.5 \end{bmatrix} X$

2.12 $\dot{X} = \begin{bmatrix} -2 & 1 \\ 0 & -2 \end{bmatrix} X + \begin{bmatrix} 0 \\ 1 \end{bmatrix} r$ $\qquad y = \begin{bmatrix} 1 & 2 \end{bmatrix} X$

2.13 $\dot{X} = \begin{bmatrix} 0 & 1 & 0 \\ 0 & 0 & 1 \\ -8 & -14 & -7 \end{bmatrix} X + \begin{bmatrix} 0 \\ 0 \\ 1 \end{bmatrix} r \qquad y = \begin{bmatrix} 2 & 1 & 3 \end{bmatrix} X$

2.14 $\dot{X} = \begin{bmatrix} -7 & 1 & 0 \\ -14 & 0 & 1 \\ -8 & 0 & 0 \end{bmatrix} X + \begin{bmatrix} 3 \\ 1 \\ 2 \end{bmatrix} r \qquad y = \begin{bmatrix} 1 & 0 & 0 \end{bmatrix} X$

2.15 $\dot{X} = \begin{bmatrix} -1 & 0 & 0 \\ 0 & -2 & 0 \\ 0 & 0 & -4 \end{bmatrix} X + \begin{bmatrix} 1 \\ 1 \\ 1 \end{bmatrix} r \qquad y = \begin{bmatrix} \dfrac{7}{3} & -7 & \dfrac{23}{3} \end{bmatrix} X$

2.16 $\dot{X} = \begin{bmatrix} -2 & 0 & 0 \\ 0 & -1 & 1 \\ 0 & 0 & -1 \end{bmatrix} X + \begin{bmatrix} 1 \\ 0 \\ 1 \end{bmatrix} r \qquad y = \begin{bmatrix} 15 & 6 & -11 \end{bmatrix} X$

2.17 $\dot{X} = \begin{bmatrix} -3 & 3 \\ 0 & -2 \end{bmatrix} X + \begin{bmatrix} 10 \\ 10 \end{bmatrix} r \qquad y = \begin{bmatrix} 1 & 0 \end{bmatrix} X \qquad \dfrac{Y(s)}{R(s)} = \dfrac{10(s+5)}{(s+2)(s+3)}$

2.18 $\dot{X} = \begin{bmatrix} 0 & 0 & -1 \\ 10 & -5 & 0 \\ 0 & 1 & -1 \end{bmatrix} X + \begin{bmatrix} 1 \\ 0 \\ 0 \end{bmatrix} r \qquad y = \begin{bmatrix} 0 & 1 & 0 \end{bmatrix} X$

2.19 $\dot{X} = \begin{bmatrix} 0 & 1 & 0 \\ 0 & 0 & 1 \\ -6 & -11 & -6 \end{bmatrix} X + \begin{bmatrix} -7 \\ 24 \\ -72 \end{bmatrix} r \qquad y = \begin{bmatrix} 1 & 0 & 0 \end{bmatrix} X + 2r$

2.20 $\phi(t) = e^{At} = \begin{bmatrix} 2e^{-2t} - e^{-4t} & \dfrac{1}{2}e^{-2t} - \dfrac{1}{2}e^{-4t} \\ -4e^{-2t} + 4e^{-4t} & -e^{-2t} + 2e^{-4t} \end{bmatrix}$

2.21 $[a = 6, \; b = 16, \; k_1 = 5, \; k_2 = 15]$

Chapter 3

3.1 e^{-t}

3.2 $e^{-t} - e^{-3t}$

3.3 $e^{-t} - \dfrac{1}{3} + \dfrac{1}{3} e^{-3t}$

3.4 $3e^{-3t}$; $1 - 4e^{-3t} + 3e^{-4t}$; $1 - e^{-3t} + 3e^{-4t}$

3.5 $7e^{-2t} \cos t - 11 e^{-2t} \sin t$

3.6 $\dfrac{1}{5} + \dfrac{1}{5} e^{-2t} \cos t + \dfrac{14}{5} e^{-2t} \sin t$

3.7 $y_{\text{ZIR}} = 2e^{-2t} + te^{-2t}$; $y_{\text{ZSR}} = 2 - 2e^{-2t} - te^{-2t}$; $y = 2$

3.8 $\dfrac{1}{6} - \dfrac{3}{2} e^{-t} + \dfrac{7}{2} e^{-2t} - \dfrac{7}{6} e^{-3t}$

3.9 $\begin{bmatrix} e^{-t} & 0 \\ 0 & e^{-2t} \end{bmatrix}$; $\dfrac{3}{2} + \dfrac{1}{2} e^{-2t}$

3.10 $-6e^{-t} + 10e^{-2t}$

3.11 e^{-2t}

3.12 $\dfrac{2}{3} + \dfrac{2\sqrt{2}}{3} e^{-t} \sin \sqrt{2} t - \dfrac{2}{3} e^{-t} \cos \sqrt{2} t$

3.13 $\begin{bmatrix} e^{-t} & e^{-t} - e^{-2t} \\ 0 & e^{-2t} \end{bmatrix}$

3.14 $\begin{bmatrix} e^{-t}(\cos 2t + \dfrac{1}{2} \sin 2t) & -\dfrac{5}{2} e^{-t} \sin 2t \\ \dfrac{1}{2} e^{-t} \sin 2t & e^{-t}(\cos 2t - \dfrac{1}{2} \sin 2t) \end{bmatrix}$

3.15
$$\begin{bmatrix} e^{-2t} - 2te^{-2t} & te^{-2t} \\ -4te^{-2t} & e^{-2t} + 2te^{-2t} \end{bmatrix}$$

3.16
$$\begin{bmatrix} 2e^{-2t} - e^{-4t} & -4e^{-2t} + 4e^{-4t} \\ \dfrac{1}{2}e^{-2t} - \dfrac{1}{2}e^{-4t} & -e^{-2t} + 2e^{-4t} \end{bmatrix}$$

3.17 $\dfrac{1}{8}\begin{bmatrix} -6 & 8 \\ -1 & 0 \end{bmatrix}$

3.18
$$\begin{bmatrix} 1 & 3 & 0 \\ -1 & -\dfrac{5}{2} & 0 \\ 0 & 0 & -\dfrac{1}{3} \end{bmatrix}$$

3.19 $-47A - 95I$

3.20 (a)
$$\begin{bmatrix} \sin(-1) & \sin(-1) - \sin(-2) \\ 0 & \sin(-2) \end{bmatrix}$$

3.20 (b)
$$\begin{bmatrix} \cos(-1) & \cos(-1) - \cos(-2) \\ 0 & \cos(-2) \end{bmatrix}$$

3.21
$$\begin{bmatrix} -1 & -1 & -1 \\ 0 & -1 & -1 \\ 0 & 0 & -1 \end{bmatrix}$$

3.22
$$\begin{bmatrix} \cos 3t & -\sin 3t & 0 \\ \sin 3t & \cos 3t & 0 \\ 0 & 0 & e^{-t} \end{bmatrix}$$

Chapter 4

4.1 $-2, -4;$ $\begin{bmatrix} 3 \\ -3 \end{bmatrix}, \begin{bmatrix} 1 \\ -3 \end{bmatrix}$

4.2 $-1, -3;$ $\begin{bmatrix} 3 \\ -1 \end{bmatrix}, \begin{bmatrix} 1 \\ -1 \end{bmatrix}$

4.3 $-1, -2, -3;$ $\begin{bmatrix} 1 \\ -1 \\ -1 \end{bmatrix}, \begin{bmatrix} 2 \\ -4 \\ 1 \end{bmatrix}, \begin{bmatrix} 1 \\ -3 \\ 3 \end{bmatrix}$

4.4 $\begin{bmatrix} -e^{-2t} + 2e^{-3t} & 2e^{-2t} - 2e^{-3t} \\ -e^{-2t} + e^{-3t} & 2e^{-2t} - e^{-3t} \end{bmatrix}$

4.5 $\begin{bmatrix} 2e^{-t} - e^{-2t} & e^{-t} - e^{-2t} \\ -2e^{-t} + 2e^{-2t} & -e^{-t} + 2e^{-2t} \end{bmatrix}$

4.6 $-2, -2;$ $\begin{bmatrix} -2 \\ -4 \end{bmatrix}, \begin{bmatrix} 1 \\ 0 \end{bmatrix}$

4.7 $-2, -2, -3;$ $\begin{bmatrix} 1 \\ 5 \\ 6 \end{bmatrix}, \begin{bmatrix} -1 \\ -4 \\ -3 \end{bmatrix}, \begin{bmatrix} 1 \\ 4 \\ 4 \end{bmatrix}$

4.8 $0, -1, -1;$ $\begin{bmatrix} 4 \\ 1 \\ 12 \end{bmatrix}, \begin{bmatrix} 0 \\ 0 \\ 16 \end{bmatrix}, \begin{bmatrix} 4 \\ 0 \\ -4 \end{bmatrix}$

4.9 $P = \begin{bmatrix} 1 & -1 \\ 1 & 1 \end{bmatrix};$ $\dot{Z} = \begin{bmatrix} -2 & 0 \\ 0 & -4 \end{bmatrix} Z + \begin{bmatrix} 0.5 \\ -1.5 \end{bmatrix} r$ $y = \begin{bmatrix} 3 & 1 \end{bmatrix} Z$

$e^{At} = \dfrac{1}{2} \begin{bmatrix} e^{-2t} + e^{-4t} & e^{-2t} - e^{-4t} \\ e^{-2t} - e^{-4t} & e^{-2t} + e^{-4t} \end{bmatrix}$

4.10 $P = \begin{bmatrix} 1 & 1 \\ 0 & -1 \end{bmatrix};$ $\dot{Z} = \begin{bmatrix} -2 & 0 \\ 0 & -3 \end{bmatrix} Z + \begin{bmatrix} 2 \\ -1 \end{bmatrix} r$ $y = \begin{bmatrix} 2 & 3 \end{bmatrix} Z$

$$e^{At} = \begin{bmatrix} e^{-2t} & e^{-2t} - e^{-3t} \\ 0 & e^{-3t} \end{bmatrix}$$

4.11 $P = \begin{bmatrix} -4 & 1 \\ -8 & 0 \end{bmatrix};$ $\dot{Z} = \begin{bmatrix} -2 & 1 \\ 0 & -2 \end{bmatrix} Z + \begin{bmatrix} -0.125 \\ -0.5 \end{bmatrix} r$ $y = \begin{bmatrix} 12 & 1 \end{bmatrix} Z$

$$e^{At} = \begin{bmatrix} e^{-2t} - 4te^{-2t} & 2te^{-2t} \\ -8te^{-2t} & e^{-2t} + 4te^{-2t} \end{bmatrix}$$

4.12 $P = \begin{bmatrix} 2 & 1 & 0 \\ 0 & -0.5 & 0 \\ 2 & 3 & 4 \end{bmatrix};$ $\dot{Z} = \begin{bmatrix} -1 & 0 & 0 \\ 0 & -2 & 0 \\ 0 & 0 & -3 \end{bmatrix} Z + \begin{bmatrix} 1 \\ -2 \\ 1.25 \end{bmatrix} r$ $y = \begin{bmatrix} 2 & 0.5 & 0 \end{bmatrix} Z$

4.13 $P = \begin{bmatrix} 0 & 4 & 0 \\ 0 & 0 & 1 \\ 16 & -4 & 0 \end{bmatrix};$ $\dot{Z} = \begin{bmatrix} -2 & 1 & 0 \\ 0 & -2 & 1 \\ 0 & 0 & -2 \end{bmatrix} Z + \begin{bmatrix} 0.25 \\ 0.25 \\ 2 \end{bmatrix} r$ $y = \begin{bmatrix} -16 & 8 & 2 \end{bmatrix} Z$

4.14 $P = \begin{bmatrix} 0 & 2 & -8 \\ 0 & 4 & 0 \\ -2 & 1 & -6 \end{bmatrix};$ $\dot{Z} = \begin{bmatrix} -1 & 1 & 0 \\ 0 & -1 & 0 \\ 0 & 0 & -5 \end{bmatrix} Z + \begin{bmatrix} 0.1875 \\ 0.75 \\ 0.0625 \end{bmatrix} r$ $y = \begin{bmatrix} 2 & 1 & -2 \end{bmatrix} Z$

4.15 $P = \begin{bmatrix} 1 & 1 \\ -2 & -4 \end{bmatrix};$ $\dot{Z} = \begin{bmatrix} -2 & 0 \\ 0 & -4 \end{bmatrix} Z + \begin{bmatrix} 0.5 \\ -0.5 \end{bmatrix} r$ $y = \begin{bmatrix} 4 & 6 \end{bmatrix} Z$

4.16 $P = \begin{bmatrix} 1 & 0 \\ -3 & 1 \end{bmatrix};$ $\dot{Z} = \begin{bmatrix} -3 & 1 \\ 0 & -3 \end{bmatrix} Z + \begin{bmatrix} 0 \\ 1 \end{bmatrix} r$ $y = \begin{bmatrix} -2 & 1 \end{bmatrix} Z$

4.17 $P = \begin{bmatrix} 1 & 0 \\ -5 & 1 \end{bmatrix};$ $\dot{Z} = \begin{bmatrix} -5 & 1 \\ 0 & -5 \end{bmatrix} Z + \begin{bmatrix} 2 \\ 11 \end{bmatrix} r$ $y = \begin{bmatrix} 11 & -2 \end{bmatrix} Z$

4.18 $P = \begin{bmatrix} 1 & 1 & 1 \\ -1 & -2 & -3 \\ 1 & 4 & 9 \end{bmatrix};$ $\dot{Z} = \begin{bmatrix} -1 & 0 & 0 \\ 0 & -2 & 0 \\ 0 & 0 & -3 \end{bmatrix} Z + \begin{bmatrix} 1 \\ -2 \\ 1 \end{bmatrix} r$ $y = \begin{bmatrix} 1 & 2 & 5 \end{bmatrix} Z$

4.19 $P = \begin{bmatrix} 1 & 0 & 0 \\ -2 & 1 & 0 \\ 4 & -4 & 1 \end{bmatrix}$; $\dot{Z} = \begin{bmatrix} -2 & 1 & 0 \\ 0 & -2 & 1 \\ 0 & 0 & -2 \end{bmatrix} Z + \begin{bmatrix} 0 \\ 0 \\ 1 \end{bmatrix} r$ $y = \begin{bmatrix} -3 & 2 & 0 \end{bmatrix} Z$

4.20 $P = \begin{bmatrix} 1 & 1 \\ 4 & 1 \end{bmatrix}$; $\dot{Z} = \begin{bmatrix} -1 & 0 \\ 0 & -4 \end{bmatrix} Z + \begin{bmatrix} 1/3 \\ 2/3 \end{bmatrix} r$ $y = \begin{bmatrix} 1 & 1 \end{bmatrix} Z$

4.21 $P = \begin{bmatrix} 1 & 0 \\ 3 & 1 \end{bmatrix}$; $\dot{Z} = \begin{bmatrix} -3 & 1 \\ 0 & -3 \end{bmatrix} Z + \begin{bmatrix} 2 \\ -1 \end{bmatrix} r$ $y = \begin{bmatrix} 1 & 0 \end{bmatrix} Z$

4.22 $P = \begin{bmatrix} 1 & 1 & 1 \\ 8 & 6 & 4 \\ 15 & 5 & 3 \end{bmatrix}$; $\dot{Z} = \begin{bmatrix} -1 & 0 & 0 \\ 0 & -3 & 0 \\ 0 & 0 & -5 \end{bmatrix} Z + \begin{bmatrix} 0.5 \\ -4.5 \\ 6 \end{bmatrix} r$ $y = \begin{bmatrix} 1 & 1 & 1 \end{bmatrix} Z$

4.23 $P = \begin{bmatrix} 1 & 0 & 1 \\ 6 & 1 & 2 \\ 5 & 5 & 1 \end{bmatrix}$; $\dot{Z} = \begin{bmatrix} -1 & 1 & 0 \\ 0 & -1 & 0 \\ 0 & 0 & -5 \end{bmatrix} Z + \begin{bmatrix} 1.3125 \\ 1.25 \\ 2.3125 \end{bmatrix} r$ $y = \begin{bmatrix} 1 & 0 & 1 \end{bmatrix} Z$

Chapter 5

5.1 Not completely controllable; completely observable.

5.2 Not completely controllable; not completely observable.

5.5 For complete controllability, $k_1 \neq 0$ or $k_1 \neq k_2(\beta - \alpha)$

For complete observability, $c_2 \neq 0$ or $c_2 \neq c_1(\alpha - \beta)$

5.6 Completely controllable and completely observable.

5.8 Completely controllable; completely observable.

5.10 Not completely controllable; not completely observable.

5.11 Completely controllable; not completely observable.

5.14 Completely controllable; not completely observable.

5.15 $P = \begin{bmatrix} 8 & 1 \\ -8 & 2 \end{bmatrix}$ $\dot{Z} = \begin{bmatrix} 0 & 1 \\ -8 & -6 \end{bmatrix} Z + \begin{bmatrix} 0 \\ 1 \end{bmatrix} Z$ $y = \begin{bmatrix} 8 & 4 \end{bmatrix} Z$

5.16 $P = \begin{bmatrix} 5 & 1 \\ 7 & 2 \end{bmatrix}$ $\dot{Z} = \begin{bmatrix} 0 & 1 \\ -8 & -6 \end{bmatrix} Z + \begin{bmatrix} 0 \\ 1 \end{bmatrix} Z$ $y = \begin{bmatrix} -9 & -3 \end{bmatrix} Z$

5.17 $P = \begin{bmatrix} 1 & 0 \\ 5 & 1 \end{bmatrix}$ $\dot{Z} = \begin{bmatrix} 0 & 1 \\ -6 & -5 \end{bmatrix} Z + \begin{bmatrix} 0 \\ 1 \end{bmatrix} Z$ $y = \begin{bmatrix} -7 & -2 \end{bmatrix} Z$

5.18 $P = \begin{bmatrix} 5 & 1 \\ 6 & 2 \end{bmatrix}$ $\dot{Z} = \begin{bmatrix} 0 & 1 \\ -9 & -6 \end{bmatrix} Z + \begin{bmatrix} 0 \\ 1 \end{bmatrix} Z$ $y = \begin{bmatrix} -7 & -3 \end{bmatrix} Z$

5.19 $P = \dfrac{1}{7} \begin{bmatrix} -9 & -2 \\ 8 & 1 \end{bmatrix}$ $\dot{Z} = \begin{bmatrix} -5 & 1 \\ -4 & 0 \end{bmatrix} Z + \begin{bmatrix} 1 \\ -3 \end{bmatrix} Z$ $y = \begin{bmatrix} 1 & 0 \end{bmatrix} Z$

5.20 $P = \dfrac{1}{3} \begin{bmatrix} 5 & -2 \\ 1 & -1 \end{bmatrix}$ $\dot{Z} = \begin{bmatrix} -6 & 1 \\ -8 & 0 \end{bmatrix} Z + \begin{bmatrix} -3 \\ -9 \end{bmatrix} Z$ $y = \begin{bmatrix} 1 & 0 \end{bmatrix} Z$

5.21 Not completely controllable and not completely observable.

Chapter 6

6.1 [−10/3 −10/3]

6.4 [6.5 2.5]

6.7 [30 2]

6.10 [10 −3 −1]

6.12 [10000 1164 55]

6.14 [3 1.5]

6.17 [0 5 −1]

6.20 [364 44 3]

6.21(a) [5/3 −7/3]; 6.21(b) $[-2 \quad 14]^T$

6.23 $[5 \quad 44]^T$

6.24 [17.5 −1.5 −1]

6.26 [61 1428 35730]

6.27 $[11 \quad 878]^T$

6.28 [16 396]

6.29 [1 −15 15]

6.31 $[1233 \quad 64 \quad 23569]^{T}$

6.32 $[-9.5 \quad -240.5]^{T}$

6.33 $-2 \pm j3$

6.34 $-5, -1 \pm j2$

6.35 $[6 \quad 9]$

6.36 $[-1 \quad 3 \quad -3]$

6.37 $[8 \quad 1 \quad 8]$

6.38 $[18 \quad 50]^{T}$

6.39 $[17 \quad 498]^{T}$

6.40 $\begin{bmatrix} 60 & 1169 & 9970 \end{bmatrix}^{T}$

6.41 $\begin{bmatrix} 21 & 221 \end{bmatrix}^{T}$

Chapter 7

7.1 $[-2.55 \sin \sqrt{10}\, t; \text{ stable}]$

7.2 $[-0.85 \sin \sqrt{15}\, t ; \text{ stable}]$

7.3 $[-1.91 \sin \sqrt{35}\, t ; \text{ stable}]$

7.4 $[k > 3, \quad \text{stable}]$

7.5 $[-6.26 \sin 2\, t, \quad \text{stable}]$

7.6 $[k > 0.4]$

7.7 $[-2.475 \sin \sqrt{50}\, t\,]$

7.8 $[-2.475 \sin t\,]$

7.9 $[k < 4.9]$

7.10 $[-0.5 \sin 10.95\, t]$

7.11 $[-2.1864 \sin 20\, t]$

7.12 $[-2.13 \sin \sqrt{3}\, t]$

7.13 $[-(\pi a / 2N)]$

7.14 $[a < 2.12]$

7.15 $[N>3.14]$

7.16 $[-5\sin 2t, \text{ stable}]$

7.17 [No limit cycle]

7.18 $[N>3.49]$

7.19 $[k<2; c(t)=-0.85\sin 5t]$

7.20 $[P - \text{unstable}, Q - \text{stable}]$

Chapter 8

8.1 $(x_1^2/25)+(x_2^2/100)=1$, ellipse

8.3 $x_2^2+4x_1^2-2x_1=16$

8.5 $x_2^2+4x_1=16$

8.7 $x_2+4x_1=7$

8.9 $x_2+2x_1+2\ln\left(\dfrac{2-x_2}{2}\right)=8$

8.10 $\quad$ ——

8.11 $x_2=\dfrac{-1}{m+0.25}x_1$

8.12 $x_2=\dfrac{-1}{m+0.25}x_1+\dfrac{2}{m+0.25}$

8.13 $g(x_2)=2x_2 \qquad f(x_1)=x_1$

8.14 $g(x_2)=0.5x_2 \quad f(x_1)=0.5x_1-1$

8.15 $f(x_1,x_2)=0.25x_2+0.5x_1^3; k=1; \delta_0=5$

8.16 $f(x_1,x_2)=0.5x_2+0.5x_1x_2-1; \; k=0.25; \; \delta_0=8$

8.17 For $x_1 > 0, x_2 = \dfrac{-1}{m+1}x_1 + \dfrac{1}{m+1}$, For $x_1 < 0, x_2 = \dfrac{1}{m+1}x_1 + \dfrac{1}{m+1}$

8.18 For $x_2 > 0, x_2 = \dfrac{1}{m+1}$, For $x_2 < 0, x_2 = \dfrac{1}{m-1}$

8.19 $\dot{x}_1 = e; \dot{x}_2 = e;$ For $x_1 > 0, \; x_2 = \dfrac{-1}{m+2};$ For $x_1 < 0, \; x_2 = \dfrac{1}{m+2}$

8.20 $\dot{x}_1 = e; \dot{x}_2 = e;$ For $x_1 > 1, \; x_2 = \dfrac{-1}{m+2};$ For $x_1 < 1, \; x_2 = \dfrac{1}{m+2};$

For $-1 < x_1 < 1, x_2 = 0$

8.21 $\dot{x}_1 = e; \dot{x}_2 = e;$ For $x_1 > 0.5, \; x_2 = \dfrac{-2}{m+3}x_1 + \dfrac{2r-1}{m+3};$

For $x_1 < 0.5, \; x_2 = \dfrac{-2}{m+3}x_1 + \dfrac{2r+1}{m+3}$ For $-0.5 < x_1 < 0.5, x_2 = \dfrac{-4}{m+3}x_1 + \dfrac{2r}{m+3}$

8.22 $\dot{x}_1 = e; \dot{x}_2 = e;$ For $x_1 > 1, \; x_2 = \dfrac{-4}{m+1}x_1 + \dfrac{4}{m+1};$

For $x_1 < 1, \; x_2 = \dfrac{-4}{m+1}x_1 - \dfrac{4}{m+1}$ For $-1 < x_1 < 1; x_2 = 0$

8.23 $\dot{x}_1 = e; \dot{x}_2 = e;$ For $x_2 < 0, \; x_2 = \dfrac{-2}{m+2}x_1 + \dfrac{2}{m+2}$

For $x_2 > 0, \; x_2 = \dfrac{-2}{m+2}x_1 - \dfrac{2}{m+2}$

8.24 $\dot{x}_1 = e; \dot{x}_2 = e;$ For $x_2 < 1; \; x_2 = \dfrac{-2}{m+1}x_1 + \dfrac{2}{m+1}$

For $x_2 > 1; x_2 = \dfrac{-2}{m+1}x_1 - \dfrac{2}{m+1}$ For $-1 < x_2 < 1; x_2 = \dfrac{-2}{m+1}x_1$

8.25 $\begin{bmatrix} x_{1e} \\ x_{2e} \end{bmatrix} = \begin{bmatrix} 0 \\ 0 \end{bmatrix}$; Centre or vortex $\begin{bmatrix} x_{1e} \\ x_{2e} \end{bmatrix} = \begin{bmatrix} 1 \\ 0 \end{bmatrix}$; saddle point

8.26 $\begin{bmatrix} x_{1e} \\ x_{2e} \end{bmatrix} = \begin{bmatrix} 0 \\ 0 \end{bmatrix}$;Stable node $\begin{bmatrix} x_{1e} \\ x_{2e} \end{bmatrix} = \begin{bmatrix} \dfrac{\pi}{6} \\ -\dfrac{\pi}{6} \end{bmatrix}$; Saddle point $\begin{bmatrix} x_{1e} \\ x_{2e} \end{bmatrix} = \begin{bmatrix} \dfrac{7\pi}{6} \\ \dfrac{7\pi}{6} \end{bmatrix}$; Saddle point

8.27 $\begin{bmatrix} x_{1e} \\ x_{2e} \end{bmatrix} = \begin{bmatrix} \dfrac{\pi}{2} \\ 0 \end{bmatrix}$;Saddle point $\begin{bmatrix} x_{1e} \\ x_{2e} \end{bmatrix} = \begin{bmatrix} \dfrac{3\pi}{2} \\ 0 \end{bmatrix}$; Stable focus

8.28 $\begin{bmatrix} x_{1e} \\ x_{2e} \end{bmatrix} = \begin{bmatrix} 0 \\ 0 \end{bmatrix}$;Stable node

8.29 $\begin{bmatrix} x_{1e} \\ x_{2e} \end{bmatrix} = \begin{bmatrix} 0 \\ 0 \end{bmatrix}$;Stable focus $\begin{bmatrix} x_{1e} \\ x_{2e} \end{bmatrix} = \begin{bmatrix} -1 \\ 0 \end{bmatrix}$; Saddle point

8.30 $\dot{x}_1 = e; x_2 = \dot{e};$

For $x_2 < 0,$ $\begin{bmatrix} x_{1e} \\ x_{2e} \end{bmatrix} = \begin{bmatrix} 1 \\ 0 \end{bmatrix}$;Stable focus $x_2 > 0,$ $\begin{bmatrix} x_{1e} \\ x_{2e} \end{bmatrix} = \begin{bmatrix} -1 \\ 0 \end{bmatrix}$;Stable focus

Chapter 9

9.1 Indefinite   9.2 Negative definite;   9.3 Indefinite

9.4 Indefinite   9.5 Negative semi-definite   9.6 Positive semi-definite

9.7 Asymptotically stable   9.8 Asymptotically stable   9.9 Stable

9.10 Stable if $k_1 = k_2 c$ 9.11 Asymptotically stable;   9.12 Asymptotically stable if $c = 1$

9.13 Asymptotically stable;   9.14 Asymptotically stable   9.15 Stable if $a = (1/3)$

9.16 Asymptotically stable;   9.17 Asymptotically stable

9.18 Stability cannot be assessed using Krasovskii's theorem

9.19 $V(X) = \dfrac{1}{2} x_1^2 + \dfrac{1}{2} x_2^2;$ $x_1 x_2 < \dfrac{1}{2}$

9.20 $V(X) = \dfrac{1}{2} x_1^2 + \dfrac{1}{2} x_2^2;$ $x_1 x_2 < 1$

9.21 Stable; $P = \begin{bmatrix} \dfrac{1}{4} & \dfrac{1}{16} \\ \dfrac{1}{16} & \dfrac{9}{32} \end{bmatrix}$

9.22 Stable; $P = \begin{bmatrix} \dfrac{7}{4} & \dfrac{5}{8} \\ \dfrac{5}{8} & \dfrac{3}{8} \end{bmatrix}$

9.23 Stable; $P = \begin{bmatrix} \dfrac{3}{2} & \dfrac{1}{2} \\ \dfrac{1}{2} & 1 \end{bmatrix}$

9.24 Unstable; $P = \begin{bmatrix} -\dfrac{1}{2} & \dfrac{3}{2} \\ \dfrac{3}{2} & -2 \end{bmatrix}$

9.25 Stable; stability cannot be assessed using Krasovskii's theorem

9.26 Stability cannot be assessed using Krasovskii's theorem

9.27 Asymptotically stable

Chapter 10

10.1 $u = -2x_1 - 3x_2 \quad J_{min} = 4$

10.2 $u = -2x_1 - \sqrt{2}\,x_2 \qquad J_{min} = \sqrt{2}$

10.3 $u = -0.35y$

10.4 $K = \begin{bmatrix} 1 & 0.236 \end{bmatrix}$

10.5 $K = \begin{bmatrix} -1 & 4 \end{bmatrix}$

10.6 $K = \begin{bmatrix} -1 & 4.12 \end{bmatrix}$

10.7 $K = [0.225 \quad\quad 0.3345]$

10.8 $K = \begin{bmatrix} 0.415 & \quad 0.415 \end{bmatrix}$

10.9 $K = \begin{bmatrix} 0.414 & 0.1973 \end{bmatrix}$

10.10 $K = \begin{bmatrix} 2 & \quad 2 \end{bmatrix}$

10.12 $K = \begin{bmatrix} 1 & \quad 1 \end{bmatrix}$

10.13 $K = [1 \quad\quad 1.646]$

10.14 $K = [1.732 \quad\quad 1.732]$

Appendices

APPENDIX – A
Matrix Fundamentals

Diagonal Matrix - A square matrix, the diagonal elements of which are non-zero and off-diagonal elements are zeros.

$$A = \begin{bmatrix} a_{11} & 0 & 0 & 0 \text{---} 0 \\ 0 & a_{22} & 0 & 0 \text{---} 0 \\ 0 & 0 & a_{33} & 0 \text{---} 0 \\ - & - & - & - - - - \\ 0 & 0 & 0 & 0 \text{---} a_{nn} \end{bmatrix}$$

Identity Matrix - A diagonal matrix of which the diagonal elements are unity.

$$I = \begin{bmatrix} 1 & 0 & 0 & 0 \text{---} 0 \\ 0 & 1 & 0 & 0 \text{---} 0 \\ 0 & 0 & 1 & 0 \text{---} 0 \\ - & - & - & - - - \\ 0 & 0 & 0 & 0 \text{---} 1 \end{bmatrix}$$

Singular Matrix

A square matrix is called a singular matrix if the associated determinant is zero; otherwise it is non-singular.

Symmetric Matrix

A square matrix A is said to be symmetric if every element $a_{ij} = a_{ji} ; i \neq j$.

Minor of a Matrix

The minor of an element of a square matrix is denoted by M_{ij} and is obtained as the determinant of the matrix after removing the ith row and j^{th} column.

Cofactor of a Matrix

The cofactors of a square matrix is obtained as

$$C_{ij} = (-1)^{i+j} M_{ij}$$

Adjoint of a Matrix

The adjoint of a square matrix is the matrix obtained after replacing the elements of the matrix by the corresponding cofactors and the taking its transpose.

Inverse of a Matrix

The inverse of a square matrix A is denoted as A^{-1} such that $AA^{-1} = A^{-1}A = I$ and $|A|$ is nonzero.

$$A^{-1} = \frac{Adjoint\ of\ A}{|A|}; |A| \neq 0$$

Inverse of a square matrix exists if and only if it is non-singular.

If $A = \begin{bmatrix} a & b \\ c & d \end{bmatrix}$ then $A^{-1} = \frac{1}{(ad-bc)} \begin{bmatrix} d & -b \\ -c & a \end{bmatrix}$, provided $(ad-bc) \neq 0$.

Rank of a Matrix

A matrix A is said to have a rank m if there exists an $m \times m$ sub-matrix such that its determinant is nonzero and the determinant of every other matrix of higher size of A is zero. The rank of A is the largest nonsingular matrix contained in it.

1. Rank of $A = \begin{bmatrix} 1 & -2 \\ -1 & 2 \end{bmatrix}$ is 1

2. Rank of $A = \begin{bmatrix} 1 & 3 & 1 \\ 3 & 9 & 2 \end{bmatrix}$ is 2

3. Rank of $A = \begin{bmatrix} 1 & 3 & 0 \\ 3 & 9 & 2 \\ 2 & 6 & 1 \end{bmatrix}$ is 2

4. Let $A = \begin{bmatrix} 1 & 2 & 3 & 4 \\ 0 & 1 & -1 & 0 \\ 1 & 0 & 1 & 2 \\ 1 & 1 & 0 & 2 \end{bmatrix}$

$$|A| = \begin{vmatrix} 1 & 2 & 3 & 4 \\ 0 & 1 & -1 & 0 \\ 1 & 0 & 1 & 2 \\ 1 & 1 & 0 & 2 \end{vmatrix} = \begin{vmatrix} 1 & -1 & 0 \\ 0 & 1 & 2 \\ 1 & 0 & 2 \end{vmatrix} + \begin{vmatrix} 2 & 3 & 4 \\ 1 & -1 & 0 \\ 1 & 0 & 2 \end{vmatrix} - \begin{vmatrix} 2 & 3 & 4 \\ 1 & -1 & 0 \\ 0 & 1 & 2 \end{vmatrix}$$

$$= 1(2) + 1(-2) + 2(-2) - 1(6) + 1(4) - [2(-2) - (6-4)] = 0$$

The rank of A is less than 4.

Consider the sub-matrix of size 3×3 of A.

$$A_1 = \begin{bmatrix} 1 & 2 & 3 \\ 0 & 1 & -1 \\ 1 & 0 & 1 \end{bmatrix} \qquad |A_1| = \begin{vmatrix} 1 & 2 & 3 \\ 0 & 1 & -1 \\ 1 & 0 & 1 \end{vmatrix} = 1 + (-2 - 3) = -4$$

The determinant of A_1 is not zero. Hence the rank of A is 3.

5. Consider $A = \begin{bmatrix} 1 & 2 & 0 & 2 \\ 1 & 2 & 0 & 1 \\ 0 & 1 & 3 & 0 \end{bmatrix}$

Consider a square sub-matrix $A_1 = \begin{bmatrix} 1 & 2 & 0 \\ 1 & 2 & 0 \\ 0 & 1 & 3 \end{bmatrix} |A_1| = 1(6) - (6) = 0$

Consider another square sub-matrix $A_2 = \begin{bmatrix} 2 & 0 & 2 \\ 2 & 0 & 1 \\ 1 & 3 & 0 \end{bmatrix} \qquad |A_2| = 2(-3) + 2(6) = 6$

Hence the rank of A is 3.

Eigen Values

The Eigen values λ_i of square matrix A are obtained by solving the characteristic equation

$$|\lambda_i I - A| = 0$$

Eigen Vectors

If A is a square matrix, X is a nonzero vector and λ is a scalar variable satisfying the equation

$$AX = \lambda X$$

then X is said to be an Eigen vector. There is an Eigen vector associated with each of the Eigen values of A. The Eigen vectors X of a square matrix A are obtained by solving the equation

$$AX = \lambda X \qquad or \qquad [\lambda I - A]X = [0]$$

It can also be determined from the cofactors of any row of $[\lambda I - A]$.

APPENDIX – B
LAPLACE TRANSFORMS

No.	Signal	Transform	Comment
1	$\delta(t)$	1	Unit impulse signal
2	$u(t)$	$\dfrac{1}{s}$	Unit step signal
3	e^{-at}	$\dfrac{1}{s+a}$	Exponentially Decaying signal
4	$\sin(\omega_0 t)u(t)$	$\dfrac{\omega_0}{s^2+\omega_0^2}$	Sinusoidal signal
5	$\cos(\omega_0 t)u(t)$	$\dfrac{s}{s^2+\omega_0^2}$	Cosinusoidal signal
6	$e^{-at}\sin(\omega_0 t)u(t)$	$\dfrac{\omega_0}{(s+a)^2+\omega_0^2}$	Exponentially damped sinusoid
7	$e^{-at}\cos(\omega_0 t)u(t)$	$\dfrac{(s+a)}{(s+a)^2+\omega_0^2}$	Exponentially damped cosinusoid

APPENDIX – C
RESPONSE IN THE S-DOMAIN

Zero Input Response: $Y_{ZIR}(s) = C[sI-A]^{-1}X(0)$

Zero State Response: $Y_{ZSR}(s) = C[sI-A]^{-1}B\ R(s)$

Total Response: $Y(s) = Y_{ZIR}(s) + Y_{ZSR}(s) = C[sI-A]^{-1}[X(0)+B\ R(s)]$

Impulse response $Y(s) = C[sI-A]^{-1}[X(0)+B]$

Unit step response $Y(s) = C[sI-A]^{-1}\left[X(0)+\dfrac{1}{s}B\right]$

$$= C\left[\dfrac{1}{s}(sI-A)^{-1}\right][sX(0)+B]$$

Response in the time domain

Zero Input Response $\qquad y_{ZIR}(t) = C\, e^{At}\, X(0)$

Zero State response $\qquad y_{ZSR}(t) = \int_0^t C e^{A(t-\theta)}\, B\, r(\theta)\, d\theta$

Impulse response $\qquad y_{ZSR}(t) = C\, e^{At}\, B$

Unit step response $\qquad y_{ZSR}(t) = C A^{-1}[e^{At} - I]B$

APPENDIX – D
COMPUTATION OF AN ANALYTIC FUNCTION OF A SQUARE MATRIX A

$$F(A) = \alpha_0\, I + \alpha_1\, A + \alpha_2\, A^2 + ---- + \alpha_{n-1}\, A^{n-1}$$

The corresponding scalar equation is

$$f(\lambda) = \alpha_0 + \alpha_1\, \lambda + \alpha_2\, \lambda^2 + ---- + \alpha_{n-1}\, \lambda^{n-1}$$

This represents n simultaneous equations corresponding to n different Eigen values and solving these, all the α coefficients $\alpha_0, \alpha_1, \alpha_2, ---, \alpha_{n-1}$ can be determined.

APPENDIX – E
STATE FEEDBACK CONTROLLER DESIGN

Transformation to Controllable Canonical Form of State Model

$$\dot{X} = AX + Bu \qquad\qquad y = CX \qquad\qquad u = -KX$$

$$\dot{Z} = (P^{-1}AP)Z + (P^{-1}B)u \qquad y = (CP)Z$$

For a third order system, the characteristic equation is

$$s^3 + a_2 s^2 + a_1 s + a_0 = 0$$

Characteristic equation of the model in controllable canonical form is

$$s^3 + a_2 s^2 + a_1 s + a_0 = 0$$

$$K_z = \left[(\alpha_0 - a_0) \quad (\alpha_1 - a_1) \quad (\alpha_2 - a_2) \right]$$

$$P = M\, M_z^{-1} \qquad\qquad K = K_z P^{-1}$$

M is the controllability matrix of the given state model

M_z is the controllability matrix of the controllable canonical form of state model

Ackermann's Formula

The characteristic equation

$$s^n + \alpha_{n-1}s^{n-1} + \alpha_{n-2}s^{n-2} + - - - + \alpha_1 s + \alpha_0 = 0$$

$$\phi(A) = A^n + \alpha_{n-1}A^{n-1} + \alpha_{n-2}A^{n-2} + - - - + \alpha_1 A + \alpha_0 I$$

$$K = [0 \quad 0 \quad 0 - - - 0 \quad 1] M^{-1} \phi(A)$$

STATE OBSERVER DESIGN

Transformation to Observable Canonical Form of State Model

$$\dot{X} = AX + Bu \qquad\qquad y = CX \qquad\qquad \dot{E}_X = [A - LC]E_X$$

$$\dot{Z} = (P^{-1}AP)Z + (P^{-1}B)u \qquad\qquad y = (CP)Z$$

For a third order system, the characteristic equation is

$$s^3 + a_2 s^2 + a_1 s + a_0 = 0$$

Characteristic equation of the model in observable canonical form is

$$s^3 + a_2 s^2 + a_1 s + a_0 = 0$$

$$L_Z = \left[(\alpha_2 - a_2) \quad (\alpha_1 - a_1) \quad (\alpha_0 - a_0)\right]^T$$

$$P = Q^{-1}Q_Z \qquad\qquad L = P L_Z$$

    is the observability matrix of the given state model.

    Q_Z is the observability matrix of the observable canonical form of state model

Ackermann's Formula

The characteristic equation

$$s^n + \alpha_{n-1}s^{n-1} + \alpha_{n-2}s^{n-2} + - - - + \alpha_1 s + \alpha_0 = 0$$

$$\phi(A) = A^n + \alpha_{n-1}A^{n-1} + \alpha_{n-2}A^{n-2} + - - - + \alpha_1 A + \alpha_0 I$$

$$L = \phi(A) Q^{-1} \begin{bmatrix} 0 & 0 & 0 - - - 0 & 1 \end{bmatrix}^T$$

APPENDIX – F
OPTIMIZATION OF QUADRATIC PERFORMANCE MEASURE

$$J = \int_0^\infty (X^T Q X + U^T R U)\, dt$$

Lyapunov Equation

$$[A - BK]^T P + P[A - BK] + K^T R K = -Q$$

Riccati Equation

$$A^T P + PA - PBR^{-1}B^T P = -Q$$

APPENDIX – G
NEWTON-RAPHSON METHOD

To solve a nonlinear or transcendental equation $f(x) = 0$.

Let an initial guess root be $x = x_0$. The value of the root $x = x_1$ after the first iteration is

$$x_1 = x_0 - \frac{f(x_0)}{f'(x_0)}$$

The value of the root at the end of kth iteration is

$$x_{k+1} = x_k - \frac{f(x_k)}{f'(x_k)}$$

The iteration is terminated when the difference between two successive values of x is negligibly small or less than a permissible error.

References

1. Ogata K., Modern Control Engineering, Prentice-Hall of India Ltd., 1994.

2. Ogata K., State Space Analysis of Control Systems, Prentice-Hall Inc., NJ., 1967.

3. Brogan W.L., Modern Control Theory, Prentice-Hall, Upper Saddle River, NJ., 1991.

4. Nise N.S., Control System Engineering, Wiley India (P) Ltd., 2007.

5. Friedland B., Control System Design: An Introduction to State Space, Mc Graw Hill, New York, 1987.

6. Timothy L.K., and Bona B.E., State Space Analysis: An Introduction, Mc Graw Hill, New York, 1968.

7. Furuta K., Sano A., and Atherton D., State Variable Methods in Automatic Control, John Wiley and Sons, 1988.

8. Kirk D.E., Optimal Control Theory: An Introduction, Prentice-Hall of India, 1970.

9. Naidu D.S., Optimal Control Systems, CRC Press, 2003.

10. Philips, C.L., and Harbor, R.D., Feedback Control systems, Prentice Hall, Englewood Cliffs, N.J., 1988.

Index